Evolutionary History of Bats
Fossils, Molecules and Morphology

Advances in morphological and molecular methods continue to uncover new information on the origin and evolution of bats. Presenting some of the most remarkable discoveries and research involving living and fossil bats, this book explores their evolutionary history from a range of perspectives.

Phylogenetic studies based on both molecular and morphological data have established a framework of evolutionary relationships that provides a context for understanding many aspects of bat biology and diversification. In addition to detailed studies of the relationships and diversification of bats, the topics covered include the mechanisms and evolution of powered flight, evolution and enhancement of echolocation, feeding ecology, population genetic structure, ontogeny and growth of facial form, functional morphology and evolution of body size. The book also examines the fossil history of bats from their beginnings over 50 million years ago to their diversification into one of the most globally widespread orders of mammals living today.

Gregg F. Gunnell is Director of the Division of Fossil Primates, Duke Lemur Center, Durham, North Carolina. He has spent the last 32 years studying the origin and diversification of modern mammals, mostly focusing on the fossil record and what it can tell us about these groups of organisms.

Nancy B. Simmons is Curator-in-Charge of the Department of Mammalogy, American Museum of Natural History, New York. Her research focuses on the systematics and evolution of bats, including projects that range from higher-level phylogenetic studies to descriptions of new species. In 2008 she was awarded the Gerrit S. Miller Award from the North American Society for Bat Research.

Cambridge Studies in Morphology and Molecules: New Paradigms in Evolutionary Biology

This new Cambridge series addresses the interface between morphological and molecular studies in living and extinct organisms. Areas of coverage include evolutionary development, systematic biology, evolutionary patterns and diversity, molecular systematics, evolutionary genetics, rates of evolution, new approaches in vertebrate palaeontology, invertebrate palaeontology, palaeobotany and studies of evolutionary functional morphology. The series invites proposals demonstrating innovative evolutionary approaches to the study of extant and extinct organisms that include some aspect of both morphological and molecular information. In recent years the conflict between "molecules vs. morphology" has given way to more open consideration of both sources of data from each side making this series especially timely.

Carnivoran Evolution: New Views on Phylogeny, Form and Function
Edited by Anjali Goswami and Anthony Friscia

Evolutionary History of Bats

Fossils, Molecules and Morphology

EDITED BY

Gregg F. Gunnell

Duke University, Durham, USA

and

Nancy B. Simmons

American Museum of Natural History,
New York, USA

CAMBRIDGE
UNIVERSITY PRESS

CAMBRIDGE
UNIVERSITY PRESS

University Printing House, Cambridge CB2 8BS, United Kingdom

One Liberty Plaza, 20th Floor, New York, NY 10006, USA

477 Williamstown Road, Port Melbourne, VIC 3207, Australia

314-321, 3rd Floor, Plot 3, Splendor Forum, Jasola District Centre, New Delhi - 110025, India

103 Penang Road, #05-06/07, Visioncrest Commercial, Singapore 238467

Cambridge University Press is part of the University of Cambridge.

It furthers the University's mission by disseminating knowledge in the pursuit of education, learning and research at the highest international levels of excellence.

www.cambridge.org
Information on this title: www.cambridge.org/9780521745260

First published 2012

A catalogue record for this publication is available from the British Library

ISBN 978-0-521-76824-5 Hardback
ISBN 978-0-521-74526-0 Paperback

Additional resources for this publication at www.cambridge.org/9780521745260.

Contents

The color plates are situated between pages 340 and 341.

Contributors

Robert J. Baker, Natural Science Research Laboratory, Museum of Texas Tech University, Lubbock, TX, USA

Olaf R. P. Bininda-Emonds, Carl von Ossietzky University Oldenburg, Faculty V, Institute for Biology and Environmental Sciences (IBU), Oldenburg, Germany

Kenneth S. Breuer, Brown University, Division of Engineering, Providence, RI, USA

Nicholas J. Czaplewski, Oklahoma Museum of Natural History, Norman, OK, USA

Shannon L. Datwyler, Department of Biological Sciences, California State University, Sacramento, CA, USA

Serena Dool, School of Biology and Environmental Science, University College Dublin, Belfield, Dublin, Ireland

Thomas P. Eiting, Graduate Program in Organismic and Evolutionary Biology, University of Massachusetts Amherst, USA

Norberto P. Giannini, Facultad de Ciencias Naturales e Instituto Miguel Lillo, Tucumán, Argentina

Gregg F. Gunnell, Division of Fossil Primates, Duke Lemur Center, Durham, NC, USA

Jörg Habersetzer, Forschungsinstitut Senckenberg, Messelforschung, Frankfurt, Germany

Suzanne Hand, Biological, Earth and Environmental Sciences, University of New South Wales, Sydney, Australia

Ivan Horáček, Department of Zoology, Charles University, Praha, Czech Republic

Jose Iriarte-Díaz, Department of Organismal Biology and Anatomy, University of Chicago, IL, USA

Winston C. Lancaster, Department of Biological Sciences, California State University, Sacramento, CA, USA

Dawn J. Larkey, Department of Biological Sciences, California State University, Sacramento, CA, USA

Elodie Maitre, Departement Sciences de la Terre, Université Lyon 1, France

Hugo Mantilla-Meluk, Department of Biological Sciences, Texas Tech University, Lubbock, TX, USA

Gary S. Morgan, New Mexico Museum of Natural History and Science, Albuquerque, NM, USA

Kevin J. Olival, EcoHealth Alliance, New York, NY, USA

Scott C. Pedersen, Department of Bio-Microbiology, South Dakota State University, Brookings, SD, USA

Calvin A. Porter, Department of Biology, Xavier University of Louisiana, New Orleans, LA, USA

Daniel K. Riskin, Discovery Channel Canada, Scarborough, Ontario, Canada

Evelyn Schlosser-Sturm, Forschungsinstitut Senckenberg, Messelforschung, Frankfurt, Germany

Bernard Sigé, Institut des Sciences de l'Evolution, Université de Montpellier 2, France

Nancy B. Simmons, Department of Mammalogy, American Museum of Natural History, New York, NY, USA

Elwyn L. Simons, Division of Fossil Primates, Duke University Lemur Center, Durham, NC, USA

Thierry Smith, Department of Paleontology, Royal Belgian Institute of Natural Sciences, Brussels, Belgium

František Špoutil, Department of Zoology, Charles University, Praha, Czech Republic

Mark S. Springer, Department of Biology, University of California, Riverside, CA, USA

Gerhard Storch, Forschungsinstitut Senckenberg, Mitarbeiter Mammalogie, Frankfurt, Germany

Sharon M. Swartz, Ecology & Evolutionary Biology, Brown University, Providence, RI, USA

Emma C. Teeling, School of Biology and Environmental Science, University College Dublin, Belfield, Dublin, Ireland

Douglas W. Timm, University of Oklahoma, Norman, OK, USA

Ronald A. Van Den Bussche, Associate Dean for Research, College of Arts & Sciences, Oklahoma State University, Stillwater, OK, USA

Preface

In 2007, the editors (Gregg F. Gunnell and Nancy B. Simmons), along with graduate student Thomas P. Eiting, organized a symposium for the Society of Vertebrate Paleontology meeting in Austin, TX. This symposium was designed to explore bat evolution by using both morphological and molecular data, and to better elucidate where results gleaned from different data and methodologies agreed and disagreed. It was (and is) our contention that both sets of data are essential to fully understanding the evolutionary history of any group of organisms, and we set out to demonstrate this at the Austin meeting. The symposium was very successful and several participants expressed an interest in contributing to a symposium volume.

During this same time frame, one of the editors (GFG) was aiding in the development of a book series for Cambridge University Press that emphasizes the importance of both molecular and morphological data, and it was decided that a book on bat evolution would be appropriate for the series. Many species of bats have been sampled genetically, and bats have a long history of morphological study because of their unique ability to fly and to navigate by echolocation. Bats also have a rich fossil record extending back over 50 million years, and a geographic distribution that spans most of the globe. Several studies have explored the genetics and morphology of bats and have found much common ground (unlike in some other groups of organisms where there has been more contention than compatibility). These facts, and the enthusiastic support of our authors, led us to propose this volume.

The book draws together leading experts in the fields of phylogenetic systematics, population genetics, functional morphology, paleobiology, development, behavior and ecology, all of whom having focused much of their research on bats. Authors represent countries from across the globe including the United States, Ireland, Belgium, Germany, France, Australia, Argentina and the Czech Republic. The study of bats is an interdisciplinary undertaking, and attempting to understand the evolutionary history and unique lifestyles of bats has attracted scientists for generations. We hope that this book will be of interest and use to bat biologists around the world today, and to future generations of researchers.

As with most endeavors of this sort, many people had a hand in the final product and we are indebted to all of them. First and foremost we owe a big debt of gratitude to all of the authors, not only for the excellent chapters that they produced, but also for their thoughtful commentary, advice, encouragement and dedication. Many authors did double duty and served as primary reviewers of each other's chapters. We are also very grateful to many of our colleagues who gave freely of their time to aid in innumerable ways – we especially thank Rick Adams, Ted Macrini, Kristen Bishop, Gary McCracken, Paul Velazco, Adam Rountrey, Andrea Wetterer, Tim Strickler, Adrian Tejedor, Stephen Rossiter, Gareth Jones, Brock Fenton, Alistair Evans and Liliana Davalos for their help. We thank those that participated in the original 2007 symposium, but who were unable to contribute to this volume for a variety of reasons – included among them are Elizabeth Dumont, Steven Hoofer, James Hutcheon, Erik Seiffert, Cassandra Miller-Butterworth, Robin Beck, Trevor Worthy, Kevin Seymour, Michael Archer, Wieslaw Bogdanowicz, Marta Gajewska and Tomasz Postawa. Our Cambridge University Press editor Dominic Lewis and his assistants Zewdi Tsegai and Megan Waddington were very supportive and helpful throughout the production of this book, as was Russell L. Ciochon, our other book series editor. To all of these people we offer our most heartfelt thank you!

1

Phylogenies, fossils and functional genes: the evolution of echolocation in bats

EMMA C. TEELING, SERENA DOOL AND
MARK S. SPRINGER

1.1 Introduction

Bats are one of the most successful orders of mammals on this planet. They account for over 20% of living mammalian diversity (~1200 species), and are distributed throughout the globe, absent only from the extreme latitudes (Simmons, 2005). Bats are the only living mammals that are capable of true self-powered flight, and likewise they are the only mammals capable of sophisticated laryngeal echolocation (Macdonald, 2006). Their global success is largely attributed to these novel adaptations (Jones and Teeling, 2006). Echolocation occurs when a bat emits a brief laryngeal-generated sound that can vary in duration (0.3–300 ms) and in frequency (8–210 kHz) and interprets the returning echoes to perceive its environment (Fenton and Bell, 1981; Thomas et al., 2004). Calls and echoes can be separated either in time or in frequency (Jones, 2005). Some bats (e.g., horseshoe bats, leaf-nosed bats and mustached bats) emit long constant-frequency calls with Doppler shift compensation (CF/DSC) by taking the velocity of their flight into account and adjusting the frequency of their outgoing calls to ensure that the incoming echoes return at a specific frequency (Thomas et al., 2004; Jones, 2005). Most other bats emit low-duty-cycle frequency-modulated calls, and separate outgoing calls and incoming echoes temporally (Thomas et al., 2004; Jones, 2005).

Echolocation calls show a great diversity in shape, duration and amplitude, and are correlated with the parameters of a bat's environment (Jones and Teeling, 2006; Jones and Holderied, 2007). The auditory capabilities of bats are extraordinary. Bats produce and interpret some of the "loudest" naturally produced airborne sounds ever recorded (130 dB; Jones, 2005), and are also capable of hearing some of the "quietest" sounds of any mammal (~-20 dB; Neuweiler, 1990). Despite the magnitude and functionality of this spectacular form of sensory perception, the evolutionary history of echolocation is still controversial.

Evolutionary History of Bats: Fossils, Molecules and Morphology, ed. G. F. Gunnell and N. B. Simmons. Published by Cambridge University Press. © Cambridge University Press 2012.

This has stemmed from inconsistent and unresolved phylogenies (Simmons and Geisler, 1998; Van Den Bussche and Hoofer, 2004; Eick *et al.*, 2005; Teeling *et al.*, 2005), and an incomplete (Teeling *et al.*, 2005; Eiting and Gunnell, 2009) and differentially interpreted fossil record (Simmons *et al.*, 2008; Veselka *et al.*, 2010) that allows for alternate interpretations of gain and loss of auditory function, and lack of molecular echolocation signatures (Teeling, 2009).

1.2   Phylogenetic controversies

Traditionally bats were divided into two subordinal groups, Megachiroptera and Microchiroptera (Koopman, 1994; Simmons, 1998; Simmons and Geisler, 1998). Megachiroptera includes the Old World family Pteropodidae, and Microchiroptera contains the remaining 17 bat families (Simmons and Geisler, 1998). This division was based mainly on morphological and paleontological data, but it also highlighted the difference in the dominant mode of sensory perception used by megabats (vision) and microbats (ultrasound). Given that all microbats are capable of sophisticated laryngeal echolocation, whereas megabats are not (Jones, 2005), it was believed that laryngeal echolocation had a single origin in the common ancestor of microbats (Teeling *et al.*, 2000). The 17 families of microbats were subsequently divided into two infraorders Yinochiroptera (rhinolophids, hipposiderids, megadermatids, craseonycterids, rhinopomatids, emballonurids, nycterids) and Yangochiroptera (vespertilionids, molossids, natalids, phyllostomids, noctilionids, furipterids, thyropterids, mormoopids, mystacinids, myzopodids), based on whether their premaxillaries were moveable/absent or fused relative to their maxillaries (Koopman, 1994; Simmons and Geisler, 1998; Hutcheon and Kirsch, 2006). This arrangement was largely supported by morphological data sets (Gunnell and Simmons, 2005) and supertree consensus studies (Jones *et al.*, 2002). However, superfamilial groupings ranged in content and number between studies (Koopman, 1994; Simmons and Geisler, 1998; Jones *et al.*, 2002; Gunnell and Simmons, 2005).

From the advent of modern molecular techniques during the 1980s and 1990s, it became apparent that molecular data did not support the monophyly of Microchiroptera and consequently, did not support a single origin of laryngeal echolocation. Rather, molecular data supported a basal division between Yinpterochiroptera (rhinolophoid microbats and pteropodids) and Yangochiroptera (all other bats; Teeling *et al.*, 2000, 2005; Hutcheon and Kirsch, 2004; Van Den Bussche and Hoofer, 2004; Eick *et al.*, 2005; Miller-Butterworth *et al.*, 2007). This topology suggested that laryngeal echolocation either originated in the ancestor of all bats and was subsequently lost in the common ancestor of megabats, or originated on more than one occasion in the microbats (Teeling *et al.*, 2000). Initially immunological distance data (Pierson, 1986), single gene data sets (Stanhope *et al.*, 1992; Porter *et al.*, 1996), single-copy DNA–DNA hybridization (Hutcheon *et al.*, 1998), studies of repetitive genomic elements (Baker *et al.*, 1997) and taxonomically limited consensus studies (Liu and Miyamoto, 1999) all supported microbat paraphyly to different degrees (Jones and Teeling, 2006). However, strong support and congruence for the association of the rhinolophoid microbats with the pteropodids was only derived from large concatenated nuclear data sets with representatives from nearly all putative bat families (Eick *et al.*, 2005 – 4 kb, four nuclear introns; Teeling *et al.*, 2005 – 13.7 kb, 18 nuclear exons and UTRs; Miller-Butterworth *et al.*, 2007 – 11 kb, 16 nuclear exons and UTRs) and rare cytogenetic signature events (Ao *et al.*, 2007).

Molecular data in the form of large nuclear and mitochondrial concatenations (Teeling *et al.*, 2000, 2005; Van Den Bussche and Hoofer, 2004; Eick *et al.*, 2005; Miller-Butterworth *et al.*, 2007) provided strong support for the monophyly of four different lineages of echolocating microbat lineages:

(1) Rhinolophoidea (rhinolophids, hipposiderids, rhinopomatids, craseonycterids, megadermatids)
(2) Emballonuroidea (nycterids and emballonurids)
(3) Vespertilionoidea (vespertilionids, molossids, natalids, miniopterids)
(4) Noctilionoidea (noctilionids, phyllostomids, mormoopids, furipterids, thyropterids, mystacinids, myzopodids).

Myzopodidae was recovered with robust support as the sister taxon to other Noctilionoidea (Teeling *et al.*, 2005; Miller-Butterworth *et al.*, 2007), or as the sister taxon to Vespertilionoidea (Eick *et al.*, 2005), albeit with weak support. Two other differences between the exon + UTR tree (Teeling *et al.*, 2005; Miller-Butterworth *et al.*, 2007; Figure 1.1) and the intron tree (Eick *et al.*, 2005; Figure 1.2) are as follows (Figure 1.3):

(1) *Thyroptera* was either the sister group to *Mystacina* (Eick *et al.*, 2005) or grouped in a clade with *Noctilio* and *Furipterus* (Teeling *et al.*, 2005; Miller-Butterworth *et al.*, 2007).
(2) Emballonuroidea and Noctilionoidea were sister taxa (Eick *et al.*, 2005) or Emballonuroidea and Vespertilionoidea were sister taxa (Teeling *et al.*, 2005; Miller-Butterworth *et al.*, 2007).

Finally, the phylogenetic position of Craseonycteridae within Rhinolophoidea was robust based on exons + UTRs (Teeling *et al.*, 2005; Miller-Butterworth *et al.*, 2007), but Craseonycteridae was not included in the Eick *et al.* (2005) data set.

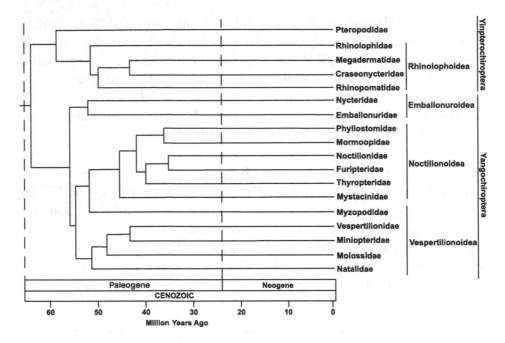

Figure 1.1 Miller-Butterworth *et al.*, 2007 bat phylogeny.

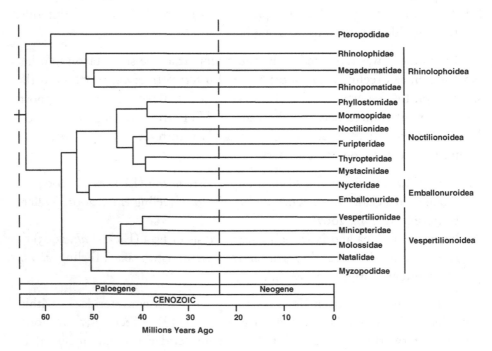

Figure 1.2 Eick *et al.*, 2005 bat phylogeny.

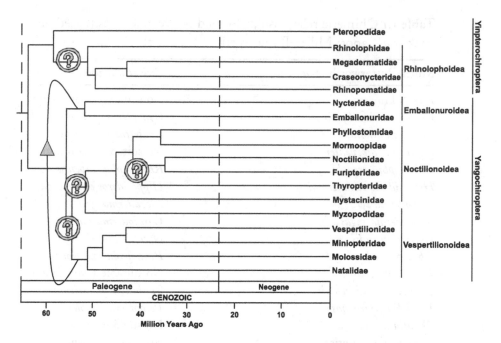

Figure 1.3 Tree depicting the questionable nodes that differ in support and familial representation between Miller-Butterworth *et al.*, 2007 and Eick *et al.*, 2005.

Here, we combine the largest published nuclear data sets for bats (Eick *et al.*, 2005; Teeling *et al.*, 2005; Miller-Butterworth *et al.*, 2007) and reconstruct phylogenetic relationships based on this concatenation. Next, we examine the evolutionary history and molecular basis of echolocation in the context of our phylogenetic results, and discuss recent fossil evidence in light of these findings. We also discuss ongoing molecular investigations into candidate genes that underlie echolocation, and describe how these studies inform the echolocation gain vs. loss debate.

1.3 Phylogenetic analysis

1.3.1 Molecular data sets

We combined nuclear gene sequences from Teeling *et al.* (2005), Miller-Butterworth *et al.* (2007) and Eick *et al.* (2005). Teeling *et al.*'s (2005) data set comprised ~13 kb of nuclear sequence (exons and UTRs) from 18 genes, and included representatives of all bat families except Miniopteridae. Miller-Butterworth *et al.*'s (2007) data set expanded on Teeling *et al.*'s (2005) data set by including two species of *Miniopterus* and an additional vespertilionid, but omitted sequences for the *ADRA2B* and *VWF* genes (these are omitted from subsequent analyses). Eick *et al.*'s (2005) data set consisted of ~4 kb from four

Table 1.1 Chimeric relationships formed between species (based on Eick *et al.*, 2005; Miller-Butterworth *et al.*, 2007).

Miller-Butterworth *et al.*, 2007	Eick *et al.*, 2005
Pteropus rayneri	No suitable taxon found
Cynopterus brachyotis	*Cynopterus sphinx*
Rousettus lanosus	*Rousettus aegytiacus*
Nyctimene albiventer	No suitable taxon found
Rhinolophus creaghi	*Rhinolophus capensis*
Hipposideros commersoni	*Hipposideros commersoni*
Megaderma lyra	*Megaderma lyra*
Macroderma gigas	*Cardioderma cor*
Nycteris grandis	*Nycteris grandis*
Rhinopoma hardwicki	*Rhinopoma hardwicki*
Emballonura atrata	No suitable taxon found
Taphozous nudiventris	*Taphozous mauritianus*
Rhynchonycteris naso	*Peropteryx kappleri*
Tonatia saurophila	No suitable taxon found
Artibeus jamaicensis	*Artibeus jamaicensis*
Desmodus rotundus	*Desmodus rotundus*
Anoura geoffroyi	*Glossophaga soricina*
Noctilio albiventris	*Noctilio albiventris*
Antrozous pallidus	No suitable taxon found
Rhogeesa tumida	*Scotophilus dinganii*
Myotis daubentoni	*Myotis tricolor*
Myzopoda aurita	*Myzopoda aurita*
Pteronotus parnellii	*Pteronotus parnellii*
Thyroptera tricolor	*Thyroptera tricolor*
Mystacina tuberculata	*Mystacina tuberculata*
Furipterus horrens	*Furipterus horrens*
Natalus stramineus	*Natalus micropus*
Tadarida brasiliensis	*Tadarida aegyptiaca*
Eumops auripendulus	*Otomops martiensseni*
Craseonycteris thonglongyai	*Craseonycteris thonglongyai*[1]
Miniopterus schreibersii	*Miniopterus natalensis*
Miniopterus fraterculus	*Miniopterus fraterculus*
Eptesicus fuscus	*Eptesicus hottentotus*

[1] Data generated in this study.

nuclear introns for 17 of the 18 bat families (missing Craseonycteridae). Our concatenation of data from Miller-Butterworth *et al.* (2007) and Eick *et al.* (2005) included several chimeric taxa (Table 1.1). When possible we concatenated the data set at the species level. When the same species was not present in both data set we concatenated the taxa at the generic level, using published phylogenies to assess intergeneric relationships (Hollar and Springer, 1997; Jones *et al.*, 2002; Baker *et al.*, 2003; Hoofer and Van den Bussche, 2003; Table 1.1). In addition to sequences from the aforementioned studies, we amplified and sequenced the missing intronic fragments for *Craseonycteris* using primers and PCR amplification conditions described in Eick *et al.* (2005) (GenBank Accession Numbers HQ231220- HQ231221). Our final data set comprised ~14 kb and consisted of exonic sequences from 12 genes (*ADORA2, ADRB2, ATP7A, BDNF, BRCA1, EDG1, PNOC, RAG1, RAG2, TITIN, TYR, ZFX*), UTR sequences from four genes (*APP, BMI1, CREM, PLCB4*) and intronic sequences from four genes (*SPTBN, PRKC1, THY, STAT5A*) for 35 taxa, of which 33 are bats and two are outgroup sequences from the laurasiatherian orders Perissodactyla and Carnivora.

1.3.2 Phylogenetic methods

The concatenated data set was aligned with the program Clustal W (Higgins and Sharp, 1988) and optimized using Se-Al (Rambaut, 1996). Insertion-deletion events were observed among taxa, and gaps were introduced (by Se-Al) to maintain the alignment. All alignment gaps were treated as missing characters in subsequent phylogenetic analyses. Alignment-ambiguous regions were identified by eye and were excluded from phylogenetic analyses. PAUP*4.0 (Swofford, 2002) was used to perform maximum parsimony (MP), maximum likelihood (ML) and minimum evolution (ME) analyses. Modeltest v3.06 (Posada and Crandall, 1998) was used to select the nucleotide substitution model that best fit the data. This was a general time reversible (GTR) model of sequence evolution with a proportion of invariant sites (I) and an allowance for a gamma (Γ) distribution of rates (GTR $+$ I $+$ Γ). Parameter estimates for ML and ME analyses were as follows: Base $=$ (0.2649 0.2456 0.2379), Nst $=$ 6, Rmat $=$ (1.1770 3.9466 0.5501 1.2454 4.3633), Rates $=$ gamma, Shape $=$ 0.7992, Pinvar $=$ 0.2626. MP analyses employed stepwise addition with ten randomized input orders. ME and ML analyses started with neighbor-joining (NJ) trees. All heuristic searches employed tree-bisection and reconnection (TBR) branch-swapping, except for ML bootstrap analyses, which employed nearest- neighbor interchange (NNI) branch-swapping. MP and ME bootstrap analyses were performed with 500 pseudoreplicate data sets; ML bootstraps were carried out with 100 pseudoreplicate data sets.

1.3.3 Phylogenetic results

Figure 1.4 shows the maximum likelihood tree with ML bootstrap percentages for the concatenated data set. The results for all bootstrap analyses are depicted in Table 1.2. The overall topology of the Miller-Butterworth *et al.*

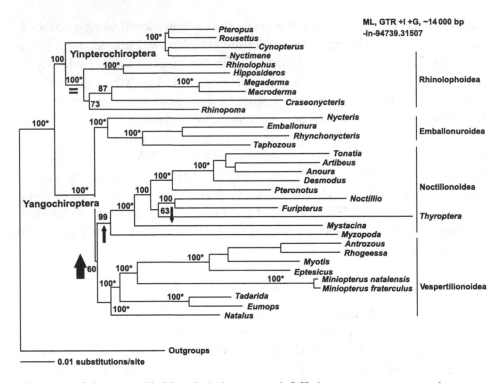

Figure 1.4 Maximum likelihood phylogram with ML bootstrap support on the nodes. Arrows indicate where bootstrap support is increased or decreased on the questionable nodes depicted in Figure 1.3. *Indicates when ML, MP and ME bootstrap analyses agree.

(2007) phylogeny is still supported and the majority of nodes received robust bootstrap support. All of the uncertain relationships depicted in Figure 1.3 were resolved in favor of the Miller-Butterworth *et al.* (2007) tree. *Myzopoda* is the sister group to other noctilionoid families, and Vespertilionoidea and Noctilionoidea are sister taxa, although the branch that groups these two superfamilies together only received moderate bootstrap support (Figure 1.4). The branch uniting the Neotropical families Noctilionidae, Furipteridae and Thyropteridae was also recovered, albeit with lower ML bootstrap support (60% vs. 91%) than in Miller-Butterworth *et al.* (2007). Analyses of the nuclear intronic data set alone do not support this node, nor do phylogenetic analyses of mitochondrial data (Van Den Bussche and Hoofer, 2004). Further phylogenetic investigations will be required to assess the validity of this clade. *Craseonycteris* still groups within Rhinolophoidea and is the sister taxon to Megadermatidae.

Table 1.2 Bootstrap values for the various clades depicted in Figure 1.4. Abbreviations are as follows: ML = maximum likelihood, MP = maximum parsimony, ME = minimum evolution.

	ML	MP	ME
Pteropodidae	100	100	100
Rhinolophidae	100	100	100
Megadermatidae	100	100	100
Megadermatidae + Craseonycteridae	87	100	100
Megadermatidae + Craseonycteridae + Rhinopomatidae	73	83	<50
Rhinolophoidea	100	100	100
Yinpterochiroptera	100	100	<50
Yangochiroptera	100	100	100
Emballonuridae	100	100	100
Nycteridae + Emballonuridae	100	56	100
Phyllostomidae	100	100	100
Phyllostomidae + Mormoopidae	100	100	100
Noctilionidae + Furipteridae	100	70	99
Noctilionidae + Furipteridae + Thyropteridae	63	<50	<50
Noctilionoidea	99	57	54
Noctilionoidea + Emballonuroidea	60	50	79
Vespertilionidae	100	100	100
Vespertilionidae + Miniopteridae	100	100	100
Miniopteridae	100	100	100
Molossidae	100	100	100
Molossidae + Miniopteridae + Vespertilionidae	100	100	100
Vespertilionoidea	100	100	100

1.4   Implications for echolocation

The suborder Yinpterochiroptera, which unites the non-echolocating Pteropodidae and the echolocating superfamily Rhinolophoidea, is still supported. This association was previously supported by independent nuclear data sets based on exon + UTRs (Teeling *et al.*, 2005; Miller-Butterworth *et al.*, 2007) and introns (Eick *et al.*, 2005), and received additional support when analyses were performed on a concatenation that included both data sets (Figure 1.4). This has direct implications for the evolution of echolocation. Two scenarios are depicted in Figure 1.5:

(A) Laryngeal echolocation was gained in the ancestor of all living bats and subsequently lost in the Pteropodidae.
(B) Laryngeal echolocation was gained independently in at least two echolocating lineages (Jones and Teeling, 2006; Teeling, 2009).

(A) Scenario 1

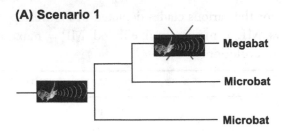

Echolocation was gained once but was lost in megabats

(B) Scenario 2

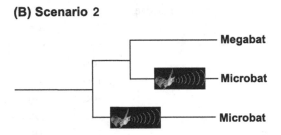

Echolocation was gained convergently at least more than once in bats

Figure 1.5 Two scenarios depicting the evolution of echolocation given the molecular phylogeny supported.

Both scenarios are equally parsimonious if we only consider the topology for living bats, although echolocation loss in Pteropodidae is more parsimonious if we consider living and fossil bats (Springer *et al.*, 2001; Teeling *et al.*, 2005). Attempts to reconstruct an ancestral type of echolocation call of extant bats given this molecular phylogeny that could indicate a single origin of echolocation failed to draw any conclusive evidence, suggesting that echolocation variation reflected environmental niche adaptation rather than shared ancestry (Jones and Teeling, 2006).

The chronogram of Teeling (2009) suggests that the four major lineages of echolocating bats originated ~58 mya, whereas basal cladogenesis within each group is in the range of 52–50 mya (Teeling *et al.*, 2005; Figure 1.1). The origin of the major lineages is coincident with the Paleocene/Eocene thermal maximum (PETM), where global temperatures increased by ~5°C. The PETM commenced at 55.8 mya and lasted ~170 ky (Woodburne *et al.*, 2009). This global warming event was associated with a significant increase in plant diversity and the peak of Tertiary insect diversity (Teeling *et al.*, 2005). The basal diversification of the echolocating lineages appears to coincide with the late Early Eocene climatic optimum (EECO, 53–50 mya; Woodburne *et al.*, 2009). This climatic episode is associated with a marked global temperature increase, resulting in the highest prolonged ocean temperatures of the Cenozoic and a continental interior mean annual temperature of 23°C. There was a significant increase in floral diversity and habitat complexity during this time (Woodburne *et al.*, 2009). Therefore, the bat chronogram implies that the major echolocating microbat lineages may have originated and radiated in response to an increase in prey diversity and roost sites (Teeling *et al.*, 2005), perhaps indicating that laryngeal echolocation in these lineages had multiple adaptive origins. One way to elucidate which hypothesis is more probable is to examine the bat fossil record to infer how echolocation evolved in the basal branches and stem taxa.

1.5   Echolocation

1.5.1   Echolocation and the fossil record

Bats have one of the most impoverished fossil records among mammals (Teeling *et al.*, 2005). Estimates of the completeness of the bat fossil record based on phylogeny-dependent (Teeling *et al.*, 2005) and -independent methods (Eiting and Gunnell, 2009) have suggested that over 70% of the bat fossil record is missing. The incompleteness of the bat fossil record stems partly from the globally poor preservation of delicate bat skeletons, leaving mainly cranial and dental remains. The latter are difficult to distinguish from those of other small insectivores that also have primitive dilambdodont molars (Gunnell and Simmons, 2005). However, even though the bat fossil record generally is patchy, incomplete and difficult to interpret, there also are fantastic Lagerstätten from the Eocene that preserve nearly complete skeletons (e.g., *Icaronycteris, Palaeochiropteryx, Archaeonycteris, Hassianycteris, Tachypteron*). Unfortunately, whole skeletons from the Oligocene, Miocene and Pliocene are rare (Gunnell and Simmons, 2005). Therefore, it is possible to identify morphological echolocation characteristics in Eocene bats, but it is difficult to track the evolution of these characters throughout the rest of the Tertiary.

1.5.2   Eocene fossils and echolocation

Up until 2008 the most primitive bat fossil found was *Icaronycteris index*, which is known from Eocene Green River Formation deposits in Wyoming, USA that have been dated at ~52.2 mya (Simmons and Geisler, 1998). It has been commonly accepted that this primitive bat was capable of both flight and laryngeal echolocation, given that it possessed wings similar to modern bats and also showed evidence of three cranial characteristics that were considered diagnostic of echolocation capabilities (Simmons and Geisler, 1998).

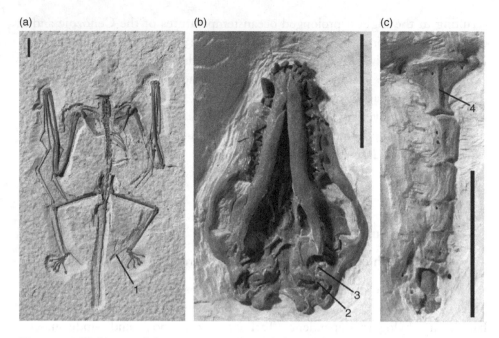

Figure 1.6 Holotype of *Onychonycteris finneyi* (adapted from Simmons *et al.*, 2008). (a) Skeleton in dorsal view. (b) Skull in ventral view. (c) Sternum in ventral view. Scale bars, 1 cm. Features labeled: 1, calcar; 2, cranial tip of stylohyal; 3, orbicular apophysis of malleus; 4, keel on manubrium of sternum. See also color plate section.

Extant bats that echolocate are shown to possess three diagnostic characters: a large cochlea, an enlarged orbicular apophysis on the malleus and a stylohyal element with an expanded, paddle-like cranial tip (Novacek, 1985; Simmons and Geisler, 1998). Indeed, in most of the Eocene bat fossils, it was possible to detect the morphological structures of wings, and all were shown to have the three echolocation diagnostic characters, suggesting that echolocation and flight had evolved in bats no later than the Early Eocene. Phylogenetic analyses of morphological data sets that enforced a molecular phylogeny scaffold suggested that *Icaronycteris* was basal to all extinct and extant bats and that under the most parsimonious scenario, echolocation evolved in the ancestor of all bats and was subsequently lost in the pteropodid lineage (Springer *et al.*, 2001). However, there is still no fossil evidence (a primitive pteropodid with the osteological characteristics of an echolocator) or conclusive molecular evidence of this loss (an echolocation gene showing evidence of loss of function).

In 2008, a new fossil bat, *Onychonycteris finneyi*, was described as a new species, genus and family (Figure 1.6; Simmons *et al.*, 2008). This fossil was found in the same Fossil Butte Member of the Early Eocene Green River formation in Wyoming, USA as *Icaronycteris*. *O. finneyi* differs from all other

Eocene bats in its larger size, possession of claws on all wing digits and retention of more primitive limb and basicranial characteristics (Simmons *et al.*, 2008). Although it has long wing-like digits and other postcranial characters suggesting that *O. finneyi* was capable of flight, this species did not possess the three osteological characteristics that are diagnostic of echolocation capabilities (Simmons *et al.*, 2008). Phylogenetic analyses placed this fossil as the most basal lineage within Chiroptera and therefore Simmons *et al.* (2008) concluded that flight must have evolved before echolocation and that echolocation must have evolved on the branch leading to the other Eocene and crown group bats.

Recently, Simmons *et al.*'s (2008) conclusions were questioned by the discovery of an additional morphological character that is shared by all echolocating bats that have so far been examined. Using advanced microcomputed tomography of fluid preserved bat skulls, Veselka *et al.* (2010) developed a three-dimensional image of the structure of the stylohyal and its connections to other elements. They found that in all echolocating bats the stylohyal lies across the tympanic bones and is either closely articulated with or fused to the tympanic bones, whereas in all pteropodids, which are either non-echolocating or are only capable of emitting tongue-click echolocation calls (e.g., *Rousettus*), there is no hard connection between the stylohyal and the tympanic bones (Veselka *et al.*, 2010; Figure 1.7). On examination of *O. finneyi*, Veselka *et al.* (2010) found that the stylohyal and tympanic bones were in contact, leading them to conclude that *O. finneyi* was capable of echolocation, despite not possessing the other three hallmark characteristics of echolocation (Veselka *et al.*, 2010). They argued that among other mammals that do not echolocate, some possess a large orbicular apophysis, and although the cochlea of *O. finneyi* was more similar in size to that of the non-echolocating pteropodids, cochlea size could vary greatly across bats; therefore these characteristics were not good indicators of echolocation capabilities (Veselka *et al.*, 2010).

Simmons *et al.* (2010) later agreed that an enlarged orbicular process is not informative of echolocation capabilities, but cochlea size is. Simmons *et al.* (2010) argued that stylohyal/tympanic articulation + modified stylohyal is only one feature as it is correlated in all echolocating bats examined and both features are absent in the pteropodids. As *O. finneyi* did not possess a modified cranial end of the stylohyal and an enlarged cochlea, they concluded that this species was not capable of laryngeal echolocation. They argue that stylohyal/tympanic connection described by Veselka *et al.* (2010) occurred as the result of the *O. finneyi* holotype being crushed and flattened with consequent damage to the skull, therefore this characteristic alone could not be used to infer echolocation capabilities in *O. finneyi* (Simmons *et al.*, 2010).

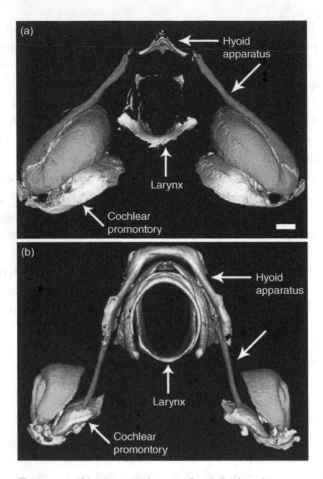

Figure 1.7 Shows articulation of stylohyal and tympanic bones in an
echolocating versus non-laryngeal echolocating bat (adapted from Veselka *et al.*,
2010). a-b, Caudal view comparing stylohyal and tympanic bones in a laryngeal
echolocating bat (a, *Desmodus rotundus*) and a pteropodid that lacks laryngeal
echolocation (b, *Rousettus aegyptiacus*). The stylohyal bone is depicted in
turquoise, the tympanic bone in yellow. See also color plate section.

Regardless of the conclusion of echolocation capabilities in *O. finneyi*, the
evolution of echolocation in bats occurred early in their evolutionary history.
Both Veselka *et al.* (2010) and Simmons *et al.* (2008) agree that the other
Eocene bats, *Icaronycteris*, *Palaeochiropteryx*, *Archaeonycteris*, *Hassianycteris* and
Tanzanycteris (Gunnell *et al.*, 2003), were more than likely capable of
echolocation. Perhaps *O. finneyi* represents a transitional fossil, and is on an
evolutionary trajectory leading to the development of all the characteristics
needed for modern echolocation capabilities. The elucidation of a timeline for
the evolution of echolocation, including the temporal sequence of acquisition

of the two echolocation characters needed for echolocation (i.e., fusion of a modified stylohyal to the tympanic bones + enlargement of the cochlea) need to be investigated. This will require the discovery of more Paleocene and Eocene fossils and the use of advanced technologies (e.g., micro CT-scanning) for imaging fossil specimens.

However, although the finding of this fossil enabled the question of flight and echolocation to be addressed, it did not enlighten the question of gain or loss of echolocation in pteropodids any further than what was already known. The gain of echolocation either convergently occurred in multiple basal and extant bat lineages or was gained once in the ancestor leading to all non-onychonycterid bats and subsequently lost in the pteropodids. Although the most parsimonious interpretation of echolocation suggests that loss in pteropodids is the most plausible explanation and optimization of echolocation on to congruent phylogenetic trees also supports this hypothesis (Springer et al., 2001; Simmons et al., 2008), the fossil and molecular data are still not present to convincingly rule out the possibility of convergent echolocation acquisition in bats.

Another way to untangle the interplay of these diagnostic characters in relation to echolocation capabilities, and shed light on the gain/loss debate, is to examine how they change in bat lineages that do not use laryngeal echolocation. In all pteropodids examined the stylohyal is not expanded nor is it fused to the tympanic bones (Simmons et al., 2010). The absence of this articulation is either a derived state in pteropodids and evolved in conjunction with the loss of echolocation, or is the ancestral state that also occurs in all other non-echolocating laurasiatherian mammals (Veselka et al., 2010). A study of stem pteropodids would shed light on this issue. However, of all bat families the non-echolocating pteropodids have perhaps the worst fossil record and are missing up to 98% of their fossil history, including ~32 mya of the pteropodid stem (Teeling et al., 2005). In the absence of older pteropodid fossils, it will be difficult to determine if laryngeal echolocation was lost in stem Pteropodidae or was never present in this lineage.

1.6   Echolocation genes

1.6.1   Functional genes and echolocation

Sequencing of the human genome (Lander et al., 2001; Venter et al., 2001) has facilitated great advances in our understanding of the genes and molecular mechanisms that underlie phenotypes and characteristics (Hayden et al., 2010). Laryngeal echolocation is a complex process that relies on the interplay of morphological and physiological adaptations. To echolocate, an organism must generate ultrasound, perceive the returning echoes and interpret this information to make an acoustic image of its environment. Two genes have directly been associated with echolocation capabilities: *FOXP2* (the '*vocalization*' gene; Li *et al.*, 2007) and *Prestin* (the '*hearing*' gene; Li *et al.*, 2008).

1.6.2 Vocalization and hearing

The *FOXP2* gene codes for a transcription factor that is implicated in the development and control of orofacial movements and thus vocalization capabilities in mammals (Li *et al.*, 2007). Li *et al.* (2007) amplified and sequenced this entire gene in 13 bat species and compared their results with 22 other mammals and 2 non-mammalian vertebrates. They further amplified, sequenced and examined the two highly variable exons, 7 and 17, in 64 species representing 42 bats (10 families), 18 cetaceans, 2 birds and a reptile (Li *et al.*, 2007). They found that this gene was completely conserved in most mammals, however in bats *FOXP2* is highly variable and this variation appears to correlate with differences in echolocation call type (Li *et al.*, 2007). The study of Li *et al.* (2007) employed excellent taxonomic coverage, and *FOXP2* is a good candidate echolocation gene, but the authors were not able to ascertain from their analyses if laryngeal echolocation was secondarily lost in Pteropodidae after originating in the common ancestor of Chiroptera, or was independently acquired in multiple echolocating lineages. Exon 17 of *FOXP2* was under high purifying selection in pteropodids, similar to other mammals, whereas exon 7 showed a similar level of non-synonymous variation and accelerated mutation rates in the pteropodids as in the other echolocating lineages. This indicates that both echolocation hypotheses may be still possible given these data: while exon 7 may be a 'smoking gun' suggesting that pteropodids could once echolocate and still retained the ancestral signatures of echolocation, exon 17 refutes this hypothesis and suggests that pteropodids have retained the ancestral state. However, when Li *et al.* (2007) reconstructed the ancestral amino acids on the stem bat branch they found no evidence of adaptive change that would indicate that echolocation gain had occurred on that branch. If *FOXP2* is important for echolocation then these results suggest that:

(1) Echolocation evolved independently in at least two microbat lineages, based on the absence of shared changes in Chiroptera, or
(2) Primitive laryngeal echolocation evolved in the common ancestor of bats, but *FOXP2* was not important in the early evolution of echolocation and was only recruited later.

The *Prestin* gene codes for a transmembrane motor protein that controls the electromotility of the outer hair cells in the cochlea (Li *et al.*, 2008). *Prestin* is considered the mammalian cochlear "amplifier" and is directly responsible for cochlear sensitivity in the mammalian ear. Therefore, changes in *Prestin* may drive the auditory specializations that are seen in echolocating bats (Li *et al.*, 2008). Li *et al.* (2008) amplified, sequenced and analyzed the complete coding sequence of this gene in 22 mammals, including 12 species of bats. Phylogenetic analyses resulted in a basal split between Megachiroptera and Microchiroptera, which is similar to the traditional morphological topology (Simmons and Geisler, 1998). Interestingly, this support derived from the variable amino-acid sites that were contained in the loops and extracellular termini of the protein. Selection tests revealed that the exposed extracellular loops and termini, which are functionally important in the protein, are under positive selection in echolocating bats. This suggests that these parts of the protein sequence were subject to convergent evolution in the echolocating lineages. Phylogenetic analyses of amino acids found in the transmembrane regions alone supported the molecular species tree (Figure 1.4) in which the echolocating bats are paraphyletic (Li *et al.*, 2008). Two recent studies have reported startling evidence of convergent evolution of amino-acid sites present in this gene that are convergently found in the CF/DSC echolocating Old World rhinolophids and hipposiderids and in the echolocating cetaceans (Li *et al.*, 2010; Liu *et al.*, 2010). Both studies concluded that these convergent amino-acid mutations are directly related to echolocation and are driven by natural selection.

Within Pteropodidae there was little evidence of the relaxed selection (Li *et al.*, 2008) that would be expected if this gene experienced loss of function due to loss of echolocation capabilities. However, perhaps this is not surprising as this gene is necessary for correct auditory functioning in mammals and therefore will always be subjected to some level of purifying selection (Teeling, 2009). Interestingly, there are shared amino-acid sites that appear to be under the same selective pressure in pteropodids as in their sister taxa the rhinolophoids, which perhaps is a relict signature of echolocation (Teeling, 2009).

1.6.3   Deafness genes

With two "echolocation genes" providing conflicting signals for the multiple gain and single-gain–single-loss hypotheses of bat echolocation, it is clear that two genes alone are not enough to elucidate the molecular control of this complex process. Echolocation is vertically integrated at the molecular, anatomical, neurophysiological and behavior levels, and also exhibits extensive taxonomic variability. Given this complexity, it seems inescapable that echolocation is underpinned by many genes (Teeling, 2009). Candidate echolocation genes could be genes directly involved in the hearing process in mammals. Known genes that control hearing in mammals are genes that when disrupted or "knocked out" cause deafness. To date ~46 genes are known to be involved in inherited non-syndromic hearing loss (Hilgert *et al.*, 2009). An examination of 15 of these genes across mammals revealed that two of these genes, *Myosin 15a* and *TECTA*, showed evidence of positive selection in bats (Kirwan, 2010), suggesting that these may be good candidate echolocation genes. However, all of the genes examined were functional in both echolocating and non-echolocating bats and the level of positive selection within the echo-locating lineages was not that strong (Kirwan, 2010). This is not surprising as these genes are required for correct auditory functioning in mammals and are under high levels of purifying selection. Perhaps this method is not optimal for finding the "smoking-gun" evidence of echolocation loss in pteropodids. However, further research into this area needs to be carried out before reaching final conclusions.

1.6.4   Where to look next for echolocation genes?

Another method to find genes involved in echolocation is to examine the genomes of closely related individuals or subspecies in a population that show evidence of divergent echolocation capabilities. This type of genomic screening can be achieved by sequencing entire genomes and finding regions that show evidence of divergent selection in echolocating versus non-echolocating lineages. Indeed the Genome 10K consortium of scientists has proposed to sequence a representative species from every bat genus (Genome 10K, 2009), and the results of this endeavor promise to greatly enhance our understanding of the genomic mechanisms that could underlie echolocation. Another search method would be to complete an extensive neutral marker screen (microsatel-lites) across the genome to elucidate if there is any evidence of divergent selection acting on these markers similar to the divergent echolocation states observed within a population (Via, 2009). The genomic regions that contain the selected markers may also contain the echolocation genes or regulatory regions acting on these genes that could directly be involved in echolocation. These types of studies, in concert with functional genomic screens, will greatly enlighten our understanding of the molecular diversity present in the bat genome and may enable us to identify more "echolocation genes" by comparing the genomes of taxa with different sensory capabilities.

However, our interpretation of the evolution of echolocation, even if every bat genome is sequenced, depends on our understanding of the evolutionary history of these species and our interpretation of the fossil record. Efforts to reconstruct the evolutionary history of bats are currently underway. The Tree

Of Life Consortium (http://mammaltree.informatics.sunysb.edu/) is reconstructing evolutionary relationships and divergence times for all families of mammals, including bats, based on large molecular and morphological databases. Robust phylogenies provide an essential framework for reconstructing the evolutionary history of echolocation. Fossils are also critical because of their unique combinations of morphological character states. The recent discovery of *Onychonycteris finneyi* (Simmons *et al.*, 2008), for example, has reopened a long-standing debate (Teeling *et al.*, 2000) on the timing of the origin of flight vs. echolocation (Veselka *et al.*, 2010). New fossil discoveries of stem pteropodids, which may or may not preserve osteological "smoking guns" that are indicative of laryngeal echolocation, are needed to discriminate between the multiple gain hypothesis, which postulates that echolocation evolved independently in at least two microbat lineages, and the single-gain–single-loss hypothesis, which postulates that echolocation evolved once in the common ancestor of living bats and was subsequently lost in the ancestor of Pteropodidae. Ongoing studies of key morphological characters related to echolocation are also allowing paleontologists to evaluate the echolocation capabilities of fossil species (Simmons *et al.*, 2010; Veselka *et al.*, 2010) and are advancing our understanding of when laryngeal echolocation arose in the history of bats. As discussed above, molecular evolutionary and genomic studies are providing additional insights into the evolutionary history of echolocation (Teeling, 2009). However, no one field alone is capable of elucidating the complete evolutionary history of this trait, and it is only by integrating the information contained in phylogenies, fossils and functional genes that we can hope to understand the evolution of laryngeal echolocation.

1.7   Acknowledgments

This study was supported by a Science Foundation Ireland PIYRA, 06/YI3/B932 award to ECT. We would like to thank Nancy B. Simmons and Gregg F. Gunnell for constructive comments on early versions of this manuscript.

1.8   REFERENCES

Ao, L., Mao, X., Nie, W. *et al.* (2007). Karyotypic evolution and phylogenetic relationships in the order Chiroptera as revealed by G-banding comparison and chromosome painting. *Chromosome Research*, **15**, 257–267.

Baker, R. J., Longmire, J. L., Maltbie, M., Hamilton, M. J. and Van Den Bussche, R. A. (1997). DNA synapomorphies for a variety of taxonomic levels from a cosmid library from the New World bat *Macrotus waterhousii*. *Systematic Biology*, **46**, 579–589.

Baker, R. J., Hoofer, S. R., Porter, C. A. and Van Den Bussche, R. A. (2003). Diversification among New World Leaf-Nosed Bats: an evolutionary hypothesis and classification inferred from digenomic congruence of DNA sequence. *Occasional Papers, Museum of Texas Tech University*, **230**, 1–32.

Eick, G. N., Jacobs, D. S. and Matthee, C. A. (2005). A nuclear DNA phylogenetic perspective on the evolution of echolocation and historical biogeography of extant bats (Chiroptera). *Molecular Biology and Evolution*, **22**, 1869–1886.

Eiting, T. P. and Gunnell, G. F. (2009). Global completeness of the bat fossil record. *Journal of Mammalian Evolution*, **16**, 151–173.

Fenton, M. B. and Bell G. P. (1981). Recognition of species of insectivorous bats by their echolocation calls. *Journal of Mammalogy*, **62**, 233–243.

Genome 10K Community of Scientists. (2009). Genome10K: a proposal to obtain whole-genome sequence for 10,000 vertebrate species. *Journal of Heredity*, **100**, 659–674.

Gunnell, G. F. and Simmons, N. B. (2005). Fossil evidence and the origin of bats. *Journal of Mammalian Evolution*, **12**, 209–246.

Gunnell, G. F., Jacobs, B. F., Herendeen, P. S. *et al.* (2003). Oldest placental mammal from sub-Saharan Africa: Eocene microbat from Tanzania – evidence for early evolution of sophisticated echolocation. *Palaeontologia Electronica*, **5**, 10 pages.

Hayden, S., Bekaert, M., Crider, T. A. *et al.* (2010). Ecological adaptation determines functional mammalian olfactory subgenomes. *Genome Research*, **20**, 1–9.

Higgins, D. G. and Sharp, P. M. (1988). CLUSTAL: a package for performing multiple sequence alignment on a microcomputer. *Gene*, **73**, 237–244.

Hilgert, N., Smith, R. J. H. and Van Camp, G. (2009). Forty-six genes causing nonsyndromic hearing impairment: which ones should be analyzed in DNA diagnostics? *Mutation Research/Reviews in Mutation Research*, **681**, 189–196.

Hollar, L. J. and Springer, M. S. (1997). Old World fruitbat phylogeny: evidence for convergent evolution and an endemic African clade. *Proceedings of the National Academy of Sciences, USA*, **94**, 5716–5721.

Hoofer, S. R and Van Den Bussche, R. A. (2003). Molecular phylognetics of the chiropteran family Vespertilionidae. *Acta Chiropterologica*, **5**, 1–63.

Hutcheon, J. M. and Kirsch, J. A. W. (2004). Camping in a different tree: results of molecular systematic studies of bats using DNA-DNA-hybridization. *Journal of Mammalian Evolution*, **11**, 17–47.

Hutcheon, J. M. and Kirsch, J. A. W. (2006). A moveable face: deconstructing the Microchiroptera and a new classification of extant bats. *Acta Chiropterologica*, **8**, 1–10.

Hutcheon, J. M., Kirsch J. A. W. and Pettigrew, J. D. (1998). Base compositional biases and the bat problem. III. The question of microchiropteran monophyly. *Philosophical Transactions of the Royal Society of London B*, **353**, 607–617.

Jones, G. (2005). Echolocation. *Current Biology*, **15**, R484–R488.

Jones, G. and Holderied, M. W. (2007). Bat echolocation calls: adaptation and convergent evolution. *Proceedings of the Royal Society of London B*, **276**, 905–912.

Jones, G. and Teeling, E. C. (2006). The evolution of echolocation in bats. *Trends in Ecology and Evolution*, **21**, 149–156.

Jones K. E., Purvis, A., MacLarnon, A., Bininda-Emonds, O. R. P. and Simmons, N. B. (2002). A phylogenetic supertree of the bats (Mammalia: Chiroptera). *Biological Reviews*, **77**, 223–259.

Kirwan, J. (2010). The molecular evolution of hearing in mammals. Unpublished M.S. thesis, University College Dublin, Ireland.

Koopman, K. F. (1994). Chiroptera: systematics. In *Handbook of Zoology*, vol. 8, pt. 60. *Mammalia*. Berlin, Germany: Walter de Gruyter.

Lander, E. S., Linton, L. M., Birren, B. *et al.* (2001). Initial sequencing and analysis of the human genome. *Nature*, **409**, 860–921.

Li, G., Wang, J., Rossiter, S. J., Jones, G. and Zhang, S. (2007). Accelerated FoxP2 evolution in echolocating bats. *PLoS ONE*, **2**, e900.

Li, G., Wang, J., Rossiter, S. J. *et al.* (2008). The hearing gene Prestin reunites echolocating bats. *Proceedings of the National Academy of Sciences, USA*, **105**, 13959–13964.

Li, Y., Liu, Z., Shi, P. and Zhang, J. (2010). The hearing gene Prestin unites echolocating bats and whales. *Current Biology*, **20**, R55–R56.

Liu, F. R. and Miyamoto, M. M. (1999). Phylogenetic assessment of molecular and morphological data for eutherian mammals. *Systematic Biology*, **48**, 54–64.

Liu, Y., Cotton, J. A., Shen, B. *et al.* (2010). Convergent sequence evolution between echolocating bats and dolphins. *Current Biology*, **20**, R53–R54.

Macdonald, D. W. (2006). *The Encyclopedia of Mammals*. Oxford: Oxford University Press.

Miller-Butterworth, C. M., Murphy, W. J., O'Brien, S. J. *et al.* (2007). A family matter: conclusive resolution of the taxonomic position of the Long-fingered Bats, *Miniopterus*. *Molecular Biology and Evolution*, **24**, 1553–1561.

Neuweiler, G. (1990). Auditory adaptations for prey capture in echolocating bats. *Physiological Reviews*, **70**, 615–641.

Novacek, M. J. (1985). Evidence for echolocation in the oldest known bats. *Nature*, **315**, 140–141.

Pierson, E. D. (1986). Molecular systematics of the Microchiroptera: higher taxon relationships and biogeography. Unpublished Ph.D. thesis, University of California, Berkeley.

Porter, C. A., Goodman, M. and Stanhope, M. J. (1996). Evidence on mammalian phylogeny from sequences of exon 28 of the von Willebrand Factor gene. *Molecular Phylogenetics and Evolution*, **5**, 89–101.

Posada, D. and Crandall, K. A. (1998). Modeltest: testing the model of DNA substitution. *Bioinformatics*, **14**, 817–818.

Rambaut, A. E. (1996). Se-Al: sequence alignment editor. Available via tree.bio.ed.ac.uk/software/seal/.

Simmons, N. B. (1998). A reappraisal of interfamilial relationships of bats. In *Bat Biology and Conservation*, ed. T. H. Kunz and P. A. Racey. Washington, DC: Smithsonian Institution Press, pp. 3–26.

Simmons, N. B. (2005). Order Chiroptera. In *Mammal Species of the World: A Taxonomic and Geographic Reference*, vol. 1, 3rd edn., ed. D. E. Wilson and D. M. Reeder. Baltimore, MD: Johns Hopkins University Press, pp. 312–529.

Simmons, N. B. and Geisler J. H. (1998). Phylogenetic relationships of *Icaronycteris, Archaeonycteris, Hassianycteris*, and *Palaeochiropteryx* to extant bat lineages, with comments on the evolution of echolocation and foraging strategies in Microchiroptera. *Bulletin of the American Museum of Natural History*, **235**, 1–182.

Simmons, N. B., Seymour K. L., Habersetzer J. and Gunnell G. F. (2008). Primitive early Eocene bat from Wyoming and the evolution of flight and echolocation. *Nature*, **451**, 818–821.

Simmons, N. B, Seymour, K. L, Habersetzer, J. and Gunnell, G. F. (2010). Inferring echolocation in ancient bats. *Nature*, **466**, E8–E10.

Springer, M. S., Teeling, E. C., Madsen, O., Stanhope, M. J. and de Jong, W. W. (2001). Integrated fossil and molecular data reconstruct bat echolocation. *Proceedings of the National Academy of Sciences, USA*, **98**, 6241–6246.

Stanhope, M. J., Czelusniak, J., Si, J. S., Nickerson, J. and Goodman, M. (1992). A molecular perspective on mammalian evolution from the gene encoding interphotoreceptor retinoid binding protein, with convincing evidence for bat monophyly. *Molecular Phylogenetics and Evolution*, **1**, 148–160.

Swofford, D. L. (2002). *PAUP*. Phylogenetic Analysis Using Parsimony (*and Other Methods)*. Version 4. Sunderland, MA: Sinauer Associates.

Teeling, E. C. (2009). Hear, hear: the convergent evolution of echolocation in bats? *Trends in Ecology and Evolution*, **24**, 351–354.

Teeling, E. C., Scally, M., Kao, D. J. *et al.* (2000). Molecular evidence regarding the origin of echolocation and flight in bats. *Nature*, **403**, 188–192.

Teeling, E. C., Springer, M. S., Madsen, O. *et al.* (2005). A molecular phylogeny for bats illuminates biogeography and the fossil record. *Science*, **307**, 580–584.

Thomas, J., Moss, C. and Vater, M. (2004). *Echolocation in Bats and Dolphins*. Chicago, IL: University of Chicago Press.

Van Den Bussche, R. A. and Hoofer, S. R. (2004). Phylogenetic relationships among recent chiropteran families and the importance of choosing appropriate out-group taxa. *Journal of Mammalogy*, **85**, 321–330.

Venter, J. C., Adams, M. D., Myers, E. W. *et al.* (2001). The sequence of the human genome. *Science*, **291**, 1304–1351.

Veselka, N., McErlain, D. D., Holdsworth, D. W. *et al.* (2010). A bony connection signals laryngeal echolocation in bats. *Nature*, **463**, 939–942.

Via, S. (2009). Natural selection in action during speciation. *Proceedings of the National Academy of Sciences, USA*, **106**, 9939–9946.

Woodburne, M. O., Gunnell, G. F. and Stucky, R. K. (2009). Climate directly influences Eocene mammal faunal dynamics in North America. *Proceedings of the National Academy of Sciences, USA*, **106**, 13399–13403.

2

Systematics and paleobiogeography of early bats

THIERRY SMITH, JÖRG HABERSETZER, NANCY
B. SIMMONS AND GREGG F. GUNNELL

2.1 Introduction

The phylogenetic and geographic origins of most extant mammalian orders are still poorly documented. Many first appear in the fossil record during the Paleocene-Eocene Thermal Maximum (PETM) at the beginning of the Eocene epoch about 55.5 million years ago (Smith *et al.*, 2006). However, three prominent orders are exceptions to this pattern. Rodents first appeared in North America about 0.5–1.0 million years before the PETM, but probably had an Asian origin like other Glires (Meng *et al.*, 2003). Bats and whales are not known with any certainty before Middle Ypresian, about 54 mya.

The earliest known bats are small, insectivorous forms that are preserved in both terrestrial and lacustrine fossil faunas. Their phylogenetic and geographic origins are still unknown, although the absence of clear transitional forms in the fossil record suggests that bat origins are potentially either quite ancient or their evolution from non-volant mammals was quite rapid. Although morphological evidence has generally supported an origin from within Euarchontoglires, sequence data from multiple genes strongly supports an origin of bats from within Laurasiatheria (Springer *et al.*, 2003; Gunnell and Simmons, 2005).

The oldest known fossil bats are early-middle Early Eocene taxa, and the first members of modern bat families and superfamilies seem to appear in the fossil record in the Middle Eocene (Gunnell and Simmons, 2005). We thus here restrict the term "early bats" to the species known from the Early and early-middle Middle Eocene (Ypresian and Lutetian, and global equivalents, encompassing European mammalian reference levels MP7 through MP13). These early bats mainly include "eochiropterans" (Eochiroptera Van Valen 1979 is a controversial paraphyletic group composed of primitive taxa; see Simmons and Geisler, 1998 for an overview) and a few taxa belonging to the first members of modern families.

Evolutionary History of Bats: Fossils, Molecules and Morphology, ed. G. F. Gunnell and N. B. Simmons. Published by Cambridge University Press. © Cambridge University Press 2012.

The fossil record for early bats is mainly composed of cranial and dental remains, with the exception of two world-famous lagerstätten that have yielded many complete skeletons – the Early Eocene Green River Formation in Wyoming and the Middle Eocene Messel Formation in Germany. However, these two areas account for only a portion of the known diversity of Early and Middle Eocene taxa. In this chapter we review the earliest records of bats from all continents and provide updated diagnoses and discussion based on the fossil material presently available.

The diverse and complex record of Middle Eocene through Lower Oligocene bats from Quercy in France has recently been reviewed by Maitre (2008) and Maitre *et al.* (2008), and will not be repeated here. Updated taxonomy of Middle Eocene Quercy bats (including *Vespertiliavus*, *Hipposideros* (*Pseudorhinolophus*), *Palaeophyllophora*, *Stehlinia*, *Carcinipteryx* and *Necromantis*) can be found in those resources. Further discussion of *Necromantis* can also be found in Chapter 6 in this volume.

2.2   Acronyms for museum collections

AMNH	American Museum of Natural History, New York, NY
BMNH M	The Natural History Museum (British Museum of Natural History), London, UK
CB	Chambi Collection, Tunisia
CM	Carnegie Museum of Natural History, Pittsburgh, PA
CUZ	Université de Montpellier II, Cuzal Locality Collections, France
FDN	Université de Montpellier II, Fordones Locality Collections, France
FNR	Université de Montpellier II, Fournes Locality Collections, France
GMH	Geiseltal Museum, Halle, Germany
GPIMUH	Geologisch-Paläontologisches Institut und Museum, Universität Hamburg, Germany
GU/RSR/VAS	Garhwal University, Srinigar, India
HLMD	Hessisches Landesmuseum, Darmstadt, Germany
IITR/SB/VLM	Indian Institute of Technology, Roorkee, India
IRSNB M	Institut Royal des Sciences Naturelles de Belgique, Brussels, Belgium
IVPP V	Institute of Vertebrate Paleontology and Paleoanthropology, Beijing, China
LIEB-PV	Laboratorio de Investigaciones en Evolución y Biodiversidad, Universidad Nacional de la Patagonia "San Juan Bosco," Argentina

MHNB	Muséum d'Histoire Naturelle de Bâle, Switzerland
MHNG	Muséum d'Histoire Naturelle de Genève, Switzerland
MNHN	Muséum National d'Histoire Naturelle, Paris, France
QM	Queensland Museum, Australia
QP, QW	MNHN, Quercy Collections, France
ROM	Royal Ontario Museum, Toronto, Canada
SMF Me	Senckenberg Museum, Messel Department, Frankfurt, Germany
SMNH	Saskatchewan Museum of Natural History, Regina, Canada
SMNK	Staatliches Museum fur Naturkunde, Karlsruhe, Germany
TNM MP	Tanzanian National Museum, Dar es Salaam, Tanzania
UM2-BRE	Université de Montpellier II, Montpellier, France
UNLSNC	Universidade Nova de Lisboa, Portugal, Silveirinha new collection
UW	University of Wyoming Museum, Laramie, WY
YPM-PU	Princeton University (collections now housed in the Yale Peabody Museum, New Haven, CT)

2.3   Diversity and systematics of early bats

2.3.1   Order Chiroptera Blumenbach, 1779

2.3.2   Family Onychonycteridae Simmons *et al.*, 2008

Synonymy – Eppsinycterididae Hooker, 2010, p. 43.

Contents – *Onychonycteris* Simmons *et al.*, 2008; *Eppsinycteris* Hooker, 1996; *Ageina* Russell *et al.*, 1973; *Honrovits* Beard *et al.*, 1992.

Emended diagnosis – Differs from other bat families by the posteriorly tilted coronoid process of the dentary; P3/p3 and P4/p4 more reduced than in Icaronycteridae, Archaeonycteridae, Hassianycteridae and Palaeochiropterygidae; M1–2 of relatively square shape, especially M1, which is nearly as long as wide with centrocrista that does not reach the labial border; paraconule present and metaconule minute or absent; ectoflexus relatively wide and shallow; p4 metaconid very small to absent; m1–3 with short cristid obliqua and hypoconulid well developed and nearly median on the talonid.

Genus *Onychonycteris* Simmons *et al.*, 2008

Type species – *Onychonycteris finneyi* Simmons *et al.*, 2008.

Type specimen – ROM 55351A, B (Figure 2.1A, C, E–H), part and counterpart of a nearly complete, articulated skeleton including skull with lower jaws, from Finney Quarry, Sandwich Beds (F-2 Layer), Green River Formation, Lincoln County, Wyoming, USA, Late Wasatchian (Wa7).

Included species – Monotypic, includes only *O. finneyi*.

Age and distribution – Late Early Eocene of North America.

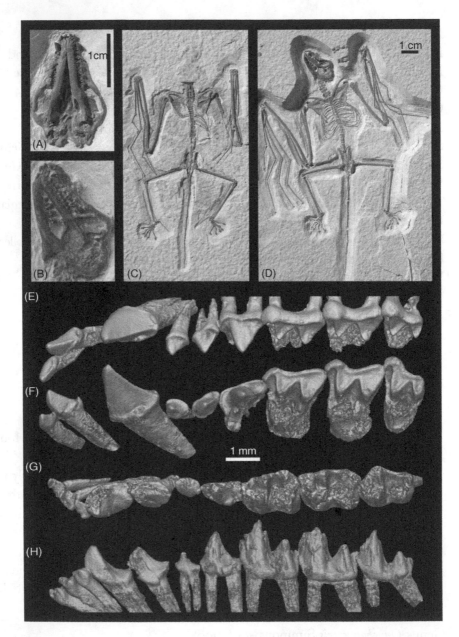

Figure 2.1 Onychonycteridae: *Onychonycteris finneyi* from Green River Formation, USA, Early Eocene, (A) skull in ventral view, (C) skeleton in dorsal view, (E-F) upper dentition in labial and occlusal views, (G-H) lower dentition in occlusal and labial views, all of holotype skeleton (ROM 55351A, B), (B) skull in ventrolateral view and (D) skeleton in ventral view, both of paratype skeleton (unaccessioned).

Emended diagnosis – Differs from all other known bats in having large claws on wing digits I, II and III, and small but distinct claws on digits IV and V; also differs in having relatively longer hindlimbs and relatively shorter forelimbs than any other known bat; differs from *Eppsinycteris* in being larger, in having a more reduced p3 and p4 and by absence of a metaconid on p4; differs from *Ageina* in having extoflexus much shallower on M2 and by absence of a metaconid on p4; differs from *Honrovits* in being larger, in having a P4 wider labiolingually, almost no ectoflexus on M1, p2 somewhat smaller than p4 (nearly identical in size and morphology in *Honrovits*), p4 paracristid oriented mesiodistally instead of angled labially with entire labial margin of p4 more mesiodistally oriented, p4 relatively larger compared to m1 (relatively much smaller in *Honrovits*), lower molar hypoflexids deeper, especially on m1–2, wider and shorter talonid on m3.

Dental characteristics – Primitive chiropteran dental formula I2/i3, C1/c1, P3/p3, M3/m3; I1 smaller than I2 with I2 being nearly caniniform; upper canines dagger-like and bilaterally compressed with weak lingual cingulum, slightly stronger posterior cingulum and no labial cingulum; P2 single-rooted, pointed, simple; P3 double-rooted; P4 triple-rooted, non-molariform; upper molars dilambdodont with parastylar hook, tall and acute cusps, sharply defined ectoloph crests, continuous labial cingulum (protofossa does not extend to cingulum), essentially no ectoflexus on M1 and ectoflexus shallow on M2 and no mesostyles; M1–2 paracone and metacone of equal height; M3 labial margin angled lingually, but with both paracone and metacone present; lower incisors small and bicuspid; lower premolar size = p4 > p2 > p3; p4 non-molariform; lower molars with strong labial cingulids and tall and acute cusps.

Other characteristics – Cochlea and orbicular apophysis relatively small; stylohyal a simple rod with small, rounded cranial tip; metacarpal formula (shortest to longest) I:II:III:IV:V; second phalanx longer than first phalanx in wing digits II–IV, first phalanx longer than second in digits I and V; radius relatively short and humerus relatively long; manubrium with keel-like ventral process; hind legs long and robust with complete fibula; feet with digit I shorter than digits II–V; calcar present; tail long.

Comments – The skeleton of *Onychonycteris finneyi* (Figure 2.1C–D) clearly indicates an animal fully capable of powered flight (Simmons *et al.*, 2008). The relatively small size of the cochlea and the lack of an expanded cranial tip of the stylohyal indicate that *O. finneyi* was probably incapable of echolocation (see Simmons *et al.*, 2010 and Veselka *et al.*, 2010 for further discussion). If true this suggests that flight preceded the ability to echolocate in the evolutionary history of bats.

Genus *Eppsinycteris* Hooker, 1996

> *Type species* – *Eppsinycteris anglica* Cooper, 1932.
>
> *Type specimen* – BMNH M13776 (Figure 2.2K), fragment of a dentary with p3–m3, from Abbey Wood, Blackheath Beds, London Basin, England, Middle Ypresian (MP8+9).
>
> *Included species* – Monotypic, includes only *E. anglica.*
>
> *Age and distribution* – Middle Early Eocene (MP8+9), Europe.
>
> *Diagnosis* – p3 three-rooted, somewhat smaller than p4; p4 trenchant with mesiodistally oriented paracristid, paraconid indistinct, small, high metaconid continuous with distinct crest extending to talonid, bilobed; lower molars with transverse protocristid, trigonid open lingually, large mesially projecting paraconid, trigonids increasingly more compressed from m1 to m3, entoconid very low and crestiform, hypoconulids distinctly lingual of center; horizontal ramus straight; coronoid process high and expanded anteroposteriorly; anterior slope of ascending ramus relatively shallow; angular process slender and slightly deflected laterally.
>
> *Comments* – Hooker (1996) cited a labially bilobed and exodaenodont p4, straight horizontal ramus and increasingly compressed molar trigonids as potentially derived character states shared by *Eppsinycteris*, *Vespertiliavus* (Eocene-Oligocene emballonurid) and extant emballonurids. Previously Cooper (1932) and Russell (1964) had considered *Eppsinycteris* to be either an amphilemurid or a geolabidid, respectively. Storch *et al.* (2002) noted the high and posteriorly angled coronoid process, the high position of the articular condyle and differences in ventral extension of the labial trigonid and talonid bases as being more similar to erinaceomorph lipotyphlans than bats. These authors also cited the relatively weak labial cingulids and the triple-rooted p3 as uncharacteristic of emballonurids (the weak cingulids are almost certainly due to wear, as noted by Hooker and a triple-rooted p3 is uncommon among bats). The family Eppsinycteridingidae recently erected by Hooker (2010) is here considered as a junior synonym of the family Onychonycteridae based on the shared characters with the dentary of *Onychonycteris* (see Figure 2.1).

Genus *Ageina* Russell *et al.*, 1973

> *Type species* – *Ageina tobieni* Russell *et al.*, 1973.
>
> *Type specimen* – MNHN Louis-409 Mu (Figure 2.2G), from Mutigny, northeast of Epernay, Marne, France, Early Eocene (MP8+9).
>
> *Included species* – Monotypic, includes only *A. tobieni.*

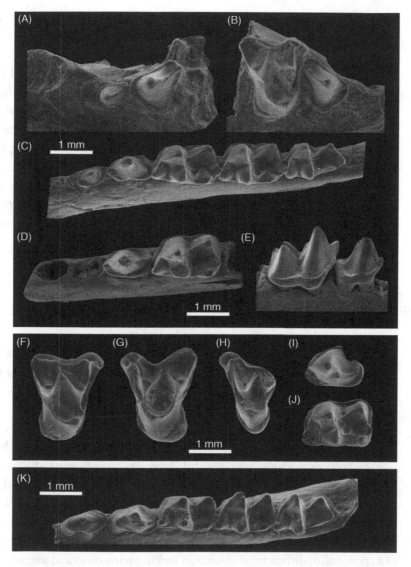

Figure 2.2 Onychonycteridae: *Honrovits tsuwape* from Wind River Formation, USA, Early Eocene, (A-B) left P3–4 and right P4-M1 (CM 62654) in occlusal view, (C) right dentary with p3-m3 (holotype, CM 62641) in occlusal view. "*Hassianycteris*" *joeli* from Belgium, Early Eocene, (D-E), right dentary with p4-m1 (holotype, IRSNB M1567) in occlusal and labial views. *Ageina tobieni* from France, Early Eocene, (F) right M1 (MNHN Louis-375Mu), (G) left M2 (holotype, MNHN Louis-409Mu), (H) left M3 (MNHN Louis-474Mu), (I) right p4 (MNHN Louis-481Mu), and (J) right m1 (MNHN Mu 5112), all in occlusal view. *Eppsinycteris anglica* from Blackheath Beds, England, Early Eocene, (K) right dentary with p3-m3 (holotype, BMNH M13776), in occlusal view.

Age and distribution – Middle-late Early Eocene, Europe; late Early Eocene, North America.

Diagnosis – Differs from other Eocene bats by the combination of: M1–2 relatively long anteroposteriorly, mesostyle absent, paracone and metacone broad labially, ectoflexus between para- and metacone relatively wide and shallow, posterior wall of metacone visible in occlusal view, paraconule and metaconule small or absent, postprotocrista connects directly to base of metacone, trigon basin wide anteroposteriorly, hypocone absent, hypocone lobe weak.

Comments – *Ageina* was considered to be a sister taxon to *Honrovits* by Beard *et al.* (1992) and those authors regarded this clade as the sister group to extant natalids. Simmons and Geisler (1998) considered both of these taxa to be nataloids, but declined to designate a family. Stucky (1984a, 1984b) mentioned the presence of *Ageina* in the late Early Eocene of North America in two separate faunal lists, but provided no further details.

Genus *Honrovits* Beard *et al.*, 1992

Type species – *Honrovits tsuwape* Beard *et al.*, 1992.

Type specimen – CM 62641 (Figure 2.2C), right dentary with p3–m3; from CM locality 1040 Quarry 6, Wind River Formation, Wyoming, USA, Late Wasatchian (Wa7).

Included species – Monotypic, includes only *H. tsuwape*.

Age and distribution – Late Early Eocene, North America.

Diagnosis – Differs from other Chiroptera (see Figure 2.2A–C) in having a distally projecting hypoconulid lobe on m3; further differs from Archaeonycteridae, Palaeochiropterygidae and Hassianycteridae in having a less transverse P4 with reduced lingual lobe, strong mesial and distal crests on p2, p4 simplified and low, centrocrista not reaching labial border of upper molars, lower molars with more nearly median, cuspidate hypoconulid; differs from *Ageina* in lacking paraconid and metaconid on p4; differs from natalids in having upper molars with centrocrista not reaching labial border and with postprotocrista joining lingual base of metacone; further differs from other natalids except possibly *Ageina*, *Chadronycteris* and *Chamtawaria* in having strongly reduced P3/p3.

Dental characteristics – Dentary apparently has a high, posteriorly recurved coronoid process (although this was not documented by any figures in Beard *et al.*, 1992); talonid on m3 unusually elongated due to a strong and distally projecting hypoconulid; p4 lacks paraconid and metaconid.

Comments – *Honrovits* was recognized as a natalid by Beard *et al.* (1992), based on having simplified and relatively homodont lower canine and premolars and relatively simple upper premolars. However, Simmons and Geisler (1998) placed the enigmatic *Honrovits* in Nataloidea *incertae sedis* because its familial affinities were ambiguous. Its unusually strong, median, cuspidate hypoconulid is a primitive character shared with many mammals. This character is unlike that of true natalids, in which the hypoconulids are reduced in size and situated low at the posterolingual corner of the lower molar talonids (Czaplewski *et al.*, 2008).

"*Hassianycteris*" *joeli* from the Late Ypresian of Belgium (Smith and Russell, 1992) differs from other species of *Hassianycteris* by its small size, reduced p4, narrow m1, relatively centrally placed hypoconulid and weak development of nyctalodonty. These characters are similar to those found in *Honrovits* (see Figure 2.2D–E), suggesting that "*Hassianycteris*" *joeli* may prove to be more closely related to the *Honrovits* clade than to any other known group.

2.3.3   Family Icaronycteridae Jepsen, 1966

Contents – monotypic, includes only *Icaronycteris* Jepsen, 1966.
Diagnosis – as for genus.

Genus *Icaronycteris* Jepsen, 1966

Type species – *Icaronycteris index* Jepsen, 1966.
Type specimen – YPM-PU 18150 (Figure 2.3A–B), a nearly complete skeleton from quarry 4 miles southeast of Fossil Butte National Monument, 18-Inch Layer (F-1 Beds), Green River Formation, Lincoln County, Wyoming, USA, Late Wasatchian (Wa7).
Included species – *I. index* Jepsen, 1966 (North America), *I. menui* Russell *et al.*, 1973 (Europe) and *I. sigei* Smith *et al.*, 2007 (India).
Age and distribution – Late Early Eocene of North America (Wa7), middle Early Eocene (MP8+9) of Europe, middle-late Early Eocene of India.
Emended diagnosis – Differs from North American contemporary *Onychonycteris finneyi* in lacking complete claws on hand digits III–V, in having forelimb proportions within the range of all other bats and in having P2/p2 smaller than P3/p3; differs from most bats by having a hypocone present on P4 (absent in *I. menui*). Upper molars with centrocrista reaching the labial margin, but no mesostyle – crista are continuous labially (except perhaps on M3 where the crista may be slightly separated); P4 with strong parastyle; M1 with distinct hypocone cusp.
Dental characteristics – Primitive chiropteran dental formula I2/i3, C1/c1, P3/p3, M3/m3; dentary shallow with narrow symphysial surface and two mental foramina, below i2–3 and below p2; p4-m3 with

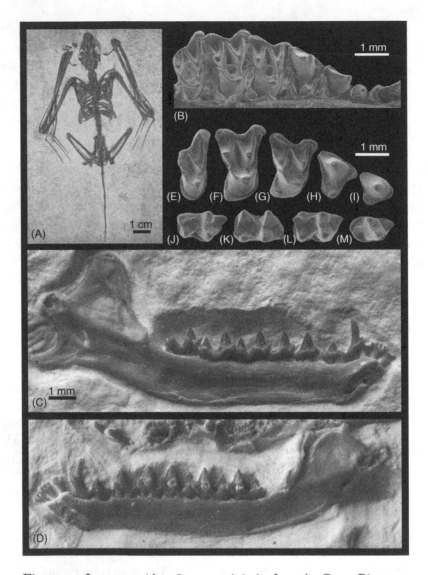

Figure 2.3 Icaronycteridae: *Icaronycteris index* from the Green River Formation, Early Eocene, (A-B) skeleton and occlusal view of right maxilla with C1-M3 (holotype, YPM-PU 18150), (C-D) left and right dentaries with i1-m3 (AMNH 125000) in lingual views. *I. menui* Russell *et al.*, 1973 from France, early Eocene, (E) left M3 reversed (MNHN Louis-387Mu), (F) left M2 reversed (MNHN Louis-416Mu), (G) right M1 (MNHN Mu-6523), (H) right P4 (MNHN Louis-109Mu), (I) right P3 (MNHN Louis-461Mu), (J) right m3 (MNHN Louis-117Mu), (K) left m2 (MNHN Louis-357Mu), (L) right m1 (MNHN Louis-353Mu) and (M) right p4 (MNHN Louis-384Mu), all in occlusal view.

well-developed paraconids and p4 with very prominent metaconid; hypoconulid of m1–3 lingually displaced and connected to entoconid by a crest.

Comments – The possible presence of the genus *Icaronycteris* in the Late Paleocene has been reported from Wyoming, North America (Gingerich, 1987; Bloch and Bowen, 2001). The specimens have been reported under the name Cf. *Icaronycteris* and come from Late Clarkforkian (Cf3) localities in the Northern Bighorn Basin, where the genus *Wyonycteris* is well represented. The latter, a small nyctitheriid-like mammal, has similar size and general morphology to bats, but it does not belong to the order Chiroptera (Hand *et al.*, 1994; Smith, 1995). Czaplewski *et al.* (2008) commented on the fact that *Icaronycteris* cf. *I. index*, a partial skeleton (UW 2244) from the Gosiute Lake of Green River Formation (Novacek, 1987), may represent a second species of *Icaronycteris* from a younger level (Bridgerian) than *I. index*. *Icaronycteris* sp. is also mentioned from late-early and early-middle Eocene localities (Wa7 and Br1a) in Wyoming (see Czaplewski *et al.*, 2008).

As far as its dental morphology is presently known, *Icaronycteris* is characterized by only plesiomorphic features and there is no evidence that it is monophyletic. Species have been referred to the genus primarily because of their age, size, geographic origin and the fact that they lack shared apomorphies with other Eocene bat genera. However, each of the species currently referred to the genus is diagnosable (see Figure 2.3E–M, 2.4A–C). The development of a partial lingual cingulum on the p4 in *I. menui* apparently represents an autapomorphy since *I. index* and *I. sigei* lack this feature. *I. sigei* is 12–20% larger than *I. index* and *I. menui*. The paraconid of p3 is more reduced in *I. sigei* than in *I. index*. The presence of a hypocone on the P4 of *I. index* may be autapomorphic since it is apparently absent in *I. menui* and all other early bats. The symphysis is more elongate posteriorly and the dentary is deeper posterior to m3 at the base of the ascending ramus in *I. sigei* compared to *I. index*. Whether or not these species form a reciprocally monophyletic group relative to other bats remains to be determined. It is possible that *Icaronycteris* as we define it here is representative of a grade of chiropteran evolution rather than a single lineage.

2.3.4 Family Archaeonycteridae Revilliod, 1917

Contents – *Archaeonycteris* Revilliod, 1917; *Protonycteris* Smith *et al.*, 2007.

Emended diagnosis – Ventral process of manubrium oriented approximately 90° to body of manubrium; posterior xiphisternum narrow, without lateral flare; dorsal ischial tuberosity present; M1–2 ectoflexus deep and broad; upper molars lacking mesostyles and hypocones; paraconule very

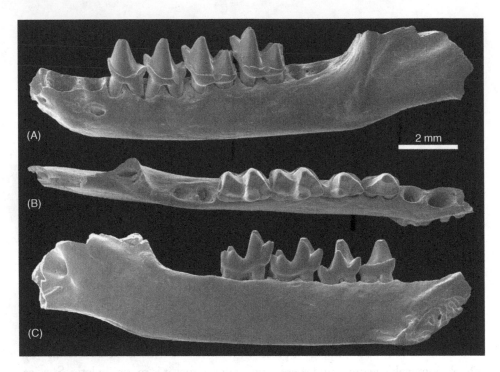

Figure 2.4 Icaronycteridae: *I. sigei* from the Cambay Formation, India, Early Eocene, (A-C) left dentary with p3-m2 (holotype, GU/RSR/VAS 137), in labial, occlusal and lingual views.

weak, metaconule absent; differs from Icaronycteridae in lacking hypocone on upper molars, in having P4 high-crowned with a low protocone shelf that is not lingually expanded and in having p4 with a less well-developed paraconid, metaconid and talonid.

Genus *Archaeonycteris* Revilliod, 1917

Type species – *Archaeonycteris trigonodon* Revilliod, 1917.

Lectotype – HLMD 1398 – Me 33a, b (Figure 2.5) part and counterpart of a very fragmentary partial skeleton including thorax, vertebral column, pelvis, tail, fragments of forelimbs and hindlimbs, from Grube Messel near Darmstadt, Messel Formation, Hesse, Germany, early Lutetian (MP11).

Included species – *A. trigonodon* Revilliod, 1917 (Germany), *A. pollex* Storch and Habersetzer, 1988 (Germany), *A. brailloni* Russell *et al.*, 1973 (France), *A. relicta* Harrison and Hooker, 2010 (England).

Age and distribution – Middle Early Eocene (MP8+9) to late Middle Eocene (MP16?) of Europe; middle-late Early Eocene of India.

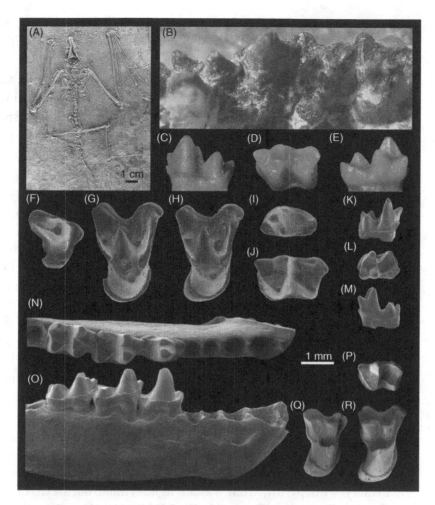

Figure 2.5 Archaeonycteridae: *Archaeonycteris trigonodon* from the Messel Formation, Germany, Middle Eocene, (A) skeleton (SMF Me-80/1379 in ventral view, (B) left maxilla with P2-M2 (SMF Me-663), in occlusal view, (C-E) left m1 (GPIMUH-244), in labial, occlusal and lingual views. *A. brailloni*, France, Early Eocene, (F) right P4 (MNHN Bn-617 Av), (G) right M1 (holotype, MNHN Av-4564), (H) right M2 (MNHN Bn-274 Av), (I) right p4 (MNHN Bn-751 Av), (J) left m1 (MNHN Bn-757 Av), all in occlusal views. *Archaeonycteris? praecursor* from Portugal, Early Eocene, (K-M) right m1 or m2 (holotype, UNLSNC-447), in labial, occlusal and lingual views. *Archaeonycteris? storchi* from Cambay Formation, India, Early Eocene, (N-O) right dentary with p4-m2 (holotype, GU/RSR/VAS 140), in occlusal and labial views. cf. *Archaeonycteris* sp. from Meudon, France, Early Eocene, (P) left m1 or m2 (MNHN 16065 Me), (Q) right M2 (MNHN 15988 Me) and (R) right M2 (MNHN 16379 Me), all in occlusal view. Figure A courtesy of the Senckenberg Forschungsinstitut, Frankfurt.

Emended diagnosis – Differs from *Protonycteris* in having narrower lower molars that have complete labial cingulids and a hypoconulid that is more lingually located on the distal aspect of the tooth.

Dental characteristics – Primitive chiropteran dental formula I2/i3, C1/c1, P3/p3, M3/m3; p4 with relatively small paraconid, metaconid and talonid; lower molars with a cristid obliqua that joins the postvallid relatively labially; upper molars lacking mesostyles, with large protocones and a deep and wide ectoflexus positioned between the parastyle and metastyle on M1–2 (see Figure 2.5).

Comments – The genus *Archaeonycteris* is well represented in Europe with the recently described *Archaeonycteris*? *praecursor* (Tabuce *et al.*, 2009) from Silveirinha (MP7), Portugal being possibly the oldest species (Figure 2.5K–M). The youngest species is *Archaeonycteris relicta* from the late Middle Eocene of Dorset, Southern England (Harrison and Hooker, 2010). The locality of Meudon (MP8+9) in the Paris Basin, France has yielded a few isolated chiropteran teeth that belong to a primitive archaeonycterid probably representing *Archaeonycteris* (Figure 2.5P–R; Smith and Gunnell, unpublished data). *Archaeonycteris*? *storchi* (Smith *et al.*, 2007) from the Early Eocene of India is the only species presently known outside Europe. *A. trigonodon* has p2/P2 and p3/P3 low crowned in comparison with *A. pollex* and *Protonycteris gunnelli*.

Genus *Protonycteris* Smith *et al.*, 2007

Type species – *Protonycteris gunnelli* Smith *et al.*, 2007.

Type specimen – GU/RSR/VAS 436 (Figure 2.6A–C), nearly complete left dentary with p3–m3, from Vastan lignite mine northeast of Surat, Cambay Formation, Gujarat, India, Middle-Late Ypresian.

Included species – monotypic, includes only *P. gunnelli*.

Age and distribution – Middle-Late Early Eocene, India.

Diagnosis – Similar in size to *Archaeonycteris trigonodon*, but differs from that taxon in having wider lower molars with an incomplete labial cingulum and a hypoconulid that is more centrally located on the distal aspect of the tooth.

Dental characteristics – Primitive chiropteran lower dental formula i3, c1, p3, m3, upper dentition as yet unknown; lower molars relatively short and wide, with posterolabially incomplete ectocingulum, strong hypoconid, relatively weak entoconid, slightly lingually shifted hypoconulid and short and weak cristid obliqua; p4 with small, low paraconid and prominent metaconid with long posterior crest; two mental foramina present, one below i3 and one below p2; narrow symphysis.

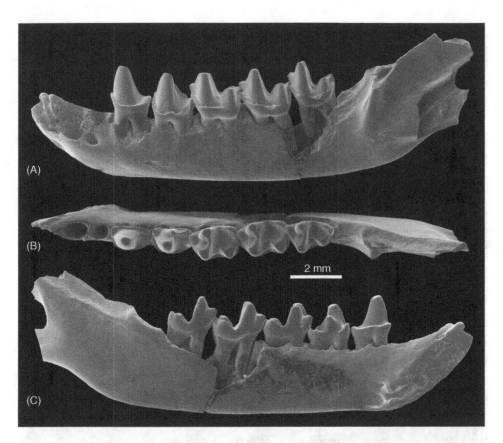

Figure 2.6 *Protonycteris gunnelli* from the Cambay Formation, India, Early Eocene, (A-C) left dentary with p3-m3 (holotype, GU/RSR/VAS 436), in labial, occlusal and lingual views.

Comments – *Protonycteris* is referred to Archaeonycteridae because of the weak entoconid and cristid obliqua that joins the postvallid relatively labially on the lower molars, and a well-developed metaconid on p4. *Protonycteris* is plesiomorphic in the position of the hypoconulid.

2.3.5   Family Palaeochiropterygidae Revilliod, 1917

Contents – Includes *Palaeochiropteryx* Revilliod, 1917; *Cecilionycteris* Heller, 1935; *Matthesia* Sigé and Russell, 1980; *Lapichiropteryx* Tong, 1997; *Microchiropteryx* Smith *et al.*, 2007 and probably *Stehlinia* Revilliod, 1919.

Diagnosis – P3 relatively less reduced than in archaeonycterids (see Figure 2.7); length of postparacrista equals length of premetacrista on M3; where known, upper molars with either one or two acute, narrow ectoflexi, parastyle

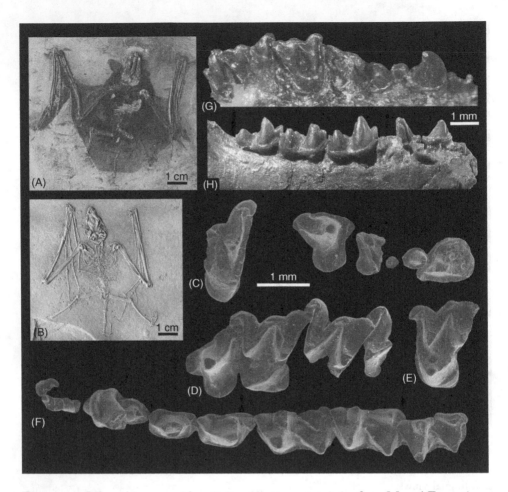

Figure 2.7 Palaeochiropterygidae: *Palaeochiropteryx tupaiodon* from Messel Formation, Germany, Middle Eocene, (A) skeleton (IRSNB M 2014) showing outline of soft tissues, in ventral view, (B) skeleton (SMF ME 10), in dorsal view, (C) right maxilla with CI-P4, M3 (lectotype, HLMD 655 [Me 25]), in occlusal view, (D, F) left maxilla P4-M3 and left dentary i1-m3 (HLMD 4271 [Me 26]), (E) left M2 (HLMD 717 [Me 43]), all in occlusal view. *P. spiegeli* from Messel Formation, (G) right maxillary with I2-M3, in occlusal view and (H) right dentary with m1-m3, in labial view joined with left dentary with p3-p4 in lingual view, of lectotype (HLMD 853 [Me 32]). Figure B courtesy of the Senckenberg Forschungsinstitut, Frankfurt.

strongly hooked, paraconules absent, metaconules minute; p4 metaconid small to nearly absent; lower molar entocristid anteroposteriorly aligned with hypoconulid, forms a 90° angle with postcristid; ventral accessory process present on cervical vertebra 5; infraspinous fossa of scapula relatively broad; ventral projection present on anteromedial flange of scapula.

Genus *Palaeochiropteryx* Revilliod, 1917

Type species – *Palaeochiropteryx tupaiodon* Revilliod, 1917.

Type specimen – HLMD 655 – Me 25 (Figure 2.7C–F), partial skeleton including skull, forelimbs, hindlimbs and tail, from Grube Messel near Darmstadt, Messel Formation, Hesse, Germany, early Lutetian (MP11).

Included species – *P. tupaiodon* Revilliod, 1917 and *P. spiegeli* Revilliod, 1917.

Age and distribution – late Early Eocene (MP10) to early Middle Eocene (MP11), Europe.

Diagnosis – Differs from *Microchiropteryx* in being larger and in having relatively broader molars with hypoconulids slightly more labially placed and a less labially directed cristid obliqua; differs from *Cecilionycteris* by having a weaker mesostyle on upper molars, in lacking double ectoflexi on M1–2 and in having a weaker p4 metaconid; differs from *Matthesia* by having a relatively smaller and less robust upper canine, in retaining small metaconules on upper molars, in having a p4 metaconid and in having sharper, less bulbous cusps.

Comments – A left m1 of *Palaeochiropteryx* cf. *tupaiodon* has been reported from the French locality of Grauves (MP10, late Early Eocene) in the Paris Basin by Russell *et al.* (1973). If correctly allocated, this would constitute the earliest record of *Palaeochiropteryx*.

Genus *Microchiropteryx* Smith *et al.*, 2007

Type species – *Microchiropteryx folieae* Smith *et al.*, 2007.

Type specimen – GU/RSR/VAS 96 (Figure 2.8D–F), left dentary fragment with m3, from Vastan lignite mine northeast of Surat, Cambay Formation, Gujarat, India, Middle-Late Ypresian.

Included species – Monotypic, includes only *M. folieae*.

Age and distribution – Middle-Late Early Eocene, India.

Emended diagnosis – Lower molars very small, long and narrow, and strongly nyctalodont (Figure 2.8A–C); m3 with well-basined, square talonid, entocristid parallel to the long axis of the tooth and postcristid perpendicular to the long axis of the tooth; differs from other palaeochiropterygids in having the hypoconulid of m1 less lingually positioned and from *Palaeochiropteryx tupaiodon* and *P. spiegeli* in having narrower lower molars and more labial cristid obliqua.

Comments – *Microchiropteryx folieae* is the oldest and smallest known palaeochiropterygid, and the smallest Early Eocene bat yet discovered.

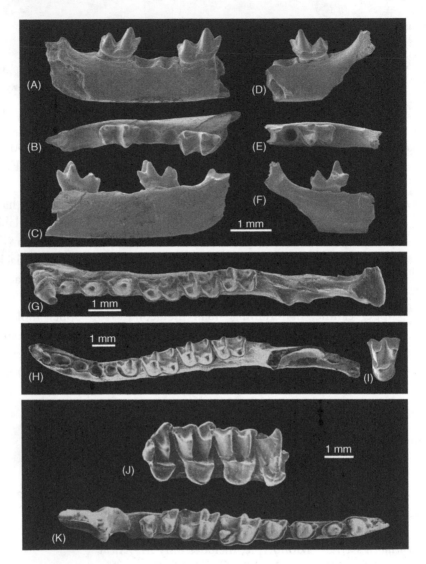

Figure 2.8 Palaeochiropterygidae: *Microchiropteryx folieae* from Cambay
Formation, India, Early Eocene, (A-C) right dentary with m1, m3 (GU/RSR/
VAS 459), in labial, occlusal and lingual views, (D-F) left dentary with m3
(holotype, GU/RSR/VAS 96), in labial, occlusal and lingual views.
Palaeochiropterygidae indet., from Geiseltal Obere Mittelkohle, Germany,
Middle Eocene, (G) left dentary with c1-m3 (GMH XLI-334), in occlusal
view. *Stehlinia* sp. from Quercy, France, Middle Eocene, (H) left dentary with
p4-m3 (CUZ-392), in occlusal view, (I) left M2 (CUZ-385), in occlusal view.
Lapichiropteryx xiei from Shanxi, China, Middle Eocene, (J-K) right
maxillary with P4-M3 and right dentary with c1-m3 (holotype, IVPP 10204,
adapted from Tong, 1997), in occlusal view. Figures H-I courtesy of
Elodie Maitre.

Genus *Cecilionycteris* Heller, 1935

Type species – *Cecilionycteris prisca* Heller, 1935.

Lectotype – GMH 3965, right dentary with c–m3 and laterally compressed palate, from Geiseltal Obere Mittelkohle, Germany, Middle Lutetian (MP13).

Included species – Monotypic, includes only *C. prisca*.

Age and distribution – Middle Middle Eocene (MP13), Europe.

Emended diagnosis – Differs from all other palaeochiropterygids in having double ectoflexi on upper molars; differs from *Palaeochiropteryx* and *Matthesia* in having a distinct, tall metaconid on p4; differs from *Lapichiropteryx* in lacking lower premolar diastemata and in having relatively longer lower premolars; differs from *Matthesia* in having more acute cusps; differs from *Microchiropteryx* in being larger and in having broader molars.

Genus *Matthesia* Sigé and Russell, 1980

Type species – *Matthesia germanica* Sigé and Russell, 1980.

Lectotype – GMH 3940, palate with right I1–M2 and left I1–P3, from Geiseltal Obere Mittelkohle, Germany, Middle Lutetian (MP13).

Included species – *M. germanica* Sigé and Russell 1980, *M? insolita* Sigé and Russell 1980.

Age and distribution – Middle Middle Eocene (MP13), Europe.

Diagnosis – differs from other palaeochiropterygids in having robust, bulbous cusps and a very large upper canine; differs from *Palaeochiropteryx* and *Cecilionycteris* in lacking a metaconid on p4; differs from *Lapichiropteryx* in lacking double ectoflexi on upper molars and in lacking lower premolar diastemata; differs from *Palaeochiropteryx* in lacking any trace of upper molar metaconules.

Comments – Both *Cecilonycteris* and *Matthesia* are only known from Geiseltal. It is possible once dental variation within *Palaeochiropteryx* is better understood that one or both of these genera may turn out to be junior synonyms of that genus.

2.3.6 Family Palaeochiropterygidae?

Genus *Stehlinia* Revilliod, 1919

Type species – *Stehlinia gracilis* Revilliod, 1922.

Type specimen – QP 632, right maxillary fragment with P4–M3, from unknown locality, old Quercy collection, France, early Late Eocene (MP17).

Included Lutetian species – *S. pusilla* Revilliod, 1922 and *S. rutimeyeri* Revilliod, 1922. Two additional Lutetian species of *Stehlinia* recognized by Maitre (2008) await formal description. In all, *Stehlinia* is represented by as many as nine species in Quercy deposits (Maitre, 2008).

Age and Distribution – Middle Middle Eocene (MP13) to early Late Oligocene (MP25), Europe.

Comments – The first occurrence of the genus *Stehlinia* is in the middle Middle Eocene of Cuzal (MP13, Figure 2.8H–I, Marandat *et al.*, 1993; Maitre, 2008). *Stehlinia* has been recently assigned to the Palaeochiropterygidae by Maitre (2008) (see also Gunnell *et al.*, Chapter 7, this volume).

Genus *Lapichiropteryx* Tong, 1997

Type species – *Lapichiropteryx xiei* Tong, 1997.

Holotype – IVPP V10204 (Figure 2.8J–K), left and right dentaries with c-m2 and c-m3 respectively, fragments of left and right maxillaries with P4–M3 and P3–M1 respectively, isolated M2 and M3, and three upper incisors, from Tuqiaogou, east of Zhaili village, Yuanqu County, Shanxi province, China.

Included species – Monotypic, includes only *L. xiei*.

Age and distribution – Late Middle Eocene (Bartonian), China. ·

Diagnosis – Differs from other palaeochiropterygids in having lower premolars that are short, relatively wide and separated by diastemata; upper molars lacking ectoflexi and with poorly developed metastylar shelves.

Comments – *Lapichiropteryx* was placed in Palaeochiropterygidae by Tong (1997). It appears to share a straight m3 entocristid with small hypoconulid positioned posterior to entoconid with *Palaeochiropteryx* although this area is somewhat broken in the type specimen. The shallower upper molar ectoflexi, the lack of upper molar hypocones or hypocone shelves, the poorly developed upper molar metastylar region, the presence of diastemata separating lower premolars, and the presence of short and wide lower premolars are somewhat atypical for a palaeochiropterygid. However, most of these characters are shared with *Stehlinia* from middle Middle Eocene of Cuzal in France. Moreover, both present a similar morphology and reduction of p4 (Smith, unpublished data).

2.3.7   Family Hassianycteridae Habersetzer and Storch, 1987

Contents – *Hassianycteris* Smith and Storch, 1981; *Cambaya* Bajpai *et al.*, 2005a.

Diagnosis – Radius very long and curved; trochiter of humerus extends well beyond humeral head; digit V metacarpal short relative to metacarpals III and IV; dentary robust and relatively deep anteriorly; P2 tiny or absent; p4 with strong (*Cambaya*) to nearly absent (*Hassianycteris*) metaconid and a short talonid; m1–2 with relatively very high and robust protoconid and hypoconid; M1–2 transversely broad and relatively short; upper molars lack an ectoflexus, but labial margin may have shallow indentation between parastyle and mesostyle; hypocone absent or weak; M3 reduced with very long preparacrista extending into strong parastyle hook.

Genus *Hassianycteris* Smith and Storch, 1981

> *Type species* – *Hassianycteris messelensis* Smith and Storch, 1981.
>
> *Type specimen* – HLMD Me 7480a,b, part and counterpart of a partial skeleton (torso and hindlimbs not well preserved), including rostral portion of skull with nearly complete lower jaws, from Grube Messel near Darmstadt, Messel Formation, Hesse, Germany, early Lutetian (MP11).
>
> *Included species* – *H. messelensis* Smith and Storch, 1981 (Germany), *H. magna* Smith and Storch, 1981 (Germany), *H. revilliodi* Russell and Sigé, 1970 (Germany), *H. kumari* Smith *et al.*, 2007 (India).
>
> *Age and distribution* – Early Middle Eocene (MP11) of Europe, middle-late Early Eocene of India.
>
> *Diagnosis* – Protoconid and hypoconid on molar teeth exceptionally tall and robust (Figure 2.9); p4 not molariform, metaconid absent and talonid short; anterior-most upper premolar reduced to tiny peg or absent; mandible deep and thickened dorsoventrally; radius unusually long and strongly curved; trochiter on humerus extends proximally well beyond level of humeral head; metacarpal of digit V relatively short compared to metacarpals III and IV.
>
> *Dental characteristics* – The genus *Hassianycteris* presents derived characters including tall premolariform P_4, tall and straight lower canine, nyctalodont lower molars with wide talonid and well-developed entoconid.
>
> *Comments* – Until recently *Hassianycteris* was essentially restricted to the Middle Eocene German locality of Messel, and was the largest bat known from that locality. *Hassianycteris* has now also been reported from the Early Eocene (Middle-Late Ypresian) of India (Smith *et al.*, 2007). Previously described *Hassianycteris joeli* from the Late Ypresian of Belgium (Smith and Russell, 1992) is here removed from the genus *Hassianycteris* and is tentatively allied with *Honrovits* (see above).

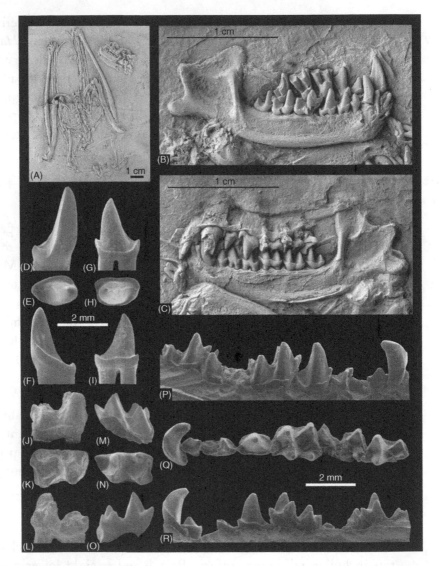

Figure 2.9 Hassianycteridae: *Hassianycteris messelensis* from Messel Formation, Germany, Middle Eocene, (A) skeleton (SMF ME 1414a), (B-C) right and left dentaries and maxillae in occlusion (SMF ME1539a), in labial views. *H. revilliodi* from Messel Formation, (P-R) right dentary c1-m3 (holotype, HLMD 4294 [Me 16]), in labial, occlusal and lingual views. *H. kumari* from the Cambay Formation, India, Early Eocene, (D-F) right c1 (GU/RSR/VAS 61), (G-I) left p4 (holotype, GU/RSR/VAS 59), (J-L) right m1 or m2 (GU/RSR/VAS 56), and M-O) left m1 or m2 (GU/RSR/VAS 561), all in labial, occlusal and lingual views. Figure A courtesy of Senckenberg Forschungsinstitut, Frankfurt.

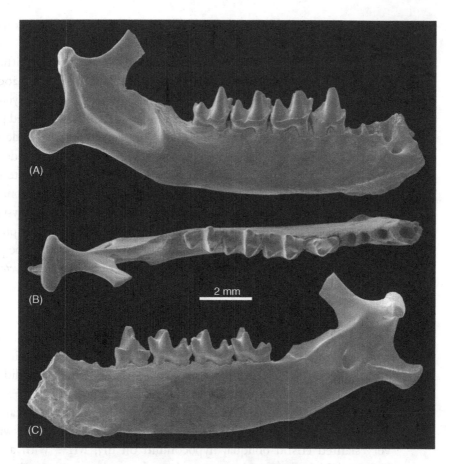

Figure 2.10 *Cambaya complexus* from Cambay Formation, India, Early Eocene, (A–C) right dentary with p4–m3 (GU/RSR/VAS 435), in labial, occlusal and lingual views.

Genus *Cambaya* Bajpai *et al.*, 2005a

Type species – *Cambaya complexus* Bajpai *et al.*, 2005a.

Type specimen – IITR/SB/VLM 508 (Figure 2.10A–C), isolated left p4, from Vastan lignite mine northeast of Surat, Cambay Formation, Gujarat, India, Middle-Late Ypresian.

Included species – Monotypic, includes only *C. complexus.*

Age and distribution – Middle-Late Early Eocene, India.

Diagnosis – Dentary relatively deep, with wide, anteroventrally expanded symphysis and large mental foramen situated between c1 and p2; p4 with very low paraconid, long crest at back of prominent metaconid, short talonid and incomplete lingual cingulum; lower molars narrow

and nyctalodont with long cristid obliqua, relatively low hypoconulid,
high entoconid on m1–2 and complete ectocingulum.

Comments – *Cambaya* was originally described as the first nyctitheriid
insectivore from India based on an isolated p4 (Bajpai *et al.*, 2005a).
Smith *et al.* (2007) referred it to Chiroptera, possibly as a hassianycterid.
Cambaya complexus presents several derived features such as reduced paraconid and talonid on p4, nyctalodont lower molars and a deeper dentary
than in all other early bats, except perhaps *Hassianycteris*. These derived
characters exclude allocation to Icaronycteridae or Archaeonycteridae.
In some features *Cambaya* resembles Palaeochiropterygidae, but the
morphology of m3 and the deep dentary with expanded symphysis are
not typical of that family. The derived characters of *C. complexus* are present
in Hassianycteridae, so we provisionally refer it to that family. The presence of a metaconid on p4 distinguishes *Cambaya* from *Hassianycteris*.

2.3.8   Family Mixopterygidae Maitre *et al.*, 2008

Contents – *Mixopteryx* Maitre *et al.*, 2008; *Carcinipteryx* Maitre *et al.*, 2008.

Diagnosis – Family associating dental characters usually observed either in
emballonurids or hipposiderids; p4 with sinuous labial cingulid and
small lingual cusps at the anterior and posterior extremities; m1–2 with
nyctalodont pattern; median entoconid on the lingual edge of talonid;
marked lingual hypoconulid, forming a crest and projected backwards;
very slanted cristid obliqua; hypoconulid on m3; M1–2 with a heel;
low-cut labial edge between parastyle and prominent mesostyle; more
marked external inclination of the angular process and of the top of
ramus with respect to dentary axis than in emballonurids.

Genus *Carcinipteryx* Maitre *et al.*, 2008

Type species – *Carcinipteryx trassounius* Sigé, 1988.

Type specimen – UM2-BRE2–750, right M1, from Le Bretou (MP 16),
Tarn-et-Garonne Dept., Quercy Phosphorites, France.

Included Lutetian species – *C. maximinensis* Maitre *et al.*, 2008. An additional unnamed species of *Carcinipteryx* is recognized from European
reference level MP13 (Figure 2.11D; Maitre, 2008). In all, four species
of *Carcinipteryx* are known from Quercy deposits.

Age and distribution – Middle Middle Eocene (MP 13) to Early Oligocene
(MP 20–21), Europe.

Diagnosis – Dental formula of I?/i2 C1/c1 P?/p2 M3/m3 (Figure 2.11).
Genus characterized by very small size; long lower molars with

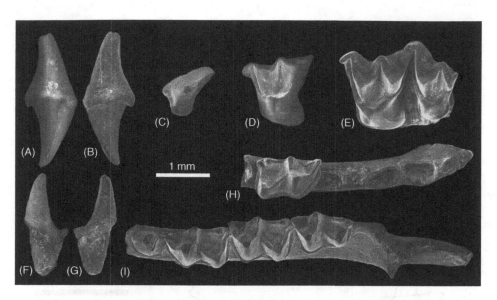

Figure 2.11 Mixopterygidae: *Carcinipteryx maximinensis* from Gard Phosphorites, France, Middle Eocene, (A-B) right C1/ (SMXG2–263) in labial and lingual views, (C) left P4 (SMXCm C.1.10), in occlusal view, (E) left maxillary with M2–3 (holotype, SMXG2–264), in occlusal view, (F-G) left c1 (SMXCm 7), in labial and lingual views, (H) right dentary with m1 (SMXCm A.1.3), in occlusal view and (I) left dentary with m1–3 (SMXCm A.1.5), in occlusal view. *Carcinipteryx* sp., (D) left M1 (SMXGI-309) in occlusal view. All figures reproduced by permission of Schweizerbart Publishers.

nyctalodont pattern, open trigonid and a crest-shaped hypoconulid, projected backwards and located on the lingual edge; cristid obliqua linked more lingually to the trigonid than on the molars of hipposiderids; M3/m3 slightly reduced; complete C1 cingulum with a notch on the lingual side; posterolingual cusp present; quadrate-shaped M1–2, pre- and postprotocrista linked to the pre- and postcingulum, low-cut labial edge between parastyle and mesostyle.

Comments – The genus *Mixopteryx* is not recorded before the Late Eocene (MP17a) and only *Carcinipteryx* is known from the Middle Eocene.

The family Mixopterygidae shows a mosaic of dental characters, some of them being typical of emballonurids (especially *Vespertiliavus*), and others of hipposiderids (Maitre *et al.*, 2008). Emballonurid-like characters include the triangular P4 in occlusal view with developed parastyle, large paracone, lingual lobe-shaped cingulum; the large upper molars with labial notch anterior to the mesostyle, well-developed parastyle, open trigon, short postprotocrista not linked to the metacone base, presence of a small hypocone and rotation of the ramus with respect to the horizontal axis. Hipposiderid-like characters include upper molars

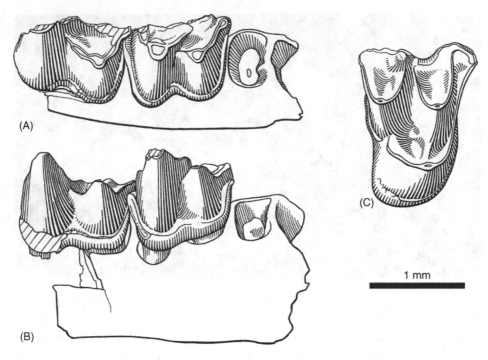

Figure 2.12 Philisidae: *Dizzia exsultans* from Chambi, Tunisia, Early Eocene, (A) left dentary with broken m1–2 (CB 1–15), in occlusal and labial views, (C) right M1 or M2 (holotype, CB 1–17) in occlusal view. A–C reproduced by permission of Schweizerbart Publishers.

with quadrate occlusal contour, postprotocrista linked to the postcingulum and a protruding mesostyle; M3 slightly reduced as in *Vaylatsia*, trigon with four crests, or else very reduced as in *Palaeophyllophora*, no mesostyle and trigon with two crests; m1 with rather narrow trigonid; m3 with hypoconulid and wide trigonid.

2.3.9   Family Tanzanycteridae Gunnell *et al.*, 2003

Contents – Monotypic, includes only *Tanzanycteris* Gunnell *et al.*, 2003.
Diagnosis – As for genus.

Genus *Tanzanycteris* Gunnell *et al.*, 2003

Type species – *Tanzanycteris mannardi* Gunnell *et al.*, 2003.
Type specimen – TNM MP-207 (Figure 2.12A–B), partial skeleton including skull, axial skeleton anterior to sacrum, partial left and right humeri, and partial left radius from Mahenge west of Singida, north-central Tanzania, Lutetian.
Included species – Monotypic, including only *T. mannardi*.

Age and distribution – Middle Middle Eocene, Africa.

Diagnosis – Differs from *Icaronycteris*, *Onychonycteris*, *Palaeochiropteryx*, *Archaeonycteris* and *Hassianycteris* in having proportionally a much larger cochlea; differs from *Onychonycteris*, *Icaronycteris* and *Palaeochiropteryx* in having a broad first rib; differs from all but *Hassianycteris* in having the trochiter of humerus extending proximally well beyond the humeral head; further differs from *Onychonycteris*, *Icaronycteris*, *Archaeonycteris* and *Hassianycteris* in having anterior laminae on ribs and from all but *Hassianycteris* in having a manubrium with a bilaterally compressed ventral keel; further differs from *Palaeochiropteryx* in having a narrow scapular infraspinous fossa.

Comments – *Tanzanycteris* shares several apparently derived character states with *Hassianycteris*, including a compressed ventral process of the manubrium, presence of posterior laminae on ribs, presence of a dorsal scapular articular facet and a trochiter that extends beyond the humeral head. Only the last of these is shared exclusively with *Hassianycteris* however, while the others are also found in *Palaeochiropteryx*. *Tanzanycteris* has a greatly enlarged cochlea that is shared only with living rhinolophids and a single species of the mormoopid *Pteronotus* (Gunnell *et al.*, 2003). The phylogenetic affinities of *Tanzanycteris* remain ambiguous; it may nest within crown clade Chiroptera (with other taxa possessing a very large cochlea, which is indicative of high duty-cycle echolocation) or represent another chiropteran stem group.

2.3.10 Family ?Philisidae Sigé, 1985

Genus *Dizzya* Sigé, 1991

Type species – *Dizzya exsultans* Sigé, 1991.

Type specimen – CB 1–17 (Figure 2.13C), M1 or M2, from Chambi, Tunisia, Late Ypresian.

Included species – Monotypic, includes only *D. exsultans*.

Age and distribution – Late Early Eocene, Africa.

Diagnosis – Small size; metacone slightly more developed than paracone; weak paraloph and metaloph; postprotocrista and postcingulum weakly continuous; lingual cingulum weak; talonid broadest in middle; hypoconid moderately high; nyctalodont (Figure 2.13A, B).

Comments – *Dizzya* was described based on a single upper molar and a lower dentary with two broken teeth (Sigé, 1991). The upper molar is distinctly philisid-like in having long and parallel postpara- and premetacrista that extend to the labial margin but do not join, leaving the labial portion of the protofossa open. Other features of the upper

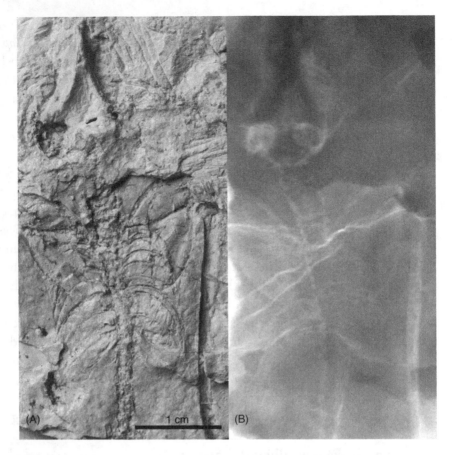

Figure 2.13 Tanzanycteridae: *Tanzanycteris mannardi* from Mahenge, Tanzania, Middle Eocene, skeleton (holotype, TNM MP-207), in ventral view, (A) photograph, (B) x-ray of specimen.

molar that resemble philisids include a strongly hooked parastyle and a small but distinct mesostyle placed just posterior to the labial terminus of the premetacrista. However the referred lower jaw is distinctly unlike philisids in being nyctalodont and in having relatively short and straight to modestly lingually angled cristid obliqua. It is possible that the upper tooth and lower jaw do not represent the same taxon, but more material is needed to make this determination (see also Gunnell *et al.*, Chapter 7, this volume).

2.3.11   Family Emballonuridae Gervais, 1855

Genus *Tachypteron* Storch *et al.*, 2002

Type species – Tachypteron franzeni Storch *et al.*, 2002.

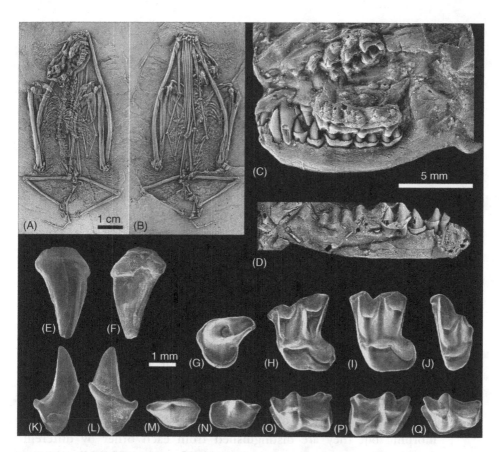

Figure 2.14 Emballonuridae: *Tachypteron franzeni* from the Messel Formation, Germany, Middle Eocene, (A-B) skeleton (holotype, IRSNB M1944a,b) in dorsal and ventral views, (C-D) left and right dentaries and maxillae (paratype, SMNH Me 676a+b), in labial views. *Vespertiliavus schlosseri* from Quercy, France, Middle Eocene, (E-F) right C1 (SLI VsD.4.8), (G) left P4 (SLI VsD.1.4), (H) left M1 (SLI VsD.1.3), (I) left M2 (SLI VsD.4.17), (J) left M3 (SLI VsD.2.25), (K-L) right c1 (SLI VsD.3.25), (M) left p2 (SLI VsD.3.1), (N) right p4 (SLI VsD.4.7), (O) left m1 (SLI VsD.2.6), (P) right m2 (SLI VsD.1.19), (Q) left m3 (SLI VsD.2.22). E-F and K-L in labial and lingual views, all others in occlusal view. Figures A-D courtesy of the Senckenberg Forschungsinstitut, Frankfurt. Figures E-Q courtesy of Elodie Maitre.

Type specimen – IRSNB M 1944a,b (field number BE4–119a+b) (Figure 2.14A, B), part and counterpart of a complete, articulated skeleton, from Grube Messel near Darmstadt, Messel Formation, Hesse, Germany, early Lutetian (MP11).

Included species – Monotypic, includes only *T. franzeni*.

Age and distribution – Early Middle Eocene (MP11), Europe.

Diagnosis – Differs from *Vespertiliavus* (Figure 2.14) in having only two upper premolars and retaining three lower premolars that progressively increase in size from anterior to posterior; differs from *Vespertiliavus* and extant emballonurids (except *Emballonura*) in having two upper incisors; differs from extant emballonurids in retaining three lower premolars and from extant *Taphozous* and *Saccolaimus* in retaining three lower incisors; differs from extant emballonurids and *Vespertiliavus* in lacking a postorbital process on the skull; postcranially differs from *Taphozous* in having a longer tail and a shorter calcar; differs from *Vespertiliavus* in having a slightly stronger styloid process on the distal humerus.

Other characteristics – Upper and lower canines are moderately high, labially swollen and lack labial cingular structures; facial part of the skull long; postcranial skeletal morphology and proportions essentially as in living *Taphozous*.

Comments – *Tachypteron franzeni* is the oldest known emballonurid and the oldest representative of an extant clade. The holotype is extraordinarily well preserved and shows perfectly most of the bones and articulations indicating a highly specialized rapid-flight bat with strikingly narrow wings (Storch *et al.*, 2002). Among emballonurids, the dentitions of *T. franzeni* and *Vespertiliavus* are plesiomorphic but they are distinguished from each other by different apomorphies suggesting an early diversification of Middle Eocene emballonurids.

Genus *Vespertiliavus* Schlosser, 1887

Type species – *Vespertiliavus bourguignati* Filhol, 1877.

Type specimen – MHNG unnumbered syntypes, figured by Filhol (1876, Pl. 7, Fig. 5–8) from several localities, Lot, Phosphorites du Quercy, France, age indeterminate (probably MP17 following the reference population of La Bouffie *sensu* Maitre (2008).

Included Lutetian species – *V. lapradensis* Sigé 1990. In all, *Vespertiliavus* is represented by as many as nine species in Quercy deposits (Maitre, 2008).

Age and distribution – Middle Middle Eocene (MP13) to early Late Oligocene (MP25), Europe.

Comments – The genus *Vespertiliavus* (Figure 2.14E-Q) first occurs in the late Middle Eocene (Late Lutetian) of Cuzal with the species *V. lapradensis* (Marandat *et al.*, 1993).

2.3.12   Family Hipposideridae, in Flower and Lydekker, 1891

Genus *Palaeophyllophora* Revilliod, 1917

Type species – *Palaeophyllophora oltina* Delfortrie, 1873.

Type specimen (Lectotype) – MHNG unnumbered, a skull figured by Delfortrie (1873, fig. 1, p.712) from Ste-Néboule de Béduer, Lot, Phosphorites du Quercy, France, Late Eocene (MP18).

Included Lutetian species – *Palaeophyllophora* is represented by two unnamed species in European reference level MP13 (Maitre, 2008). In all, *Palaeophyllophora* is represented by as many as seven species in Quercy deposits (Maitre, 2008).

Age and distribution – Middle Middle Eocene (MP13) to middle Early Oligocene (MP23), Europe.

Comments – The genus *Palaeophyllophora* is present at Cuzal (MP13) based on a p4 that has a morphology close to the one of *P. oltina* and *P. quercyi* (Maitre, 2008).

Genus *Hipposideros* Gray, 1831

Type species – *Hipposideros* (*Pseudorhinolophus*) *morloti* Pictet, in Pictet et al., 1855.

Type specimen – MHNB unnumbered, syntypes, figured by Pictet (1857, Pl. 6, Fig. 1–10) from Les Alleveys, Mormont, Switzerland, Late Eocene (MP19).

Included Lutetian species – A single indeterminate species of *Hipposideros* has been recognized from MP13 by Maitre (2008). In all, as many as ten species of *Hipposideros* may be represented in Quercy deposits (Maitre, 2008).

Age and distribution – Middle Middle Eocene (MP13) to Middle Miocene, Europe.

Comments – The genus *Hipposideros* is mentioned from Chamblon (MP13), Switzerland, based on three isolated teeth (Maitre, 2008).

2.3.13   Family indeterminate

Genus *Necromantis* Weithofer 1887

Type species – *Necromantis adichaster* Weithofer, 1887.

Type specimen – QW 627 (see Chapter 6 for figures of *Necromantis*), dentary fragment, from unknown locality, old Quercy collection, France, Late Eocene (MP17).

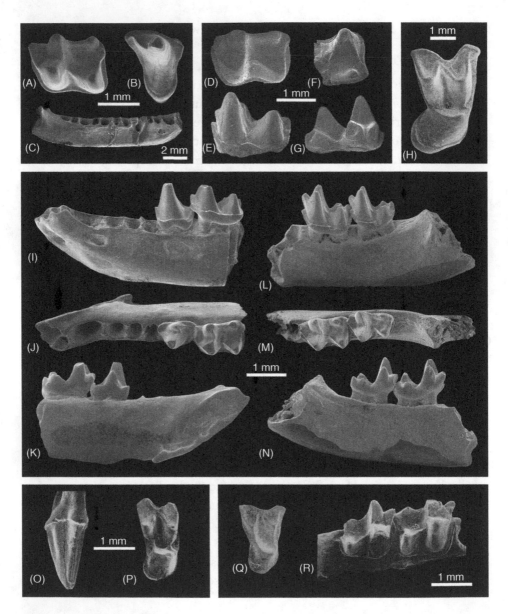

Figure 2.15 Bats unidentified at the family level. *Australonycteris clarkae*, from
Australia, Early Eocene, (A) left m2? (holotype, QM F19147), in occlusal view, (B)
right P4 (QM F 19148), in occlusal view, (C) right edentulous dentary (QM F19149), in
lateral view. Unidentified bat from Laguna Fría, Argentina, Early Eocene (Tejedor
et al., 2005), (D-E, G) left m1 or m2 (LIEB-PV 999), in occlusal, labial and lingual
views, (F) right lower molar talonid (LIEB-PV 1000) in occlusal view. *Necromantis* sp.
from Quercy, France (Marandat *et al.*, 1993), Middle Eocene, (H) right M1 or M2 (CUZ-
383), in occlusal view. *Jaegeria cambayensis* from the Cambay Formation, India, Early
Eocene, (I-K) left dentary with p4-m1 (GU/RSR/VAS 458) in labial, occlusal and lingual

Included Lutetian species – Two new species of *Necromantis* have been described by Hand *et al.* (see Chapter 6, this volume). Three species of *Necromantis* are now represented in Quercy deposits (Maitre, 2008).

Age and distribution – Middle Middle Eocene (MP13) to middle Late Eocene (MP17), Europe.

Comments – The presence of the genus *Necromantis* is documented from the middle Middle Eocene locality of Cuzal by Marandat *et al.* (1993) and Maitre (2008) (see Figure 2.15H).

Genus *Jaegeria* Bajpai *et al.*, 2005b

Type species – *Jaegeria cambayensis* Bajpai *et al.*, 2005b.

Type specimen – IITR/SB/VLM/585, left m1 or m2, from Vastan lignite mine northeast of Surat, Cambay Formation, Gujarat, India, Middle-Late Ypresian.

Included species – Monotypic, includes only *J. cambayensis.*

Age and distribution – Middle-Late Early Eocene, India.

Emended diagnosis – Dentary shallow (Figure 2.15I–N) with two mental foramina, one below i3 and the other below the posterior end of the canine alveolus; p4 with low paraconid, well-developed metaconid and short, rounded talonid; m1–3 nyctalodont with very lingual hypoconulid just posterior to entoconid and long cristid obliqua forming an acute angle with the posterior crest; m3 with somewhat reduced talonid basin.

Comments – *Jaegeria* was originally described as a herpetotheriine marsupial by Bajpai *et al.* (2005b), based on a single tooth and some unassociated tooth fragments. Smith *et al.* (2007) subsequently referred additional material to *Jaegeria* and recognized its true chiropteran affinities. However, the material now known to represent this genus is insufficient to definitively ascertain its relationships.

Caption for Figure 2.15 (*cont.*) views, (L-N) left dentary with m2–3 (GU/RSR/VAS 100), in labial, occlusal and lingual views. Chiroptera indet. 1 from Fordones, France (Marandat, 1991), Early Eocene, (O) right C1 (FDN-267), in lingual view, (P) right M1 or M2 (FDN-265), in occlusal view. Chiroptera indet. 2 from Fournes, France (Marandat, 1991), Early Eocene, (Q) right M1 or M2 (FNR-46), in occlusal view and (R) right dentary with m2–3 (FNR-02) in occlusal view. Figures A-G © Copyright 2011 The Society of Vertebrate Paleontology. Reprinted and distributed with permission of the Society of Vertebrate Paleontology. Figure H reproduced from Marandat *et al.* (1993), Geobios, Copyright (1993), with permission from Elsevier. Figures O-R reproduced from Marandat (1991) and used with permission of Palaeovertebrata (Henri Cappetta).

Genus *Australonycteris* Hand *et al.*, 1994

> *Type species* – *Australonycteris clarkae* Hand *et al.*, 1994.
>
> *Type specimen* – QM F19147 (Figure 2.15A), left m2, from Main Quarry, Tingamarra property, Boat mountain region, Murgon, Murgon Formation, southeast Queensland, Australia.
>
> *Included species* – Monotypic, includes only *A. clarkae.*
>
> *Age and distribution* – Early Eocene, Australia.
>
> *Diagnosis* – Long and slender dentary (Figure 2.15A–C); lower molars with cusps distinct, separate and conical; talonid cusps relatively short; hypoconulid lingually displaced, but without crest connecting to entoconid; P4 wider than long with lingual lobe slender and lingually expanded.
>
> *Comments* – A fragmentary periotic was described by Hand *et al.* (1994) as possibly representing *Australonycteris*. The periotic includes such chiropteran-like features as a flattened promontorium with thin bone enclosing the labyrinth, the presence of an anteromedial flange on the promontorium, a large basal cochlear turn and the presence of well-developed lamina ossea spiralis primaria and secundaria. All of these characteristics suggest that the periotic represents a chiropteran and most probably an echolocating taxon. Additional evidence is needed to confirm that this periotic does represent *Australonycteris* and to test the hypothesis that *A. clarkae* was capable of echolocation.

The age of *Australonycteris clarkae* was cited as 54.6 Ma by Hand *et al.* (1994). If this age is correct it would make *A. clarkae* one of the oldest known bats discovered to date.

2.3.14   Chiroptera *incertae sedis*

Genus *Wallia* Storer, 1984

> *Type species* – *Wallia scalopidens* Storer, 1984.
>
> *Type specimen* – SMNH P1654.312, right M1, Saskatchewan, Canada, Late Uintan (Ui3).
>
> *Included species* – Monotypic, includes only *W. scalopidens.*
>
> *Age and distribution* – Late Middle Eocene, North America.
>
> *Diagnosis* – Upper molars strongly dilambdodont, oblique postprotocrista, indistinct para- and metaconules, large, low hypocone (on M1–2), strong preprotocrista forming anterior shelf extending to distinct parastylar hook; lower molar with anteroposteriorly compressed, oblique trigonid and talonid, well-developed labial and posterior cingulid, and hypoconulid as high as postcristid.

Comments – *Wallia* was originally considered to be a proscalopid lipotyphlan (Storer, 1984), but Legendre (1985) referred it to Molossidae based on the presence of shelf-like anterior cingulum continuous with the preprotocrista and the presence of a distinct hypocone. Storer (1984) made careful comparisons with a variety of chiropterans and was struck by the resemblances between the upper molars of *Wallia* and those of palaeochiropterygids, especially *Cecilonycteris*. However, Storer (1984) also noted that the referred lower molar was unlike any chiropteran and therefore felt that the affinities of *Wallia* must lie elsewhere. The upper molars are also similar to those of some nyctitheriid mammals such as *Pontifactor* and *Wyonycteris*. These taxa are characterized by a long centrocrista reaching the labial border, a strong mesostyle and the presence of a hypocone (Smith, 1995; Smith, unpublished data). More complete specimens are required to confirm the taxonomic status of *Wallia*.

2.3.15   Other early bats

The discovery of two lower molars of bats at Laguna Fría, an Early Eocene locality in northwestern Chubut Province, Argentina (Figure 2.15D–G), has been reported by Tejedor *et al.* (2005): a well-preserved left m?2 (LIEB-PV 999) and a right talonid (LIEB-PV 1000). Before that, the oldest South American record of a potential bat was a broken lower molar of Middle or Late Eocene age from the Santa Rosa locality (?Yahuarango Formation), in Peru (Czaplewski and Campbell, 2004). None of these specimens has been assigned to any chiropteran family, living or extinct, because they are too incomplete.

Other bat specimens have been reported from the Early Eocene of Fordones and Fournes (Figure 2.15O–R) in southern France (Marandat, 1991). Undescribed bat teeth are known from Meudon in France (see Figure 2.5P–R). Tong (1997) recorded the possible presence of *Icaronycteris*, an indeterminate archaeonycterid, and indeterminate chiropterans from late Middle Eocene of Shanxi and Henan Provinces in China, but details of these finds have never been provided. The upper premolar described as a possible P3 of *Icaronycteris* resembles a P3 of a palaeochiropterygid by the rounded outline. Moreover, its size matches that of *Lapichiropteryx*. An archaic bat with a hypoconulid-entoconid cristid has also been recently recorded from the Early Eocene El Kohol area in Algeria (Ravel *et al.*, 2010). In addition to *Ageina*, Stucky (1984a, 1984b) reported the presence of *Icaronycteris* sp. from the late Early (Wa7) and early Middle (Br1a) Eocene in Wyoming.

2.4   Discussion

Recent work on the "hearing gene" (Li *et al.*, 2008) *Prestin* has shown that bats and cetaceans have convergently developed amino-acid mutations that may be directly related to echolocation (Li *et al.*, 2010; Liu *et al.*, 2010; also see Teeling *et al.*, Chapter 1, this volume). There may be some biogeographical similarities between bats and whales as well, because of their locomotor specializations (flying or swimming). It is possible that bats and whales may have been able to undergo relatively widespread geographic dispersal, whereas other mammals may have been less mobile and thus less likely to have dispersed rapidly from their sites of origin. However, there is no hard evidence to indicate early colonization by either bats or whales, which could suggest that they either have a later phylogenetic divergence or a different geographic origin, or their fossil record is less complete than that of other mammals (which is certainly the case for bats, see Teeling *et al.*, 2005 and Eiting and Gunnell, 2009).

Teeling *et al.* (2005) utilized a molecular phylogenetic approach to examine biogeographic patterns within Chiroptera. Based on this analysis, these authors hypothesized that bats originated near the K-T boundary in Laurasia, perhaps in North America, although this is far from certain. Further, they suggested that three of the major microbat clades (emballonuroids, vespertilionoids and rhinolophoids) originated in Laurasia, whereas noctilionoids were more likely to have originated in Gondwana, perhaps in South America. Our review of the earliest fossil records of bats from all relevant continents supports the likely Laurasian origin of most major microbat clades.

Since the description of *Icaronycteris index* from the late Early Eocene of Wyoming (Jepsen, 1966), the search for older and/or more primitive bats in North America has been ongoing (Gingerich, 1987; Novacek, 1987; Beard *et al.*, 1992; Bloch and Bowen, 2001). Despite extensive field collecting in Paleocene deposits, no bats are known with certainty before Wasatchian Zone Wa7 (late Early Eocene) in North America. Three genera of certain bats are known from the Early Eocene in North America, *Icaronycteris*, *Onychonycteris* and *Honrovits*. The recently described *Onychonycteris* is restricted to North America and is the most primitive bat yet discovered based on skeletal morphology (Simmons *et al.*, 2008). It belongs with *Honrovits* in the family Onychonycteridae that is also present in Europe (Figure 2.16). *Icaronycteris* is known from Europe and the Indian subcontinent, as well as North America, although the affinities of the Old World species remain somewhat uncertain. There are no certain bats in the Middle Eocene in North America, with *Wallia* being the only potential chiropteran; however its affinities are doubtful.

The diversity of Early Eocene bats is relatively high in Europe, despite being restricted to just a few localities. At least six genera belonging to four families

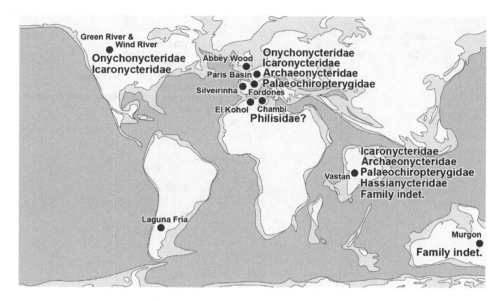

Figure 2.16 Paleogeographic map of the Early Eocene (modified from Scotese, 2006) showing the distribution of bat records.

are recognized and several other Early Eocene bat taxa have yet to be described (Marandat, 1991; Smith, unpublished data). Moreover, bats are well represented from the middle Early Eocene in Europe. No bats are recorded from the earliest Eocene (Paleocene-Eocene boundary), but some early-middle Early Eocene localities (Silveirinha, Meudon) have yielded bat teeth.

By the Middle Eocene, at least 12 genera belonging to at least six families are present in Europe (Figure 2.17) including several modern bat groups (Gunnell and Simmons, 2005). Nearly half of this diversity comes from the oldest localities (MP13–16) of the Quercy area (Maitre *et al.*, 2008).

The bat fauna from Vastan Mine in India is the earliest known from the Indo-Pakistan subcontinent and also from Asia. Surprisingly, it also represents the most diverse known Early Eocene bat fauna. However, none of the seven Vastan bats seems to belong to an extant family, with the possible exception of *Jaegeria cambayensis*, which has not been definitively identified at the family level (Smith *et al.*, 2007). Where known, all Vastan bats have the primitive chiropteran lower dental formula of 3.1.3.3 and also possess an elongate angular process of the mandible (Simmons and Geisler, 1998). Of the seven species, three are particularly primitive (*Icaronycteris sigei*, *Protonycteris gunnelli* and *Archaeonycteris? storchi*), based on having an unreduced metaconid on p4 and tribosphenic molars (Russell and Sigé, 1970).

The new genus *Protonycteris* is phylogenetically significant because it has an incomplete ectocingulum and a nearly central hypoconulid on the talonid basin.

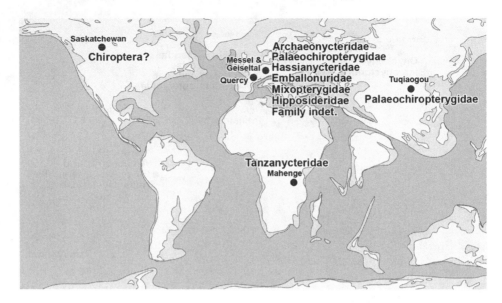

Figure 2.17 Paleogeographic map of the Middle Eocene (modified from Scotese, 2006) showing the distribution of bat records.

The presence of an ectocingulum and lingual displacement of the hypoconulid have been considered important synapomorphies of bat molars, and their incomplete development in *Protonycteris* suggests both that this species might be very plesiomorphic and that these features may not represent useful synapomorphies of the entire bat clade. The four other species known from the Vastan Mine (*Hassianycteris kumari*, *Cambaya complexus*, *Jaegeria cambayensis* and *Microchiropteryx folieae*) are more derived in having nyctalodont teeth. *J. cambayensis* and *M. folieae* have fully developed nyctalodonty with the hypoconulid located lingual and just posterior to the entoconid. The Vastan fauna thus comprised species both seemingly quite primitive and relatively derived in terms of dental morphology.

The resemblance of the Indian bat fauna from Vastan to Eocene European bat faunas is not surprising given that much of the Early Eocene mammalian assemblage from Indo-Pakistan shows affinities with Europe (Rose *et al.*, 2006, 2009; Gunnell *et al.*, 2008; Rana *et al.*, 2008). It is especially the case that the Vastan bats share connections with those from the late Early Eocene of the Paris Basin (Russell *et al.*, 1973) and the early Middle Eocene of the German Messel locality (Habersetzer and Storch, 1987; Storch *et al.*, 2002).

In East Asia, Early Cenozoic bats are known only from the late Middle Eocene of the Yuanqu Basin in China (Tong, 1997) represented by *Lapichiropteryx*, a distinctive potential palaeochiropterygid that is somewhat similar to *Stehlinia* from the middle Middle Eocene of Cuzal in France. Other Asian bat genera are also present, but too fragmentary to be identified with certainty. Pre-Late Eocene African bats are poorly represented by *Dizzya* in the Early Eocene and *Tanzanycteris* in the Middle Eocene. Both genera belong to monogeneric families that are restricted to Africa. An archaic bat has also been recorded from the Early Eocene of Algeria, but is too fragmentary to be determined (Ravel *et al.*, 2010). Bats are also present in Australia during the Early Eocene represented by the primitive *Australonycteris*, perhaps representing the oldest bat known (Hand *et al.*, 1994). Finally, bats are also present in the Early Eocene of South America with two unidentified lower molars that clearly are chiropterans (Tejedor *et al.*, 2005).

In summary, the best known Early Cenozoic bat faunas come from North America, Europe and the Indo-Pakistan subcontinent. Diversity is highest in Europe and Indo-Pakistan with several genera shared in common between the two areas, suggesting the existence of a Euro-west-south-Asian faunal province, maybe including North Africa. This province would approximately cover the territory of the Tethys. Many of the bats from the Vastan Mine appear to exhibit very primitive dental morphology suggesting that Indo-Pakistan potentially could be the source area for the initial radiation of bats. However, by far the most primitive known bat skeletal morphology is exhibited by *Onychonycteris* from North America. Nevertheless, older onychonycterids with more primitive dental morphology such as *Ageina* and *Eppsinycteris* are restricted to Europe. New fossil bats are being discovered at an accelerated rate, so more evidence will soon be available to address the questions of the ultimate phylogenetic and geographic origins of Chiroptera.

2.5 Acknowledgments

We thank Richard Smith (RBINS), Elodie Maitre and Bernard Sigé (University of Lyon), Claire Sagne and Pascal Tassy (MNHN, Paris), Jerry Hooker (BMNH, London), Philip Gingerich (UM, Ann Arbor), Chris Beard (CM, Pittsburgh), John Flynn and Michael Novacek (AMNH, New York), Gerhard Storch, Evelyn Schlosser-Sturm and Stephan Schaal (SMF, Frankfurt), Kevin Seymour (ROM, Toronto) and Norbert Micklich (HLMD, Darmstadt) for giving access to comparative material, providing casts and pictures or discussions about fossil bats; Annelise Folie (RBINS) for help with manuscript and figure preparation, Julien Cillis, Pieter Missiaen and Eric De Bast (RBINS), and Katrin Krohmann, Elvira Brahm and Anika Vogel (SMF, Frankfurt) for producing SEM and optical photographs, Claude Desmedt (RBINS) for map drawings and Eric Dewamme (RBINS) for casting. This work was supported by Research Project MO/36/020 of the Belgian Federal Science Policy Office to TS and US National Science Foundation grants DEB-0629811 and 0949859 to NBS.

2.6   REFERENCES

Bajpai, S., Kapur, V. V., Das, D. P. *et al.* (2005a). Early Eocene land mammals from Vastan Lignite Mine, District Surat (Gujarat), western India. *Journal of the Palaeontological Society of India*, **50**, 101–113.

Bajpai, S., Kapur, V. V., Thewissen, J. G. M., Tiwari, B. N. and Das, D. P. (2005b). First fossil marsupial from India: Early Eocene *Indodelphis* n. gen. and *Jaegeria* n. gen. from Vastan lignite mine, District Surat, Gujarat. *Journal of the Palaeontological Society of India*, **50**, 147–151.

Beard, K. C., Sigé, B. and Krishtalka, L. (1992). A primitive vespertilionoid bat from the early Eocene of central Wyoming. *Comptes Rendus de L'Academie des Sciences, Serie II*, **314**, 735–741.

Bloch, J. I. and Bowen, G. J. (2001). Paleocene–Eocene microvertebrates in freshwater limestones of the Willwood Formation, Clarks Fork Basin, Wyoming. In *Eocene Biodiversity: Unusual Occurrences and Rarely Sampled Habitats*, ed. G. F. Gunnell. New York: Kluwer Academic/Plenum, pp. 95–129.

Castelnau, F. de (1855). *Expédition dans les parties centrales de l'Amérique du Sud de Rio de Janeiro à Lima, et de Lima au Para. 7ème partie. [Zoologie] Animaux nouveaux ou rares recueillis pendant l'expédition* [t.I, 2]. Paris: Mammifères/par Paul Gervais.

Cooper, C. F. (1932). On some mammalian remains from the Lower Eocene of the London Clay. *Annals and Magazine of Natural History, London*, **10**, 458–467.

Czaplewski, N. J. and Campbell, K. E. Jr. (2004). A possible bat (Mammalia: Chiroptera) from the ?Eocene of Amazonian Peru. In *The Paleogene Mammalian Fauna of Santa Rosa, Amazonian Peru*, ed. K. E. Campbell, Jr. Natural History Museum of Los Angeles County, Science Series, **40**, pp. 141–144.

Czaplewski, N. J., Morgan, G. S. and McLeod, S. A. (2008). Chiroptera. In *Evolution of Tertiary Mammals of North America Volume 2*, ed. C. M. Janis, G. F. Gunnell and M. D. Uhen. Cambridge: Cambridge University Press, pp. 174–197.

Delfortrie, M. (1873). Un singe de la Famille des lémuriens dans les phosphates de chaux quaternaires du département du Lot. *Actes de la Société linnéenne de Bordeaux*, **29**, 87–95.

Eiting, T. P. and Gunnell, G. F. (2009). Global completeness of the bat fossil record. *Journal of Mammalian Evolution*, **16**, 151–173.

Filhol, H. (1877). Recherches sur les phosphorites du Quercy. Étude des fossiles qu'on y rencontre, et spécialement des Mammifères. *Annales des Sciences Géologiques, Paris*, **7** (7), 1–220.

Flower, W. H. and Lydekker, R. (1891). *An Introduction to the Study of Mammals, Living and Extinct.* London: Adam and Charles Black.

Gervais, F. L. P. (1855). *Mammifères. Animaux nouveaux, ou rares, recueillis pendant l'expédition dans les parties centrales de l'Amérique du Sud.* Paris: P. Bertrand.

Gingerich, P. D. (1987). Early Eocene bats (Mammalia, Chiroptera) and other vertebrates in freshwater limestones of the Willwood Formation, Clark's Fork Basin, Wyoming. *Contributions from the Museum of Paleontology, University of Michigan*, **27**, 275–320.

Gray, J. E. (1831). Descriptions of some new genera and species of bats. *Zoological Miscellany*, **1831**, 37–38.

Gunnell, G. F. and Simmons, N. B. (2005). Fossil evidence and the origin of bats. *Journal of Mammalian Evolution*, **12**, 209–246.

Gunnell, G. F., Jacobs, B. F., Herendeen, P. S. *et al.* (2003). Oldest placental mammal from sub-Saharan Africa: Eocene microbat from Tanzania – evidence for early evolution of sophisticated echolocation. *Palaeontologica Electronica*, **5**, 1–10.

Gunnell, G. F., Gingerich, P. D., ul-Haq, M. *et al.* (2008). New euprimates (Mammalia) from the early and middle Eocene of Pakistan. *Contributions from the Museum of Paleontology, University of Michigan*, **32**, 1–14.

Habersetzer, J. and Storch, G. (1987). Klassifikation und funktionelle Flügelmorphologie paläogener Fledermäuse (Mammalia, Chiroptera). *Courier Forschungsinstitut Senckenberg*, **91**, 117–150.

Hand, S., Novacek, M., Godthelp, H. and Archer, M. (1994). First Eocene bat from Australia. *Journal of Vertebrate Paleontology*, **14**, 375–381.

Harrison, D. L. and Hooker, J. J. (2010). Late Middle Eocene bats from the Creechbarrow Limestone Formation, Dorset, South England with description of a new species of Archaeonycteris (Chiroptera: Archaeonycteridae). *Acta Chiropterologica*, **12**, 1–18.

Heller, F. (1935). Fledermäuse aus der eozänen Braunkohle des Geiseltales bei Halle a.S. *Nova Acta Leopoldina*, **2**, 301–314.

Hooker, J. J. (1996). A primitive emballonurid bat (Chiroptera, Mammalia) from the earliest Eocene of England. *Palaeovertebrata*, **25**, 287–300.

Hooker, J. J. (2010). The mammal fauna of the early Eocene Blackheath Formation of Abbey Wood, London. *Monograph of the Palaeontological Society*, **165**, 1–162.

Jepsen, G. L. (1966). Early Eocene bat from Wyoming. *Science*, **154**, 1333–1339.

Legendre, S. (1985). Molossidés (Mammalia, Chiroptera) cénozoiques de l'Ancien et du Nouveau Monde: statut systématique; intégration phylogénique de données. *Neues Jahrbuch für Geologie und Paläontologie Abhandlungen*, **170**, 205–227.

Li, G., Wang, J., Rossiter, S. J. *et al.* (2008). The hearing gene Prestin reunites echolocating bats. *Proceedings of the National Academy of Sciences, USA*, **105**, 13959–13964.

Li, Y., Liu, Z., Shi, P. and Zhang, J. (2010). The hearing gene Prestin unites echolocating bats and whales. *Current Biology*, **20**, R55–R56.

Liu, Y., Cotton, J. A., Shen, B. *et al.* (2010). Convergent sequence evolution between echolocating bats and dolphins. *Current Biology*, **20**, R53–R54.

Maitre, E. (2008). Les Chiroptères paléokarstiques d'Europe occidentale, de l'Eocène moyen à l'Oligocène inférieur, d'après les nouveaux matériaux du Quercy (SW France): Systématique, Phylogénie, Paléobiologie. Unpublished Ph.D. thesis, Université Claude Bernard – Lyon 1.

Maitre, E., Sigé, B. and Escarguel, G. (2008). A new family of bats in the Paleogene of Europe: systematics and implications for the origin of emballonurids and rhinolophoids. *Neues Jahrbuch für Geologie und Paläontologie Abhandlungen*, **250**, 199–216.

Marandat, B. (1991). Mammifères de l'Ilerdien moyen (Eocène inférieur) des Corbières et du Minervois (Bas-Languedoc, France) Systématique, biostratigraphie, corrélations. *Palaeovertebrata*, **20** (2–3), 55–144.

Marandat, B., Crochet, J-Y., Godinot, M. *et al.* (1993). Une nouvelle faune à Mammifères d'âge Eocène moyen dans les Phosphorites du Quercy. *Géobios*, **26**, 617–623.

Meng, J., Hu, Y. and Li, C. (2003). The osteology of *Rhombomylus* (Mammalia, Glires): implications for phylogeny and evolution of Glires. *Bulletin of the American Museum of Natural History*, **275**, 1–247.

Novacek, M. J. (1987). Auditory features and affinities of the Eocene bats *Icaronycteris* and *Palaeochiropteryx* (Microchiroptera, incertae sedis). *American Museum Novitates*, **2877**, 1–18.

Pictet, F. J., Gaudin, C. and de la Harpe, P. (1855). Mémoire sur les animaux vertébrés trouvés dans le terrain sidérolithique du Canton de Vaud. *Matériaux pour la paléontologie Suisse, Genève*, **1855**, 27–115.

Rana, R. S., Kumar, K., Escarguel, G. *et al.* (2008). An ailuravine rodent from the lower Eocene Cambay Formation at Vastan, western India, and its palaeobiogeographic implications. *Acta Palaeontologica Polonica*, **53**, 1–14.

Ravel, A., Marivaux, L, Tabuce, R. and Mahboubi, M. (2010). Oldest bat (Chiroptera, Eochiroptera) from Africa: Early Eocene from El Kohol (Algeria). *Journal of Vertebrate Paleontology*, **28**, 149A.

Revilliod, P. A. (1917). Fledermäuse aus der Braunkohle von Messel bei Darmstadt. *Abhandlungen der Großherzoglich Hessischen Geologischen Landesanstalt zu Darmstadt*, **7**, 157–201.

Revilliod, P. (1919). L'etat actuel de nos connaissances sur les Chiropteres fossiles (Note préliminaire). *Compte rendu des séances de la Société de physique et d'histoire naturelle de Genève*, **36**, 93–96.

Revilliod, P. (1922). Contribution a l'étude des Chiroptères des terrains Tertiaires. Troisième partie et fin. *Mémoires de la Société Paléontologique Suisse*, **45**, 133–195.

Rose, K. D., Smith, T., Rana, R. S. *et al.* (2006). Early Eocene (Ypresian) continental vertebrate assemblage from India, with description of a new anthracobunid (Mammalia, Tethytheria). *Journal of Vertebrate Paleontology*, **26**, 219–225.

Rose, K. D., Rana, R. S., Sahni, A. *et al.* (2009). Early Eocene primates from Gujarat, India. *Journal of Human Evolution*, **56**, 366–404.

Russell, D. E. (1964). Les Mammifères Paléocènes D'Europe. *Mémoires du Muséum National D'Histoire Naturelle*, **8**, 1–324.

Russell, D. E. and Gingerich, P. D. (1981). Lipotyphla, Proteutheria(?), and Chiroptera (Mammalia) from the Early-Middle Eocene Kuldana Formation of Kohat (Pakistan). *Contributions from the Museum of Paleontology, University of Michigan*, **25**, 277–287.

Russell, D. E. and Sigé, B. (1970). Révision des chiroptères lutétiens de Messel (Hesse, Allemagne). *Palaeovertebrata*, **3**, 83–182.

Russell, D. E., Louis, P. and Savage, D. E. (1973). Chiroptera and Dermoptera of the French Early Eocene. *University of California Publications in Geological Sciences*, **95**, 1–57.

Schlosser, M. (1887). Die Affen, Lemuren, Chiropteren, Insectivoren, Marsupialier, Creodonten un Carnivoren des europäischen Tertiärs und deren Beziehungen zu ihren lebenden und fossilen aussereuropäischen Verwandten. *Beiträge zur Paläontologie Österreich-Ungarns und des Orients*, **6** (I–II), 1–224.

Scotese, C. R. (2006). PALEOMAP Project, www.scotese.com.

Sigé, B. (1985). Les Chiroptères Oligocènes du Fayum, Egypte. *Geologica et Palaeontologica*, **19**, 161–189.

Sigé, B. (1988). Le gisement du Bretou (Phosphorites du Quercy, Tarn-et-Garonne, France) et sa faune de vertébrés de l'Eocène supérieur. IV. Insectivores et Chiroptères. *Palaeontographica Abteilung A*, **205**, 69–102.

Sigé, B. (1990). Nouveaux chiroptères de l'Oligocène moyen des phosphorites du Quercy, France. *Comptes Rendus de l'Academie des Sciences, Paris*, **310**, 1131–1137.

Sigé, B. (1991). Rhinolophoidea et Vespertilionoidea (Chiroptera) du Chambi (Eocène inférieur de Tunisie). Aspects biostratigraphique, biogéographique et paléoécologique de l'origine des chiroptères modernes. *Neues Jahrbuch für Geologie und Paläontologie Abhandlungen*, **182**, 355–376.

Sigé, B. and Russell, D. E. (1980). Compléments sur les chiroptères de l'Eocène moyen d'Europe. Les genres *Palaeochiropteryx* et *Cecilionycteris*. *Palaeovertebrata, Mémoire Jubilaire en Hommage René Lavocat*, **1980**, 81–126.

Simmons, N. B. and Geisler, J. H. (1998). Phylogenetic relationships of *Icaronycteris*, *Archaeonycteris*, *Hassianycteris*, and *Palaeochiropteryx* to extant bat lineages, with comments on the evolution of echolocation and foraging strategies in Microchiroptera. *Bulletin of the American Museum of Natural History*, **235**, 1–182.

Simmons, N. B., Seymour K. L., Habersetzer, J. and Gunnell, G. F. (2008). Primitive early Eocene bat from Wyoming and the evolution of flight and echolocation. *Nature*, **451**, 818–822.

Simmons, N. B., Seymour K. L., Habersetzer, J. and Gunnell, G. F. (2010). Inferring echolocation in ancient bats. *Nature*, **466**, E8.

Smith, J. D. and Storch, G. (1981). New Middle Eocene bats from the "Grube Messel" near Darmstadt, W Germany (Mammalian: Chiroptera). *Senckenbergiana Biologica*, **61**, 153–168.

Smith, R. and Russell, D. E. (1992). Mammifères (Marsupialia, Chiroptera) de l'Yprésien de la Belgique. *Bulletin Institut Royal des Sciences Naturelle de Belgique, Sciences de la Terre*, **62**, 223–227.

Smith, T. (1995). Présence du genre *Wyonycteris* (Mammalia, Lipotyphla) à la limite Paléocène-Eocène en Europe. *Comptes Rendus de l'Académie des Sciences, Paris, Série IIa*, **321**, 923–930.

Smith, T., Rose, K. D. and Gingerich, P. D. (2006). Rapid Asia-Europe-North America geographic dispersal of earliest Eocene primate *Teilhardina* during the Paleocene-Eocene Thermal Maximum. *Proceedings of the National Academy of Sciences, USA*, **103**, 11223–11227.

Smith, T., Rana, R. S., Missiaen, P. *et al.* (2007). High bat (Chiroptera) diversity in the Early Eocene of India. *Naturwissenschaften*, **94**, 1003–1009.

Springer, M. S., Murphy, W. J., Eizirik, E. and O'Brien, S. J. (2003). Placental mammal diversification and the Cretaceous-Tertiary boundary. *Proceedings of the National Academy of Sciences, USA*, **100**, 1056–1061.

Storch, G. and Haberstezer, J. (1988). *Archaeonycteris pollex* (Mammalia, Chiroptera), eine neue Fledermaus aus dem Eozän der grube Messel bei Darmstadt. *Courier Forschungsinstitut Senckenberg*, **107**, 263–273.

Storch, G., Sigé, B. and Habersetzer, J. (2002). *Tachypteron franzeni* n. gen., n. sp., earliest emballonurid bat from the Middle Eocene of Messel (Mammalia, Chiroptera). *Paläontologische Zeitschrift*, **76**, 189–199.

Storer, J. E. (1984). Mammals of the Swift Current Creek local fauna (Eocene: Uintan), Saskatchewan. *Natural History Contributions (Saskatchewan Culture and Recreation)*, **7**, 1–158.

Stucky, R. K. (1984a). Revision of the Wind River faunas, Early Eocene of central Wyoming. Part 5. Geology and biostratigraphy of the upper part of the Wind River Formation, northeastern Wind River Basin. *Annals of Carnegie Museum*, **53**, 231–294.

Stucky, R. K. (1984b). The Wasatchian-Bridgerian Land Mammal Age boundary (Early to Middle Eocene) in western North America. *Annals of Carnegie Museum*, **53**, 347–382.

Tabuce, R., Antunes, M. T. and Sigé, B. (2009). A new primitive bat from the earliest Eocene of Europe. *Journal of Vertebrate Paleontology*, **29**, 627–630.

Teeling, E. C., Springer, M. S., Madsen, O. *et al.* (2005). A molecular phylogeny for bats illuminates biogeography and the fossil record. *Science*, **307**, 580–584.

Tejedor, M. F., Czaplewski, N. J., Goin, F. J. and Aragon, E. (2005). The oldest record of South American bats. *Journal of Vertebrate Paleontology*, **25**, 990–993.

Tong, Y.-S. (1997). Middle Eocene small mammals from Liguanqiao Basin of Henan province and Yuanqu Basin of Shanxi province, Central China. *Palaeontologica Sinica* 18, *New Series C*, **26**, 1–256.

Van Valen, L. (1979). The evolution of bats. *Evolutionary Theory*, **4**, 103–121.

Veselka, N., McErlain, D. D., Holdsworth, D. W. *et al.* (2010). A bony connection signals laryngeal echolocation in bats. *Nature*, **463**, 939–942.

Weithofer, A. (1887). Zur Kenntniss der fossilen Cheiropteren der französischen Phosphorite. *Sitzungsberichte der Mathematisch-Naturwissenschaftlichen Classe der Kaiserlichen, Akademie der Wissenschaften*, pp. 341–360.

3

Shoulder joint and inner ear of *Tachypteron franzeni*, an emballonurid bat from the Middle Eocene of Messel

JÖRG HABERSETZER, EVELYN
SCHLOSSER-STURM, GERHARD STORCH
AND BERNARD SIGÉ

3.1 Introduction

Over 600 Middle Eocene bat specimens have been excavated from the Messel pit (Grube Messel, near Darmstadt, Germany), and seven species have been described thus far. Many of the fossils are preserved as complete skeletons, often with soft body outlines and gut contents. Six of the bat species represent three extinct families, whereas *Tachypteron franzeni* can be assigned to the extant family Emballonuridae (Storch *et al.*, 2002). *T. franzeni* is known only from two specimens; however, these are extraordinarily well preserved, including the shoulder joints and inner ears, so this had already been recognized in the original description of *T. franzeni*, and these close resemblances to extant emballonurids led to the conclusion that *T. franzeni* had already evolved similar bioacoustic specializations and a similar flight style to modern taxa.

The shoulder joints of bats are sophisticated structures showing remarkable morphological variation. Miller's (1907) investigations on the differentiations of the shoulder within the Microchiroptera were continued by the studies of other authors (Vaughan, 1970; Strickler, 1978; Hermanson and Altenbach, 1983).

Three different types of shoulder joint can be distinguished within the Chiroptera: the primitive morphology of the shoulder joint with a globular humeral head and corresponding glenoid cavity, as seen in Megachiroptera and Rhinopomatidae; a derived shoulder joint with an oblong humeral head and a single trough-like articular surface on the scapula, found in members of the superfamilies Emballonuroidea, Rhinolophoidea and Noctilionoidea; a derived shoulder joint with a secondary articulation between the tuberculum majus and a secondary articular facet on the dorsal side of the scapula, as seen in the

Evolutionary History of Bats: Fossils, Molecules and Morphology, ed. G. F. Gunnell and N. B. Simmons. Published by Cambridge University Press. © Cambridge University Press 2012.

remaining families. Their distribution within the order gives evidence of parallel evolution of the derived types (Schlosser-Sturm and Schliemann, 1995). The morphological modifications of the derived joints are interpreted as a functional response to a biomechanical demand connected with flight (Norberg, 2002), i.e., to limit pronation of the humerus during the downstroke of the wing beat cycle, realized in two different mechanical ways (Schlosser-Sturm, 1982; Altenbach, 1987; Schliemann and Schlosser-Sturm, 1999). Because movement restriction was described for the primitive type as well (Bergemann, 2003), functional interpretations are still a matter of controversy.

The inner – i.e., trabecular – structure of the bones of fossil and extant bat shoulder girdles is so far largely unknown. Empirical scaling studies of transverse sections of the humeral shaft, distal to the joint region, showed homology of trabecular elements within and between the vespertilionids *Eptesicus fuscus* and *Myotis lucifugus* (Swartz *et al.*, 1998). Mammalian bones and their trabecular architecture were studied recently with regard to mechanical properties, loads and stresses (Ito *et al.*, 2002: Lim *et al.*, 2006; Stauber *et al.*, 2006) and by means of two-dimensional or enhanced three-dimensional methods (Kirkpatrick, 1994; Fajardo and Muller, 2001).

The consensus is that the pattern of trabeculae is always a compromise reflecting all normal, average stresses acting on a bone, and consequently we expect to see the following in response to transformations of the ancestral bat shoulder joint configuration:

(1) change of the axis of rotation in those joints with a secondary articular facet for the abducted wing, because novel joint elements become mechanically active;
(2) different lever arms and sizes of the tuberculae accompanied by differently specialized spongiosa; and
(3) functional adaptations in the trabecular substructure of the articular facets caused by different modes of joint operation of the two specialized joint types.

In this study the outer shape and internal structure of the shoulder of *Tachypteron franzeni* is compared with articulated shoulder joints of the black-bearded tomb bat *Taphozous melanopogon*, which has previously been used as an extant model for this Eocene Messel bat (Storch *et al.*, 2002), and additionally with the black mastiff bat *Molossus ater*. *T. melanopogon* and *M. ater* represent the two modern types of derived shoulder joints.

The inner-ear size of bats varies to a very high degree. Non-echolocating pteropodids have the smallest ears; microchiropteran species, which do echolocate but also use information other than acoustics as important cues for prey catching (olfactory, visual, noises of the prey), have ears that are larger than in non-echolocators, but smaller than in aerial hawking bats that only rely on their echolocation system. The latter are surpassed in cochlear size (oblique diameter) by bats using long, constant-frequency ultrasonic sounds. Finally, the largest cochlear sizes are displayed by those specialists that echolocate with very long, constant-frequency ultrasonic sounds, including Doppler shift compensation and have an expanded frequency scale on their basilar membrane, a so-called acoustic fovea (Habersetzer and Storch, 1992). Considering these relationships between form and function, the size and the longitudinal distribution of the volume of the cochlea of *T. franzeni* is of interest. Studies of the internal morphology of the cochlea and of the trabecular microstructure of the elongated tuberculum majus of the humerus are possible, due to the excellent preservation of this fossil bat. Thus, previous conclusions about flight and echolocation of *T. franzeni* can be re-examined by additional data and by refined methods in this study.

X-ray studies of Messel bats have been undertaken for almost 30 years, but microcomputed tomography (= μCT) was very rarely applied to Messel fossils because they are prepared on large resin plates. The relatively large size of these plates is a hindrance to the study of details, because the complete plate has to be placed into the X-ray beam that projects onto the sensor. Thus, the maximal possible resolution is limited not only by the X-ray apparatus itself, but also by the relation of the plate size compared to the region of interest (ROI) in the specimen to be studied. For bats with a typical plate size of 15 cm, the typical resolution is 150 μm, although the same μCT equipment actually could have a 10× to 30× higher resolution, given its specifications, if the total size of the object were correspondingly smaller. In the 2007 SVP bat symposium (Habersetzer *et al.*, 2007) it was demonstrated for the first time that these limitations could be successfully overcome with new ROI-μCT algorithms (Kastner *et al.*, 2006) or two-axis-μCT (Habersetzer, 1998), both of which deal with a particular ROI of the samples. Thus another aim of this chapter is to compare different radiographic and tomographic methods.

3.2   Material and methods

For the study of shoulders we used the holotype of *Tachypteron franzeni* preserved on plates A and B (Figures 3.1, 3.2) and the prepared left shoulders of an adult *Taphozous melanopogon* (Figure 3.4) and an adult *Molossus ater* (Figure 3.3). For the prepared specimens of extant bats, the humerus, scapula and clavicle were left articulated with all ligaments and scanned in an abducted position. Directional terms in the morphological description of the bat shoulder refer to a spatial position in which the wings are fully extended, like in the middle of the downstroke of the wingbeat cycle. Thus the upper side of the

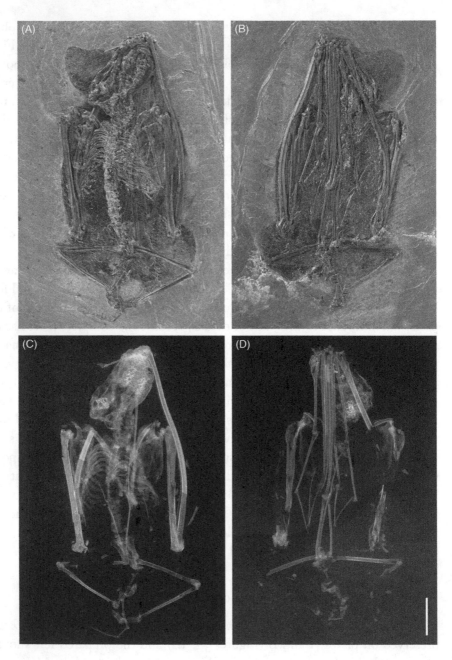

Figure 3.1 (A) and (B) photographs of part and counterpart of holotype (IRSNB M1944a+b, formerly BE 4–119a+b) of *Tachypteron franzeni*. (C) and (D) contact-microradiographs of same. Scale is 10 mm.

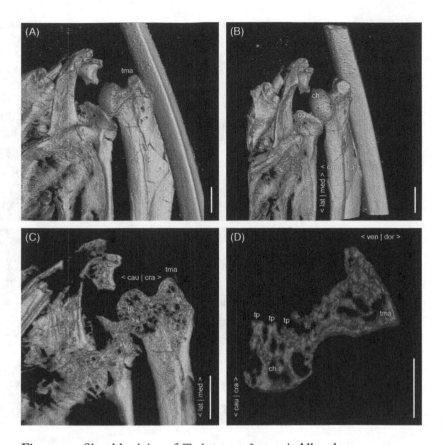

Figure 3.2 Shoulder joint of *Tachypteron franzeni*. All scales are 1 mm.
Directional terms: (cau)dal, (cra)nial, (dor)sal, (lat)eral, (med)ial, (ven)tral;
these refer to orientation of the humerus exclusively. (A) General dorsal view
of the right shoulder joint. The forearm is adducted. The humeral head is
preserved with a strong tuberculum majus (tma). (B) Shoulder joint with
glenoid cavity (cg) in a slightly rotated view shows elongated humeral head
(ch). Note the narrow width in the proximal part compared to the broad distal
part and the notch (no) for muscle attachment at the tuberculum majus.
(C) Moderately enlarged detail of joint (compare with A). The image includes
nine aligned slices with a total thickness of 149 μm from the middle of the
preserved bones. Honeycomb-like trabeculae (ht) support the border of
the cavitas glenoidalis (middle), aligned trabecular plates (tp) the center of
the head above. Note small trabecula-supported cavities in the tuberculum
majus (tma). (D) Virtual dorsoventral cut through the proximal part of the
humerus (plane diagonal to the surface of the fossil plate). This single slice is
20 μm thick and is perpendicular to shaft axis. This right humerus mirrors
corresponding details in Figures 3.3D and 3.4D. Note that parts of the proximal
epiphysis beyond the diagonal fracture zone are missing on the fossil plate.
Below: humeral head (ch) with parallel trabecular plates (tp) in the center;
right: tuberculum majus (tma) with small trabecula-supported cavities.

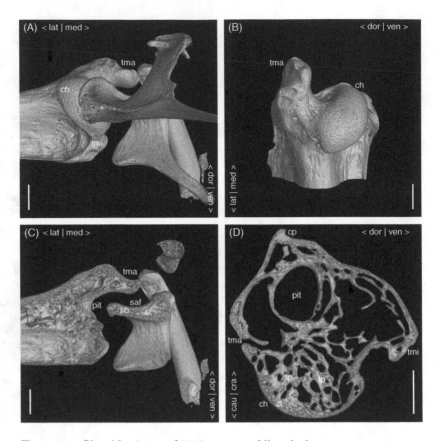

Figure 3.3 Shoulder joint of *Molossus ater*. All scale-bars are 1 mm.
Directional terms as for Figure 3.2. (A) Abducted left shoulder joint, caudal
view. Humerus positioned like in flight (mid-downstroke). Gaps in images
due to joint cartilage (not shown). Tuberculum majus (tma) is directed
medially and extends well beyond the head, tuberculum minus (tmi) is very
small and close to the caput humeri (ch). (B) Proximal epiphysis of left
humerus, oriented as in adducted right forearm in Figure 3.2B. The diagonally
oriented oval head (ch) and the separated tuberculum majus (tma) aligned
parallel to the shaft axis are visible. (C) Left shoulder joint seen, as in (A), but
virtually cut vertically in cranial part of joint. Portions of locked secondary
articulations – between tuberculum majus (tma) and the secondary articular
facet (saf) and between the jointed tuberositas supraglenoidalis (st) and pit on
the proximal humerus (pit) – are visible. Note dense sandwich (sb) bone in the
lateral part of the tuberositas supraglenoidalis, compared to porous spongiosa
in the tuberculae and shaft. (D) Vertical cut through proximal humerus,
perpendicular to long axis. In the image eight aligned single slices are added to
a total thickness of 80 μm. Top: crista pectoralis (cp); bottom: humeral head
(ch) with trabecular plates (tp) aligned towards the base of the tuberculum
majus (tma) and on the left side dense zone of transition (zt). Left hollow space:
base of tuberculum majus; round hollow area at middle: cranial pit of the
humerus (pit); right hollow space with few trabeculae: tuberculum minus (tmi).

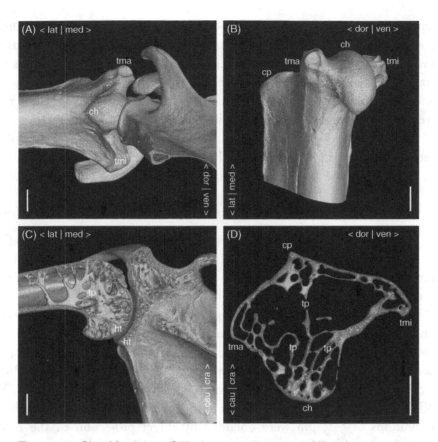

Figure 3.4 Shoulder joint of *Taphozous melanopogon*. All scales are 1 mm. Directional terms as in Figure 3.2. (A) General view of the abducted left shoulder joint seen caudally. Humerus stretched out horizontally in flight position, posture like Figure 3.3A. The distal oval part of the elongated caput humeri (ch) is visible. The tuberculum majus (tma) is shorter than tuberculum minus (tmi) and they are directed dorso- and ventromedially but neither surpasses the head. (B) Isolated proximal epiphysis of left humerus, oriented comparable to adducted right forearm of fossil in Figure 3.2B. Note small elongated head (ch) oriented in the shaft axis, the large crista pectoralis (cp) and the size and orientation of the tuberculae (tma, tmi). (C) Dorsal view of horizontally cut joint showing large bracing trabecular plate (tp) in the middle of the humeral shaft, humeral head and the opposed honeycomb-like spongiosa (ht) in the border of the cavitas glenoidalis. This typical tissue underneath joint surfaces can also be recognized in the fossil, Figure 3.2C. (D) Vertical cut through the proximal part of humerus, perpendicular to long axis. The image consists of eight aligned single slices with a total thickness of 88 μm. The spongiosa is not as dense as in *Molossus* (Figure 3.3D). Top: crista pectoralis (cp); below: humeral head (ch), note the large bracing trabecular plate (tp) in the middle and three smaller parallel trabecular plates. Left: base of tuberculum majus (tma); right: tuberculum minus (tmi) with large hollow space below.

wing is dorsal, the lower ventral; the crista pectoralis of the humerus is directed craniad, the humeral head caudad. The complete shoulder girdles are illustrated in caudal view (Figures 3.3A and 3.4A). The resolution of μCT-series (voxel size) is 16.5 μm for *Tachypteron franzeni*, 11.0 μm for *Taphozous melanopogon* and 10.0 μm for *Molossus ater*. For the study of the inner ears (Figures 3.5–3.8) we used CT-series with slightly different resolutions, namely 10.4 μm for *T. franzeni* and 15.5 μm for prepared skulls of *T. melanopogon*. Histological morphometric data were additionally provided by A. Fiedler, Frankfurt, for *T. melanopogon*, *M. ater* and the Indian false vampire bat *Megaderma lyra*.

3.2.1   Synopsis of different X-ray and CT techniques for fossils

Due to the fundamental importance of different X-ray methods for the study of Messel bats and other fossils preserved on large plates, we tested and extensively compared two different radiographic and three (A–C) different tomographic methods.

In the first radiographic method, contact microradiographs served as overview images for comparative purposes, because the resolution is limited to 25 μm by the storage screen (SR-HD-IP, Fuji, Japan). This screen was exposed to a conventional X-ray source (Faxitron 43804 X-ray cabinet, Hewlett Packard, USA) and processed with a laser scanning digitizer (HD-CR 35 NDT, Duerr-NDT, Germany). In the second method, microfocus-microradiographs with resolutions of 5–15 μm were obtained by direct projection of the enlarged specimen onto the storage screen (see above) or alternatively onto a real-time digital sensor (C7942 CK12, version modified for small bones, Hamamatsu, Japan) using a microfocus X-ray tube (FXT 100.52, Feinfocus/Yxlon, Germany).

The first two μCT methods are similar and can be united under the term "limited-angle-μCT." However, they require two different apparatuses and procedures (A, B described below). Both methods were developed in order to avoid serious artifacts in imaging when large parts of the fossil plate enter and leave the field of view, which typically occurs twice during a 360° rotation in both conventional "standard" CT and μCT.

(A) PμCT: In planar microcomputed tomography image data are acquired by an *xy*-translation of the probe which is synchronous to a similar but proportionally larger *xy*-translation of the sensor. The region of interest is geometrically enlarged on the sensor, as in all these μCT methods. Movements of both probe and sensor are executed in relation to a stationary X-ray source. This results in a 360° rotation of the sensor in a plane parallel to the surface of the sample. One of the advantages is that image projections are already oriented planar to the probe surface and thus can

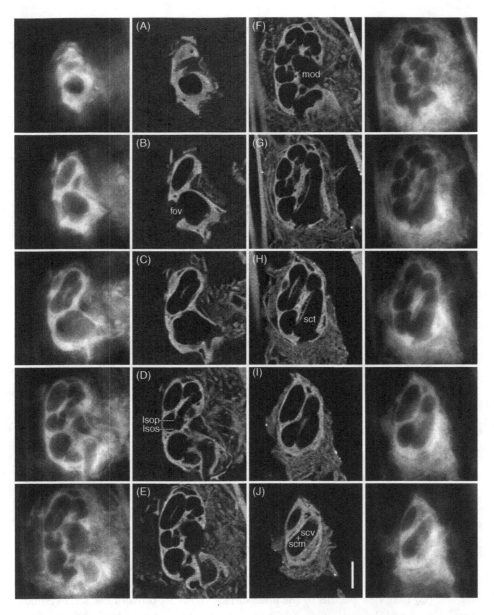

Figure 3.5 Special μCT of the inner ear of *Tachypteron franzeni*. Scale bar is 1 mm. Ten equidistant slices of the right cochlea (plate B, Figure 3.1) are compared for two different μCT methods. The two center columns A–J show images acquired using the ROI-μCT technique, from the top the deepest layer of the fossil plate approaching the most superficial layer. The two marginal rows show the same areas acquired using the P-μCT technique. Morphological details labeled: fenestra ovale (fov), lamina spiralis ossea primaria (lsop), lamina spiralis ossea secundaria (lsos), modiolus (mod), scala tympani (sct), scala vestibuli (scv), scala media (scm).

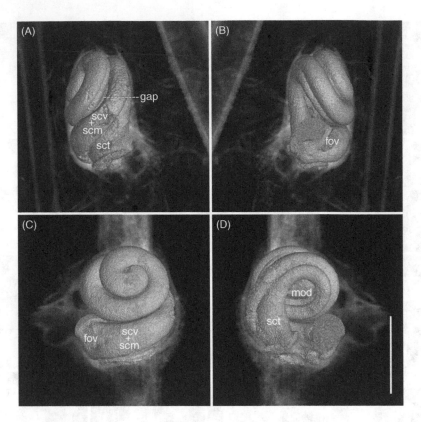

Figure 3.6 Virtual cochlear endocast of *Tachypteron franzeni*. Scale bar is 2 mm. Each background is a microradiograph computed from the ROI-µCT-data set and shows long bone elements of the wing, bone fragments and vestibular ampullae. (A) Front view of right cochlea as preserved on plate B (see also Figure 3.1). Note irregular shape of the helix between the first and the second turn (gap). The scala vestibuli (scv) and the scala media (scm) form a common lumen and are separated from the scala tympani (sct) by bone impression of the secondary spiral lamina (see also Figure 3.7).
(B) Hind view of the cochlea, which in reality is buried in the plate. The oval window (fenestra ovale = fov) indicates the former position of the stapes.
(C) Left view parallel to the plate surface shows 2.5 turns of the scala vestibuli (scv) and the scala media (scm) from the basal fenestra ovale (fov) to the apex.
(D) Right view to plate B provides insight to the modiolar axis (mod).
The rough surface of the scala tympani (sct) indicates less bone preservation of the basal part of the cochlea compared to other areas in (A)–(C).

be directly processed by tomosynthesis. This procedure, described below (B), is used in conventional clinical tomography and has been adapted to µ-tomography with microfocus-X-ray tubes by special µ-mechanical setup and software (µ-3D Visualizer, Feinfocus/Yxlon, Germany; software by

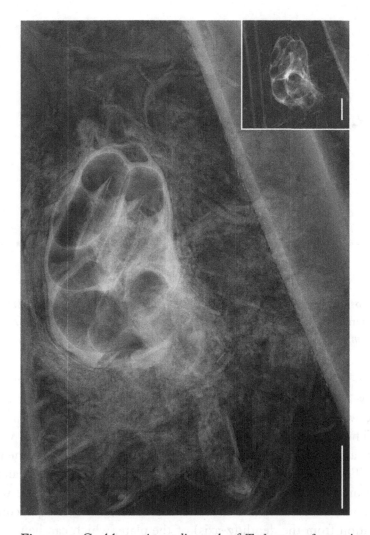

Figure 3.7 Cochlear microradiograph of *Tachypteron franzeni* made with the microfocus X-ray tube of the two-axis-μCT apparatus showing microstructural details. The inset image in the upper right is a microradiograph from the ROI-μCT data set (see Figure 3.6, upper left), which shows a much lower resolution compared with the large image, even though this larger image is already scaled down by 0.35×.

Fraunhofer Gesellschaft, Germany). One of the disadvantages of this method is that the X-ray beam, which projects at an angle of 45° from the center beam of the microfocus tube, has a low radiation density in relation to its maximum at the center. Due to the low signal/noise ratio in the images, a 2×2 binning of the pixel matrix was necessary, and thus the theoretically excellent properties of PμCT cannot fully be realized.

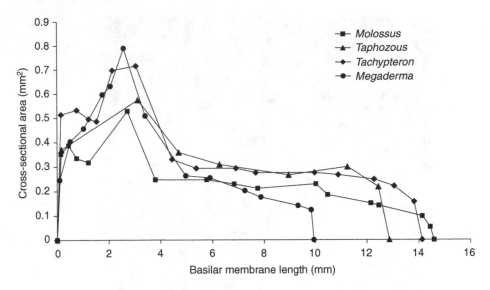

Figure 3.8 Cross-sectional area of the cochlear canal (canalis spiralis cochleae) vs. basilar membrane length of *Molossus ater, Taphozous melanopogon, Megaderma lyra* and *Tachypteron franzeni*. For details, see text.

(B) Two-axis-μCT: The fossil plate is simultaneously rotated around the x- and y-axes, both of which refer to the width and the height of the sample, whereas the X-ray tube and the digital X-ray sensor remain static. These movements result in a series of projection images with a constant angle of 45° in relation to the (perpendicular) z-axis to the plate surface. The projection images cover a 360° circle in the xy-plane and are focused on a small region of interest (e.g., the inner ear) of the whole plate. Thus, because of these object movements, the images contain information from the depth (z-axis) of the plate, which can be used for building a series of tomograms, even though the plate never passes completely into the field of view (Habersetzer, 1998). A cosine function is used to compensate for the distortion to each image introduced by the 45° angle of view; then these images are used to calculate an image stack of tomograms by tomosynthesis. In tomosynthesis, the same set of digital projection images with different xy-offsets for specific (z-) steps in depth of the sample is repetitively processed. For the complete two-step procedure the prototype software Tomoimage (by Mark Peinl, Morphisto GmbH at Senckenberg, Germany) was used. To complete this brief description, the following essential details are also appended because they are originally described only in the German patent (Habersetzer, 1998).

1st step: rotating specimen in *x*- and *y*-axis, while *x/y* rotations are phaseshifted by constantly 90° (phaselocked), recording a minimum of 48 (of 360) positions at an angle of 45° with respect to the *z*-axis, related to the isocenter of ROI. This provides coplanar images from different angles of view around the isocenter.

2nd step: tomosynthesis by *n*-fold (*n* dertermines the number of slices) overlay of images under *xy*-shifts (*xy*-shift dertermines the thickness of slices) of the isocenter which generates a *z*-image stack of a series of *n* tomograms.

The limited-angle-µCT methods introduced here for comparison (PµCT and two-axis-µCT) both use digital tomosynthesis and thus could also be described as tomosynthesis-µCT. Both clearly differ from the "standard" computed tomography and are based to some extent on older traditional tomography procedures, e.g., conventional laminography, which provides only a single slice per measurement. However, now the reconstruction of several slices from one set of projections by tomosynthesis is possible by sophisticated digital imaging procedures. Due to the principally lower resolution in *z*-axis compared to the (mono-) axial "standard" computed tomography, this application is mainly recommended for objects whose geometry prevents them from being studied by means of "standard" CT. Many efforts have been made to extend the range of application of "standard" CT and µCT to these kinds of difficult objects with some progress, e.g., with an oblique radiation geometry instead of an orthogonal radiation geometry.

(C) ROI-CT and ROI-µCT are the third new technical approach suitable for the purposes of this study. ROI-µCT allows large parts of the fossil plate to enter and leave the field of view during data acquisition without strong artifacts in the resulting tomograms. For this particular situation special algorithms were optimized (Kastner *et al.*, 2006). The microfocus X-ray tube provides a maximal resolution of 4–5 µm in "standard" µCT-mode when sample size (or plate size) does not exceed 4–5 mm. In the case of *Tachypteron franzeni* the plate size was 130 mm, thus allowing a maximum resolution of only 130 µm. ROI-µCT, however, allows an increase in the resolution to 15 µm, which is a gain of 8.7× compared to the "standard" situation above. Following the virtual pixel matrix concept (Kastner *et al.*, 2006) this would mean that a normal CT could theoretically produce similar images with a sensor/reconstruction size of 8700-pixel side-length instead of a normally 1000-pixel sensor side-length, because the ratio of resolution to object size is 1: 8700.

However, the artifacts of these algorithms are a progressive fusion and loss of contrast of bones in the images as the relative size of the region of interest decreases. Thus, compromises must be found between loss of density information and gain in resolution. ROI-μCT images were obtained using an industrial Micro CT System (RayScan 200 XE, RayScan Technologies, Germany, Micro-CT-lab of AUDI, Germany). The same system was also used with a "standard" μCT-mode, e.g., for the shoulder preparations of extant bats, which entirely fit into the field of view of the X-ray beam projecting on the sensor.

Raw CT-data were processed on a special workstation (Dell Precision T 7400) in the X-ray lab of Senckenberg Research Institute with VGStudio MAX 2.0.1 (Volume Graphics, Germany).

3.3   Results

The holotype of *Tachypteron franzeni* is preserved on two plates showing the dorsal view on plate A and the ventral view of the fossil on plate B (Figure 3.1A, B). A detailed description of this and the paratype can be found in Storch *et al.* (2002). Additional information is depicted on the contact-microradiographs of both plates. Many details visible on the photo of plate B, such as the right radius, are preserved on this plate only as a mold; however, the same bones are completely preserved on plate A (Figure 3.1C, D). It is also visible on the radiographs that the right inner ear is completely preserved and sheltered between strong wing bones on plate B, and the right shoulder joint is mainly preserved on plate A. Both are the main subjects of this chapter.

The shape and functional parameters of bat wings have been studied for extant bats (Habersetzer, 1986; Norberg and Rayner, 1987) and were compared with the data of six of the Messel bats (Habersetzer and Storch, 1987). *Hassianycteris magna* and *H. messelensis* were interpreted as high and rapidly flying species; *Archaeonycteris pollex* and *A. trigonodon* were considered unspecialized, hunting at a middle flight altitude and *Palaeochiropteryx spiegeli* and *P. tupaiodon* supposedly flew slowly and with high maneuverability close to the ground or close to the water surface of the former Messel lake. The functional wing morphology of *Tachypteron franzeni* suggests that this species hunted at high altitude, above the tree canopy (Storch *et al.*, 2002), thus completing the ecological scenario of flight and hunting corridors. The updated version is shown here in Figure 3.9.

3.3.1   Shoulder joint

In the following description, two high and rapid flying extant bats, *Molossus ater* and *Taphozous melanopogon*, are compared with *Tachypteron franzeni*.

Figure 3.9 Bats from the Messel pit. From top to bottom: *Tachypteron franzeni, Hassianycteris magna, Hassianycteris messelensis, Archaeonycteris pollex, Archaeonycteris trigonodon, Palaeochiropteryx spiegeli, Palaeochiropteryx tupaiodon*. The order from top to bottom correlates with decreasing aspect ratio and wing loading.

Molossus ater – The outer shape of the shoulder joint of *Molossus ater* – as in other vespertilionoid bats – is dominated by an extremely well-developed secondary articulation. Figure 3.3A shows the left joint from caudal view with the wing fully stretched. The tuberculum majus is very prominent and extends well proximal to the head, while the tuberculum minus is inconspicuous, not rising proximal to, and lying close to the irregularly oval humeral head. The tuberculum majus and humeral head are separated by a distinct neck (Figure 3.3B) which passes into a pit cranially. This species has, compared to other bats studied by the authors, an average-sized crista pectoralis, no crista tuberculi majoris, but a well-developed crista tuberculi minoris. On the scapula two clearly defined and connected articular facets, and a prominent tuberositas supraglenoidalis can be seen (Figure 3.3C). The cavitas glenoidalis – caudally round and flat, cranially becoming narrow and passing into the big cone of the tuberositas supraglenoidalis, is in permanent contact with the humeral head. The secondary articulation, however, that between the tuberculum majus and the triangular secondary articular facet of the scapula, is only functional when the humerus is abducted with the arm fully stretched. This is typically the case through the middle of the downstroke of the wing beat (Figure 3.3C) as described for *Desmodus* (Altenbach, 1979), *Glossophaga* (v. Helversen, 1986) and *Plecotus* (Norberg, 1976). Also in this phase of the wing beat cycle the tuberositas supraglenoidalis fits into the pit cranial to the humeral head. Thus the joint becomes fixed, and pronatory movement impossible (Schlosser-Sturm and Schliemann, 1995). This interpretation from prepared and separated joint elements of *Molossus ater* is here strongly supported by the natural functional posture of the articulated joint in Figures 3.3A and C, in which the complete joint capsule is made invisible by digital imaging of the CT data.

The spongiosa, i.e., cancellous bone with large, often marrow-filled spaces between the bony trabeculae, is quite dense in the proximal part of the humerus, in the scapula underneath the articular facets, in the tuberositas supraglenoidalis (Figure 3.3C) and in the strengthened bony frame close to the joint area. There are fewer trabeculae, isolated plates and struts, lying in the marrow cavity of the humeral shaft distal to the joint. The head is completely filled with a dense mesh of spongiosa (Figure 3.3D). The number and connectivity of trabeculae is apparently greater and the robustness in the center less distinct than in the other two species described below. In the center of the head trabeculae are longer, often plate-like, connected with small but robust plates and directed mainly towards the caudal wall of the shaft (perpendicular to the plane of Figure 3.3D). Close to the base of the tuberculum majus (Figure 3.3D, left and bottom) there is a zone of transition between the corticalis and the honeycomb-like spongiosa lying directly underneath the joint surface. Proximal plates run towards the base of the tuberculum majus. All in all the spongiosa is more robust on the dorsomedial side of the head than on the ventrolateral side. The tuberculum minus is mainly hollow with few trabeculae parallel to the corticalis and underneath the attachment of M. subscapularis. The tuberculum majus is largely hollow too, braced by few robust plates. Strong sandwich bone is found in its tip (Figure 3.3C). In this bone tissue, very short trabeculae firmly bind two parallel cortical plates.

In the scapula spongiosa can be found mainly around the glenoid cavity. Few trabeculae are in the acromion and coracoid processes, in the margo and angulus cranialis or in the margo lateralis. The first layer underneath the corticalis of the articular facet of the cavitas glenoidalis is a zone of transition, followed by honeycomb-like trabeculae. Fan-shaped buttresses anchor the spongiosa within the shoulder blade. Three layers can be seen in the secondary articular facet as well: a zone of transition, honeycomb-like trabeculae and, next to the ventral corticalis, a bundle of strong, parallel trabecular plates lying perpendicular to the longitudinal axis of the tuberositas supraglenoidalis, forming the sandwich bone in its tip (Figure 3.3C).

Taphozous melanopogon – The shoulder joints of examined extant emballonurid bats have a single articulation but are derived in a different way. In *Taphozous melanopogon* the proximal epiphysis of the humerus is small. Figure 3.4A shows the left joint in caudal view. The head is elongate, in the direction of the humeral shaft, proximally narrow with a ridge (Figure 3.4B), and distally broader and oval. The width of the head is *c.* 30% larger when measured 1 mm distal to the base of the tuberculum majus compared to 1 mm proximal of the same structure. The dorsal side of the head is rounded, the ventral side flattened, and is not well separated from the tuberculum majus. The bigger tuberculum minus, on the other hand, is clearly separated from the head by a shallow notch. Neither extends proximally beyond the head, and no cranial pit is developed. The most prominent outer structure of the humerus is the big, dorsally curved crista pectoralis. No crista tuberculi majoris is visible, but a small crista tuberculi minoris is (Figures 3.4A, B).

The cavitas glenoidalis of the scapula has a single articular facet, which is elongate and nearly the same width both cranially and caudally. It looks groove-like, concave in both directions because of big cartilaginous callosities (labrum glenoidale) on its craniodorsal and cranioventral aspect. The tuberositas supraglenoidalis is small and inconspicuous, and there is no cone developed. When the arm is abducted, the narrow cranial part of the humeral head becomes wedged between the callosities of the labrum glenoidale, and pronatory movement of the arm is blocked (Schliemann and Schlosser-Sturm, 1999).

The most remarkable internal structure revealed by μCT in the shaft is a big, strong and multiply bracing plate, which lies horizontally in the middle of the proximal part of the shaft and the humeral head. It consists of a transverse bony perforated plate which is solidly anchored to the shaft on the caudal side and with V-shaped or tree-like struts on the cranial side. This structure is continued throughout the whole humeral head (Figure 3.4C).

The head is filled with three different kinds of spongiosa. Robust honey-comb-like trabeculae are found directly underneath the articular surface, dorsally more so than ventrally. In the inner volume, porous spongiosa with relatively big cavities dominates. Trabecular plates are common, whereas struts are very rare. Two central structures can be discriminated. The first is a network of small plates in the dorsal part, as shown in Figure 3.4C, which is a continu-ance of the big central bracing plate of the shaft, and the second structure is a single strong ventral plate that is connected with the corticalis of the caudal side of the tuberculum minus. The trabecular bone of the tuberculum majus mostly consists of plate-like structures too. These are shown in cross section in Figure 3.4D (left side of image) with fewer trabeculae from dorsal (left) to caudal (bottom). Considering all slices through the tuberculum majus, two big plates are on the dorsal side parallel to the corticalis separating the tuberculum from the shaft. Between these parallel plates the trabeculae also form few vesicular structures. Underneath the muscle attachment (approx. 0.5 mm prox-imal to the plane of Figure 3.4D, see upper facet in Figure 3.4A), the cavity on the cranial side (Figure 3.4D) no longer exists, being replaced by a single plate and thin but predominant honeycomb-like spongiosa.

Trabecular bone can be seen in the tip of the tuberculum minus (Figure 3.4D, right side), on the cranial side parallel to the corticalis. A big cavity is situated in the middle of the tuberculum. This cavity is crossed only by a single delicate trabecular strut distally from the slice shown in the figure. Approximately 1 mm distal at the smooth transition to the shaft, the tuberculum minus is supported by a big plate. Most of the other trabeculae are plate-like and aligned towards the tip. Honey-comb-like spongiosa underneath the muscle attachments can hardly be found.

Within the studied area of the scapula, trabecular bone lies only around the glenoid cavity and, less commonly, in the tip of acromion and coracoid pro-cesses, the margo and angulus cranialis, and the margo lateralis. In the center of the coracoid and acromion processes there are no trabeculae, but instead thin longitudinal tubular bone walls. Underneath the corticalis of the glenoid cavity there is first a thin layer of honeycomb-like spongiosa (Figure 3.4C) and then some plates aligned dorsoventrally parallel to the joint surface. Below that, longer plates radiate in the plane of the shoulder blade towards the margo and angulus cranialis, and the base of the acromion process.

Tachypteron franzeni – Most of the shoulder joints of *Tachypteron franzeni* are preserved on part A of the fossil (Figure 3.2A), and some important ventral fragments of both joints appear on the counterpart B (Figures 3.1B, D). The humeral head of *T. franzeni* is elongated and curved, describing a plane in which the shaft axis also lies. Proximally the head is narrow and broader distally. The width of the head measured above the neck of tuberculum majus is about 30% smaller than the width below the neck (Figure 3.2B). The head is dorsally rounded; some flat fragments visible only on the fossil plate B suggest the ventral surface was flat. The tuberculum majus of the humerus is well developed, but does not rise above the head or articulate with the scapula, as also described previously (Storch *et al.*, 2002). It is clearly separated from the head by a neck (Figure 3.2B); a cranial pit is not detectable. The tuberculum minus on the fossil part B is large; the relative position to the head is not preserved. The shaft is broken and compressed, showing fragments of a large crista pectoralis, but no internal details (Figures 3.1D, 3.2D). No crista tuberculi majoris is developed, and a crista tuberculi minoris on part B is not detectable.

The glenoid fossa of the scapula is apparently single and without a secondary articular surface along its dorsocranial rim (Figure 3.2A). The distal part is round. The proximal part, however, is slightly deformed, so that a tuberositas supraglenoidalis could not be distinguished.

Trabeculae are preserved in the humeral head, tuberculum majus (Figures 3.2C, D) and scapula (Figure 3.2C). In the humeral head, the layer directly underneath the articular facet consists of honeycomb-like trabeculae. In the middle of the head, few but robust trabeculae can be seen coursing orthogonally towards the articular facet. They show few interconnections (Figures 3.2C, D).

In a different three-dimensional orientation of the slices, trabeculae can be seen taking course from the middle of the shaft towards the articular facet of the dorsal part of the head. In the middle of the tuberculum majus is a central cavity braced by trabecular plates. Other plates are preserved under the notch (Figures 3.2A, B) where the M. infraspinatus attached, and here they separate three small cavities (Figure 3.2D). This enlarged detail of the right humerus is different in orientation compared to Figures 3.2A–C. The plane of this view intersects the plane of the image in Figure 3.2B at an angle of approx. 45° along the axis of the humeral neck and thus mirrors corresponding details in Figure 3.3D and Figure 3.4D, with the exception of non-preserved parts of the humerus, e.g., tuberculum minus and crista pectoralis. Thus, *T. franzeni* compares substantially in detail with *Taphozous*.

In the scapula the outermost layer of honeycomb-like trabeculae can clearly be recognized (Figure 3.2C). More trabeculae can be identified in the shoulder

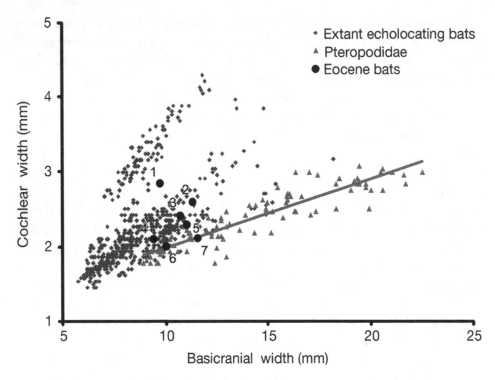

Figure 3.10 Cochlear size of Eocene bats compared to cochlear size of extant microbats and extant Pteropodidae: 1 *Tanzanycteris mannardi*, 2 *Hassianycteris messelensis*, 3 *Tachypteron franzeni*, 4 *Palaeochiropteryx tupaiodon*, 5 *Archaeonycteris trigonodon*, 6 *Icaronycteris index*, 7 *Onychonycteris finneyi*.

of *Tachypteron*, but cannot definitely be assigned to a certain area of the flattened fossil.

3.3.2 Inner ear

Cochlear size was estimated in a broad sample of extant bat species as its oblique diameter across the second half-turn in relation to the basicranial width by Habersetzer and Storch (1992), and the same measurements from microradiographs were taken for Eocene bats. This method of comparing sizes of the cochlea was also used by other authors (Simmons and Geisler, 1998; Gunnell *et al.*, 2003) and the data set was recently expanded by Simmons *et al.* (2008). This expanded data set is represented in Figure 3.10, slightly modified by indicating particular Eocene bat species compared. *Tanzanycteris mannardi* (Gunnell *et al.*, 2003) has a huge rhinolophoid-like cochlea and surpasses in relative cochlear size all other Eocene bats and most extant aerial hawking insectivorous bats (Figure 3.10, 1). *Hassianycteris messelensis* (Figure 3.10, 2) has the largest cochleae of all Messel bats and was also assumed to have a similar echolocation system to emballonurid bats, because based on its flight apparatus it was also adapted for hunting in the open above the tree canopy. *Tachypteron franzeni* (Figure 3.10, 3) has a relatively large cochlea too, and together with *Hassianycteris* and *Palaeochiropteryx tupaiodon* (Figure 3.10, 4) these three extinct species had cochlear measurements within the range of extant echolocating bats. *Archaeonycteris trigonodon* (Figure 3.10, 5), the most primitive of the Messel bats, has a relatively smaller cochlea compared to the other three species. Its cochlear size is close to that of non-echolocating pteropodids (with a few exceptions), as are the Green River bats *Icaronycteris index* (Figure 3.10, 6) and the non-echolocating insectivorous *Onychonycteris finneyi* (Figure 3.10, 7; Simmons *et al.*, 2008).

The holotype of *Tachypteron franzeni* has a cochlear size of 2.42 mm, and the paratype 2.40 mm (Storch *et al.*, 2002). We remeasured the cochlea of the holotype of *T. franzeni* and confirm these measurements on the basis of the new two-dimensional and three-dimensional data sets. We can also confirm the older statement that the above procedure of oblique size measurement provides reliable data independent of the three-dimensional orientation of the cochlear axis (modiolar axis) in the skull, because we obtained the same results (within $\pm$0.012 mm) when we measured the width of the virtually isolated cochlea of plate B (see also Figure 3.1, Figures 3.5–3.7) in different angles of view ($\pm$45°) with VGStudio MAX. We also compared the relation of cochlea size/basicranial width of *T. franzeni* (0.22 and 0.23) with those of 15 species of extant emballonurids and found that *Tachypteron* falls within the range for the extant species (Table 3.1). But some extant species have a slightly larger relative cochlear size.

Internal two-dimensional (Figure 3.5) and three-dimensional (Figure 3.6) details of the completely preserved right cochlea of *Tachypteron franzeni* are described below. The cochlea was studied with two differing methods of microcomputed tomography. Using normal µCT, the resolution would be limited to approximately that shown in Figure 3.1D for the whole plate B, and this is not suitable for showing inner details of the cochlea. As already pointed out, mechanical preparation of the cochlea of the holotype can hardly be considered; two special µCT methods, in contrast, allow the specimen to be scanned as it lies on the plate. The two center rows in Figure 3.5 show ten equidistant slices obtained with the ROI-µCT technique. The distance between the selected slices is 230 µm, and the thickness of a single slice is 10.44 µm. The two marginal rows show the same areas recorded with the P-µCT technique. The slices are in parallel orientation to the plate surface, as is also shown at low magnification in Figure 3.1. The corresponding three-dimensional view of Figure 3.6A provides a guide for orientation and comparison of the slices.

Table 3.1 Basicranial width, cochlear size and relative cochlear size (cochlear size/basicranial width) of *Tachypteron franzeni* and 15 extant emballonurid species. IRSNB = Royal Belgian Institute of Natural Sciences Brussels; LNK = Landessammlungen für Naturkunde Karlsruhe; SMF = Senckenberg Museum Frankfurt.

Genus	Species	Collection	Basicranial width (mm)	Cochlear size (mm)	Relative cochlear size
Tachypteron	*franzeni* (holotype)	IRSNB M1944	10.70	2.42	0.22
Tachypteron	*franzeni* (paratype)	LNK Me 676	10.80	2.40	0.23
Taphozous	*affinis*	SMF 18358	12.47	3.08	0.25
Taphozous	*australis*	SMF 17795	10.40	2.75	0.26
Taphozous	*georgianus*	SMF 17435	10.36	2.63	0.25
Taphozous	*hamiltoni*	SMF 46545	11.50	2.83	0.25
Taphozous	*hildegardeae*	SMF 39393	11.34	2.75	0.24
Taphozous	*hilli*	SMF 68728	10.28	2.71	0.26
Taphozous	*longimanus*	SMF 55573	11.13	2.63	0.24
Taphozous	*mauritianus*	SMF 32790	10.85	2.75	0.25
Taphozous	*melanopogon*	SMF 46333	11.13	2.67	0.24
Taphozous	*nudiv. kachhensis*	SMF 59567	12.63	2.83	0.22
Taphozous	*nudiventris magnus*	SMF 74086	13.28	2.87	0.22
Taphozous	*perforatus*	SMF 33108	10.77	2.71	0.25
Taphozous	*philippinensis*	SMF 29063	10.77	2.83	0.26
Taphozous	*sudani*	SMF 18945	10.20	2.67	0.26
Taphozous	*theobaldi*	SMF 63096	12.31	2.91	0.24

The selected series of CT slices starts with the deepest layer of the fossil plate (Figure 3.5A), which shows the dorsolateral side of the cochlea in its natural and undisturbed position in the unbroken skull and ends with the most superficial slice (Figure 3.5J, bottom) at the ventromedial side of the cochlea. The morphological description below refers to the central images (Figure 3.5A–J). At the upper part of image (B) the dorsal part of the second turn (canalis spiralis cochleae) becomes visible, and in (C) the helical spiral lamina (lamina spiralis ossea primaria, lsop) intrudes into the second turn as a thin lunate shadow, the outer spiral lamina (lamina spiralis ossea secundaria, lsos) becomes visible at the inner bone wall of the canalis cochleae as a small protrusion (upper right). In (D) the second turn is sliced across the canalis spiralis modioli, which houses the nerve cells of the spiral ganglion, and the lsop is shown with its junction to the modiolus. Above is the apical half-turn and below is the hook region of the first half-turn. In (E) the second turn is sliced across the modiolus and the bony margins of the spiral canal of the modiolus with the distinct double bony layers of the lsop visible for the first and the second turn. The opposite lsop and lsos are at best recognized at the first turn. In (F) the modiolus displays the large conical inner space for the acoustic nerve (canalis centralis modioli); the oblique cross-sectional area of the scala tympani of the first turn increases (the left side of the image) and in (G) it is completely fused with the cross-sectional area of the scala tympani at the right side of the image, which also indicates the end of the first half-turn. From (G) to (H) the central column of the cochlea (modiolus) is cut progressively more peripherally, first passing through the spiral canal of the modiolus (G) and then the primary spiral lamina (H), whose two layers are connected by perpendicular bony elements. In Figure 3.5I, the previously separated sections of the upper canal becomes joined, and in (J) this common bony canal for the scala vestibuli and the scala media is the last sliced part of the basal turn. All morphological details are more clearly shown by ROI-μCT (center images) compared to the P-μCT (marginal images). At present, ROI-μCT is the superior method for these kinds of objects, for specimens on a slab.

A virtual cochlear endocast of *Tachypteron franzeni* (Figure 3.6) is used to show the three-dimensional shape of the cochlear canal (canalis spiralis cochleae) from different angles of view. Each part of Figure 3.6 consists of a three-dimensional reconstruction superimposed on a background microradiograph computed from the ROI-μCT data set for the respective angle of view. The long bone elements of the wing, which include the radius and the fifth finger, can serve for comparison with Figure 3.1. In addition to the bone fragments of the skull in the background, two vestibular ampullae are visible, and the whole vestibular organ is represented only by a few fragments. The three-dimensional endocast was closed at its vestibular end with a smoothed, gray-shaded convex surface just behind the oval window (fenestra ovale cochleae) and shortly after the end of the basilar membrane. The oval window is also marked with a gray-shaded convex surface which protrudes slightly. The distinct longitudinal groove on the endocast is the impression of the secondary spiral lamina, which partly divides the upper canal with its two scalae from the lower canal with its single scala tympani. The upper scala vestibuli and scala media cannot be discriminated because the delicate Reissner membrane is not fossilized and so no information for the endolymphatic space is available. The basilar membrane was between the primary and secondary osseous spiral lamina, and the length of this membrane can be determined by following this helical structure along the two and a half turns with a curve-measuring tool in VGStudio MAX. The length of the basilar membrane is 12.9 mm.

The cochlea of *T. franzeni* is comparable to those of emballonurids not only in size, but also in the shape and distribution of inner cavities like the cochlear and modiolar canal which have similar slopes and spatial orientation. For this we compared high-resolution microradiographs of *Taphozous nudiventris kachhensis*, *T. melanopogon* and *T. hilli* and a ROI-μCT image series of *Taphozous melanopogon*. Besides the highly oblique orientation of the cochlear modiolar axis (ventral component approximately 40°), a striking feature of the cochlea of these three *Taphozous* species is a large gap between the end of the first half-turn and the neighboring end of the third half-turn. On account of this gap, the cochlear canal is briefly ventrally deflected. *Tachypteron* shares this character, as can be seen in the virtual three-dimensional endocasts (Figure 3.6A).

Although virtual three-dimensional endocasts represent a remarkable advance for studying bat fossils on large plates, we were suspicious that not all preserved structures in the fossil specimen could be adequately depicted by this method. The ROI-μCT can deal with large plates, but is still limited by the plate size through CT geometry because the plate has to pass in front of the X-ray tube while being rotated. This limits the minimal distance between the ROI of the fossil to the tube surface and thus predetermines the required distance from the ROI to the X-ray detector for a particularly high magnification. The X-ray beam intensity is always low for microfocus tubes, and more so as focus size is calibrated close to its theoretical minimum for high magnification, which is necessary for sharp images of small regions of interest. With very low X-ray beam intensity, the detector sensitivity becomes a limiting factor. Detector sensitivity limits the maximal possible distance of the detector from the X-ray source and thus the maximal possible magnification. Compromises have to be found between high magnification and low signal at the sensor.

3.4   Comparison of imaging methods

All endocast images in Figure 3.6 show a coarsely grained surface structure which does not represent the real bone structure, but originates from the low signal/noise ratio in the gray values of the ROI-tomograms. At present there is no applicable filter or smoothing function available for removing these particular artifacts, because the producer (see Table 3.2) still keeps the algorithm secret. In contrast, the microradiographs provide a unique richness of detail: smaller and more delicate structures of the interior walls of the cochlear canal and the thin bony layers of the lamina spiralis are made visible only by this method (Figure 3.7). The cochlear microradiograph of

Table 3.2 Main characteristics of the special μCT-procedures compared with normal μCT.

	P-μCT	Two-axis-μCT	ROI-μCT	standard μCT
Technical feature	*xy*-translation (P = planar)	*xy*-rotation (synchronous)	*y*-rotation (ROI=region of interest)	*y*-rotation
Size relation of specimen/ ROI	no limit by size (*xy*), only by thickness (*z*)	up to 50	up to 10	< 1 (half rotation <2)
Other advantages	highest possible enlargements	also suited for single axis μCTs	also suited for normal μCT	–
Main disadvantages	only suited for this particular procedure	highest demands in μ-mechanics	probe size limits enlargement	probe size limits enlargement
Promoters and producers	Fraunhofer Feinfocus	Senckenberg Fraunhofer	Wällischmiller, Ray Scan Technologies	many companies

Tachypteron franzeni was made with the microfocus X-ray tube of the two-axis-μCT apparatus with only 7 mm distance from the tube, compared to 60 mm distance shown in the upper right inset image, which is a microradiograph from the ROI-μCT data set. This inset image has a much lower resolution compared to the large image, which in fact has already been scaled down by the factor 0.35×. These differences exemplify very clearly which of the details are visible in the microradiograph and which details are lost using the ROI-μCT. It is obvious what future potential in quality gain can be expected from technical upgrades of P-μCT and two-axis-μCT, for the methods do not require full rotation of the specimen and thus allow a very short distance between the ROI and the tube.

At present ROI-μCT is best suited for three-dimensional, and microradiography for two-dimensional. The two-dimensional cross-sectional areas shown in Figure 3.7 were used to acquire the data for Figure 3.8, which shows the cross-sectional area of the cochlear canal (canalis spiralis cochleae) in relation to basilar membrane length in *Tachypteron franzeni* and certain other species. For the three extant echolocating bats *Molossus ater*, *Taphozous melanopogon* and

Table 3.3 Number of hair cells on the basilar membrane. Hair cells numbers of three extant bats were counted from histological series and used in calculation for *Tachypteron franzeni*.

Number of hair cells	Megaderma lyra	Tachypteron franzeni	Taphozous melanopogon	Molossus ater
Total number	5831	7598		
hair cells/mm	589	589		
Total number		**7688**	8460	
hair cells/mm		**596**	596	
Total number		7663		8673
hair cells/mm		594		594
	Counted	Calculated	Counted	Counted

Megaderma lyra (Fiedler, 1983; Fiedler, personal communication), histological data were used. The basal hook region and the first half-turn are poorly documented (technical problems discussed by Fiedler, 1983); however, the upper turns show relatively smooth curves and species-specific differences. Kraus (1983) also found such differences in three other bat species. Our purpose was to test this kind of analysis for the first time in fossil material. The data plots of *Taphozous melanopogon* and *Tachypteron franzeni* resemble each other compared to the other two species. Thus, cross-sectional area of the cochlear turn is another character in which both these species are similar.

From the basilar membrane length it is possible to reconstruct the number of hair cells, because the relative number of hair cells per mm is almost constant (approximately 590) in examined living species. In Table 3.3 the data of three extant bats are given (Fiedler, 1983; Fiedler, personal communication) with absolute hair cell numbers counted from histological preparations and relative number per mm of the basilar membrane. These data were used to calculate both values for *Tachypteron franzeni*. It is obvious that *T. franzeni* compares more closely with the high and rapid hunting bats *Taphozous melanopogon* and *Molossus ater* than with the Indian false vampire bat *Megaderma lyra*, which hunts close to the ground (Marimuthu *et al.*, 1995) and does not produce loud and long-ranging ultrasonic sounds (like *Taphozous* or *Molossus*), but only makes faint sounds, often interrupted by simply listening to the noise of prey.

3.5   Discussion and conclusions

For a long time it was not possible to study Messel fossils or other specimens preserved on large plates by means of computed tomography. This is because of the disadvantages of the high specimen/ROI size ratio for fossils preserved on large plates, as already pointed out in the Introduction. A critical evaluation of the applied X-ray- and μCT-methods can be made from the figures in the Results section and from the technical descriptions in Material and Methods. Our evaluation, which is summarized comprehensively in Table 3.2, is based on these results and also on various other specimens from Messel and the Green River Formation. This evaluation also includes our experiences with the application of ROI-μCT for the newly described primate *Darwinius masillae* (Franzen *et al.*, 2009) and a Messel lizard (Smith *et al.*, 2008), both of which were preserved on large plates.

On account of ease of use, standard μCT is the first choice whenever possible. If the probe is only slightly larger than the region of interest, the 360° rotation of the probe may be reduced to 180° and the rotation axis may be shifted to the right or left field of view to exclude some parts of the specimen from being projected into the raw image data. When the specimen/ROI ratio is 2 or larger, the region of interest in the specimen would have to be cut out mechanically. If this is possible (usually not for unique fossil specimens), the full resolution of the apparatus can be used. Using synchrotron radiation (SR) for μCT of prepared inner ears of extant bats provides superb image quality, and the use of the full size of the detector only for this ROI isolated from the skull by preparation even makes it possible to zoom into details of the cochlea (Habersetzer *et al.*, 2004). For prepared human cochleae, it was recently shown that even histological details of the corti organ can be studied by SR-μCT (Lareida *et al.*, 2009).

In this chapter, *Tachypteron franzeni* exemplifies the case in which normal μCT and SR-μCT cannot be applied for the investigation of the desired small ROIs. As shown in Figure 3.5, ROI-μCT provides better results compared to P-μCT (and thus also better results than the two-axis-μCT). The three-dimensional images of the ROI-μCT allow a complete reconstruction of the course of the cochlear scalae (Figure 3.6), but also show a lot of grain-like artifacts on the surface. These artifacts do not reflect the preservation of the fossil itself, but originate from the low signal/noise ratio of the tomograms, as can be proven by the microradiograph (Figure 3.7). This was also the reason why cross-sectional areas were taken from the microradiographs and not from the tomographic images. With these limitations of ROI-μCT (which are even more pronounced in P-μCT), we are presently not able to analyze trabecular thickness, degree of anisotropy or trabecular bone volume (Odgaard, 2001; Fajardo *et al.*, 2002), which, however, can be done without difficulty for prepared samples of extant bats. The next upgrade of ROI-μCT will have increased resolution by a factor of 2, and thus will increase the number of voxels (cubic pixels) eightfold. P-µCT was discontinued by the producer, but will be relaunched with fourfold resolution, which will result in a 64-fold increase in the number of voxels. Very likely the presently limiting signal/noise ratios will be improved by new detector types, and image quality can be additionally improved by voxel-binning when the resolution is much higher. All this will allow an effective reduction of the artifacts, as displayed in Figure 3.5 and may also constitute a break-through for two-axis-µCT, which has the highest demands on µ-mechanics, but conversely, is the method with the highest versatility (see Table 3.2: other advantages). Two-axis-µCT will be especially interesting in combination with synchrotron radiation, because it is the practical alternative to P-µCT, which cannot be executed with the geometry of SR. However, it remains to be seen what the above-presented µCT-methods will provide for the best collection of quantitative data in the near future.

The topics discussed here are of principal importance for studying microscopic details in Messel fossils preserved on large plates. Impressive, state-of-the-art studies of cochlear endocasts of the apatemyid *Carcinella sigei* from the Phosphorites du Quercy, France (Koenigswald *et al.*, 2009) and of *Henkelotherium guimarotae* from Guimarota, Portugal (Ruf *et al.*, 2009) were recently published. These studies also illustrate the enormous difficulties in approaching real microscopic resolution of cochlear endocasts by µCT imaging, which arise even when the fossil material is not preserved on large plates. The results depicted from the CT images and microradiographs of *Tachypteron franzeni* are, despite the limitations in some quantitative aspects discussed above, best suited to study specializations of the emballonurid shoulder joint and the general shape and internal detail of the emballonurid cochlea.

The shoulder joint of bats works in two different operational modes. When the arm is adducted, the distal parts of the humeral head and glenoid cavity are in contact. The joint is then disposed for the needs of quadrupedal movement: the distal parts of the joint are more generalized, the joint surfaces of the head and scapula are rounded and the resulting ball-and-socket articulation presents no restriction to movement. This is observed in the two extant bats in our sample. The specialized parts of the joint are found proximally on the humerus and adjacent scapula. These parts are in contact when the arm is fully abducted, as in flight, and the forces of lift and propulsion must be transmitted to the body without loss.

Here a broad and detailed description of the general bone morphology of the shoulder joint is given for the extant bats, even if some of the findings cannot be applied to the fossil specimen under consideration in this study. A summary comparison and interpretation is given in Table 3.4.

Table 3.4 Description and interpretation of the shoulder joint.

Structure	Species		
	Molossus ater	*Taphozous melanopogon*	*Tachypteron franzeni*
Humeral head			
– Outer shape	Irregularly oval head, oriented diagonally to shaft axis	Head small, elongated in direction of shaft axis with proximal rim, distally oval, dorsally rounded, ventrally flattened	Head elongated in direction of shaft axis, proximal narrow, distally broader and oval, dorsally rounded, ventrally flat fragments visible (part B)
– Inner layers			
– Zone of transition	Present at dorsal part	Not present	Not present
– Honeycomb-like trabeculae	Under joint surface or zone of transition	Under joint surface	Under joint surface
– Center of head	Mesh of spongiosa plates central and proximal, directed to caudal wall of shaft	Porous, robust spongiosa; central robust plates big, strong horizontal plate in head and shaft	Porous, robust spongiosa; central robust plates not present (probably due to preservation)
Interpretation	Humeral head slides from flight operation mode (blocked joint) into crawling mode (free movement) Head has no blocking function by itself. Axis of rotation only if humerus is adducted.	Humeral head works distally as a ball-and-socket joint, the proximal rim acts like a wedge that blocks pronation. Part of axis of rotation	Humeral head works distally as a ball-and-socket joint, the proximal rim acts like a wedge that blocks pronation. Part of axis of rotation. Wedge is not as distinct as in *Taphozous*
Tuberculum majus			
– Outer shape	Prominent, rising well above the head, separated from head by a neck ending cranially in a distinct pit	Not rising above the head, not separated from head by a neck, no cranial pit	Well developed, but not rising above the head, separated from head by a neck, no pit visible

Table 3.4 (cont.)

Structure	Species		
	Molossus ater	Taphozous melanopogon	Tachypteron franzeni
– Inner layers			
– Sandwich bone	In the tip	Not present	Not present
– Zone of transition	At the base of tuberculum	Not present	Not present
– Honeycomb-like trabeculae	Not present	Few in the tip	Not present
– Center	Big hollow braced by plates	Small cavities braced by horizontal plates	Not present or preserved Small cavities braced by plates, likely horizontal
Interpretation	Horizontal lever arm for dorsal scapular muscles Blocking mechanism Axis of rotation if abducted	Vertical lever arm for dorsal scapular muscles No blocking mechanism No axis of rotation	Vertical lever arm for dorsal scapular muscles No blocking mechanism No axis of rotation
Tuberculum minus			
– Outer shape	Inconspicuous, not separated but close to head, not rising above the head	Bigger than tub. majus, clearly separated from head, not rising above the head	Big, relative position to head not preserved
– Inner layers			
– Honeycomb-like trabeculae	Not present	Few	Could not be identified
– Center	Mainly hollow, few trabeculae	Strong plate separating from shaft	
Interpretation	Weak bony structure No additional lever arm	Big vertical lever arm for M. subscapularis developed	Big vertical lever arm for M. subscapularis developed

The two different tissues encountered underneath the covering cortical bone are as follows:

(1) First there may be a zone of transitional structure between cortical and trabecular bone. This zone of transition consists of tight bone with few and tiny pores. This tissue was found in our sample only in *Molossus* in the dorsal part of the humeral head at the base of the tuberculum majus and in its secondary articular facet (Figure 3.3D). Its function is similar to that of the honeycomb-like spongiosa, to transmit strong compressive forces immediately and without deformation from the corticalis to deeper parts of the bone (Currey, 2006).

(2) Four different kinds of the second tissue, trabecular bone (cf. spongiosa), can be distinguished in the three specimens. They vary in form and function. In honeycomb-like spongiosa, short trabeculae are cross-linked in all directions; many of them are plate-like. All bats analyzed here show a layer of honeycomb-like spongiosa directly underneath the bony joint surface in the humeral head and the cavitas glenoidalis (Figure 3.4C, Figure 3.2C); in *Molossus* this tissue also exists beneath the secondary articular facet, and in *Taphozous* there is comparatively little in the tip of the tuberculum majus and minus. This kind of tissue is found universally underneath joint surfaces; it transmits loads directly from the thin subchondral bone to the dense cortex and thus helps to protect the cartilage when loaded during impact (Currey, 2006). Underneath the attachment of muscles and tendons this kind of spongiosa works to distribute traction (Currey, 1984).

Plate-like trabeculae can be found in *Molossus* in the proximal part of the shaft, in the tuberculum majus, in the center and the proximal parts of the head (Figure 3.3D) and beneath the secondary articular facet, where a bundle of strong, parallel plates lie perpendicular to the long axis of the tuberositas supraglenoidalis and also form sandwich bone (see below) in its tip. In *Taphozous* this kind of spongiosa is predominant. The most remarkable structure of the proximal humerus is the big, strong, horizontal and multiply bracing plate. It connects the caudal and the cranial corticalis, close to the base of the crista pectoralis (Figure 3.4C), and consists of a transverse, bony, perforated plate that is solidly anchored to the shaft on the caudal side and with V-shaped or branch-like struts on the cranial side. This structure is continued throughout the whole humeral head. It is not found in *Molossus*, nor in *Noctilio leporinus*, another species with a derived shoulder joint with a single articular facet (data not shown). Both instead have a lot of nearly parallel, smaller plates, as also described by Swartz *et al.* (1998). Plate-like trabeculae carry loads in any direction parallel to the plate. They are found "underneath loaded surfaces, particularly where the pattern of stress is reasonably constant" (Currey, 1984, p. 33). The plate-like structure reflects high mechanical stress (Ding *et al.*, 2002). For *Tachypteron* only plate-like structures in the humeral head can be compared because unfortunately the shaft of the fossil is broken.

Long, delicate, single trabecular struts are seen in *Molossus* and *Taphozous* crossing hollow parts of the head or tuberculae, forming short branching struts that anchor plates to the cortical bone (Figure 3.3C, Figure 3.4C). In the flattened *Tachypteron*, struts cannot definitely be identified. Struts or rod-like structures reflect low mechanical stress (Ding *et al.*, 2002). Cylindrical struts with no preferred orientation are found deep in bones, well away from any loaded surface (Currey, 1984). They are rarely cross-linked, only to prevent buckling. Rogers and LaBarbera (1993) determined that these inconspicuous struts are in fact very important for bone mechanics and that bone is destabilized immediately if they are destroyed. Thus, trabecular struts in *Molossus* and *Taphozous* play an important form-stabilizing role in deeper bone regions with low mechanical stress by saving bone mass and consequently body weight.

Sandwich bone is seen in our sample only in *Molossus* in the tip of tuberculum majus and tuberositas supraglenoidalis (Figure 3.3C), and beneath the secondary articular facet. This kind of spongiosa is found in thin bones that are exposed to high mechanical stresses, especially bending forces. It resembles I-beams in engineering (Currey, 1984).

Our hypothetical expectations expressed in the Introduction are discussed below on the basis of the new results:

(1) Only in *Molossus* is the axis of rotation shifted in abduction from the humeral head and cavitas glenoidalis to the secondary articulation and tuberositas supraglenoidalis. In the emballonurid bats, the axis of rotation remains in the middle of head and shaft.

(2) Different size and lever arms of the tuberculae are accompanied by typically developed spongiosa. In the forearm of *Tachypteron*, the outer shape of the three lever arms, tuberculae majus, minus and crista pectoralis, is likely comparable to *Taphozous* as is the spongiosa of the tuberculum majus, in which spatial orientation is preserved as well. In the abducted forearm of *Taphozous*, the tuberculum majus, the lever arm of Mm. supra- and infraspinatus, is directed dorsomediad. The tuberculum minus, the lever arm of M. subscapularis, is elongated and directed ventromediad. Both pull the proximal rim of the humeral head into the narrow cavitas glenoidalis and maintain it there. Inside, both tuberculae are braced by parallel plates.

The crista pectoralis, directed dorsocraniad, is the large and long lever arm of the M. pectoralis. Its pronatory forces *are transfe*rred by the strong medial trabecular plate crossing from the shaft to the head and into the wedged rim. In the a*bducted fo*rearm of *Molossus*, the tuberculum majus, lever arm of dorsal Mm. supra- and infraspinatus, is elongated horizontally in the *proximal d*irection. The M. subscapularis pulls the tiny tuberculum minus ventrally tight to the scapula. The crista pectoralis is of average size, so the M. pectoralis is attached close to the humeral shaft. Its pronatory forces are transmitted to the corticalis and resisted by the bony lock between the pressure points in the trabeculae, which are strengthened in: (1) the dorsal part of caput humeri and the dorsal border of the cavitas glenoidalis, (2) the tuberositas supraglenoidalis and the pit proximal on the caput humeri and (3) the ventral side of the tuberculum majus and the secondary articular facet.

(3) Spatial variation in the internal construction of the two specialized joint types:

- In *Tachypteron*, the humeral head is small, elongate, proximally narrow, distally oval and *strengthened* by trabecular plates in the middle. The spongiosa is mainly porous but robust. The inner structure of the tuberculum majus is built similarly as in *Taphozous*.
- In *Taphozous*, the humeral head is small, elongate with a proximal rim and distally oval. It is strengthened by a big, trabecular plate lying horizontally in the middle of the head and shaft. The spongiosa is mainly porous but robust, with parallel horizontal plates.
- In *Molossus*, the humeral head is irregularly oval; the spongiosa in the head is mainly a dense mesh. There are only isolated plates and struts lying in the shaft distal to the joint. The strengthened parts are found dorsally, in the tuberculum majus, in the dorsomedial part of the humeral head and on the dorsal side of the scapula.

The arrangement in layers underneath the articular facets shows little taxonomic variation. In *Tachypteron* the same substructure can be found as in the extant bats of our sample. In descending order: cortical bone, honeycomb-like trabeculae and deeper layers of plates. In *Molossus*, a zone of transition is additionally found between cortical bone and honeycomb-like trabeculae in the head, the tuberculum majus and the two articular facets. That suggests exceptional forces acting on those parts of the bones.

Thus, the shoulder joint of *Tachypteron franzeni* is of the derived type with a single articulation. The outer shape of the head is not at all globular, but rather elongated, distally broader and proximally narrow, though not as narrow as in

Taphozous melanopogon. The size and orientation of the tuberculae and the large crista pectoralis are also comparable and typical for derived joints with a single articulation. The trabecular bracing structures in the middle of the shaft and in the humeral head support this interpretation. No external or internal characteristics typical for *Molossus ater* can be detected in *Tachypteron franzeni.* It has been shown that the joint function is very similar to that presented for *Taphozous melanopogon.*

The cochlea of *Tachypteron franzeni* is characterized by its size (Figures 3.5, 3.10), endocast shape (Figure 3.6) and basilar membrane length (Figures 3.6, 3.7). All these compare well with the extant emballonurid species *Taphozous melanopogon.* The shape of the scalae cochleae (cross-sectional area vs. basilar membrane length: see Figures 3.7–3.8) of *T. franzeni* and *T. melanopogon* are very similar, but different than *Molossus ater,* another extant bat with similar ecological specializations to rapid hunting flights at high altitudes. Finally, there are large differences in the shape of scalae cochleae and the basilar membrane length when *T. franzeni* is compared with *Megaderma lyra* (Figure 3.8), an extant bat with a very different echolocating system, hunting behavior and flight style (Marimuthu *et al.*, 1995).

The internal details made visible by special microradiography and μCT-methods in this chapter provide evidence that the echolocation performance and flight style of *Tachypteron franzeni* was very similar compared to extant emballonurid bats. These new data support the interpretation of Storch *et al.* (2002), whose conclusions were based only on the cochlear size and on other morphological characters, such as wing shape and area, wing loading and aspect ratio. Taking together new and previously published facts, we conclude that the view of *Tachypteron franzeni* as showing a rapid and constant flight style and using an echolocation system adapted for hunting in the open air above the forest canopy is now very strongly supported. This unique fossil extends the previously described ecological spectrum of various foraging strategies in the Messel bat community to the extreme type of high-altitude flight foraging (Figure 3.9). It seems now less speculative to us to interpret this fossil as a highly derived and specialized emballonurid similar to many bats of today. The sympatry in the Messel bat fauna of "old-timers" like *Archaeonycteris trigonodon* and "modern specialists" like *Tachypteron franzeni* is all the more intriguing with the discovery of the most primitive known bat, *Onychonycteris finneyi* (Simmons *et al.*, 2008), in the Green River Formation. This insectivorous but non-echolocating bat occurred together with *Icaronycteris index,* a close relative of *Archaeonycteris trigonodon* from Messel, but both Green River bats were only about 5 myr older than the Messel bats. Obviously, bats of widely differing evolutionary stages were present in a very narrow time interval. This makes further efforts to collect quantitative data about the internally preserved morphological details even more promising for broader comparative studies including more representatives of Early and Middle Eocene bats.

3.6   *Tachypteron franzeni* summary

Tachypteron franzeni is the only one of seven bat species known from Grube Messel (near Darmstadt, Germany) which can be assigned to an extant family. The extraordinarily well-preserved holotype shows trabecular micro-structures of the elongated tuberculum majus of the humerus and also internal microscopic details of the cochlea which compare well to emballonurid species (e.g., *Taphozous melanopogon, T. kachhensis*), but less to other extant bats with similar morphological and ecological specializations (e.g., *Molossus ater*).

The shoulder joint and inner ear were examined as a whole by means of various μCT-methods and high-resolution radiographs. Several bone tissues could be identified in the shoulder joint region; three-dimensional images of functional units of trabecular structures are compared and interpreted. These details made visible by μCT, in addition to other morphological characters, strongly support the view of a rapid and constant flight style for *Tachypteron franzeni* in combination with an echolocation system adapted for hunting in the open air above the forest canopy. Thus, this type of ecological adaptation extends the known spectrum of foraging strategies in the Messel bat community to the extreme of high-altitude flight foraging.

3.7   Acknowledgments

We thank Dr. T. Smith, Brussels, for the loan of holotype specimen, Mark Peinl (Morphisto/Senckenberg) and Dipl. Ing. R. Gipser (Gipser Consulting) for their technical contributions for Tomoimage-C++-program and two-axis-μCT, Juliane Eberhardt (Senckenberg) for technical assistance, Dr. A. Fiedler (Frankfurt) for providing histological data of extant bats and Dr. K. Smith (Senckenberg) for revising the English. The chapter was improved by many suggestions by two referees, one of them Dr. T. Macrini, a second anonymous. Dr. F. Beckmann (Synchroton μCT, DESY, Hamburg), Dr. R. Reinhold (Xylon-Comet/Feinfocus P-μCT, Hannover), Dr. M. Brodmann and Dr. M. Sindel (AUDI-CT-Lab/Wällischmiller ROI-μ CT, Neckarsulm,) Dr. M. Goebbels (BAM, Berlin) and the team of Volume Graphic (VGStudio Max, Heidelberg) were helpful with many discussions and proposals. We are grateful to Ermann Foundation (Ermann Stiftung, Frankfurt), Fraunhofer Gesellschaft (Pat.-Stelle f.d.dt. Forschung, München) and Deutsche Forschungsgemeinschaft (DFG, Bonn) for financial support.

3.8   REFERENCES

Altenbach, J. S. (1979). Locomotor morphology of the vampire bat, *Desmodus rotundus*. *Special Publication, American Society of Mammalogists*, **6**, 1–137.

Altenbach, J. S. (1987). Bat flight muscle function and the scapulo-humeral lock. In *Recent Advances in the Study of Bats*, ed. M. B. Fenton, P. Racey and J. M. V. Rayner. Cambridge: Cambridge University Press, pp. 100–118.

Bergemann, Y. (2003). Funktionsmorphologische Untersuchungen am Schultergelenk der Chiroptera mit besonderer Berücksichtigung der Megachiroptera und der Rhinopomatidae. Unpublished Ph.D. thesis, Hamburg: University of Hamburg.

Currey, J. D. (1984). *The Mechanical Adaptations of Bones*. Princeton, NJ: Princeton University Press.

Currey, J. D. (2006). *Bones: Structure and Mechanics*. Princeton, NJ: Princeton University Press.

Ding, M., Odgaard, A., Danielsen, C. C. and Hvid, I. (2002). Mutual associations among microstructural, physical and mechanical properties of human cancellous bone. *Journal of Bone and Joint Surgery – British Volume*, **84B**, 900–907.

Fajardo, R. J. and Muller, R. (2001). Three-dimensional analysis of nonhuman primate trabecular architecture using micro-computed tomography. *American Journal of Physical Anthropology*, **115**, 327–336.

Fajardo, R. J., Ryan, T. M. and Kappelman, J. (2002). Assessing the accuracy of high-resolution X-ray computed tomography of primate trabecular bone by comparisons with histological sections. *American Journal of Physical Anthropology*, **118**, 1–10.

Fiedler, J. (1983). Vergleichende Cochlea-Morphologie der Fledermausarten *Molossus ater, Taphozous nudiventris, T. kachensis* und *Megaderma lyra*. Unpublished Ph.D. thesis, Frankfurt am Main: Johann Wolfgang Goethe-Universität.

Franzen, J. L., Gingerich, P. D., Habersetzer, J. *et al.* (2009). Complete primate skeleton from the middle Eocene of Messel in Germany: morphology and paleobiology. *PLoS ONE*, **4**, e5723, 1–27.

Gunnell, G. F., Jacobs, B. F., Herendeen, P. S. *et al.* (2003). Oldest placental mammal from Sub-Saharan Africa: Eocene microbat from Tanzania – evidence from early evolution of sophisticated echolocation. *Palaeontologia Electronica*, **5**.

Habersetzer, J. (1986). Vergleichende flügelmorphologische Untersuchungen an einer Fledermausgesellschaft in Madurai. In *BIONA Report 5: Bat Flight – Fledermausflug*, ed. W. Nachtigall. Stuttgart: Gustav Fischer Verlag, pp. 75–104.

Habersetzer, J. (1998). Tomographieverfahren und Anordnung zur Erzeugung von großflächigen Tomogrammen (Aktenzeichen 19542762.9–52). Forschungsinstitut Senckenberg, Senckenberganlage 25, 60325 Frankfurt. München: Deutsches Patentamt, pp. 1–18.

Habersetzer, J. and Storch, G. (1987). Klassifikation und funktionelle Flügelmorphologie paläogener Fledermäuse (Mammalia, Chiroptera). Forschungsergebnisse zu den Grabungen in der Grube Messel bei Darmstadt. *Courier Forschungsinstitut Senckenberg*, **91**, 117–150.

Habersetzer, J. and Storch, G. (1992). Cochlea size in extant Chiroptera and Middle Eocene microchiropterans from Messel. *Naturwissenschaften*, **79**, 462–466.

Habersetzer, J., Scherf, H., Beckmann, F. and Seidel, R. (2004). 3-D-Animation knöcherner Gesamtskelette und mikro-tomographischer Skelettdetails von Fossilien aus der Grube Messel. *Courier Forschungsinstitut Senckenberg*, **252**, 237–241.

Habersetzer, J., Storch, G., Schlosser-Sturm, E. and Sigé, B. (2007). Shoulder joints and inner ears of *Tachypteron franzeni*, an emballonurid bat from the Middle Eocene of Messel. *Journal of Vertebrate Paleontology*, **27** (Suppl. 3), 85A.

Helversen, O. V. (1986). Blütenbesuch bei Blumenfledermäusen: Kinematik des Schwirrfluges und Energiebudget im Freiland. In *BIONA Report 5: Bat Flight – Fledermausflug*, ed. W. Nachtigall. Stuttgart: Gustav Fischer Verlag, pp. 107–126.

Hermanson, J. W. and Altenbach, J. S. (1983). The functional anatomy of the shoulder of the pallid bat, *Antrozous pallidus*. *Journal of Mammalogy*, **64**, 62–75.

Ito, M., Nishida, A., Koga, A. *et al.* (2002). Contribution of trabecular and cortical components to the mechanical properties of bone and their regulating parameters. *Bone*, **31**, 351–358.

Kastner, J., Heim, D., Salaberger, D., Sauerwein, C. and Simon, M. (2006). Advanced applications of computed tomography by combination of different methods. *European Congress of Non-Destructive-Testing*, pp. 1–8.

Kirkpatrick, S. J. (1994). Scale effects on the stresses and safety factors in the wing bones of birds and bats. *Journal of Experimental Biology*, **190**, 195–215.

Koenigswald, W. V., Ruf, I. and Gingerich, P. D. (2009). Cranial morphology of a new apatemyid, *Carcinella sigei* n. gen. n. sp. (Mammalia, Apatotheria) from the late Eocene of southern France. *Palaeontographica – Beiträge zur Naturgeschichte der Vorzeit*, **288**, 53–91.

Kraus, H. J. (1983). Vergleichende und funktionelle Cochlea-Morphologie der Fledermausarten *Rhinopoma hardwickei*, *Hipposideros speoris* und *Hipposideros fulvus* mit Hilfe einer Computerunterstützten Rekonstruktionsmethode. Unpublished Ph.D. thesis, Frankfurt am Main: Johann Wolfgang Goethe-Universität.

Lareida, A., Beckmann, F., Schrott-Fischer, A. *et al.* (2009). High-resolution X-ray tomography of the human inner ear: synchrotron radiation-based study of nerve fibre bundles, membranes and ganglion cells. *Journal of Microscopy*, **234**, 95–102.

Lim, D. Y., Seliktar, R., Wee, J. Y., Tom, J. and Nunes, L. (2006). The effect of the loading condition corresponding to functional shoulder activities on trabecular architecture of glenoid. *Journal of Biomechanical Engineering – Transactions of the ASME*, **128**, 250–258.

Marimuthu, G., Habersetzer, J. and Leippert, D. (1995). Active acoustic gleaning from the water surface by the Indian false vampire bat, *Megaderma lyra*. *Ethology*, **99**, 61–74.

Miller, G. S. (1907). The families and genera of bats. *Bulletin of the United States National Museum*, **57**, 1–282.

Norberg, U. M. (1976). Aerodynamics, kinematics and energetics of horizontal flapping flight in the long-eared bat *Plecotus auritus*. *Journal of Experimental Biology*, **65**, 179–212.

Norberg, U. M. (2002). Structure, form, and function of flight in engineering and the living world. *Journal of Morphology*, **252**, 52–81.

Norberg, U. M. and Rayner, J. M. V. (1987). Ecological morphology and flight in bats: wing adaptations, flight performance, foraging strategy and echolocation. *Philosophical Transactions of the Royal Society of London B*, **316**, 335–427.

Odgaard, A. (2001). Quantification of cancellous bone architecture. In *Bone Mechanics Handbook*, ed. S. C. Cowin. Boca Raton, FL: CRC Press, pp. 14.01–14.19.

Rogers, R. R. and LaBarbera, M. (1993). Contribution of internal bony trabeculae to the mechanical properties of the humerus of the pigeon (*Columba livia*). *Journal of Zoology*, **230**, 433–441.

Ruf, I., Luo, Z. X., Wible, J. R. and Martin, T. (2009). Petrosal anatomy and inner ear structures of the Late Jurassic *Henkelotherium* (Mammalia, Cladotheria, Dryolestoidea): insight into the early evolution of the ear region in cladotherian mammals. *Journal of Anatomy*, **214**, 679–693.

Schliemann, H. and Schlosser-Sturm, E. (1999). The shoulder joint of the Chiroptera – morphological features and functional significance. *Zoologischer Anzeiger*, **238**, 75–86.

Schlosser-Sturm, E. (1982). Zur Funktion und Bedeutung des sekundären Schultergelenks der Microchiropteren. *Zeitschrift für Säugetierkunde*, **47**, 253–255.

Schlosser-Sturm, E. and Schliemann, H. (1995). Morphology and function of the shoulder joint of bats (Mammalia: Chiroptera). *Journal of Zoological Systematics and Evolutionary Research*, **33**, 88–98.

Simmons, N. B. and Geisler, J. H. (1998). Phylogenetic relationships of *Icaronycteris*, *Archaeonycteris*, *Hassianycteris*, and *Palaeochiropteryx* to extant bat lineages, with comments on the evolution of echolocation and foraging strategies in Microchiroptera. *Bulletin of the American Museum of Natural History*, **235**, 4–182.

Simmons, N. B., Seymour, K. L., Habersetzer, J. and Gunnell, G. F. (2008). Primitive early Eocene bat from Wyoming and the evolution of flight and echolocation. *Nature*, **451**, 818–821.

Smith, K. T., Rieppel, O. and Habersetzer, J. (2008). A complete necrosaur (Squamata: Anguimorpha) from the middle Eocene lagerstätte of Messel, Germany. *Journal of Vertebrate Paleontology*, **28** (Suppl. 3), 144A.

Stauber, M., Rapillard, L., Vanlenthe, G. H., Zysset, P. and Muller, R. (2006). Importance of individual rods and plates in the assessment of bone quality and their contribution to bone stiffness. *Journal of Bone and Mineral Research*, **21**, 586–595.

Storch, G., Sigé, B. and Habersetzer, J. (2002). *Tachypteron franzeni* n. gen., n. sp., earliest emballonurid bat from the Middle Eocene of Messel (Mammalia, Chiroptera). *Paläontologische Zeitschrift*, **76**, 189–199.

Strickler, T. L. (1978). *Functional Osteology and Myology of the Shoulder in the Chiroptera*. Basel and New York: Karger Verlag.

Swartz, S. M., Parker, A. and Huo, C. (1998). Theoretical and empirical scaling patterns and topological homology in bone trabeculae. *Journal of Experimental Biology*, **201**, 573–590.

Vaughan, T. A. (1970). The skeletal system. In *Biology of Bats 1*, ed. W. A. Wimsatt. New York: Academic Press, pp. 97–138.

4

Evolutionary history of the Neotropical Chiroptera: the fossil record

GARY S. MORGAN AND NICHOLAS J. CZAPLEWSKI

4.1 Introduction

Recent improvements in the fossil record of bats in the Americas provide new raw material to bolster interpretations of bat evolution and paleobiogeography. Herein we review the fossil record of bats from North America and South America and provide a narrative explanation of the historical development of the Neotropical bat fauna from a paleontological and biogeographical perspective. The phylogeography of bats in the Western Hemisphere as it pertains to the development of the modern Neotropical biogeographic region may also be important in developing strategies for the conservation of the tropical American chiropteran fauna.

Throughout most of the Cenozoic Era there were three widely separated tropical regions in the New World: the island continent of South America, excluding temperate latitudes in Argentina and Chile; the West Indies and Middle America, including Central America and central and southern Mexico, and at times extending northward to encompass the Gulf Coastal Plain from Texas east to Florida. North America and South America possessed very different mammalian faunas prior to the Great American Biotic Interchange, which began with a limited exchange of large non-volant mammals in the Late Miocene (~9 Ma), presumably by overwater dispersal, and then became a fully fledged interchange following the formation of the Panamanian Isthmus joining these two continents in the Early Pliocene (~5 Ma). From the Pliocene onward, there was extensive intermingling between North and South American biotas, and the tropical portions of these two continents now share a common Neotropical mammalian fauna that extends northward to about the Tropic of Cancer (~23°N latitude) in central Mexico and southward to about the Tropic of Capricorn (~23°S latitude) in southern Brazil, Paraguay and northern Argentina. The tropical American bat fauna has become so extensively

Evolutionary History of Bats: Fossils, Molecules and Morphology, ed. G. F. Gunnell and N. B. Simmons. Published by Cambridge University Press. © Cambridge University Press 2012.

intermixed since the Interchange, particularly in Middle America, that it is very difficult to determine the origin of most groups that lack a fossil record. By studying pre-Pliocene bats we can establish the pre-Interchange chiropteran faunas of North and South America, thereby determining which groups of bats were involved in the Interchange and where they may have originated. The Tertiary record of New World bats is primarily limited to temperate North America and tropical South America, with two localities from Patagonia in temperate South America. The Middle American Tertiary chiropteran record is limited to the extinct vespertilionid genus †*Plionycteris* from two Early Pliocene sites in Mexico (Lindsay and Jacobs, 1985; Carranza-Castañeda and Walton, 1992). (A family, genus, or species name preceded by a dagger (†) designates an extinct taxon.) Tertiary bats are unknown from the West Indies, although the large number of endemic genera and species among the chiropteran fauna, particularly in the Phyllostomidae, Mormoopidae and Natalidae, suggests a long period of isolation and autochthonous speciation, probably since the Miocene if not earlier.

Pre-Interchange land mammal faunas in Middle America and southern North America from sites now within the Neotropical region include an Early to Middle Miocene (Late Hemingfordian/Early Barstovian NALMAs) fauna from Panama (Whitmore and Stewart, 1965; MacFadden, 2006; Kirby *et al.*, 2008) and Middle Miocene (Barstovian NALMA) faunas from the states of Oaxaca and Chiapas in southern Mexico (Ferrusquia-Villafranca, 2003). These sites share many mammals with faunas of similar age from the western United States and Florida, but bear no resemblance to faunas of Miocene age from South America. These Tertiary mammal faunas do appear to have a tropical component among the non-volant mammals (e.g., protoceratid artiodactyls; Webb *et al.*, 2003; Woodburne *et al.*, 2006). Small mammals, including bats, are very poorly known from the Tertiary of tropical Middle America, although this clearly represents a collecting bias. Surely, both Middle America and the West Indies supported a tropical chiropteran fauna during the Tertiary, distinct from the vespertilionid-dominated bat fauna of temperate North America, but fossil evidence of this fauna is lacking.

4.2   Materials and methods

Many of the chiropteran fossils upon which this review is based have been examined by one or both of us, including almost all of the fossils from the Tertiary of Florida (GSM, NJC) and South America (NJC). Chiropteran fossils from the Florida Tertiary are primarily housed in the vertebrate paleontology collection of the Florida Museum of Natural History (FLMNH), University of Florida (UF), with a small sample from the Miocene Thomas

Farm site, including the holotypes of †*Miomyotis floridanus* and †*Suaptenos whitei*, in the Museum of Comparative Zoology (MCZ), Harvard University. The fossils from La Venta, Colombia, collected by field teams from Duke University, Instituto Nacional de Investigaciones en Geociencias, Minería y Química (INGEOMINAS), Oklahoma Museum of Natural History (OMNH) and Kyoto University, are cataloged in the Museo Geológico of INGEOMINAS (IGM) in Bogotá, Colombia, with the holotype and other specimens of †*Notonycteris magdalenensis* from La Venta in the University of California Museum of Paleontology (UCMP) at Berkeley. Several Tertiary bat fossils from Mexico are housed in the Instituto de Geología, Ciudad Universitaria (IGCU), Universidad Nacional Autónoma de Mexico, México City. Tertiary bats from South America are housed in several museums, including: Natural History Museum of Los Angeles County (LACM), California, USA; Museu Nacional do Rio de Janeiro (MNRJ), Brazil; Museu Geológico Valdemar Lefevre, Instituto Geológico (IGG), São Paulo, Brazil; Museo de La Plata (MLP), Argentina; and Laboratorio de Investigaciones en Evolución y Biodiversidad, Universidad Nacional de la Patagonia "San Juan Bosco" (LIEB-PV), Esquel, Argentina. Other abbreviations used are: Great American Biotic Interchange (GABI or Interchange), Local Fauna (LF), North American land mammal age (NALMA) and South American land mammal age (SALMA).

We follow the chapters in Woodburne (2004a) for the North American Cenozoic land mammal biochronology, including: Robinson *et al.* (2004) for the Eocene (Wasatchian, Bridgerian, Uintan and Duchesnean NALMAs), Prothero and Emry (2004) for the latest Eocene through the Early Oligocene (Chadronian, Orellan and Whitneyan NALMAs), Tedford *et al.* (2004) for the Late Early Oligocene through the earliest Pliocene (Arikareean, Hemingfordian, Barstovian, Clarendonian and Hemphillian NALMAs) and Bell *et al.* (2004) for the Early Pliocene through the Late Pleistocene (Blancan, Irvingtonian and Rancholabrean NALMAs). We follow Madden *et al.* (2005), Pascual (2006) and Cione *et al.* (2007) for the biochronology of South American Cenozoic mammals. The Pliocene–Pleistocene boundary recently was moved from 1.8 Ma to 2.6 Ma (Gibbard *et al.*, 2010), reflecting the timing of the onset of Northern Hemisphere glaciation and the beginning of the Ice Age (Sosdian and Rosenthal, 2009). Since this change has not yet gained widespread acceptance in the geological and paleontological communities, we feel it would be confusing to change the ages of sites long regarded as Pliocene (Tertiary) to Pleistocene (Quaternary). Therefore, we follow the old Pliocene–Pleistocene boundary at about 1.8 Ma, with the understanding that sites from 1.8–2.6 Ma in age we consider Late Pliocene (Late Blancan) would be regarded as Early Pleistocene by other workers. This change affects sites 48–51 from western North America

Table 4.1 Tertiary bat localities in North America, exclusive of Florida. Extinct taxa indicated by †; type locality for species indicated by *. Abbreviations: indet. (indeterminate, fossils are too incomplete or fragmentary for positive identification); unid. (unidentified, fossils may be identifiable with further study).

Locality and state	Age and NALMA	Taxa present	References
1. Fossil Lake Wyoming	Early Eocene Wasatchian	†Icaronycteridae (†*Icaronycteris index**) †Onychonycteridae (†*Onychonycteris finneyi**)	Jepsen (1966, 1970) Simmons *et al.* (2008)
2. Wind River Basin Wyoming	Early Eocene Wasatchian	Chiroptera incertae sedis (†*Honrovits tsuwape**)	Beard *et al.* (1992)
3. Elderberry Canyon Nevada	Middle Eocene Bridgerian	Chiroptera (two species present, unid.)	Emry (1990)
4. Tabernacle Butte Wyoming	Middle Eocene Bridgerian	Molossidae (?)	McKenna *et al.* (1962) Legendre (1985)
5. Powder Wash, Uinta Basin Utah	Middle Eocene Bridgerian	Chiroptera incertae sedis (cf. †*Ageina* sp.)	Krishtalka and Stucky (1984)
6. Clarno Nut Beds Oregon	Middle Eocene Bridgerian	Chiroptera (unid.)	Brown (1959) Hanson (1996)
7. Lake Uinta Colorado	Middle Eocene Uintan	Chiroptera (unid.)	Grande (1984)
8. Friars Formation California	Middle Eocene Uintan	Chiroptera (large unid. sample)	This chapter
9. Swift Current Creek Saskatchewan, Canada	Middle Eocene Uintan	Molossidae (†*Wallia scalopidens**) Chiroptera (at least one additional species, unid.)	Storer (1984) Legendre (1985)
10. Lac Pelletier Saskatchewan, Canada	Late Eocene Duchesnean	Chiroptera (unid.)	Storer (1995)
11. Raben Ranch Nebraska	Late Eocene Chadronian	Palaeochiropterygidae? (†*Stehlinia*? sp.) Chiroptera incertae sedis (†*Chadronycteris rabenae**)	Ostrander (1983, 1985)
12. Flagstaff Rim Wyoming	Late Eocene Chadronian	Chiroptera (two species; unid.)	Emry (1973)

Table 4.1 (*cont.*)

Locality and state	Age and NALMA	Taxa present	References
13. Cedar Creek Colorado	Early Oligocene Orellan	Vespertilionidae (†*Oligomyotis casementi**)	Galbreath (1962)
14. Cook Ranch Montana	Early Oligocene Orellan	Vespertilionidae (unid.)	Tabrum *et al.* (1996)
15. Ridgeview Nebraska	Late Oligocene Early Arikareean	Vespertilionidae (?†*Oligomyotis* or ?*Myotis*; two sp. present)	Czaplewski *et al.* (1999)
16. Eastend Saskatchewan, Canada	Late Oligocene Early Arikareean	Chiroptera (indet.)	Storer (2002)
17. Pollack Farm Delaware	Early Miocene Early Hemingfordian	Vespertilionidae (unid.)	Emry and Eshelman (1998)
18. Vedder California	Early Miocene Late Hemingfordian	Chiroptera indet. (originally identified as Phyllostomidae)	Hutchison and Lindsay (1974) Czaplewski *et al.* (2008)
19. Companion Nebraska	Early Miocene Late Hemingfordian	Chiroptera (indet. and unid.)	Czaplewski *et al.* (1999)
20. Barstow California	Middle Miocene Barstovian	Chiroptera (unid.)	Lindsay (1972)
21. Anceney Montana	Middle Miocene Late Barstovian	Vespertilionidae (†*Ancenycteris rasmusseni**)	Sutton and Genoways (1974)
22. Annie's Geese Cross Nebraska	Middle Miocene Late Barstovian	Vespertilionidae (†*Potamonycteris biperforatus**, cf. *Myotis* sp.)	Czaplewski (1991)
23. Myers Farm Nebraska	Middle Miocene Late Barstovian	Vespertilionidae (at least two unid. sp. present)	Corner (1976)
24. Fort Polk Louisiana	Middle Miocene Late Barstovian	Vespertilionidae (*Antrozous* sp.)	Schiebout (1997)
25. Mathews Ranch California	Middle Miocene Early Clarendonian	Chiroptera indet. (originally identified as Phyllostomidae)	James (1963) Czaplewski *et al.* (2008)
26. WaKeeney Kansas	Late Miocene Middle Clarendonian	Vespertilionidae (*Myotis* sp.)	Wilson (1968)

Table 4.1 (*cont.*)

Locality and state	Age and NALMA	Taxa present	References
27. Whisenhunt Oklahoma	Late Miocene Middle Clarendonian	Vespertilionidae (*Myotis* cf. *M. yumanensis*)	Dalquest *et al.* (1996)
28. Ashfall Nebraska	Late Miocene Middle Clarendonian	Vespertilionidae (*Myotis* sp.)	Czaplewski *et al.* (1999)
29. Pratt Slide Nebraska	Late Miocene Late Clarendonian	Vespertilionidae (*Lasiurus* sp.)	Czaplewski *et al.* (1999)
30. Black Butte Oregon	Late Miocene Late Clarendonian	Vespertilionidae (unid.)	Shotwell (1970)
31. Little Valley Oregon	Late Miocene Hemphillian	Vespertilionidae (unid.)	Shotwell (1970)
32. Rattlesnake Oregon	Late Miocene Early Hemphillian	Vespertilionidae (unid.)	Martin (1983)
33. McKay Reservoir Oregon	Late Miocene Late Hemphillian	Vespertilionidae (unid.)	Shotwell (1956)
34. Higgins Texas	Late Miocene Early Hemphillian	Vespertilionidae (*Antrozous* sp.; indet.)	Dalquest and Patrick (1989) Czaplewski (1993b)
35. Redington Arizona	Late Miocene Late Hemphillian	Vespertilionidae (cf. *Eptesicus*, *Myotis* sp.)	Czaplewski (1993a)
36. Yepómera Chihuahua, Mexico	earliest Pliocene Late Hemphillian	Vespertilionidae (†*Plionycteris trusselli**)	Lindsay and Jacobs (1985)
37. Rancho el Ocote Guanajuato, Mexico	earliest Pliocene Late Hemphillian	Vespertilionidae ("†*Plionycteris*")	Carranza-Castañeda and Walton (1992)
38. Moapa Nevada	Miocene/Pliocene Hemphillian/ Blancan	Vespertilionidae (*Antrozous pallidus*, *Myotis* sp.)	Czaplewski (1993a)
39. Beck Ranch Texas	Early Pliocene Early Blancan	Vespertilionidae (*Antrozous pallidus*, *Eptesicus fuscus*, *Lasionycteris noctivagans*, *Lasiurus borealis*)	Dalquest (1978)

Table 4.1 (*cont.*)

Locality and state	Age and NALMA	Taxa present	References
40. Fox Canyon Kansas	Early Pliocene Early Blancan	Vespertilionidae (†*Lasiurus fossilis**)	Hibbard (1950)
41. Fossil Creek Roadcut Idaho	Early Pliocene Early Blancan	Vespertilionidae (†*Lasiurus fossilis*)	Thewissen and Smith (1987)
42. Sand Point Idaho	Early Pliocene Early Blancan	Vespertilionidae (*Antrozous pallidus*)	Thewissen and Smith (1987)
43. Verde Arizona	Early Pliocene Early Blancan	Vespertilionidae (*Lasiurus* cf. *L. blossevillii*, *Myotis* sp.)	Czaplewski (1987, 1993a)
44. Old Woman Formation California	Pliocene Blancan	Vespertilionidae (*Eptesicus* cf. *E. fuscus*)	Czaplewski (1993a)
45. Anza-Borrego Desert California	Late Pliocene Late Blancan	Vespertilionidae (†*Anzanycteris anzensis**)	White (1969)
46. McRae Wash Arizona	Late Pliocene Late Blancan	Molossidae (*Eumops* cf. *E. perotis*) Vespertilionidae (indet.)	Czaplewski (1993a)
47. Wolf Ranch Arizona	Late Pliocene Late Blancan	Vespertilionidae (*Antrozous* sp.)	Harrison (1978) Czaplewski (1993a)
48. Curtis Ranch Arizona	Late Pliocene Late Blancan	Vespertilionidae (†*Simonycteris stocki**)	Stirton (1931)
49. 111 Ranch Arizona	Late Pliocene Late Blancan	Vespertilionidae (unid.)	Tomida (1987)
50. Blanco Texas	Late Pliocene Late Blancan	Vespertilionidae (indet., originally identified as *Tadarida*)	Dalquest (1975)
51. Big Springs Nebraska	Late Pliocene Late Blancan	Vespertilionidae (*Lasiurus* sp.)	Czaplewski *et al.* (1993a)

in Table 4.1 (111 Ranch and Curtis Ranch, Arizona, Blanco, Texas and Big Springs, Nebraska) and sites 14–17 from Florida in Table 4.2 (Macasphalt Shell Pit, Haile 7G, Inglis 1A, De Soto Shell Pit), all of which are Late Blancan between 1.8 and 2.6 Ma in age. Despite the boundary change these sites are still Late Blancan because the definitions of the North American land mammal ages are based on mammal faunas not chronology (Woodburne, 2004a). This

Table 4.2 Tertiary fossil sites in Florida containing bats. Extinct taxa indicated by †; type locality for species indicated by *. Abbreviations: indet. (indeterminate, fossils are too incomplete or fragmentary for positive identification); unid. (unidentified, fossils may be identifiable with further study).

Locality and county	Age and NALMA	Taxa present	References
1. I-75 Alachua County	Early Oligocene Whitneyan	Emballonuridae (two undescribed species) Mormoopidae (one undescribed species) †Speonycteridae (†*Speonycteris aurantiadens*, †*S. naturalis**) Natalidae (genus and species indet.) Vespertilionidae (genus and species indet.)	Morgan, unpublished data Morgan, unpublished data Czaplewski and Morgan (Chapter 5, this volume)
2. Brooksville 2 Hernando County	Late Oligocene Early Arikareean	Emballonuridae (two undescribed species) Mormoopidae (one undescribed species) †Speonycteridae (†*Speonycteris aurantiadens**) Molossidae (genus and species indet.)	Morgan, unpublished data Morgan, unpublished data Czaplewski and Morgan (Chapter 5, this volume)
3. White Springs Hamilton County	Late Oligocene Early Arikareean	Vespertilionidae (two species; both genus and species undet.)	MacFadden & Morgan (2003)
4. Live Oak (=SB-1A) Suwannee County	Late Oligocene Early Arikareean	Family, genus and species indet.	Frailey (1978); Morgan (1993)
5. Buda Alachua County	Early Miocene Late Arikareean	Emballonuridae (undescribed genus and species)	Frailey (1979) Morgan *et al.* (in prep.)

Table 4.2 (*cont.*)

Locality and county	Age and NALMA	Taxa present	References
6. Miller Dixie County	Early Miocene Early Hemingfordian	Vespertilionidae (two species; both genus and species unid.)	Baskin (2003); Wang (2003)
		Molossidae (genus and species unid.)	This chapter
7. Thomas Farm Gilchrist County	Early Miocene Early Hemingfordian	Emballonuridae (undescribed genus and species)	Lawrence (1943)
		Natalidae (†*Primonatalus prattae**)	Czaplewski and Morgan (2000)
		Vespertilionidae (five species, †*Karstala silva**, †*Miomyotis floridanus**, †*Suaptenos whitei**, and 2 undescribed sp.)	Czaplewski *et al.* (2003a) Morgan and Czaplewski (2003)
		Molossidae (two indet. species, near *Tadarida* or *Mormopterus*)	Morgan *et al.* (in prep.)
8. Seaboard Leon County	Early Miocene Early Hemingfordian	Vespertilionidae (genus and species unid.)	Olsen (1964)
9. Brooks Sink Bradford County	Early Miocene Late Hemingfordian	Vespertilionidae (genus and species unid.)	Morgan and Pratt (1988)
10. Bird Branch Polk County	Middle Miocene Early Barstovian	Vespertilionidae (genus and species unid.)	Hulbert and MacFadden (1991)
11. Love Bone Bed Alachua County	Late Miocene Late Clarendonian	Vespertilionidae (genus and species indet.)	Webb *et al.* (1981)
12. McGehee Farm Alachua County	Late Miocene Early Hemphillian	Vespertilionidae (genus and species indet.)	Morgan and Hulbert (2008)

Table 4.2 (*cont.*)

Locality and county	Age and NALMA	Taxa present	References
13. Tyner Farm Alachua County	Late Miocene Early Hemphillian	Vespertilionidae (genus and species unid.)	Morgan and Hulbert (2008)
14. Macasphalt Shell Pit Sarasota County	Late Pliocene Late Blancan	Molossidae (*Tadarida* sp.)	Morgan (1991) Czaplewski *et al.* (2003a)
15. Haile 7G Alachua County	Late Pliocene Late Blancan	Vespertilionidae (genus and species unid.)	Morgan and Hulbert (2008)
16. Inglis 1A Citrus County	Late Pliocene Late Blancan	Phyllostomidae (†*Desmodus archaeodaptes*) Vespertilionidae (one species each of *Antrozous*, *Corynorhinus*, *Eptesicus*, *Lasiurus*, *Myotis*, *Perimyotis*)	Morgan *et al.* (1988) Morgan (1991)
17. De Soto Shell Pit De Soto County	Late Pliocene Late Blancan	Vespertilionidae (genus and species unid.)	Morgan (1991)

change does not affect any South American chiropteran faunas because no Pliocene or early Pleistocene bats are known from that continent.

We define several geographic and biogeographic terms used throughout the chapter to eliminate any confusion regarding their meaning. North America and South America are considered the only two continents in the Western Hemisphere (i.e., we exclude West Antarctica). Mexico and Central America (together often called Middle America or Mesoamerica) and the West Indies are considered part of North America, as are the West Indian islands south to St. Vincent and Barbados. The Caribbean islands of Trinidad, Tobago and Margarita on the continental shelf north of Venezuela are considered part of South America, as are the islands of the Netherlands Antilles, including Aruba, Bonaire and Curaçao. Grenada and the Grenadines are part of the Lesser Antilles geologically, but possess a South American chiropteran fauna, and are thus considered South American with regard to their faunal affinities. Temperate and arctic North America (called the Nearctic region or realm in the biogeographic literature; Olson *et al.*, 2001) consists of Alaska, Canada, Greenland, the continental United States and northern Mexico south to the Tropic of Cancer at about 23°N latitude. Tropical North America (the northern extension of the Neotropical region in the biogeographic literature) consists of central and southern Mexico south of the Tropic of Cancer and Central America from Belize and Guatemala south to Panama, and also includes the West Indies. Although Olson *et al.* (2001) included southern peninsular Florida and the Florida Keys in the Neotropics, Morgan and Emslie (2010) considered all of Florida to be part of the Nearctic region because its vertebrae fauna is predominantly temperate. The Neotropical and Nearctic regions were defined on the content of their modern fauna and flora, and various authors define these regions and their boundaries differently (Olson *et al.*, 2001; Ferrusquía-Villafranca *et al.*, 2005; Lomolino *et al.*, 2006), even with respect to bats alone (Ortega and Arita, 1998; Proches, 2005). The boundary between the Neotropical and Nearctic regions was much different prior to the GABI than at present. Before the Pliocene, but after the Early Cenozoic (Paleocene-Eocene) final connection with Antarctica, the "Neotropical region" included the island continent of South America, with both tropical and temperate components and considerable mammalian provinciality (Kohn *et al.*, 2004; Pascual, 2006; Croft *et al.*, 2008, 2009), as well as the West Indies. The highly endemic modern and Quaternary mammalian fauna of the West Indies had its origins primarily from the Tertiary of South America (Phyllostomidae, most groups of non-volant mammals), although several Antillean bat families appear to be Nearctic in origin (Mormoopidae, Natalidae). The Middle Cenozoic pre-Pliocene Nearctic region, with periodic and relatively frequent biotic interchanges with the Palearctic region (Europe and Asia) via Greenland and Beringia, extended to the southern border of Central America, which at that time had a strictly North American fauna, at least with regard to non-volant or terrestrial mammals. Thus, throughout much of the Cenozoic, the Nearctic region also consisted of not only a temperate component in what are now the United States and arctic Canada (e.g., Hulbert and Harington, 1999; Tedford and Harington, 2003; Woodburne, 2004b), but also a tropical component in Middle America (MacFadden, 2006; Woodburne *et al.*, 2006). Following the connection of the two continents at the Panamanian Isthmus in the Early Pliocene, many species of South American tropical animals and plants moved northward into Central America and Mexico and vice versa (Webb, 2006). Tropical North America is now considered the northern extension of the Neotropical region, whereas the current Nearctic region consists of a temperate and arctic fauna and flora.

Of the 19 living families of bats worldwide, only three are widespread in temperate regions, Vespertilionidae, Molossidae, and in the Eastern Hemisphere, Rhinolophidae. This reflects in part the ability of some members of these families to hibernate and/or migrate for the cold season. All other families of bats are virtually restricted to the tropical zone, although a few have member species that extend into adjacent subtropical regions of the temperate zones.

4.3   North American Tertiary bat faunas

Although each of the Tertiary North American land mammal ages from the Wasatchian (Early Eocene) through the Blancan (Pliocene) has at least one record of Chiroptera, the Cenozoic record of bats in North America (Czaplewski *et al.*, 2008) pales in comparison to the European record (Sigé and Legendre, 1983). With the exception of several Eocene lake deposits (Green River Formation in Wyoming, Elderberry Canyon in Nevada) and ten paleokarst sites in Florida, the North American Tertiary chiropteran record mostly consists of isolated and fragmentary remains from fluvial and alluvial deposits (Czaplewski *et al.*, 2008). The North American Tertiary bat record is comparable in overall diversity and age distribution to the Tertiary chiropteran faunas from Africa and Australia and is richer than the Asian record (Gunnell and Simmons, 2005; Eiting and Gunnell, 2009).

Tables 4.1–4.3 list the bats from Tertiary (Eocene through Pliocene) sites from western North America and Florida. Figure 4.1 (North America, exclusive of Florida) and Figure 4.2 (Florida) show the location of most North American Tertiary chiropteran sites. Tables 4.1 and 4.2 include sites with bats identified to genus, but do not contain every site with specimens identified only as Chiroptera. Several Eocene sites with large unidentified samples of bats are listed because they may contain taxa relevant to the early evolution of the Neotropical Chiroptera. Most Oligocene, Miocene and Pliocene bat sites from western North America and Middle Miocene and younger sites from Florida contain only members of the Vespertilionidae. These records are listed in Tables 4.1 and 4.2, but are not discussed in the text because these taxa are not clearly related to Neotropical vespertilionids. Table 4.3 is a combined faunal list of all bats known from the Tertiary of North America.

Pleistocene records of bats with Neotropical affinities are briefly discussed in a separate section.

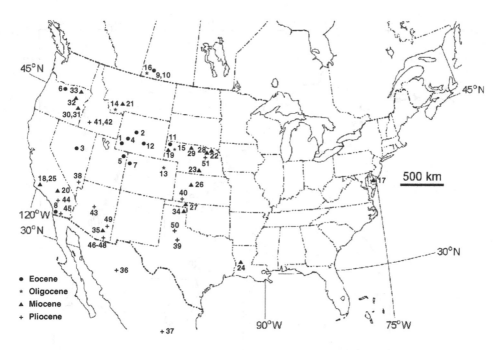

Figure 4.1 Map of North America showing location of major Tertiary chiropteran sites, exclusive of Florida. Numbers on map refer to site numbers in Table 4.1, which provides additional information on the sites, including age and NALMA, species of bats present and references. All but five of the sites are in the United States. Symbols designate different epochs: Eocene (filled circle); Oligocene (asterisk); Miocene (triangle), Pliocene (plus sign).

4.4.1   Western North America

The most complete and thoroughly studied fossil bats from North America are also the oldest, the Early Eocene (Wasatchian) †*Icaronycteris index* and †*Onychonycteris finneyi*, placed in the extinct families †Icaronycteridae and †Onychonycteridae, respectively. Both species were described from remarkably complete skeletons derived from lacustrine sediments in the Fossil Lake area, Green River Formation, southwestern Wyoming (Jepsen, 1966, 1970; Simmons *et al.*, 2008). †*Icaronycteris* and †*Onychonycteris* are critically important in understanding the early evolution of the Chiroptera; however, their relationships to the modern families of Neotropical bats are unclear (Simmons and Geisler, 1998; Simmons *et al.*, 2008). The only other described Early Eocene bat from North America, †*Honrovits tsuwape* from the Wind River Basin, Wind River Formation, Wyoming, was originally referred to the extant Neotropical family Natalidae (Beard *et al.*, 1992), but has since

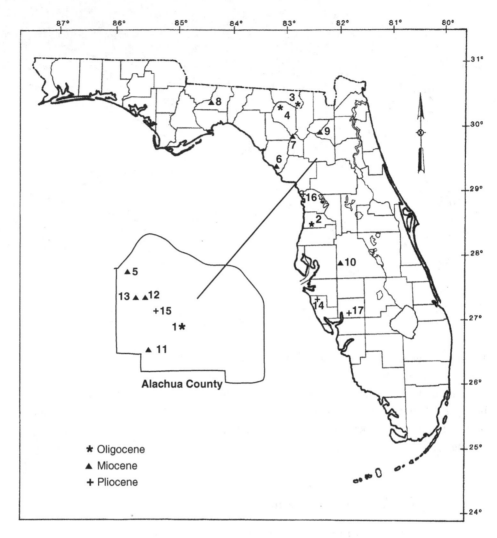

Figure 4.2 Map of Florida (and close-up view of Alachua County where many sites exist) showing location of major Tertiary chiropteran sites. Numbers on map refer to site numbers in Table 4.2, which provides additional information on the sites, including age and NALMA, species of bats present and references. Symbols designate different epochs: Oligocene (asterisk); Miocene (triangle), Pliocene (plus sign).

been removed from the Natalidae (*sensu stricto*) and considered a nataloid or indeterminate vespertilionoid (Simmons and Geisler, 1998; Morgan and Czaplewski, 2003; Czaplewski *et al.*, 2008).

There are several records of Middle Eocene (Bridgerian) Chiroptera from North America. Krishtalka and Stucky (1984) tentatively listed a lower molar of the European genus *Ageina* of early Bridgerian age from Powder Wash, Uinta

Table 4.3 Tertiary chiropteran fauna from North America. Extinct taxa indicated by †. Within each family, the genera and species are arranged alphabetically. The sites from which these bats are known, their ages and literature citations are listed in Table 4.1 (North America, exclusive of Florida) and Table 4.2 (Florida).

---

†Icaronycteridae
    †*Icaronycteris index*
†Onychonycteridae
    †*Onychonycteris finneyi*
†Palaeochiropterygidae
    †*Stehlinia?* sp.
Emballonuridae
    Two undescribed genera, three undescribed species
†Speonycteridae
    †*Speonycteris aurantiadens*
    †*Speonycteris naturalis*
Phyllostomidae
    †*Desmodus archaeodaptes*
Mormoopidae
    One undescribed genus and species
Natalidae
    †*Primonatalus prattae*
Molossidae
    Eumops cf. *E. perotis*
    Tadarida sp.
    †*Wallia scalopidens*
    indet. genus near *Tadarida* or *Mormopterus*
Vespertilionidae
    †*Ancenycteris rasmusseni*
    Antrozous pallidus
    Antrozous sp.
    †*Anzanycteris anzensis*
    Corynorhinus sp.
    Eptesicus fuscus
    cf. *Eptesicus* sp.
    †*Karstala silva*
    Lasionycteris noctivagans
    Lasiurus borealis
    †*Lasiurus fossilis*
    Lasiurus cf. *L. blossevillii*
    †*Miomyotis floridanus*
    Myotis cf. *M. yumanensis*

Table 4.3 (*cont.*)

Myotis sp.
†*Oligomyotis casementi*
Perimyotis sp.
†*Plionycteris trusselli*
†*Potamonycteris biperforatus*
†*Simonycteris stocki*
†*Suaptenos whitei*
Chiroptera incertae sedis
   cf. †*Ageina* sp.
   †*Chadronycteris rabenae*
   †*Honrovits tsuwape*

Basin, Green River Formation, Utah. A fragmentary dentary with m2 from the Bridgerian Tabernacle Butte LF, Bridger Formation, Wyoming was assigned to the Chiroptera by McKenna *et al.* (1962), and later considered a vespertilionoid, possibly a molossid, by Legendre (1985). Other records include a partial skeleton of an unidentified bat from the Late Bridgerian Clarno Formation, Oregon (Brown, 1959; Hanson, 1996) and a partial skeleton and other material in the Bridgerian Elderberry Canyon LF, Sheep Pass Formation, Nevada (Emry, 1990).

Later in the Middle Eocene (Uintan) the first bat belonging to an extant family appears in North America, the molossid †*Wallia scalopidens* from the Late Uintan Swift Current Creek LF, Cypress Hills Formation, Saskatchewan, Canada (Storer, 1984; Legendre, 1985). Excluding the possible molossid from Tabernacle Butte, †*Wallia* is the oldest member of the Molossidae (~42–44 Ma), slightly older than †*Cuvierimops* from several Late Middle Eocene sites (~39 Ma) in the Quercy phosphorites in France (Legendre, 1985). The only record of bats from the Duchesnean (Late Middle Eocene) is from Lac Pelletier, Saskatchewan, Canada, consisting of isolated teeth representing additional specimens of †*Wallia*? and other unidentified species (Storer, 1995).

The only described bat from the latest Eocene Chadronian is the enigmatic †*Chadronycteris rabenae* from the Raben Ranch and Dirty Creek Ridge LFs, Chadron Formation, northwestern Nebraska (Ostrander, 1983, 1985). †*Stehlinia*? was tentatively reported from Raben Ranch (Ostrander, 1985), although this identification needs further verification. †*Stehlinia* is otherwise known from the Eocene and Oligocene of Europe (Sigé, 1974). A Chadronian fauna from Flagstaff Rim, White River Formation, in Wyoming has produced several bats from a possible owl pellet deposit (Emry, 1973). Several maxillae with teeth from Flagstaff Rim are similar to †*Chadronycteris* and several humeri are similar to †*Oligomyotis*, the earliest North American vespertilionid from the Early Oligocene (see below).

The post-Eocene Tertiary chiropteran record from western North America consists almost entirely of Vespertilionidae; mostly extinct genera from the Oligocene and Miocene, with extant genera first appearing in the Late Miocene and Pliocene. Table 4.1 summarizes the best known of these vespertilionid faunas. Most of the sites are not discussed further here because the bats do not have obvious Neotropical affinities. The earliest New World record of the Vespertilionidae is the extinct genus and species †*Oligomyotis casementi*, described from a single distal humerus from the Early Oligocene (Orellan NALMA) Cedar Creek LF, White River Formation, northeastern Colorado (Galbreath, 1962). Remarkably, from the Oligocene through the Pliocene, there is only one record of a bat from western North America that does not belong to the Vespertilionidae, a tooth of the large molossid *Eumops* cf. *E. perotis* from the Late Pliocene (Late Blancan) McRae Wash LF in southeastern Arizona (Czaplewski, 1993a). *Eumops perotis* is a living species found in the southwestern United States, Mexico and South America.

4.4.2   Florida

The oldest land vertebrate fauna from Florida is the Early Oligocene I-75 LF from Gainesville, Alachua County (Patton, 1969). I-75 has a fairly extensive vertebrate fauna composed of about 45 species, including sharks, fish, anurans, salamanders, turtles, lizards, snakes and mammals (Patton, 1969; Holman and Harrison, 2001). The mammalian biostratigraphy of the I-75 LF indicates a Late Early Oligocene age (Late Whitneyan NALMA; 30–31 Ma; Patton, 1969; Prothero and Emry, 2004). I-75 provides the earliest record of the taxonomic diversity and community structure of North American mid-Cenozoic chiropteran faunas. About 40 specimens representing seven species of bats are known from I-75 (Table 4.2), including large and small emballonurids representing the earliest known New World members of the family, mormoopids (the earliest known record of that family), †*Speonycteris aurantiadens* and †*S. naturalis* (see Czaplewski and Morgan, Chapter 5, this volume), an indeterminate natalid and an indeterminate vespertilionid.

The vertebrate assemblage from Late Oligocene Brooksville 2 is composed of frogs, lizards, snakes and a diverse fauna of mammals. Hayes (2000) placed the Brooksville 2 LF in the Late Early Arikareean Ar2; 24–28 Ma). The similarity between the chiropteran faunas of Brooksville 2 and I-75 indicates that Brooksville probably belongs in the older portion of the Late Early Arikareean, between 26 and 28 Ma. The chiropteran sample from Brooksville 2 consists of about

200 fossils of five species: Emballonuridae (two species); Mormoopidae (one species); †*Speonycteris aurantiadens* and a large indeterminate molossid (Czaplewski *et al.*, 2003a; Czaplewski and Morgan, Chapter 5, this volume). The five bats from Brooksville 2 belong to families having Neotropical affinities, all of which are shared with I-75, except the molossid.

The Buda LF (Frailey, 1979), Arikareean (Ar3; earliest Miocene; ~22–23 Ma), has a single upper molar of an extinct emballonurid (Morgan, unpublished data). There are three species of bats from the Miller LF (Hemingfordian, He1; ~19 Ma), one species of Molossidae similar to *Tadarida*, represented by several complete humeri and a complete edentulous dentary, and two different genera of Vespertilionidae both known only from the humerus.

Thomas Farm near Bell, Gilchrist County, Florida has produced the best known Early Miocene vertebrate fauna in eastern North America, composed of about 90 species of vertebrates primarily consisting of terrestrial forms (Pratt, 1989, 1990; Morgan and Czaplewski, 2003). The mammalian biochronology of Thomas Farm indicates an early Hemingfordian age (He1, Early Miocene, 18–19 Ma; Morgan and Czaplewski, 2003; Tedford *et al.*, 2004). The Thomas Farm chiropteran fauna is composed of nine species, including four belonging to Neotropical families: Emballonuridae (one species); †*Primonatalus prattae* (Natalidae) and two species of Molossidae, similar to *Tadarida* and *Mormopterus* (Czaplewski *et al.*, 2003a; Morgan and Czaplewski, 2003; Morgan, unpublished data). The remaining five species belong to the Vespertilionidae, including three extinct forms known only from Thomas Farm, †*Karstala silva*, †*Miomyotis floridanus* and †*Suaptenos whitei* (Lawrence, 1943; Czaplewski and Morgan, 2000), and two other undescribed genera.

The latest Pliocene (Late Blancan) Inglis 1A LF has the largest Pliocene chiropteran fauna from North America comprised of seven species: the extinct vampire bat †*Desmodus archaeodaptes* and six vespertilionids, including one species each of *Antrozous*, *Corynorhinus*, *Eptesicus*, *Lasiurus*, *Myotis* and *Perimyotis* (=*Pipistrellus*) (Morgan *et al.*, 1988; Morgan, 1991). The vampire bat from Inglis is the oldest North American record of the Phyllostomidae, as well as the oldest record of *Desmodus*. Among the vespertilionids, Inglis 1A has the only record of *Antrozous* from eastern North America and the earliest records of the genera *Corynorhinus* and *Perimyotis*.

A distal humerus of a large species of the molossid *Tadarida* was identified from the Late Pliocene Macasphalt Shell Pit LF near Sarasota, Sarasota County along the southwestern Gulf Coast (Morgan, 1991; Czaplewski *et al.*, 2003b). Biostratigraphy, paleomagnetic stratigraphy and strontium isotope chronology indicate a Late Blancan age for the Macasphalt Shell Pit LF (Late Pliocene; 2.2–2.6 Ma; Jones *et al.*, 1991; Morgan, 2005).

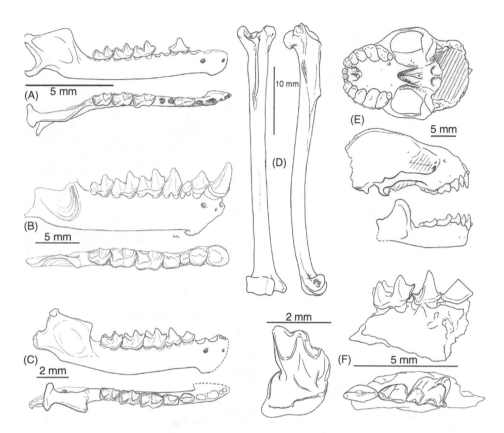

Figure 4.3 North American bat fossils. (A) †*Primonatalus prattae* (Natalidae) from Florida, USA: reconstruction of right dentary based on several specimens, in lateral and occlusal views. (B) †*Speonycteris aurantiadens* (†Speonycteridae) from Florida: reconstruction of right dentary based on several specimens, in lateral and occlusal views. (C) Undescribed Mormoopidae from Florida: reconstruction of right dentary based on several specimens, in lateral and occlusal views. (D) †*Desmodus archaeodaptes* (Phyllostomidae) from Florida: humerus in anterior and lateral views. (E) †*Cubanycteris silvai* (Phyllostomidae) from Cuba: cranium and mandible, in palatal and lateral views. (F) Undescribed Emballonuridae from Florida: M1 in occlusal view; dentary fragment with p2-m1 in labial and occlusal views.

4.5   South American Tertiary bat faunas

Czaplewski (2005) reviewed the pre-Pleistocene history of bats in South America, a record that is the poorest of all the continents despite the incredible diversity of bats in the modern Neotropical fauna. Figure 4.4 shows the location of all South American Tertiary chiropteran sites. These sites are listed in Table 4.4, together with information on their age, taxa of bats

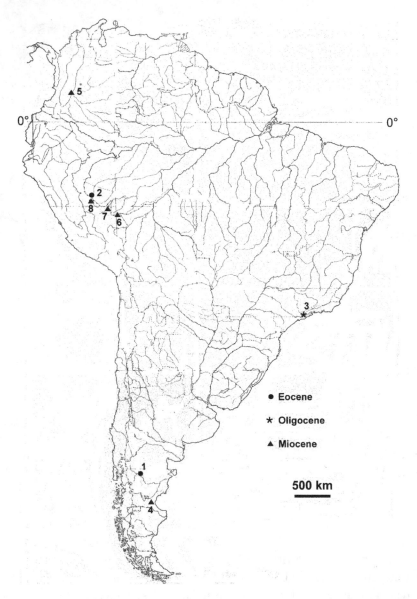

Figure 4.4 Map of South America showing the location of the major Tertiary chiropteran sites. Numbers on map refer to site numbers in Table 4.4, which provides additional information on the sites, including age and SALMA, species of bats present and references. Symbols designate different epochs: Eocene (filled circle); Oligocene (asterisk); Miocene (triangle).

Table 4.4 Tertiary bat localities from South America. Extinct taxa indicated by †; type locality for species indicated by *. Abbreviations: indet. (indeterminate, fossils are too incomplete or fragmentary for positive identification); unid. (unidentified, fossils may be identifiable with further study).

Locality and country	Age and SALMA	Taxa present	References
1. Laguna Fría Argentina	Early Eocene	Chiroptera incertae sedis	Tejedor et al. (2005, 2009)
2. Santa Rosa Peru	Late Eocene (?)	Chiroptera incertae sedis	Czaplewski and Campbell (2004)
3. Tremembé Brazil	Late Oligocene Deseadan	Molossidae (†Mormopterus faustoi*)	Paula Couto (1956) Paula Couto and Mezzalira (1971)
4. Gran Barranca Argentina	Early Miocene Colhuehuapian	Phyllostomidae (indet. genus of phyllostomine); Molossidae (†Mormopterus barrancae*, Mormopterus indet. sp.)	Czaplewski (2005, 2010)
5. La Venta Colombia	Middle Miocene Laventan	Emballonuridae (Diclidurus sp., genus indet.); Noctilionidae (Noctilio albiventris); Phyllostomidae (†Notonycteris magdalenensis*, †Notonycteris sucharadeus*, Tonatia or Lophostoma sp., †Palynephyllum antimaster*); Thyropteridae (Thyroptera lavali, Thyroptera cf. T. tricolor); Molossidae (Eumops sp., †Mormopterus colombiensis*, †Potamops mascahebenes*, genus indet.); ?Vespertilionidae (genus indet.)	Savage (1951); Czaplewski (1997) Czaplewski et al. (2003b)
6. Río Acre Peru	Late Miocene Huayquerian	Noctilionidae (†Noctilio lacrimaelunaris*)	Czaplewski (1996)
7. Río Purus Peru	Late Miocene Huayquerian	Molossidae (genus indet.)	Czaplewski (1996)
8. Río Yurúa Peru	Late Miocene Huayquerian	Emballonuridae (undescribed genus); Molossidae (genus indet.)	Czaplewski (2005) Czaplewski (unpublished data)

Table 4.5 Tertiary chiropteran fauna from South America. Extinct taxa indicated by †. Within each family, the genera and species are arranged alphabetically. The sites from which these bats are known, their ages and literature citations are listed in Table 4.4.

Emballonuridae
   Diclidurus sp.
   genus and species undescribed
   genus and species indet.
Phyllostomidae
   †*Notonycteris magdalenensis*
   †*Notonycteris sucharadeus*
   †*Palynephyllum antimaster*
   Tonatia or *Lophostoma* sp.
   genus and species indet.
Noctilionidae
   Noctilio albiventris
   †*Noctilio lacrimaelunaris*
Thyropteridae
   Thyroptera lavali
   Thyroptera cf. *T. tricolor*
Molossidae
   Eumops sp.
   †*Mormopterus barrancae*
   †*Mormopterus colombiensis*
   †*Mormopterus faustoi*
   †*Potamops mascahehenes*
   genus and species indet., near *Nyctinomops*
?Vespertilionidae
   genus and species indet.
Chiroptera
   family, genus and species indet. (at least two sp.)

---

represented and references. The composite Tertiary chiropteran fauna from South America is provided in Table 4.5. The most significant factor contributing to the meager record of fossil bats in South America relates to taphonomy, specifically the total lack of Tertiary paleokarst deposits containing bats. The Paleocene Itaboraí Fauna in southeastern Brazil consists of paleokarst deposits, but predates the first occurrence of the Chiroptera. Only in the Late Pleistocene are there karst deposits in South America containing bats, primarily caves in Brazil and Venezuela. Figure 4.5 provides illustrations of some representative Tertiary bat fossils from South America.

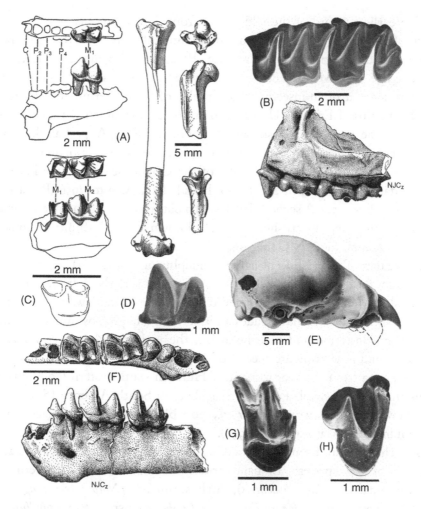

Figure 4.5 South American bat fossils. (A) †*Notonycteris magdalensensis*
(Phyllostomidae) from Colombia: holotype jaw fragment with m1; jaw
fragment with talonid of m1 and complete m2; proximal and distal ends of
humerus; proximal half of femur (from Savage, 1951). (B) †*Potamops
mascahehenes* (Molossidae) from Colombia: holotype maxilla with P4–M3;
occlusal view photo of upper teeth and lateral view drawing of maxilla (from
Czaplewski, 1997). (C) †*Palynephyllum antimaster* (Phyllostomidae) from
Colombia: holotype M1 or M2, in occlusal view. (D) †*Noctilio lacrimaelunaris*
(Noctilionidae) from Peru: holotype m1 in occlusal view. (E) †*Desmodus
draculae* (Phyllostomidae) from Venezuela: holotype cranium in lateral view
(from Morgan *et al.*, 1988). (F) †*Mormopterus barrancae* (Molossidae) from
Argentina: holotype dentary with p3-m2 in occlusal and labial views (from
Czaplewski, 2010). (G) *Thyroptera lavali* (Thyropteridae) from Colombia:
broken M2 in occlusal view. (H) *Diclidurus* sp. (Emballonuridae) from
Colombia: M3 in occlusal view (from Czaplewski, 1997).

The earliest record of the Chiroptera from South America consists of two teeth of an insectivorous bat of an indeterminate family from the Laguna Fría site in northwestern Chubut Province, Patagonia, Argentina of Late Early Eocene age (49.5 Ma; Tejedor *et al.*, 2005, 2009). The second Eocene record is a lower molar of an insectivorous bat, also of an indeterminate family, from the Santa Rosa LF in the Ucayali department, Amazonian Peru, of probable Late Eocene age (Czaplewski and Campbell, 2004). A partial skeleton of †*Mormopterus faustoi*, an extinct species of an extant genus of molossid, was described from the bituminous shales of the Oligocene Tremembé Formation (Deseadan) in the state of São Paulo, Brazil (Paula Couto, 1956; Paula Couto and Mezzalira, 1971). A second skeleton of a bat from the Tremembé Formation is more complete, but crushed and of unknown familial affinity (Leonardos, 1924; Mezzalira, 1966; NJC, personal observation).

Three bats of early Miocene age (Colhuephuapian; ~20 Ma) are known from the Gran Barranca, Chubut Province, Argentina, derived from the Puesto Almendra Member of the Sarmiento Formation (Czaplewski, 2005, 2010). The Gran Barranca fauna includes an indeterminate phyllostomine, the oldest known phyllostomid and two molossids, the extinct †*Mormopterus barrancae* and a second indeterminate species of *Mormopterus*. The Middle Miocene La Venta Fauna from the Department of Huila in southwestern Colombia contains the richest assemblage of Tertiary bats in South America (Savage, 1951; Czaplewski, 1997; Czaplewski *et al.*, 2003b). This fauna is placed in the Laventan (12–13 Ma; Kay *et al.*, 1997).

The chiropterans from La Venta (Czaplewski, 1997; Czaplewski *et al.*, 2003b) consist of 14 species: Emballonuridae (*Diclidurus* and indeterminate); Noctilionidae (*Noctilio albiventris*); Phyllostomidae (†*Notonycteris magdalenensis*, †*Notonycteris sucharadeus*, *Tonatia* or *Lophostoma* sp., †*Palynephyllum antimaster*); Thyropteridae (*Thyroptera lavali*, *Thyroptera* cf. *T. tricolor*); Molossidae (*Eumops* sp., †*Mormopterus colombiensis*, †*Potamops mascahehenes* and genus indeterminate) and ?Vespertilionidae (indeterminate). Three of the phyllostomids are referable to the subfamily Phyllostominae, whereas †*Palynephyllum antimaster* provides the oldest record of a nectar- and pollen-feeding phyllostomid. The type of †*Potamops mascahehenes* occurs in a site in the La Victoria Formation, slightly older than the remainder of the bats derived from several sites in the Villavieja Formation, including the Monkey Beds and Fish Bed.

Bats are known from three sites of Late Miocene age (Huayquerian, from the Madre de Dios Formation in the western Amazon Basin in Peru (Czaplewski, 1996, 2005): an extinct Noctilionidae (†*Noctilio lacrimaelunaris*) from Río Acre, Madre de Dios department; an indeterminate Molossidae from Río Purus, Madre de Dios department; and an undescribed Emballonuridae and an indeterminate molossid from Río Yurúa, Ucayali department. No fossil bats are yet known from the Pliocene of South America, nor are there records from the Early Pleistocene.

4.6   Pleistocene bat records

4.6.1   Emballonuridae

Emballonurids probably inhabited Mesoamerica between the Miocene and Pleistocene, but are not known there prior to two Late Pleistocene records of extant species from cave deposits in Mexico. *Balantiopteryx io* was identified from Cueva de El Abra in Tamaulipas, north of its current range in southern Veracruz (Dalquest and Roth, 1970) and *Peropteryx macrotis* is known from Gruta de Loltún in Yucatán within its modern range (Arroyo-Cabrales and Polaco, 2003, 2008). There are Late Pleistocene records of the living species *Peropteryx macrotis* from cave deposits in the states of Bahia and Minas Gerais, Brazil (Czaplewski and Cartelle, 1998; Lessa *et al.*, 2005).

4.6.2   Phyllostomidae

Nearly 30 species of phyllostomids have been identified from Pleistocene cave sites in the states of Bahia, Goiás, Mato Grosso do Sul, Minas Gerais and São Paulo in Brazil (Czaplewski and Cartelle, 1998; Lessa *et al.*, 2005; Salles *et al.*, 2006; Avilla *et al.*, 2007) and 12 species of phyllostomids are known from Quaternary cave deposits in Venezuela (Linares, 1968; Morgan *et al.*, 1988; Rincón, 1999, 2001). Shockey *et al.* (2009) identified three phyllostomids, the vampire *Desmodus* sp. and the glossophagine nectar bats *Anoura* sp. and *Platalina genovensium*, from Late Pleistocene deposits in Jatun Uchco Cave at 2150 m in elevation in the Peruvian Andes. Late Pleistocene tar pits have produced three species of phyllostomines from Venezuela (Czaplewski *et al.*, 2005) and one phyllostomine from Peru (Czaplewski, 1990). The only extinct Pleistocene phyllostomid from South America is the giant vampire bat †*Desmodus draculae*, described from Cueva del Guácharo in Venezuela (Morgan *et al.*, 1988), and since reported from caves in Brazil (Trajano and de Vivo, 1991; Cartelle and Abuhid, 1994; Czaplewski and Cartelle, 1998) and an open site of supposed Holocene age in Argentina (Pardiñas and Tonni, 2000).

The Late Pleistocene phyllostomid fauna from Middle America consists of 12 extant species from cave faunas on the Yucatán peninsula, including Gruta de Loltún and Cueva de Spukil in Mexico (Arroyo-Cabrales and Polaco, 2003, 2008) and Cebada Cave in Belize (Czaplewski *et al.*, 2003c). An extinct vampire bat, †*Desmodus stocki*, smaller than †*D. draculae*, but larger than the extant

D. rotundus, was described from San Josecito Cave, Nuevo León, northern Mexico (Jones, 1958) and is also known from Late Pleistocene sites in the Mexican states of San Luis Potosí and México (Arroyo-Cabrales and Polaco, 2008) and in the southern United States (Ray *et al.*, 1988).

Antillean Quaternary cave deposits preserve a diverse fauna of Phyllostomidae consisting of at least 21 species, including 15 living and six extinct species, comprising more extinct phyllostomids than are known from the Quaternary of the entire mainland Neotropics (Silva Taboada, 1979; Suárez and Díaz-Franco, 2003; Mancina and García-Rivera, 2005). Other extinct forms include the vampire †*Desmodus puntajudensis* from Cuba (Woloszyn and Mayo, 1974; Suárez, 2005) and the large phyllonycterine †*Phyllonycteris major* from Puerto Rico and Antigua (Anthony, 1917; Pregill *et al.*, 1988).

A fourth extinct species of vampire bat, †*Desmodus archaeodaptes*, is known only from the Late Pliocene and Early Pleistocene of Florida (Morgan *et al.*, 1988; Morgan, 1991).

4.6.3   Mormoopidae

Mormoopids have been described from Quaternary cave deposits in Cuba, †*Mormoops magna* and †*Pteronotus pristinus* (Silva Taboada, 1974), and *M. megalophylla* is known from Quaternary cave deposits in Cuba, Hispaniola, Jamaica and the Bahamas, but is now locally extinct in the West Indies (Silva Taboada, 1974; Morgan, 2001). There are also extralimital records of *M. megalophylla* from Late Pleistocene cave and karst deposits in Florida, the West Indies and Brazil (Morgan, 1991, 2001; Czaplewski and Cartelle, 1998). Mormoopids have a sparse Quaternary record in South America, consisting of Late Pleistocene records from Brazil, including *M. megalophylla* from the state of Bahia and *P. davyi* and *P. parnellii* from the states of Bahia and Goiás (Czaplewski and Cartelle, 1998; Aguiar Fracasso and Salles 2005; Lessa *et al.*, 2005). The Quaternary record of the Mormoopidae is especially rich and diverse in the West Indies, with fossils from numerous caves in Puerto Rico, as well as the Bahamas, Cayman Islands, and Anguilla, Antigua and Barbuda in the Lesser Antilles (Silva Taboada, 1979; Pregill *et al.*, 1988, 1994; Morgan, 1989, 1994, 2001).

4.6.4   Furipteridae and Noctilionidae

Furipterids have been identified from Late Quaternary deposits in the states of Bahia and Minas Gerais in Brazil (Czaplewski and Cartelle, 1998; Lessa *et al.*, 2005) and from the Atacama Desert of southern Peru (Morgan and Czaplewski, 1999).

With the exception of a canine tooth of *Noctilio* from the Middle Pleistocene at Las Grutas, Necochea in coastal Buenos Aires province, Argentina (Merino *et al.*, 2007), there are no other Pleistocene records of *Noctilio* from Middle and South America. However, there are Late Quaternary records of *N. leporinus* from Cuba, Isla de Pinos, Puerto Rico and Antigua in the West Indies (Morgan, 2001).

4.6.5 Natalidae

South American Pleistocene records of natalids are limited to *Natalus espiritosantensis* from Bahia, Goiás and Minas Gerais in Brazil (Winge, 1893; Czaplewski and Cartelle, 1998; Aguiar Fracasso and Salles, 2005). A locally extinct population of *N. tumidirostris* was reported from Late Quaternary cave deposits on Tobago (Eshelman and Morgan, 1985). The most extensive Quaternary record of the Natalidae is from the islands of the West Indies where widespread cave deposits produce an abundance of chiropteran fossils (Morgan, 2001). All eight extant Antillean natalid species are represented as fossils, mostly on islands where they still occur. However, there are several extralimital Late Quaternary records from smaller islands. Although natalids are currently unknown from the Cayman Islands, *N. major* and *Chilonatalus micropus* were reported from Late Quaternary fossil deposits on Grand Cayman (Morgan, 1994, 2001). There are also extralimital records of *Natalus* from the Bahamas, including *N. primus* from Abaco, Andros and New Providence, and *N. major* from Grand (= Middle) Caicos in the Caicos Islands (Morgan, 1989, 2001; Tejedor, 2011).

4.6.6 Molossidae

There are two genera of Late Pliocene molossids from North America, *Eumops* from McRae Wash in Arizona (Czaplewski, 1993a) and a large *Tadarida* from Macasphalt Shell Pit in Florida (Czaplewski *et al.*, 2003a). Extinct †*T. constantinei*, from Slaughter Canyon Cave (= New Cave) in New Mexico (Lawrence, 1960; Morgan, 2003) is Middle Pleistocene (Late Irvingtonian?) in age (Lundberg and McFarlane, 2006; Polyak *et al.*, 2006). *Tadarida*, perhaps conspecific with †*T. constantinei*, is known from an Irvingtonian site in Mammoth Cave, Kentucky (Jegla and Hall, 1962). *E. underwoodi* is present at Late Pleistocene Lecanto 2A LF in Florida (Morgan, 1991), far outside the modern range of this species (Hall, 1981; Simmons, 2005). Extant *E. floridanus* was first described as an extinct species, *Molossides floridanus*, from the Late Pleistocene Melbourne LF in east-central Florida (Allen, 1932) and also occurs in a Rancholabrean karst deposit in southern Florida (Martin, 1977; Morgan, 2002; Timm and Genoways, 2004). There are two extralimital Late Pleistocene records of *E. perotis* from Mexico, Cueva de El Abra in southern Tamaulipas and Lago de Chapala in Jalisco (Dalquest and Roth, 1970; Arroyo-Cabrales and Polaco, 2008). Two other molossids identified from Cueva de El Abra, *Nyctinomops aurispinosus* and *N. laticaudatus*, are Neotropical species at the northernmost limit of their ranges (Dalquest and Roth, 1970). Only three species of molossids are known from Quaternary deposits in the West Indies with several extralimital records of *T. brasiliensis* from the Bahamas (Morgan, 2001). There are Late Quaternary records of eight living species of molossids from Bahia and Minas Gerais in Brazil, all within their current ranges (Czaplewski and Cartelle, 1998; Lessa *et al.*, 2005).

4.6.7   Vespertilionidae

Temperate North America has a rich Pleistocene record of vespertilionids (Kurtén and Anderson, 1980). There are a few extralimital occurrences, including a Late Pliocene record of *Antrozous* from Florida, a genus now restricted to western North America except for an outlying population in Cuba (Morgan, 1991; Simmons, 2005). The Mesoamerican and West Indian Quaternary vespertilionid records mostly consist of modern species occurring within their current ranges (Morgan, 2001; Arroyo-Cabrales and Polaco, 2008). Two Neotropical species of vespertilionids are recorded from the Yucatan peninsula, *Eptesicus furinalis* and *Lasiurus ega* from Gruta de Loltun in Mexico and *L. ega* from Cebada Cave in Belize (Czaplewski *et al.*, 2003c; Arroyo-Cabrales and Polaco, 2008). There is a single Quaternary record of an extralimital vespertilionid from the West Indies, *Myotis* cf. *M. austroriparius* from Hole in the Wall Cave on Abaco in the Bahamas (Morgan, 2001). Vespertilionids from the Quaternary of South America include six species from Brazil (Czaplewski and Cartelle, 1998; Lessa *et al.*, 2005), three species from Peru (Cadenillas and Martinez, 2005), two species from Venezuela (Czaplewski *et al.*, 2005) and one species from the Early Holocene of Argentina (Iudica *et al.*, 2003). Significant extralimital Quaternary records include two endemic Neotropical genera, *Histiotus* from Minas Gerais (Winge, 1893) and possibly Goiás, Brazil (Aguiar Fracasso and Salles, 2005), and from Argentina (Iudica *et al.*, 2003), and *Rhogeessa* from Venezuela (Czaplewski *et al.*, 2005).

4.7   Discussion

Prior to the discoveries over the past several decades of significant Tertiary bat faunas in both North America and South America, the evolutionary history of the New World Chiroptera was based mostly on their modern geographic distributions and species-richness patterns with limited data from a few scattered fossils. The current Nearctic bat fauna consists of four families (Phyllostomidae, Mormoopidae, Molossidae, Vespertilionidae), 20 genera and 47 species (Simmons, 2005), but is dominated by vespertilionids (more than 70% of the fauna) with a few tropical intruders belonging to the three other families. Six of the seven molossids, all five of the phyllostomids and the single mormoopid known from the Nearctic region are Neotropical species whose ranges extend northward into warm temperate latitudes in northern Mexico and the southern tier of US states. The modern Neotropical bat fauna from South America, Central America, Mexico and the West Indies is remarkably diverse, with nine families (Emballonuridae, Phyllostomidae, Mormoopidae, Noctilionidae, Furipteridae, Thyropteridae, Natalidae, Vespertilionidae, Molossidae), 94 genera and 322 species (Simmons, 2005). Six of these families, five in the Noctilionoidea (Phyllostomidae, Mormoopidae, Noctilionidae, Furipteridae, Thyropteridae) and the Natalidae, are essentially endemic to the Neotropical region, with a few species of phyllostomids and a mormoopid occurring northward into the southwestern United States or the Florida peninsula. The Noctilionidae, Furipteridae, Thyropteridae and Natalidae have not been recorded from the modern fauna of temperate North America. Among the three families of New World bats that have a worldwide distribution, the Vespertilionidae occurs in both temperate and tropical regions and the Emballonuridae and Molossidae are primarily tropical, although quite a few molossids occur in warm temperate or subtropical zones around the globe. The Vespertilionidae is the only New World bat family that appears to have a temperate origin based on their modern distribution, even though there are more than twice as many species of vespertilionids in the Neotropics compared to the Nearctic region. New World Emballonuridae have a Neotropical distribution, whereas six of the 38 species of Neotropical Molossidae have ranges that extend northward into the southern United States. With the exception of the Vespertilionidae, the most parsimonious explanation for the modern distribution and species-richness patterns of the eight other New World bat families would seem to be that they originated on the island continent of South America while it was isolated from other continents in the Eocene-Miocene, and then dispersed northward into tropical habitats in Mesoamerica following the connection of South America with North America in the Early Pliocene and the onset of the Great American Biotic Interchange. The highly endemic West Indian chiropteran fauna suggests a mid-Tertiary origin from South America.

Substantial Tertiary chiropteran faunas discovered over the past 20 years in Florida and Colombia, and smaller samples of fossil bats from several other sites in western North America, Argentina and Peru, have contributed to a considerably better understanding of the evolutionary history of most of the

New World bat families. Since 1990, the number of bat families recorded from the North American Tertiary has tripled from three to nine, extinct genera have almost doubled from 10 to 19 and extinct species have more than doubled from 11 to 23 (Table 4.3). The improvement in the South American Tertiary bat record has been even more dramatic, with only two families, two genera and two species known before 1990, compared to the current record of six families, 11 genera and about 20 species (Table 4.5). This much-enhanced record clearly establishes that the Tertiary history of Neotropical bats is far more complicated than the simplistic notion of a North American origin for the New World Vespertilionidae and a South American origin for the remaining eight New World families with a primarily tropical distribution.

The evolutionary scenarios we present for the various New World bat families are based primarily on the most current information available from the fossil record. We also rely on data from recent phylogenetic analyses, current patterns of endemism, and the rapidly improving knowledge of modern bat distributions and analyses of biogeography in Latin America (e.g., Mantilla-Meluk *et al.*, 2009; Pacheco *et al.*, 2009).

4.7.1   Origins of the Neotropical chiropteran fauna

There are two conflicting hypotheses for the origin of the Neotropical Emballonuridae. The fossil evidence suggests a North American origin with dispersal of an ancestral emballonurine to South America by overwater dispersal across the Bolivar Strait from tropical North America prior to the Middle Miocene. However, a mid-Cenozoic overwater dispersal of emballonurids from Africa also has been proposed (Teeling *et al.*, 2005; Lim, 2007, 2009). Phylogenetic analyses of both recent and fossil New World Emballonuridae (Lim, 2007; Lim *et al.*, 2008; Morgan, unpublished data) indicate that all members of this group belong to the subfamily Emballonurinae, which also includes the Old World genera *Coleura*, *Emballonura* and *Mosia*. Furthermore, with the possible exception of one of the undescribed genera from the Oligocene of Florida, all other Tertiary New World emballonurids belong to the Neotropical tribe Diclidurini (Lim *et al.*, 2008; Morgan, unpublished data), including the Early Miocene undescribed taxon from Florida, the extant genus *Diclidurus* from the Middle Miocene of Colombia and an undescribed new genus from the Late Miocene of Peru. Lim (2007, 2009) proposed an African origin for the monophyletic New World diclidurines; however, this hypothesis does not adequately account for the presence of emballonurids in the Oligocene and Miocene of Florida. His evolutionary scenario would require two oceanic overwater dispersal events, first from Africa to South America and then from

South America to North America, both of which must have occurred prior to the Oligocene, much earlier than the oldest record of emballonurids in South America. The hypothesized molecular dates for the basal split of the New World emballonurids of 30 Ma (Teeling *et al.*, 2005) to 32.5 Ma (Lim, 2007) are both rather close in age to the oldest fossils of this group from the Early Oligocene of Florida (29–30 Ma), but are much older than the earliest emballonurine in South America from the Middle Miocene of Colombia (12–13 Ma). From a paleontological perspective, another problem with an African origin is that the earliest known undoubted Old World emballonurine is a record of *Coleura* from the Pliocene Omo Fauna of Ethiopia (Wesselman, 1984), whereas New World emballonurines are known from much older faunas. However, two Old World Paleogene emballonurids have been allied with emballonurines, *Vespertiliavus* from France (Sigé *et al.*, 1997) and †*Dhofarella* from Oman and Egypt (Sigé *et al.*, 1994; Gunnell *et al.*, 2008). It is possible that the North American and South American emballonurids had a dual origin, from Eurasia and Africa, respectively, and that only the South American diclidurines survived and the North American members of the family went extinct after the early Miocene. However, this hypothesis would require that the Oligocene and Miocene emballonurids from Florida and the Tertiary and modern emballonurids from South America are non-monophyletic, whereas our studies suggest that one of the undescribed taxa from Florida is a member of the Diclidurini, while the other is sister to the Diclidurini if not a member of this tribe (Morgan, unpublished data).

We propose an alternative hypothesis for the origin of the New World Emballonuridae based on the available fossil evidence. An undescribed genus from the Early Oligocene of Florida was derived from a basal Old World emballonurid (something like †*Tachypteron* or †*Vespertiliavus*, from the Eocene of Europe) that gave rise to both the Old World (Emballonurini) and New World (Diclidurini) emballonurines. An ancestral emballonurid dispersed to North America from Eurasia across Beringia sometime in the Middle to Late Eocene, giving rise to this Early Oligocene taxon which is similar to extant emballonurines in the reduction of the parastylar region on M1, but differs from living taxa in the primitive presence of a p3 in the lower jaw, a character it shares with †*Tachypteron* and †*Vespertiliavus*. The Early Oligocene form (or a similar taxon) gave rise to a second undescribed genus from the Early Miocene of Florida, which lacks a p3 and is morphologically a member of the Diclidurini. Emballonurids disappeared from what is now temperate North America (e.g., Florida) after the Early Miocene, but probably continued to inhabit tropical Mesoamerica. Sometime prior to the Middle Miocene appearance of emballonurids in Colombia, a diclidurine dispersed from Central America across the oceanic water gap (Bolivar Strait) to the then island continent of South America. Although the water gap separating North America and South America in the mid-Cenozoic was substantial, it still would have been much less than the distance across the Atlantic Ocean that separated South America from Africa at this same time. Based on molecular divergence dates, Lim (2007) proposed that the two subtribes of diclidurines (Diclidurina and Saccopterygina) differentiated in the Early Oligocene (27 Ma) and that the eight modern genera of diclidurines originated in the Early to Middle Miocene between 19 and 14 Ma. The extinct Florida Miocene emballonurid is a putative diclidurine, although the limited fossil sample available for this genus does not allow a definitive referral to either the subtribe Diclidurina or Saccopterygina. It is known from the Early Miocene (~18 Ma) during the time interval when Lim *et al.* (2008) proposed that the modern diclidurine genera first appeared. The extant *Diclidurus* first appears at the end of this time interval. Although the generic differentiation within the Diclidurini is presumed to have occurred primarily in South America, it is possible that some of the modern genera had already appeared in tropical Mesoamerica prior to their dispersal to South America. Only a Tertiary fossil record of emballonurids from Central America will answer this question. Screenwashing at Early to Middle Miocene mammal sites in Panama by Slaughter (1981) and more recently by GSM has produced a limited fauna of small mammals, but so far no bats.

The occurrence of two families of Noctilionoidea in the Early Oligocene of Florida, the extinct †Speonycteridae (two species of †*Speonycteris*) and a mormoopid, suggests this group may have had its origins in North America sometime prior to 30 Ma, rather than in South America as the modern distribution of noctilionoids would indicate. A member of the †Speonycteridae or another closely related basal noctilionoid apparently dispersed from Central America to South America in the Late Oligocene or earlier and gave rise to the remarkable radiation of noctilionoids that had their first appearance in South America in the Early Miocene of Argentina (Phyllostomidae) and Middle Miocene of Colombia (Noctilionidae, Thyropteridae). Two other noctilionoid families, Mormoopidae and Furipteridae, are unknown in South America before the Late Pleistocene. Several recent chiropteran phylogenies (e.g., Teeling *et al.*, 2005) placed two Old World bat families within the Noctilionoidea, the Myzopodidae now endemic to Madagascar and the Mystacinidae now restricted to New Zealand, but including a Miocene genus from Australia. This disjunct distribution across several continents and islands in the Southern Hemisphere among supposedly related taxa led Teeling *et al.* (2005) to propose a Gondwanan or Southern Hemisphere origin for the Noctilionoidea. Lim (2009) supported an African origin for the Neotropical noctilionoids based on his hypothesis of an ancestral area for this group determined from the molecular phylogeny of Teeling *et al.* (2005). However, there is currently no paleontological evidence to support an African origin for noctilionoids. A proposed molecular date of 52 Ma (Early Eocene) for the origin of the Noctilionoidea seems much too old (Teeling *et al.*, 2005), as this is about the age of the earliest known New World bats, †*Icaronycteris* and †*Onychonycteris*, currently placed in extinct families that are basal to the modern chiropteran superfamilies (Simmons *et al.*, 2008). The Oligocene fossils of †Speonycteridae and Mormoopidae from Florida suggest that a North American origin for the Noctilionoidea is more likely, and that earlier noctilionoids should be sought among several unstudied Middle and Late Eocene chiropteran samples from western North America.

Two families endemic to the Neotropical region, Mormoopidae and Natalidae, are now known to have their earliest and only Tertiary records in Florida: the extinct mormoopid and an indeterminate natalid from the Oligocene, and the extinct natalid †*Primonatalus* from the Early Miocene (Morgan and Czaplewski, 2003; Morgan, unpublished data). The fossils from northern Florida tentatively indicate that, although the Mormoopidae and Natalidae are now restricted to the Neotropics, both families apparently originated in what is now temperate North America in the Oligocene or earlier. Paleoclimatic records indicate a distinct glaciation beginning abruptly in the Early Oligocene about 33 Ma and lasting until the later Oligocene, about 26 Ma (Zachos *et al.*, 1992, 2008; Woodburne, 2004b). This Early Oligocene glaciation, accompanied by polar ice build-up, contributed to the initial emergence of the Florida peninsula above sea level, allowing habitation by bats and other terrestrial vertebrates by about 30 Ma. After their disappearance from Florida in the Late Oligocene (mormoopids) and Early Miocene (natalids), these two families probably survived in the West Indies and/or tropical Mesoamerica during the Miocene and Pliocene, but are currently unknown from the fossil record during this period. Both the Mormoopidae and Natalidae attain their maximum species richness and endemism in the Late Pleistocene and modern fauna of the West Indies, suggesting these two families reached the Antilles from either Florida or Mesoamerica early in their history, in the Oligocene or Miocene, and then underwent a long period of autochthonous evolution there. Dávalos (2005) has suggested based on molecular genetics and phylogenetic relationships that the ancestor of the extant natalid radiation occurred in the West Indies, although ultimately derived from continental North America, and then one genus (*Natalus*) later dispersed to Middle and South America. The Mormoopidae seem to have a more complicated history, although both the fossil record and studies of phylogenetic relationships indicate an origin for mormoopids in Florida/tropical North America, diversification in Mesoamerica and the West Indies and a later dispersal to South America (Dávalos, 2006; Morgan, unpublished data). An important feature in the evolutionary history of both mormoopids and natalids is their roosting ecology, as most species in these two families are obligate cave dwellers and prefer large cave systems with high temperature and humidity. The fossil record of mormoopids and natalids is certainly tied to caves, as almost all fossils of these two families are either from paleokarst deposits (Oligocene, Miocene and Late Pleistocene fossils from Florida) or cave deposits (Late Pleistocene fossils from the West Indies, Mexico and Brazil). One of the only non-karst fossils of these two families is a partial skeleton of the mormoopid *Pteronotus* cf. *P. parnellii* from an Early Pleistocene lake deposit in El Salvador (Webb and Perrigo, 1984). The lack of a Tertiary record, minimal endemism and limited modern species richness suggest that both mormoopids and natalids reached South America late in their history, probably in the Pliocene after the connection of North and South America at the Panamanian Isthmus.

There is a long but intermittent Tertiary record of the Molossidae in the New World, with Middle Eocene, Late Oligocene, Early Miocene and Late Pliocene records from North America, and Oligocene and Early, Middle and Late Miocene records from South America. There is little overlap at the generic level among Tertiary molossids between the two continents, with †*Wallia*, *Eumops*, *Tadarida* and several indeterminate but apparently extinct taxa from North America and †*Potamops*, *Eumops*, at least three extinct species of *Mormopterus* and a possible *Nyctinomops* from South America. The extinct genera †*Wallia* and †*Potamops* lack any obvious relationship to extant molossid taxa. The South American *Eumops* is from the Middle Miocene of Colombia (Czaplewski, 1997), while the North American record is much younger, from the Late Pliocene of Arizona (Czaplewski, 1993), suggesting a South American origin for *Eumops* and then dispersal to North America in the Pliocene during the Interchange (Morgan, 2008). The presence of *Tadarida* in the Late Pliocene of Florida and two taxa similar to *Tadarida* from the Early Miocene of Florida (Czaplewski *et al.*, 2003a), together with the current absence of a Tertiary fossil record of *Tadarida* from South America, suggest that the New World representatives of this genus may have originated in North America. *Tadarida* occurs in Oligocene and Miocene sites in Europe, supposedly including fossils referred to *Rhizomops*, the genus or subgenus that includes the only two New World species of *Tadarida*, the wide-ranging *T. brasiliensis* and the large extinct Middle Pleistocene species †*T. constantinei* from New Mexico (Lawrence, 1960; Legendre, 1984, 1985; Morgan, 2003). *Tadarida* probably immigrated from Eurasia to North America in the Oligocene or Early Miocene, reaching South

America in the Pliocene after the Interchange. The New World record of *Mormopterus* consists of at least three extinct species, †*M. faustoi* from the Oligocene of Brazil, †*M. barrancae* from the Early Miocene of Argentina and †*M. colombiensis* from the Middle Miocene of Colombia (Paula Couto, 1956, 1983; Czaplewski, 1997, 2005), as well as three extant species, two from South America and one from Cuba.

Lim (2009) also proposed an African origin for the New World Molossidae; however, the only genus of molossid from the Tertiary of Africa, *Tadarida* from the Early Miocene of Kenya (Arroyo-Cabrales *et al.*, 2002), is unknown before the Late Pleistocene in South America, but occurs in the Late Pliocene and perhaps the Early Miocene of Florida. The earliest and most speciose genus of molossid in the Tertiary of South America, *Mormopterus*, is unknown from the fossil and modern record in Africa (although there are Plio-Pleistocene and modern records in Madagascar), but it does occur in the Miocene of Europe, Asia and Australia. The lack of a well-constrained phylogeny and spotty Tertiary record hinders our attempt to present a coherent evolutionary history for the New World Molossidae. The fossil evidence suggests a possible dual origin for some New World molossids, with *Tadarida* having a Eurasian origin, *Mormopterus* a southern origin and the remaining taxa of uncertain origin. The presence of the oldest molossid in the Middle Eocene of North America (†*Wallia*) suggests the possibility of a North American origin for the family. As with the Emballonuridae, the poor fossil record of the Molossidae in Africa offers little empirical support for an African origin of the South American members of this family.

The Tertiary history of the Vespertilionidae in the New World is almost entirely limited to temperate North America, including nine extinct genera ranging in age from Early Oligocene to Late Pliocene, with extant genera first appearing in the Late Miocene (Czaplewski *et al.*, 2008). From the Early Oligocene onward in the western United States and from Middle Miocene and younger sites in Florida, the North American Tertiary chiropteran fauna is dominated by vespertilionids, much as it is today. This record probably reflects the temperate climates that characterized much of North America, beginning in the Middle Miocene, if not earlier, and continuing to the present. A tooth of an indeterminate vespertilionid from the Middle Miocene of Colombia represents the only Tertiary record of this family in South America (Czaplewski *et al.*, 2003b). Although the Neotropical vespertilionid fauna is considerably more diverse than are Nearctic vespertilionids (Simmons, 2005), this is probably a reflection of the overall higher species diversity of bats in the New World tropics compared to the Nearctic region. Moreover, only five of the 13 vespertilionid genera recorded from the Neotropics, *Eptesicus*, *Lasiurus*, *Myotis*,

Histiotus and *Rhogeessa*, occur in South America, and only the latter two are endemic to the Neotropical region. The remaining eight genera are principally Nearctic in occurrence, with ranges that extend southward into the northern-most portion of the Neotropics in Mexico and northern Central America. Six of the extant vespertilionid genera from the Neotropical region have a Tertiary record in temperate North America, further indicating a Nearctic origin.

Contrary to our proposed evolutionary scenarios based primarily on the fossil record, a Southern Hemisphere or Gondwanan origin via overwater dispersal from Africa across the Atlantic Ocean to South America (the "Out of Africa" hypothesis) has been proposed for seven of the nine families of bats that currently inhabit South America, including Emballonuridae, the five families in the Noctilionoidea and Molossidae (Teeling *et al.*, 2005; Lim, 2009). A northern origin (either dispersal from Eurasia to North America or evolution in North America) for some or all of these groups was discounted for several reasons. Foremost among the reasons for an African origin of most groups of Neotropical bats are: (1) the ancestral area of origin, based on a combination of modern distribution and molecular phylogenetic analyses, supposedly favors Africa (Lim, 2009); (2) the predominantly tropical distribution of these bats in the present fauna, with the underlying assumption that dispersal of these now-tropical forms from Eurasia into North America across the high latitude Bering land bridge would have been improbable (Teeling *et al.*, 2005; Lim, 2009); and (3) the hypothesized African origin of two other Neotropical mammal groups, the platyrrhine primates and caviomorph rodents, both of which first appeared in South America during the Late Eocene or Oligocene (Flynn and Wyss, 1998).

We suggest that the "Out of Africa" hypothesis for the origin of the Neotropical chiropteran fauna overlooks several important factors. First and most importantly, the North American fossil record establishes the presence of the Molossidae by the Middle Eocene and the Emballonuridae, Noctilionoidea (†Speonycteridae, Mormoopidae), Natalidae and Vespertilionidae by the Early Oligocene. These groups first appear much later in South America (earliest South American records in parentheses): Molossidae (Late Oligocene), Emballonuridae (Middle Miocene), Noctilionoidea (Phyllostomidae – Early Miocene; Noctilionidae and Thyropteridae – Middle Miocene; Mormoopidae and Furipteridae – Late Pleistocene), Natalidae (Late Pleistocene) and Vespertilionidae (Middle Miocene). Second, these groups have a limited to non-existent Tertiary record in Africa, their supposed continent of origin (earliest African records in parentheses from Gunnell and Simmons, 2005; Gunnell *et al.*, 2008): Molossidae (Early Miocene); Emballonuridae (Late Eocene for earliest African member of the family, but Pliocene for earliest emballonurine); Noctilionoidea (no fossil record); Natalidae (no fossil record) and Vespertilionidae (Late

Eocene). Third, there were numerous episodes of dispersal of various mamma-lian groups between Eurasia and North America during the time period when bat groups with Neotropical affinities may have arrived in North America from Eurasia, between the Late Early Eocene and Early Oligocene (~50–30 Ma; Woodburne and Swisher, 1995). Beginning in the Middle Eocene, the climate throughout much of North America began to change to more temperate, seasonal, savanna-like conditions from the warm, tropical climate of the Early Eocene (Woodburne *et al.*, 2009). Bat families now restricted to the tropics dominated Florida Oligocene faunas, suggesting that tropical climates per-sisted in Florida much later than in the remainder of the continent. We also infer that with the onset of Antarctic glaciation in the Early Oligocene, the regional climate of the southeastern United States began cooling, which even-tually drove the tropical bat families from Florida after the Early Miocene.

4.7.2   New World bats and the Great American Biotic Interchange

Because of their ability to fly, bats are less affected by water barriers than are terrestrial or non-volant mammals. It is well established that strong-flying bats occasionally cross considerable spans of open ocean water. Perhaps the best New World example is the vespertilionid *Lasiurus*, as demonstrated by the occurrence of *Lasiurus cinereus* and an extinct small species of *Lasiurus* as Quaternary fossils in lava tube caves in the remote Hawaiian Islands (A. C. Ziegler and F. G. Howarth, unpublished m.s.; personal observation), and other *Lasiurus* species in Iceland and the Galapagos Islands (Koopman and Gudmundsson, 1966; Krzanowski, 1977; Koopman and McCracken, 1998). Nonetheless, most bats do not migrate long distances, are not known for their overwater dispersal capabilities, and thus have only rarely crossed major oceanic water barriers. Most bats are only capable of dispersing moderate distances over water or may be blown between islands and landmasses by storms (Presley and Willig, 2008).

During much of the Cenozoic, the overwater dispersal ability of some bats might have contributed to the evolutionary diversification of the tropical chiropteran fauna in the Western Hemisphere. Tectonic plate movements pro-vided changing, intermittently available subcontinental blocks and island arcs between North America and South America during the Cenozoic, long before the GABI (e.g., Coates and Obando, 1996; Iturralde-Vinent and MacPhee, 1999). Figure 4.6 provides two paleogeographic sketch maps of southern North America, Middle America and the Caribbean region, and northern South America during the Late Paleogene and Early Neogene. The relative proximity of these landmasses could have facilitated dispersals between the continents and

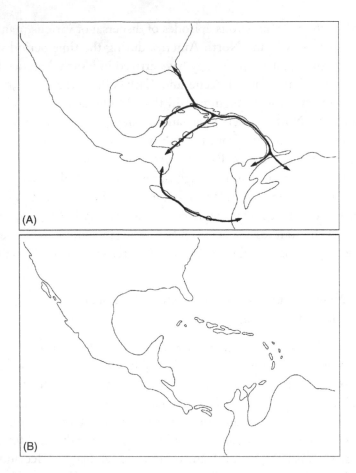

Figure 4.6 Paleogeographic sketch maps of central Western Hemisphere showing southern North America, Middle America, Caribbean region and northern South America during Late Paleogene and Early Neogene. (A) Late Eocene to Early Oligocene, with possible dispersal routes along "Gaarlandia," islands and over water. (B) Late Oligocene to Early Miocene. Maps compiled from several sources including Iturralde-Vinent (2006), Kirby *et al.* (2008), Pindell and Kennan (in press) and the websites of Iturralde-Vinent.

among the islands, and enhanced the isolation and diversification potential for bat populations. However, as noted above, many bats are incapable of crossing wide expanses of water, and until fossils can be found to establish the composition of the chiropteran fauna that occurred in Central America and the Caribbean Basin during the Tertiary, the significance of this region for the pre-Pliocene evolution of Neotropical bats remains hypothetical.

Prior to the onset of the GABI in the Late Miocene (~9 Ma), North America and South America are known to have shared at least three families of bats,

Emballonuridae, Molossidae and Vespertilionidae. This is not surprising considering that these three families had an essentially worldwide distribution in the Tertiary, also occurring in Eurasia, Africa and Australia (Gunnell and Simmons, 2005). It is instructive to compare the two most diverse Tertiary chiropteran faunas in North America and South America, the Early Miocene (~18 Ma) Thomas Farm LF from Florida and the somewhat younger Middle Miocene (~12 Ma) La Venta Fauna of Colombia. The La Venta chiropteran fauna is more diverse (14 species) than Thomas Farm (nine species), which is not unexpected considering the tropical latitude of La Venta (3°N) compared to the temperate/subtropical latitude of Thomas Farm (30°N). The nine species of bats from Thomas Farm are divided among four families: Emballonuridae (one species), Natalidae (one species), Molossidae (two species) and Vespertilionidae (five species). The La Venta fauna consists of 14 species in six families: Emballonuridae (two species), Phyllostomidae (three species of phyllostomines, one glossophagine), Noctilionidae (one species), Thyropteridae (two species), Molossidae (four species) and Vespertilionidae (one species). These two chiropteran faunas have no genera in common (unless more complete molossid material from Thomas Farm reveals the presence of *Mormopterus*, which also occurs at La Venta), but do share three families. The fact they share any families is significant. The non-volant mammals from these two faunas have no families in common and share only one other order, the Rodentia. All rodents from La Venta belong to the South American caviomorph group of African origin, whereas the three families of rodents from Thomas Farm, Sciuridae, Heteromyidae and Eomyidae, are either of North American or Eurasian origin. The other members of the Thomas Farm mammalian fauna belong to the orders Lipotyphla, Carnivora, Perissodactyla and Artiodactyla, whereas the La Venta mammals include the orders Marsupialia (*sensu lato*), Xenarthra, Primates, Litopterna and Notoungulata. Perhaps the more relevant question is, why do Thomas Farm and La Venta share any families of bats when the rest of their mammalian faunas, indeed the entire pre-Interchange non-volant mammalian faunas of the two continents, are utterly dissimilar? The shared bat families suggest limited Pre-Late-Miocene interchange of the chiropteran faunas of North and South America by overwater dispersal, across a wide and deep oceanic water barrier that prevented the exchange of non-volant mammals prior to the Late Miocene to Early Pliocene advent of the GABI. Conversely, it has been suggested that several of these families, in particular Emballonuridae and Molossidae, came from Africa, which would suggest a dual origin, with South American taxa having arrived from Africa by overwater dispersal across the Atlantic Ocean and North American taxa derived from Eurasia by an overland route across Beringia. Future fossil discoveries and phylogenies of these groups should help to confirm or refute these possible dual origins.

The participation of bats in the GABI has not been investigated thoroughly, primarily owing to their poor fossil record, leading to uncertainty about their place of origin. The classic Interchange scenario proposes two major phases of biotic exchange between North America and South America during the Late Cenozoic, the first occurring in the Late Miocene and the second in the Pliocene and Early Pleistocene (e.g., Webb, 1976). The Late Miocene phase (~7–9 Ma) involved the first appearance in temperate North America of two genera of ground sloths of South American origin and the occurrence of large procyonids in temperate South America of North American origin. More recent work in the Amazon Basin of Peru suggests that gomphotheriid proboscideans, tapirs and peccaries first appeared in northern South America in the Late Miocene (Campbell et al., 2001, 2010). This Late Miocene interchange is generally hypothesized to have involved overwater dispersal (Webb, 1976), although Campbell et al. (2000) proposed a possible Late Miocene land connection between the two continents. We are not aware of any bat dispersals that coincide with the Late Miocene phase of the GABI, lending support to the presence of a substantial water gap. The second phase of the GABI began in the Early Pliocene with the connection of North and South America at the Panamanian Isthmus. The Plio-Pleistocene phase of the GABI (~1.5–5 Ma) involved a widespread interchange of faunas and floras between the continents that has been well documented (e.g., Stehli and Webb, 1985). Among mammals, the GABI included glyptodonts, armadillos, pampatheres, several additional species of ground sloths, capybaras and porcupines of South American origin appearing in North American Pliocene faunas and cricetid rodents, leporids, carnivorans, horses, tapirs, camels, deer, proboscideans and other groups of North American origin appearing in South American Plio-Pleistocene faunas. We suggest that the Pliocene records of *Desmodus* and *Eumops* in temperate North America and the Pleistocene appearance of mormoopids and natalids in South America are both dispersal events related to the Plio-Pleistocene phase of the GABI. The evolutionary histories we propose above for several of the Neotropical bat groups, including Emballonuridae, Noctilionoidea and Vespertilionidae, and possibly Molossidae, indicate a third older phase of the Interchange involving the dispersal of these bat groups from North America to South America in the Oligocene or Early Miocene (~16–30 Ma). This hypothesized Oligocene–early-Miocene phase of the Interchange does not coincide with the dispersal of any groups of non-volant mammals.

4.8   Final comments and future research

It seems almost trite in paleontology to wish for a more complete fossil record, but the paucity of Tertiary bat fossils from North America and particularly South America has clearly hindered a better understanding of the evolutionary history of New World chiropteran fauna. However, rather than decrying the poor fossil record of bats, we point out that over the past 20 years the Tertiary record of bats in the New World has improved dramatically, with a doubling of the described taxa in North America and a ten-fold increase in the number of known taxa from South America. Much of this improvement can be attributed to the widespread use of screenwashing techniques. Nonetheless, much work remains to be done, both in the field and under the microscope.

Perhaps the most glaring information gap in the North American bat record is the Middle to Late Eocene (~34–49 Ma), a time period during which the archaic families of bats from the Early Eocene were being replaced by extant families. Although the published record of North American bats during the Mid to Late Eocene is minimal, with the only described species being the earliest molossid †*Wallia scalopidens*, much unstudied material exists. A number of sites, including the Middle Eocene Elderberry Canyon in Nevada, Swift Current Creek in Saskatchewan and Friars Formation in California and Late Eocene Lac Pelletier in Saskatchewan and Flagstaff Rim in Wyoming, contain important unstudied samples of bats that will greatly improve our understanding of this important time period in chiropteran evolution. These bats are derived from a variety of taphonomic and paleoecologic situations, mostly typical fluvial/alluvial deposits, but also including a freshwater lake deposit (Elderberry Canyon) and a possible owl roost with abundant microvertebrates (Flagstaff Rim). Large and/or well-preserved samples of Tertiary bats are scarce in western North America, primarily because of the rarity of pre-Pleistocene paleokarst deposits, although welcome exceptions are the beautifully preserved skeletons from lacustrine deposits in the Early Eocene Green River Formation in Wyoming and an impressive assemblage of bat fossils from fluvial deposits in the Middle Miocene Myers Farm site in Nebraska. Owing to the abundance of paleokarst deposits, the Florida Tertiary bat record is the richest in North America, but has its limitations, in particular the lack of a pre-Oligocene terrestrial record. Other gaps exist in the Florida record, especially the Middle to Late Miocene, during which time many modern genera of vespertilionids first appeared. The few Late Miocene bats known from Florida are mostly fragmentary specimens of vespertilionids that are not identifiable below the family level.

The South American Tertiary bat record contains many gaps, both chronologic and geographic. The highly incomplete record from the Paleogene is perhaps the most critical gap in the record. There are only two known bat localities in South America for the entire Eocene epoch, one in Argentina and one in Peru, each consisting of small samples of isolated teeth of uncertain taxonomic affinity. The Oligocene record is composed of two partial skeletons from Brazil, one of the molossid *Mormopterus* and the second of an indeterminate family. In North America and elsewhere, bat faunas from the Eocene and Oligocene document a transition from archaic to extant families, thus we can probably assume that most of the early chiropteran record is currently missing from South America. As in North America, screenwashing for microvertebrates is becoming standard in South America and has been especially productive for bats in the La Venta badlands in Colombia and in the western Amazon through the work of Kenneth Campbell and Carl Frailey. Screenwashing of South American Tertiary sites should continue to produce bats, although most of the samples discovered to date have been limited in abundance, completeness and diversity, with the obvious exception of La Venta. Another glaring gap in the South American record is the total lack of Pliocene bats. In North America, modern chiropteran faunas with many extant species first appear in the Pliocene. The most important factor limiting the Tertiary chiropteran record in South America is the paucity of known paleokarst deposits containing bats. Karst areas in northern Venezuela and southeastern Brazil have numerous caves and Late Pleistocene deposits with bat fossils, but no known fossiliferous Tertiary paleokarst deposits. Other than the well-known Paleocene deposits from the Itaboraí area in Brazil (now inaccessible) which lack bats, we are unaware of other paleokarst areas in South America with fossiliferous infillings of Tertiary age. We strongly encourage investigators in South America, Mesoamerica and the West Indies to search for paleokarst situations and fissure fillings that might contain Tertiary mammalian fossils, especially bats.

Despite the oft-cited deficiencies in the chiropteran fossil record, we are strong believers in the potential of fossils to provide important data on the evolutionary history of bats. Even the study of existing specimens can be better integrated with the molecular and morphological systematics of modern bats to refine temporal resolution of bat phylogeny and biogeography. We find the arguments of a southern Gondwanan origin of noctilionoids cogent and anticipate with enthusiasm the discovery of Paleogene or Early Neogene noctilionoid fossils in South America, but we are using a strictly empirical interpretation of the paleontological evidence. Our hypothesis is capable of being falsified (or verified) by additional fossil evidence. Fossil bats will almost certainly continue to be discovered during routine screenwashing of Tertiary – possibly even Paleocene – vertebrate sites throughout the New World. Consequently, new discoveries from ages or localities not previously known may dramatically alter previous hypotheses on the evolutionary history of certain chiropteran groups. This has certainly been true of the important bat samples from Florida and Colombia, suggesting that future discoveries may well necessitate an update of our current effort.

4.9 Acknowledgments

We are particularly grateful to G. F. Gunnell and N. Simmons, the editors of this volume and organizers of the symposium on bat evolution at the 2007 Society of Vertebrate Paleontology meeting in Austin, Texas, for inviting us to participate in both of these endeavors. Numerous people have assisted us in the field and with washing and sorting microvertebrate matrix, including R. Cifelli, J. Guerrero Díaz, R. Hitz, R. Hulbert, R. Kay, J. Krejca, R. Madden, T. Miller, T. Naeher-Stephens, A. Poyer, A. Pratt, C. Ross, M. Sanchez-Villagra, A. Serna González, B. Shockey, E. Simons, D. Steadman and A. Walton. C. Ray and the late P. Brodkorb initiated screenwashing at Thomas Farm, Florida in the late 1950s and early 1960s, first revealing the incredibly rich sample of bats from that site. G. F. Gunnell and A. Tejedor provided helpful reviews of the manuscript. R. Emry of the US National Museum of Natural History (USNM) has generously allowed us to study the Eocene bats from Elderberry Canyon, Nevada and Flagstaff Rim, Wyoming. G. Corner of the University of Nebraska State Museum has kindly permitted us to study the Miocene bats from Myers Farm, Nebraska. For loans, access to additional specimens and other assistance while using collections we are grateful to: R. Hulbert, D. Webb, B. MacFadden and L. Wilkins (FLMNH); M. Mares and J. Braun (OMNH); L. Gordon, D. Wilson, D. Schmidt, F. Grady and R. Emry (USNM); N. Simmons (American Museum of Natural History); J. Cook, J. Dunnum and W. Gannon (Museum of Southwestern Biology); P. Holroyd and the late D. Savage (UCMP); J. Patton and B. Stein (Museum of Vertebrate Zoology); S. McLeod and K. Campbell (LACM); T. Deméré (San Diego Museum of Natual History); E. Lindsay (University of Arizona Laboratory of Paleontology); B. Patterson (Field Museum of Natural History); J. Arroyo-Cabrales (Instituto Nacional de Antropologia e Historia, Mexico City); A. Rincón (Caracas, Venezuela); M. Pressinotti, P. Souza and F. Pires (IGG, São Paulo, Brazil); R. Gregorin and H. Britski (Museu Zoologia Universidade São Paulo, Brazil); A. Kellner, D. Henriques, J. de Oliveira, L. de Oliveira and L. Salles (MNRJ, Rio de Janeiro, Brazil); L. P. Bergqvist (Universidade Federal Rio de Janeiro, Brazil); C. Cartelle and G. Wilson Fernandes (Universidade Federal de Minas Gerais, Belo Horizonte, Brazil); A. Carlini and F. Goin (MLP, La Plata, Argentina); M. F. Tejedor (LIEB-PV,

Esquel, Argentina); B. Sigé (Montpellier, France); C. de Muizon (Paris, France) and M. Takai (Kyoto University, Japan). In addition to those already named, we also extend our gratitude for insights, opportunities, friendship and aid in various ways during our studies of fossil bats to V. Abuhid, L. Aguiar, E. Aragón, L. Avilla, Bambui Group speleologists, C. D. Czaplewski, J. L. Czaplewski, G. Lessa and M. Reguero.

4.10 REFERENCES

Aguiar Fracasso, M. P. and Oliveira Salles, L. (2005). Diversity of bats from Serra da Mesa (state of Goiás, Brazil). *Zootaxa*, **817**, 1–19.

Allen, G. M. (1932). A Pleistocene bat from Florida. *Journal of Mammalogy*, **13**, 256–259.

Anthony, H. E. (1917). Two new fossil bats from Porto Rico. *Bulletin of the American Museum of Natural History*, **37**, 565–568.

Arroyo-Cabrales, J. and Polaco, O. J. (2003). Caves and the Pleistocene vertebrate paleontology of Mexico. In *Ice Age Cave Faunas of North America*, ed. B. W. Schubert, J. I. Mead and R. W. Graham. Bloomington and Indianapolis, IN: Indiana University Press, pp. 273–291.

Arroyo-Cabrales, J. and Polaco, O. J. (2008). Fossil bats from Mesoamerica. *Arquivos do Museu Nacional, Rio de Janeiro*, **66**, 155–160.

Arroyo-Cabrales, J., Gregorin, R., Schlitter, D. A. and Walker, A. (2002). The oldest African molossid bat cranium (Chiroptera: Molossidae). *Journal of Vertebrate Paleontology*, **22**, 380–387.

Avilla, L. S., Winck, G. R., Rodrigues Francisco, V. M. *et al.* (2007). The fauna of fossil bats as tool for the characterization of Quaternary paleoenvironments. *Anuário do Instituto de Geociências-UFRJ*, **30**, 19–26.

Baskin, J. A. (2003). New procyonines from the Hemingfordian and Barstovian of the Gulf Coast and Nevada, including the first fossil record of Potosini. In *Vertebrate Fossils and their Context – Contributions in Honor of Richard H. Tedford*, ed. L. J. Flynn. *Bulletin of the American Museum of Natural History*, **279**, 125–146.

Beard, K. C., Sigé, B. and Krishtalka, L. (1992). A primitive vespertilionoid bat from the early Eocene of central Wyoming. *Comptes Rendus de l'Académie des Sciences, Paris*, **314**, 735–741.

Bell, C. J., Lundelius, E. L., Jr., Barnosky, A. D. *et al.* (2004). The Blancan, Irvingtonian, and Rancholabrean mammal ages. In *Late Cretaceous and Cenozoic Mammals of North America: Biostratigraphy and Geochronology*, ed. M. O. Woodburne. New York: Columbia University Press, pp. 232–314.

Brown, R. W. (1959). A bat and some plants from the upper Oligocene of Oregon. *Journal of Paleontology*, **33**, 125–129.

Cadenillas, R. and Martínez, J.-N. (2005). Additional bats from late Pleistocene Talara tar seeps (northwestern Peru): paleoenvironmental implications. *II Congreso Latino-Americano de Paleontologia de Vertebrados, Boletim de Resumos, Rio de Janeiro, Museu Nacional, Serie Livros*, **12**, 58–59.

Campbell, K. E., Jr., Frailey, C. D. and Romero Pittman, L. (2000). The late Miocene gomphothere *Amahuacatherium peruvium* (Proboscidea: Gomphotheriidae) from Amazonian Peru: implications for the Great American Faunal Interchange. *República de Perú, Sector Energía y Minas, Instituto Geológico Minero y Metalúrgico, Serie D: Estudios Regionales, Boletín*, **23**, 1–152.

Campbell, K. E., Jr., Heizler, M., Frailey, C. D., Romero-Pittman, L. and Prothero, D. R. (2001). Upper Cenozoic chronostratigraphy of the southwestern Amazon Basin. *Geology*, **29**, 595–598.

Campbell, K. E., Jr., Prothero, D. R., Romero-Pittman, L., Hertel, F. and Rivera, N. (2010). Amazonian magnetostratigraphy: dating the first pulse of the Great American Faunal Interchange. *Journal of South American Earth Sciences*, **29**, 619–626.

Carranza-Castañeda, O. and Walton, A. H. (1992). Cricetid rodents from the Rancho El Ocote fauna, late Hemphillian (Pliocene), state of Guanajuato. *Universidad Nacional Autónoma de México, Instituto de Geología, Revista*, **10**, 71–93.

Cartelle, C. and Abuhid, V. S. (1994). Chiroptera do Pleistocene final-Holoceno da Bahia. *Acta Geologica Leopoldensia*, **17**, 429–440.

Cione, A. L., Tonni, E. P., Bargo, S. *et al.* (2007). Mamíferos continentals del Mioceno tardio a la actualidad en la Argentina: cincuenta años de estudios. *Asociación Paleontológica Argentina, Ameghiniana 50 Aniversario, Publicación Especial*, **11**, 257–278.

Coates, A. G. and Obando, J. A. (1996). The geologic evolution of the Central American isthmus. In *Evolution and Environment in Tropical America*, ed. J. B. C. Jackson, A. F. Budd and A. G. Coates. Chicago, IL: University of Chicago Press, pp. 21–56.

Corner, R. G. (1976). An early Valentinian vertebrate local fauna from southern Webster County, Nebraska. M.S. thesis, University of Nebraska, Lincoln.

Croft, D. A., Flynn, J. J. and Wyss, A. R. (2008). The Tinguiririca fauna of Chile and the early stages of "modernization" of South American mammal faunas. *Arquivos do Museu Nacional, Rio de Janeiro*, **66**, 191–211.

Croft, D. A., Anaya, F., Auerbach, D., Garzione, C. and MacFadden, B. J. (2009). New data on Miocene Neotropical provinciality from Cerdas, Bolivia. *Journal of Mammalian Evolution*, **16**, 175–198.

Czaplewski, N. J. (1987). Middle Blancan vertebrate assemblage from the Verde Formation, Arizona. *Contributions to Geology, University of Wyoming*, **25**, 133–155.

Czaplewski, N. J. (1990). Late Pleistocene (Lujanian) occurrence of *Tonatia sylvicola* in the Talara Tar Seeps, Peru. *Anais Acadamie Brasileira Cien*, **62**, 235–238.

Czaplewski, N. J. (1991). Miocene bats from the lower Valentine Formation of northeastern Nebraska. *Journal of Mammalogy*, **72**, 715–722.

Czaplewski, N. J. (1993a). Late Tertiary bats (Mammalia, Chiroptera) from the southwestern United States. *Southwestern Naturalist*, **38**, 111–118.

Czaplewski, N. J. (1993b). *Pizonyx wheeleri* Dalquest and Patrick (Mammalia: Chiroptera) from the Miocene of Texas referred to the genus *Antrozous* H. Allen. *Journal of Vertebrate Paleontology*, **13**, 378–380.

Czaplewski, N. J. (1996). Opossums (Didelphidae) and bats (Noctilionidae and Molossidae) from the late Miocene of the Amazon Basin. *Journal of Mammalogy*, **77**, 84–94.

Czaplewski, N. J. (1997). Chiroptera. In *Vertebrate Paleontology in the Neotropics: The Miocene Fauna of La Venta, Colombia*, ed. R. F. Kay, R. H. Madden, R. L. Cifelli and J. J. Flynn. Washington, DC: Smithsonian Institution Press, pp. 410–431.

Czaplewski, N. J. (2005). A review of the pre-Pleistocene fossil record of bats (Chiroptera) in South America. *II Congresso Latino-Americano de Paleontologia de Vertebrados, Boletim de Resumos, Rio de Janeiro, Museu Nacional, Serie Livros*, **12**, 87–89.

Czaplewski, N. J. (2010). Colhuehuapian bats (Mammalia: Chiroptera) from the Gran Barranca, Chubut province, Argentina. In *The Paleontology of Gran Barranca: Evolution and Environmental Change through the Middle Cenozoic of Patagonia*, ed. R. H. Madden, A. A. Carlini, M. G. Vucetich and R. F. Kay. Cambridge: Cambridge University Press, pp. 240–252.

Czaplewski, N. J. and Campbell, K. E., Jr. (2004). A possible bat (Mammalia: Chiroptera) from the ?Eocene of Amazonian Peru. *Natural History Museum of Los Angeles County, Science Series*, **40**, 141–144.

Czaplewski, N. J. and Cartelle, C. (1998). Pleistocene bats from cave deposits in Bahia, Brazil. *Journal of Mammalogy*, **79**, 784–803.

Czaplewski, N. J. and Morgan, G. S. (2000). A new vespertilionid bat (Mammalia: Chiroptera) from the early Miocene (Hemingfordian) of Florida, USA. *Journal of Vertebrate Paleontology*, **20**, 736–742.

Czaplewski, N. J., Bailey, B. E. and Corner, R. G. (1999). Tertiary bats (Mammalia: Chiroptera) from northern Nebraska. *Transactions of the Nebraska Academy of Sciences*, **25**, 83–93.

Czaplewski, N. J., Morgan, G. S. and Naeher, T. M. (2003a). Molossid bats from the late Tertiary of Florida with a review of the Tertiary Molossidae of North America. *Acta Chiropterologica*, **5**, 61–74.

Czaplewski, N. J., Takai, M., Naeher, T. M., Shigehara, N. and Setoguchi, T. (2003b). Additional bats from the middle Miocene La Venta Fauna of Colombia. *Revista de la Academia Colombiana de Ciencias Exactas, Físicas, y Naturales*, **27**, 263–282.

Czaplewski, N. J., Krejca, J. and Miller, T. E. (2003c). Late Quaternary bats from Cebada Cave, Chiquibul Cave System, Belize. *Caribbean Journal of Science*, **39**, 23–33.

Czaplewski, N. J., Rincón, A. D. and Morgan, G. S. (2005). Fossil bat (Mammalia: Chiroptera) remains from Inciarte tar pit, Sierr de Perijá, Venezuela. *Caribbean Journal of Science*, **41**, 768–781.

Czaplewski, N. J., Morgan, G. S. and McLeod, S. A. (2008). Chapter 12, Chiroptera. In *Evolution of Tertiary Mammals of North America, Vol. 2: Small Mammals, Xenarthrans, and Marine Mammals*, ed. C. M. Janis, G. F. Gunnell and M. D. Uhen. Cambridge: Cambridge University Press, pp. 174–197.

Dalquest, W. W. (1975). Vertebrate fossils from the Blanco Local Fauna of Texas. *Occasional Papers, The Museum, Texas Tech University*, **30**, 1–52.

Dalquest, W. W. (1978). Early Blancan mammals of the Beck Ranch Local Fauna of Texas. *Journal of Mammalogy*, **59**, 269–298.

Dalquest, W. W. and Patrick, D. B. (1989). Small mammals from the Early and Medial Hemphillian of Texas, with descriptions of a new bat and gopher. *Journal of Vertebrate Paleontology*, **9**, 78–88.

Dalquest, W. W. and Roth, E. (1970). Late Pleistocene mammals from a cave in Tamaulipas, Mexico. *Southwestern Naturalist*, **15**, 217–230.

Dalquest, W. W., Baskin, J. A. and Schultz, G. E. (1996). Fossil mammals from a late Miocene (Clarendonian) site in Beaver County, Oklahoma. *Contributions in Mammalogy: A Memorial Volume Honoring Dr. J. Knox Jones, Jr., Museum of Texas Tech University*, **1996**, 107–137.

Dávalos, L. M. (2005). Molecular phylogeny of funnel-eared bats (Chiroptera: Natalidae), with notes on biogeography and conservation. *Molecular Phylogenetics and Evolution*, **37**, 91–103.

Dávalos, L. M. (2006). The geography of diversification in the mormoopids (Chiroptera: Mormoopidae). *Biological Journal of the Linnean Society*, **88**, 101–118.

Eiting, T. P. and Gunnell, G. F. (2009). Global completeness of the bat fossil record. *Journal of Mammalian Evolution*, **16**, 151–173.

Emry, R. J. (1973). Stratigraphy and preliminary biostratigraphy of the Flagstaff Rim area, Natrona County, Wyoming. *Smithsonian Contributions to Paleobiology*, **18**, 1–43.

Emry, R. J. (1990). *Mammals of the Bridgerian (middle Eocene) Elderberry Canyon Local Fauna of eastern Nevada*. Geological Society of America Special Paper, **243**, 187–210.

Emry, R. J. and Eshelman, R. E. (1998). The early Hemingfordian (Early Miocene) Pollack Farm Local Fauna: first Tertiary land mammals described from Delaware. In *Geology and Paleontology of the Lower Miocene Pollack Farm Fossil Site, Delaware*, ed. R. N. Benson. *Delaware Geological Survey Special Publication*, **21**, 153–173.

Eshelman, R. E. and Morgan, G. S. (1985). Tobagan Recent mammals, fossil vertebrates, and their zoogeographical implications. *National Geographic Society Research Reports*, **21**, 137–143.

Ferrusquía-Villafranca, I. (2003). Mexico's Middle Miocene mammalian assemblages: an overview. In *Vertebrate Fossils and their Context – Contributions in Honor of Richard H. Tedford*, ed. L. J. Flynn. *Bulletin of the American Museum of Natural History*, **279**, 321–347.

Ferrusquía-Villafranca, I., González Guzmán, L. I. and Cartron, J.-L. E. (2005). Northern Mexico's landscape, part I: the physical setting and constraints on modeling biotic evolution. In *Biodiversity, Ecosystems, and Conservation in Northern Mexico*, ed. J.-L. E. Cartron, G. Ceballos and R. S. Felger. New York: Oxford University Press, pp. 11–38.

Flynn, J. J. and Wyss, A. R. (1998). Recent advances in South American mammalian paleontology. *Trends in Ecology and Evolution*, **13**, 449–454.

Frailey, C. D. (1978). An early Miocene (Arikareean) fauna from northcentral Florida (the SB-1A Local Fauna). *Museum of Natural History, University of Kansas, Occasional Papers*, **75**, 1–20.

Frailey, C. D. (1979). The large mammals of the Buda Local Fauna (Arikareean: Alachua County, Florida). *Bulletin of the Florida State Museum, Biological Sciences*, **24**, 123–173.

Galbreath, E. C. (1962). A new myotid bat from the Middle Oligocene of northeastern Colorado. *Transactions of the Kansas Academy of Sciences*, **65**, 448–451.

Gibbard, P. L., Head, M. J., Walker, M. J. C. and the Subcommission on Quaternary Stratigraphy. (2010). Formal ratification of the Quaternary system/period and the Pleistocene series/epoch with a base at 2.58 Ma. *Journal of Quaternary Science*, **25**, 96–102.

Grande, L. (1984). Paleontology of the Green River Formation, with a review of the fish fauna. *Geological Survey of Wyoming Bulletin*, **63**, 1–333.

Gunnell, G. F. and Simmons, N. B. (2005). Fossil evidence and the origin of bats. *Journal of Mammalian Evolution*, **12**, 209–246.

Gunnell, G. F., Simons, E. L. and Seiffert, E. R. (2008). New bats (Mammalia: Chiroptera) from the late Eocene and early Oligocene, Fayum depression, Egypt. *Journal of Vertebrate Paleontology*, **28**, 1–11.

Hall, E. R. (1981). *The Mammals of North America*, vol. 1, 2nd edn. New York: John Wiley & Sons, Inc.

Hanson, C. B. (1996). Stratigraphy and vertebrate faunas of the Bridgerian-Duchesnean Clarno Formation, north-central Oregon. In *The Terrestrial Eocene-Oligocene Transition in North America*, ed. D. R. Prothero and R. J. Emry. Cambridge: Cambridge University Press, pp. 206–239.

Harrison, J. A. (1978). Mammals of the Wolf Ranch Local fauna, Pliocene of the San Pedro Valley, Arizona. *Occasional Papers of the Museum of Natural History, University of Kansas*, **73**, 1–18.

Hayes, F. G. (2000). The Brooksville 2 Local Fauna (Arikareean, latest Oligocene): Hernando County, Florida. *Bulletin of the Florida Museum of Natural History*, **43**, 1–47.

Hibbard, C. W. (1950). Mammals from the Rexroad Formation from Fox Canyon, Kansas. *Contributions from the Museum of Paleontology, University of Michigan*, **8**, 113–192.

Holman, J. A. and Harrison, D. L. (2001). Early Oligocene (Whitneyan) snakes from Florida (USA): remaining boids, indeterminate colubroids, summary and discussion of the I-75 Local Fauna snakes. *Acta Zoologica Cracoviensia*, **44**, 25–36.

Hulbert, R. C., Jr. and Harington, C. R. (1999). An early Pliocene hipparionine horse from the Canadian Arctic. *Palaeontology*, **42**, 1017–1025.

Hulbert, R. C., Jr. and MacFadden, B. J. (1991). Morphological transformation and cladogenesis at the base of the adaptive radiation of Miocene horses. *American Museum Novitates*, **3000**, 1–61.

Hutchison, J. H. and Lindsay, E. H. (1974). The Hemingfordian mammal fauna of the Vedder Locality, Branch Canyon Formation, Santa Barbara County, California. Part I: Insectivora, Chiroptera, Lagomorpha and Rodentia (Sciuridae). *PaleoBios*, **15**, 1–19.

Iturralde-Vinent, M. A. (2006). Meso-Cenozoic Caribbean paleogeography: implications for the historical biogeography of the region. *International Geology Review*, **48**, 791–827.

Iturralde-Vinent, M. A. and MacPhee, R. D. E. (1999). Paleogeography of the Caribbean region: implications for Cenozoic biogeography. *Bulletin of the American Museum of Natural History*, **238**, 1–95.

Iudica, C. A., Arroyo-Cabrales, J., McCarthy, T. J. and Pardiñas, U. F. J. (2003). An insect-eating bat (Mammalia: Chiroptera) from the Pleistocene of Argentina. *Current Research in the Pleistocene*, **20**, 101–103.

James, G. T. (1963). Paleontology and nonmarine stratigraphy of the Cuyama Valley Badlands, California. Part I. Geology, faunal interpretations, and systematic descriptions of Chiroptera, Insectivora, and Rodentia. *University of California Publications in Geological Sciences*, **45**, 1–154.

Jegla, T. C. and Hall, J. S. (1962). A Pleistocene deposit of the free-tailed bat in Mammoth Cave, Kentucky. *Journal of Mammalogy*, **43**, 477–481.

Jepsen, G. L. (1966). Early Eocene bat from Wyoming. *Science*, **154**, 1333–1339.

Jepsen, G. L. (1970). Bat origin and evolution. In *Biology of Bats*, vol. 1, ed. W. A. Wimsatt. New York and London: Academic Press, pp. 1–64.

Jones, D. S., MacFadden, B. J., Webb, S. D. *et al.* (1991). Integrated geochronology of a classic Pliocene fossil site in Florida: linking marine and terrestrial biochronologies. *Journal of Geology*, **99**, 637–648.

Jones, J. K., Jr. (1958). Pleistocene bats from San Josecito Cave, Nuevo León, Mexico. *University of Kansas Publications, Museum of Natural History*, **9**, 389–396.

Kay, R. F., Madden, R. H., Cifelli, R. L. and Flynn, J. J. (eds.) (1997). *Vertebrate Paleontology in the Neotropics: The Miocene Fauna of La Venta, Colombia*. Washington, DC: Smithsonian Institution Press.

Kirby, M. X., Jones, D. S. and MacFadden, B. J. (2008). Lower Miocene stratigraphy along the Panama Canal and its bearing on the Central American peninsula. *PLoS ONE*, **3**, e2791.

Kohn, M. J., Josef, J. A., Madden, R. H. *et al.* (2004). Climate stability across the Eocene-Oligocene transition, southern Argentina. *Geology*, **32**, 621–624.

Koopman, K. F. and Gudmundsson, F. (1966). Bats in Iceland. *American Museum Novitates*, **2262**, 1–6.

Koopman, K. F. and McCracken, G. F. (1998). The taxonomic status of *Lasiurus* (Chiroptera: Vespertilionidae) in the Galapagos Islands. *American Museum Novitates*, **3243**, 1–6.

Krishtalka, L. and Stucky, R. K. (1984). Middle Eocene marsupials (Mammalia) from northeastern Utah and the mammalian fauna from Powder Wash. *Annals of Carnegie Museum*, **53**, 31–45.

Krzanowski, A. (1977). Contribution to the history of bats on Iceland. *Acta Theriologica*, **22**, 272–273.

Kurtén, B. and Anderson, E. (1980). *Pleistocene Mammals of North America*. New York: Columbia University Press.

Lawrence, B. (1943). Miocene bat remains from Florida, with notes on the generic characters of the humerus of bats. *Journal of Mammalogy*, **24**, 356–369.

Lawrence, B. (1960). Fossil *Tadarida* from New Mexico. *Journal of Mammalogy*, **41**, 320–322.

Legendre, S. (1984). Étude odontologique des représentants actuels du groupe *Tadarida* (Chiroptera, Molossidae). Implications phylogéniques, systématiques et zoogéographiques. *Revue Suisse de Zoologie*, **91**, 399–442.

Legendre, S. (1985). Molossidés (Mammalia, Chiroptera) cénozoïques de l'Ancien et du Nouveau Monde; statut systématique; intégration phylogénique de données. *Neues Jahrbuch für Geologie und Paläontologie, Abhandlungen*, **170**, 205–227.

Leonardos, O. H. (1924). Os folhelhos petrolíferos do Vale do Paraíba. *Engenharia, Rio de Janeiro*, **1**, 21–24.

Lessa, G., Cartelle, C. and Fracasso, M. P. de A. (2005). The bats (Mammalia, Chiroptera) fossils of Brazil. *II Congresso Latino-Americano de Paleontologia de Vertebrados, Boletim de Resumos, Rio de Janeiro, Museu Nacional, Serie Livros*, **12**, 153–155.

Lim, B. K. (2007). Divergence times and origin of Neotropical sheath-tailed bats (Tribe Diclidurini) in South America. *Molecular Phylogenetics and Evolution*, **45**, 777–791.

Lim, B. K. (2009). Review of the origins and biogeography of bats in South America. *Chiroptera Neotropical*, **15**, 391–410.

Lim, B. K., Engstrom, M. D., Bickham, J. W. and Patton, J. C. (2008). Molecular phylogeny of New World sheath-tailed bats (Emballonuridae: Diclidurini) based on loci from the four genetic transmission systems in mammals. *Biological Journal of the Linnean Society*, **93**, 189–209.

Linares, O. (1968). Quirópteros subfósiles encontrados en las cuevas Venezolanas. Parte 1. Depósito de la Cueva de Quebrada Honda. *Boletín de la Sociedad Venezolana de Espeleología*, **1**, 119–145.

Lindsay, E. H. (1972). Small mammal fossils from the Barstow Formation, California. *University of California Publications in Geological Sciences*, **93**, 1–104.

Lindsay, E. H. and Jacobs, L. L. (1985). Pliocene small mammal fossils from Chihuahua, Mexico. *Universidad Nacional Autónoma de México, Instituto de Geologia, Paleontología Mexicana*, **51**, 1–50.

Lomolino, M. V., Riddle, B. R. and Brown, J. H. (2006). *Biogeography*, 3rd edn. Sunderland, MA: Sinauer Associates.

Lundberg, J. and McFarlane, D. A. (2006). A minimum age for canyon incision and for the extinct molossid bat, *Tadarida constantinei*, from Carlsbad Caverns National Park, New Mexico. *Journal of Cave and Karst Studies*, **68**, 115–117.

MacFadden, B. J. (2006). North American Miocene land mammals from Panama. *Journal of Vertebrate Paleontology*, **26**, 720–734.

MacFadden, B. J. and Morgan, G. S. (2003). New oreodont (Mammalia, Artiodactyla) from the late Oligocene (early Arikareean) of Florida. *Bulletin of the American Museum of Natural History*, **279**, 368–396.

Madden, R. H., Bellosi, E., Carlini, A. A. *et al.* (2005). Geochronology of the Sarmiento Formation at Gran Barranca and elsewhere in Patagonia: calibrating middle Cenozoic mammal evolution in South America. *Actas 16º Congreso Geológico Argentino, La Plata, Tomo*, **4**, 411–412.

Mancina, C. A. and García-Rivera, L. (2005). A new genus and species of fossil bat (Chiroptera: Phyllostomidae) from Cuba. *Caribbean Journal of Science*, **41**, 22–27.

Mantilla-Meluk, H., Jiménez, A. M. and Baker, R. J. (2009). Phyllostomid bats of Colombia: annotated checklist, distribution, and biogeography. *Special Publications Museum of Texas Tech University*, **56**, 1–37.

Martin, J. E. (1983). Additions to the early Hemphillian (Miocene) rattlesnake fauna from central Oregon. *Proceedings of the South Dakota Academy of Science*, **62**, 23–33.

Martin, R. A. (1977). Late Pleistocene *Eumops* from Florida. *Bulletin of the New Jersey Academy of Science*, **22**, 18–19.

McKenna, M. C., Robinson, P. and Taylor, D. W. (1962). Notes on Eocene Mammalia and Mollusca from Tabernacle Butte, Wyoming. *American Museum Novitates*, **2102**, 1–33.

Merino, M. L., Lutz, M. A., Verzi, D. H. and Tonni, E. P. (2007). The fishing bat *Noctilio* (Mammalia, Chiroptera) in the middle Pleistocene of central Argentina. *Acta Chiropterologica*, **9**, 401–407.

Mezzalira, S. (1966). Os fósseis do Estado de São Paulo. *São Paulo, Instituto Geográfico do Geológico, Boletim*, **45**, 1–132.

Morgan, G. S. (1989). Fossil Chiroptera and Rodentia from the Bahamas, and the historical biogeography of the Bahamian mammal fauna. In *Biogeography of the West Indies: Past, Present, and Future*, ed. C. A. Woods. Gainesville, FL: Sandhill Crane Press, pp. 685–740.

Morgan, G. S. (1991). Neotropical Chiroptera from the Pliocene and Pleistocene of Florida. *Bulletin of the American Museum of Natural History*, **206**, 176–213.

Morgan, G. S. (1993). Mammalian biochronology and marine-nonmarine correlations in the Neogene of Florida. In *The Neogene of Florida and Adjacent Areas*, ed. V. A. Zullo, W. B. Harris, T. M. Scott and R. W. Portell. *Florida Geological Survey, Special Publication*, **37**, 55–66.

Morgan, G. S. (1994). Late Quaternary fossil vertebrates from the Cayman Islands. In *The Cayman Islands: Natural History and Biogeography*, ed. M. A. Brunt and J. E. Davies. Dordrecht, Netherlands: Kluwer Academic Publishers, pp. 465–508.

Morgan, G. S. (2001). Patterns of extinction in West Indian bats. In *Biogeography of the West Indies: Patterns and Perspectives*, ed. C. A. Woods and F. E. Sergile. Boca Raton, FL: CRC Press, pp. 369–407.

Morgan, G. S. (2002). Late Rancholabrean mammals from southernmost Florida, and the Neotropical influence in Florida Pleistocene faunas. *Smithsonian Contributions to Paleobiology*, **93**, 15–38.

Morgan, G. S. (2003). The extinct Pleistocene molossid bat *Tadarida constantinei* from Slaughter Canyon Cave, Carlsbad Caverns National Park, New Mexico. *Bat Research News*, **44**, 159–160.

Morgan, G. S. (2005). The Great American Biotic Interchange in Florida. *Bulletin of the Florida Museum of Natural History*, **45**, 271–311.

Morgan, G. S. (2008). Vertebrate fauna and geochronology of the Great American Biotic Interchange in North America. *New Mexico Museum of Natural History and Science Bulletin*, **44**, 93–140.

Morgan, G. S. and Czaplewski, N. J. (1999). First fossil record of *Amorphochilus schnablii* (Chiroptera: Furipteridae) from the late Quaternary of Peru. *Acta Chiropterologica*, **1**, 75–79.

Morgan, G. S. and Czaplewski, N. J. (2003). A new bat (Chiroptera: Natalidae) from the early Miocene of Florida, with comments on natalid phylogeny. *Journal of Mammalogy*, **84**, 729–752.

Morgan, G. S. and Emslie, S. D. (2010). Tropical and western influences in vertebrate faunas from the Pliocene and Pleistocene of Florida. *Quaternary International*, **217**, 143–158.

Morgan, G. S. and Hulbert, R. C., Jr. (2008). Cenozoic vertebrate fossils from paleokarst deposits in Florida. In *Caves and Karst of Florida*. *National Speleological Society Convention Guidebook*, **2008**, pp. 248–271.

Morgan, G. S. and Pratt, A. E. (1988). An early Miocene (late Hemingfordian) vertebrate fauna from Brooks Sink, Bradford County, Florida. In *Heavy Mineral Mining in Northeast Florida and an Examination of the Hawthorne Formation and Post-Hawthorne Clastic Sediments*, ed. F. L. Pirkle and J. G. Reynolds. *Southeastern Geological Society Guidebook*, **29**, 53–69.

Morgan, G. S., Linares, O. J. and Ray, C. E. (1988). New species of fossil vampire bats (Mammalia: Chiroptera: Desmodontidae) from Florida and Venezuela. *Proceedings of the Biological Society of Washington*, **101**, 912–928.

Olsen, S. J. (1964). The stratigraphic importance of a lower Miocene vertebrate fauna from Florida. *Journal of Paleontology*, **38**, 477–482.

Olson, D. M., Dinerstein, E., Wikramanayake, E. D. *et al.* (2001). Terrestrial ecoregions of the world: a new map of life on Earth. *BioScience*, **51**, 933–938.

Ortega, J. and Arita, H. T. (1998). Neotropical-Nearctic limits in Middle America as determined by distributions of bats. *Journal of Mammalogy*, **79**, 772–783.

Ostrander, G. E. (1983). New early Oligocene (Chadronian) mammals from the Raben Ranch local fauna, northwest Nebraska. *Journal of Paleontology*, **57**, 128–139.

Ostrander, G. E. (1985). Correlation of the early Oligocene (Chadronian) in northwestern Nebraska. *South Dakota School of Mines and Technology, Dakoterra*, **2**, 205–231.

Pacheco, V., Cadenillas, R., Salas, E., Tello, C. and Zeballos, H. (2009). Diversity and endemism of Peruvian mammals. *Revista Peruviana de Biología*, **16**, 5–32.

Pardiñas, U. F. J. and Tonni, E. P. (2000). A giant vampire (Mammalia, Chiroptera) in the late Holocene from the Argentinean pampas: paleoenvironmental significance. *Palaeogeography, Palaeoclimatology, and Palaeoecology*, **160**, 213–221.

Pascual, R. (2006). Evolution and geography: the biogeographic history of South American land mammals. *Annals of the Missouri Botanical Garden*, **93**, 209–230.

Patton, T. H. (1969). An Oligocene land vertebrate fauna from Florida. *Journal of Paleontology*, **43**, 543–546.

Paula Couto, C. de. (1956). Une chauve-souris fossile des argiles feuilletées Pléistocènes de Tremembé, État de São Paulo (Brésil). *Actes IV Congrès International du Quaternaire, Roma-Pise*, Août-Septembre 1953, vol. 1, 343–347.

Paula Couto, C. de and Mezzalira, S. (1971). Nova conceituação geocronológica de Tremembé, Estado de São Paulo, Brasil. *Anais da Academia Brasileira de Ciências*, **43** (Suppl.), 473–488.

Polyak, V. J., Asmerom, Y. and Rasmussen, J. B. T. (2006). Old bat guano in Slaughter Canyon Cave. *New Mexico Geological Society Guidebook*, **57**, 23–24.

Pratt, A. E. (1989). Taphonomy of the microvertebrate fauna from the early Miocene Thomas Farm locality, Florida (U.S.A.). *Palaeogeography, Palaeoclimatology, and Palaeoecology*, **76**, 125–151.

Pratt, A. E. (1990). Taphonomy of the large vertebrate fauna from the Thomas Farm locality (Miocene, Hemingfordian), Gilchrist County, Florida. *Bulletin of the Florida Museum of Natural History, Biological Sciences*, **35**, 35–130.

Pregill, G. K., Steadman, D. W., Olson, S. L. and Grady, F. V. (1988). Late Holocene fossil vertebrates from Burma Quarry, Antigua, Lesser Antilles. *Smithsonian Contributions to Zoology*, **463**, 1–27.

Pregill, G. K., Steadman, D. W. and Watters, D. R. (1994). Late Quaternary vertebrate faunas of the Lesser Antilles: historical components of Caribbean biogeography. *Bulletin of the Carnegie Museum of Natural History*, **30**, 1–51.

Presley, S. J. and Willig, M. R. (2008). Composition and structure of Caribbean bat (Chiroptera) assemblages: effects of inter-island distance, area, elevation and hurricane-induced disturbance. *Global Ecology and Biogeography*, **17**, 747–757.

Proches, S. (2005). The world's biogeographical regions: cluster analysis based on bat distributions. *Journal of Biogeography*, **32**, 607–614.

Prothero, D. R. and Emry, R. J. (2004). The Chadronian, Orellan, and Whitneyan North American land mammal ages. In *Late Cretaceous and Cenozoic Land Mammals of North America: Biostratigraphy and Geochronology*, ed. M. O. Woodburne. New York: Columbia University Press, pp. 156–168.

Ray, C. E., Linares, O. J. and Morgan, G. S. (1988). Paleontology. In *Natural History of Vampire Bats*, ed. A. M. Greenhall and U. Schmidt. Boca Raton, FL: CRC Press, Inc, pp. 19–30.

Rincón, A. D. (1999). Los pequeños mamíferos subfósiles presentes en cuevas de la Sierra de Perijá, estado Zulia, Venezuela. *El Guácharo, Boletín Espeología*, **48**, 1–75.

Rincón, A. D. (2001). Quirópteros subfósiles presentes en los depósitos de guano de la Cueva de Los Murciélagos, Isla de Toas, estado Zulia, Venezuela. *Anartia, Publicacion Ocasional Museo Biología, Universidad de Zulia*, **13**, 1–13.

Robinson, P., Gunnell, G. F., Walsh, S. L. *et al.* (2004). Wasatchian through Duchesnean biochronology. In *Late Cretaceous and Cenozoic Land Mammals of North America: Biostratigraphy and Geochronology*, ed. M. O. Woodburne. New York: Columbia University Press, pp. 106–155.

Salles, L. O., Cartelle, C., Guedes, P. G. *et al.* (2006). Quaternary mammals from Serra da Bodoquena, Mato Grosso do Sul, Brazil. *Boletim do Museu Nacional (Rio de Janeiro)*, **521**, 1–12.

Savage, D. E. (1951). A Miocene phyllostomatid bat from Colombia, South America. *University of California Publications, Bulletin of the Department of Geological Sciences*, **28**, 357–366.

Schiebout, J. A. (1997). The Fort Polk Miocene microvertebrate sites compared to those from east Texas. *Texas Journal of Science*, **49**, 23–32.

Shockey, B. J, Salas-Gismondi, R., Baby, P. *et al.* (2009). New Pleistocene cave faunas of the Andes of central Perú: radiocarbon ages and the survival of low latitude, Pleistocene DNA. *Palaeontologia Electronica*, **12**, 1–15.

Shotwell, J. A. (1956). Hemphillian mammalian assemblage from northeastern Oregon. *Geological Society of America Bulletin*, **67**, 717–738.

Shotwell, J. A. (1970). Pliocene mammals of southeast Oregon and adjacent Idaho. *Museum of Natural History, University of Oregon, Bulletin*, **17**, 1–103.

Sigé, B. (1974). Données nouvelles sur le genre *Stehlinia* (Vespertilionoidea, Chiroptera) du Paléogène d'Europe. *Palaeovertebrata*, **6**, 253–272.

Sigé, B. (1997). Les remplissages karstiques polyphasés (Éocène, Oligocène, Pliocène) de Saint-Maximin (Phosphorites du Gard) et leur apport à la connaissance des faunes européennes, notamment pour l'Éocène Moyen (MP 13). In *3.B Systématique: Euthériens Entomophages*, ed. L.-P. Aguilar, S. Legendre and J. Michaux. *Actes du Congrès BiochroM'97, Mémoires et Travaux de l'Ecole Pratique des Hautes Études, Institut de Montpellier*, **21**, 737–750.

Sigé, B. and Legendre, S. (1983). L'histoire des peuplements de chiroptères du Bassin Méditerranéen: L'apport comparé des remplissages karstiques et des dépôts fluvio-lacustres. *Mémoires de Biospéologie*, **10**, 209–225.

Sigé, B., Thomas, H., Sen, S., Gheerbrant, E., Roger, J. and Al-Sulaimani, Z. (1994). Les Chiroptères de Taqah (Oligocène Inférieur, Sultanat D'Oman). Premier inventaire systématique. *Münchner Geowissenschaftliche Abhandlungen*, **26**, 35–48.

Sigé, B., Crochet, J.-Y., Sudre, J., Aguilar, J.-P. and Escarguel, G. (1997). Nouveaux sites d'âges variés dans les remplissages karstiques du Miocène inférieur de Bouzigues (Hérault, Sud de la France). Partie I: sites et faunes 1 (Insectivores, Chiroptères, Artiodactyles). *Geobios*, **20**, 477–483.

Silva Taboada, G. (1974). Fossil Chiroptera from cave deposits in central Cuba, with description of two new species (Genera *Pteronotus* and *Mormoops*), and the first West Indian record of *Mormoops megalophylla. Acta Zoologica Cracoviensia*, **19**, 33–73.

Silva Taboada, G. (1979). *Los Murciélagos de Cuba*. La Habana, Cuba: Editorial Academia.

Simmons, N. B. (2005). Order Chiroptera. In *Mammal Species of the World: A Taxonomic and Geographic Reference*, ed. D. E. Wilson and D. M. Reeder. Baltimore, MD: Johns Hopkins University Press, pp. 312–529.

Simmons, N. B. and Geisler, J. H. (1998). Phylogenetic relationships of *Icaronycteris, Archaeonycteris, Hassianycteris*, and *Palaeochiropteryx* to extant bat lineages, with comments on the evolution of echolocation and foraging strategies in Microchiroptera. *Bulletin of the American Museum of Natural History*, **235**, 1–182.

Simmons, N. B., Seymour, K. L., Habersetzer, J. and Gunnell, G. F. (2008). Primitive early Eocene bat from Wyoming and the evolution of flight and echolocation. *Nature*, **451**, 818–822.

Slaughter, B. H. (1981). A new genus of geomyoid rodent from the Miocene of Texas and Panama. *Journal of Vertebrate Paleontology*, **1**, 111–115.

Sosdian, S. and Rosenthal, Y. (2009). Deep sea temperature and ice volume changes across the Pliocene-Pleistocene climate transitions. *Science*, **325**, 306–310.

Stehli, F. G. and Webb, S. D., eds. (1985). *The Great American Biotic Interchange*. New York: Plenum Press.

Stirton, R. A. (1931). A new genus of the family Vespertilionidae from the San Pedro Pliocene of Arizona. *University of California Publications, Bulletin of the Department of Geological Sciences*, **20**, 27–30.

Storer, J. E. (1984). Mammals of the Swift Current Creek local fauna (Eocene: Uintan, Saskatchewan). *Saskatchewan Culture and Recreation, Museum of Natural History, Natural History Contribution*, **7**, 1–158.

Storer, J. E. (1995). Small mammals of the Lac Pelletier lower fauna, Duchesnean, of Saskatchewan, Canada: insectivores and insectivore-like groups, a plagiomenid, a microsyopid and Chiroptera. In *Vertebrate Fossils and the Evolution of Scientific Concepts. Writings in Tribute to Beverly Halstead by Some of his Many Friends*, ed. W. A. S. Sarjeant. United Kingdom: Gordon and Breach Publishers, pp. 595–615.

Storer, J. E. (2002). Small mammals of the Kealey Springs local fauna (Early Arikareean; late Oligocene) of Saskatchewan. *Paludicola*, **3**, 105–133.

Suárez, W. (2005). Taxonomic status of the Cuban vampire bat (Chiroptera: Phyllostomidae: Desmodontinae: *Desmodus*). *Caribbean Journal of Science*, **41**, 761–767.

Suárez, W. and Díaz-Franco, S. (2003). A new fossil bat (Chiroptera: Phyllostomidae) from a Quaternary cave deposit in Cuba. *Caribbean Journal of Science*, **39**, 371–377.

Sutton, J. F. and Genoways, H. H. (1974). A new vespertilionine bat from the Barstovian deposits of Montana. *Occasional Papers, The Museum, Texas Tech University*, **20**, 1–8.

Tabrum, A. R., Prothero, D. R. and Garcia, D. (1996). Magnetostratigraphy and biostratigraphy of the Eocene-Oligocene transition, southwestern Montana. In *The Terrestrial Eocene-Oligocene Transition in North America*, ed. D. R. Prothero and R. J. Emry. Cambridge: Cambridge University Press, pp. 278–311.

Tedford, R. H. and Harington, C. R. (2003). An Arctic mammal fauna from the early Pliocene of North America. *Nature*, **425**, 388–390.

Tedford, R. H., Albright, L. B., III, Barnosky, A. D. *et al.* (2004). Mammalian biochronology of the Arikareean through Hemphillian interval (late Oligocene through earliest Pliocene epochs). In *Late Cretaceous and Cenozoic Mammals of North America: Biostratigraphy and Geochronology*, ed. M. O. Woodburne. New York: Columbia University Press, pp. 169–231.

Teeling, E. C., Springer, M. S., Madsen, O. *et al.* (2005). A molecular phylogeny for bats illuminates biogeography and the fossil record. *Science*, **307**, 580–584.

Tejedor, A. (2011). Systematics of funnel-eared bats (Chiroptera: Natalidae). *Bulletin of the American Museum of Natural History*, **353**, 1–140.

Tejedor, A., Tavares, V. da C. and Silva-Taboada, G. (2005). A revision of the extant Great Antillean bats of the genus *Natalus*. *American Museum Novitates*, **3493**, 1–22.

Tejedor, M. F., Czaplewski, N. J., Goin, F. J. and Aragón, E. (2005). The oldest record of South American bats. *Journal of Vertebrate Paleontology*, **25**, 990–993.

Tejedor, M. F., Goin, F. J., Gelfo, J. N. *et al.* (2009). New early Eocene mammalian fauna from western Patagonia, Argentina. *American Museum Novitates*, **3638**, 1–43.

Thewissen, J. G. M. and Smith, G. R. (1987). Vespertilionid bats (Chiroptera, Mammalia) from the Pliocene of Idaho. *Contributions from the Museum of Paleontology, University of Michigan*, **27**, 237–245.

Timm, R. M. and Genoways, H. H. (2004). The Florida bonneted bat, *Eumops floridanus* (Chiroptera: Molossidae): distribution, morphometrics, systematics, and ecology. *Journal of Mammalogy*, **85**, 862–865.

Tomida, Y. (1987). *Small Mammal Fossils and Correlation of Continental Deposits, Safford and Duncan Basins, Arizona, USA*. Tokyo: National Science Museum, pp. 1–141.

Trajano, E. and Vivo, M. de. (1991). *Desmodus draculae* Morgan, Linares, and Ray, 1988, reported for southeastern Brasil, with paleoecological comments (Phyllostomidae, Desmodontinae). *Mammalia*, **55**, 456–460.

Wang, X. (2003). New material of *Osbornodon* from the early Hemingfordian of Nebraska and Florida. In *Vertebrate Fossils and their Context – Contributions in Honor of Richard H. Tedford*, ed. L. J. Flynn. *Bulletin of the American Museum of Natural History*, **279**, 163–176.

Webb, S. D. (1976). Mammalian faunal dynamics of the Great American Interchange. *Paleobiology*, **2**, 220–234.

Webb, S. D. (2006). The Great American Biotic Interchange: patterns and processes. *Annals of the Missouri Botanical Garden*, **93**, 245–257.

Webb, S. D. and Perrigo, S. C. (1984). Late Cenozoic vertebrates from Honduras and El Salvador. *Journal of Vertebrate Paleontology*, **4**, 237–254.

Webb, S. D., MacFadden, B. J. and Baskin, J. A. (1981). Geology and paleontology of the Love Bone Bed from the late Miocene of Florida. *American Journal of Science*, **281**, 513–544.

Webb, S. D., Beatty, B. L. and Poinar, G., Jr. (2003). New evidence of Miocene Protoceratidae including a new species from Chiapas, Mexico. *Bulletin of the American Museum of Natural History*, **279**, 348–367.

Wesselman, H. B. (1984). The Omo micromammals – systematics and paleoecology of early man sites from Ethiopia. In *Contributions to Vertebrate Evolution*, ed. M. K. Hecht and F. S. Szalay. Basel: Karger, pp. 1–219.

White, J. A. (1969). Late Cenozoic bats (subfamily Nyctophylinae) from the Anza-Borrego Desert of California. *University of Kansas Museum of Natural History, Miscellaneous Publications*, **51**, 275–282.

Whitmore, F. C., Jr. and Stewart, R. H. (1965). Miocene mammals and Central American seaways. *Science*, **148**, 180–185.

Wilson, R. L. (1968). Systematics and faunal analysis of a lower Pliocene vertebrate assemblage from Trego County, Kansas. *Contributions from the Museum of Paleontology, University of Michigan*, **22**, 75–126.

Winge, H. (1893). Jordfunde og nulevende Flagermus (Chiroptera) fra Lagoa Santa, Minas Gerais, Brasilien. *Med udsigt over Flagermusenes undbyrdes Slaegstkab. E Museo Lundii*, **2**, 1–92.

Woloszyn, B. W. and Mayo, N. A. (1974). Postglacial remains of a vampire bat (Chiroptera: *Desmodus*) from Cuba. *Acta Zoologica Cracoviensia*, **19**, 253–265.

Wooburne, M. O., ed. (2004a). *Late Cretaceous and Cenozoic Land Mammals of North America: Biostratigraphy and Geochronology*. New York: Columbia University Press.

Woodburne, M. O. (2004b). Global events and the North American mammalian biochronology. In *Late Cretaceous and Cenozoic Land Mammals of North America: Biostratigraphy and Geochronology*, ed. M. O. Woodburne. New York: Columbia University Press, pp. 315–343.

Woodburne, M. O. and Swisher, C. C., III. (1995). Land mammal high-resolution geo-chronology, intercontinental overland dispersals, sea level, climate, and vicariance. *Society of Economic Paleontologists and Mineralogists, Special Publication*, **54**, 335–364.

Woodburne, M. O., Cione, A. L. and Tonni, E. P. (2006). Central American provincialism and the Great American Biotic Interchange. *Universidad Autónoma de México, Instituto de Geología and Centro de Geociencias, Publicación Especial*, **4**, 73–101.

Woodburne, M. O., Gunnell, G. F. and Stucky, R. K. (2009). Climate directly influences Eocene mammal faunal dynamics in North America. *Proceedings of the National Academy of Sciences*, USA, **106**, 13399–13403.

Zachos, J. C., Breza, J. R. and Wise, S. W. (1992). Early Oligocene ice-sheet expansion on Antarctica. *Geology*, **20**, 569–573.

Zachos, J. C., Dickens, G. R. and Zeebe, R. E. (2008). An early Cenozoic perspective on greenhouse warming and carbon-cycle dynamics. *Nature*, **451**, 279–283.

5

New basal noctilionoid bats (Mammalia: Chiroptera) from the Oligocene of subtropical North America

NICHOLAS J. CZAPLEWSKI AND GARY S. MORGAN

5.1 Introduction

Bats (Chiroptera) are generally considered to be monophyletic based on morphological and molecular data (Simmons, 1998; Gunnell and Simmons, 2005; Teeling *et al.*, 2005), but the relationships among the families, especially extinct families, are not well resolved (Simmons and Geisler, 1998; Gunnell and Simmons, 2005). Recent molecular phylogenetic work suggests that one group of bats, the Noctilionoidea, consists of a monophyletic clade including at least the families Mystacinidae, Mormoopidae, Noctilionidae and Phyllostomidae (Pierson *et al.*, 1986; Kirsch *et al.*, 1998; Kennedy *et al.*, 1999; Van Den Bussche and Hoofer, 2000; Teeling *et al.*, 2003; Hutcheon and Kirsch, 2004), and probably also the families Thyropteridae, Furipteridae and Myzopodidae (Hoofer *et al.*, 2003; Van Den Bussche and Hoofer, 2004; Teeling *et al.*, 2005; Miller-Butterworth *et al.*, 2007), although Hoofer *et al.* (2003) explicitly excluded Myzopodidae. Gunnell and Simmons (2005) found morphological data supporting a more restricted Noctilionoidea composed of the first four families, Mystacinidae, Noctilionidae, Phyllostomidae and Mormoopidae, but which is sister to a clade composed of Myzopodidae, Thyropteridae, Furipteridae and Natalidae. Early fossils of noctilionoid bats are scarce; reviewed below are some pre-Pleistocene records of noctilionoids and putative noctilionoids as fossils.

The oldest Paleogene bat fossils known from South America are two isolated teeth from the Early Eocene of Chubut, Argentina, that could potentially represent a noctilionoid (Tejedor *et al.*, 2005, 2009), but the specimens are actually insufficient to realize the phylogenetic affinities of the taxon they exemplify. A possible ?bat represented by a single broken tooth of uncertain but possibly Eocene age from Santa Rosa, Peru, has a dental character rather

Evolutionary History of Bats: Fossils, Molecules and Morphology, ed. G. F. Gunnell and
N. B. Simmons. Published by Cambridge University Press. © Cambridge University Press 2012.

similar to one that is unique to noctilionids, but this specimen, too, is unsubstantial (Czaplewski and Campbell, 2004).

The distal humerus of a bat from the Fayum Depression (Late Eocene-Early Oligocene) in Egypt, named *Vampyravus orientalis* by Schlosser (1910, 1911), was considered by him to represent a bat irrefutably with a relationship to the Phyllostomatoidea (= Noctilionoidea). The bone was nicely redescribed and compared in detail by Sigé (1985), who determined that the type humerus is "not sufficient for sure classification, nor could it [prove] an African origin of the Phyllostomatoidea." Gunnell *et al.* (2008) suggested instead that the *Vampyravus* humerus may represent the same taxon as the large Fayum bat *Philisis sphingis*, a philisid not a noctilionoid; even more recently, Gunnell *et al.* (2009) determined that *Vampyravus* was distinct from all other Fayum bats, including philisids, and proposed that it might represent one of two extant bat families, Emballonuridae or Rhinopomatidae.

The endemic New Zealand family Mystacinidae presently consists of only one surviving species (another species is recently extinct), but Hand (Hand *et al.*, 1998, 2001, 2005, 2007) described three fossil species of an extinct genus *Icarops* from the Early Miocene, and an indeterminate mystacinid from the Late Oligocene of Australia, plus at least one mystacinid from the Early Miocene of New Zealand.

The Mormoopidae are represented in the pre-Pleistocene by one genus and species from the Oligocene of Florida, in the I-75 (Whitneyan) and Brooksville 2 (Early Arikareean) local faunas (Morgan, unpublished data). These are by far the earliest record of this family, and the fossils represent a Paleogene mormoopid that is in several ways intermediate between the two extant genera of the family, *Pteronotus* and *Mormoops*.

The oldest certain Phyllostomidae are from the Neogene of South America. The earliest record is a single tooth of a phyllostomine from the Early Miocene (Colhuehuapian) of Argentina at about 20.2–20.0 Ma (Czaplewski, 2010). In the Middle Miocene (Laventan) of Colombia, several phyllostomids are represented in the La Venta fauna; these include three Phyllostominae, *Notonycteris* (two species) and a *Tonatia* or *Lophostoma*, as well as a nectar-feeding phyllostomid *Palynephyllum* (Savage, 1951; Czaplewski, 1997; Czaplewski *et al.*, 2003b). Three previously reported Neogene "?phyllostomids" from California, USA (James, 1963; Hutchison and Lindsay, 1974) are taxonomically indeterminate. One of these, University of California Museum of Paleontology (UCMP) 54572, is from Big Cat Quarry (UCMP locality V5847; Clarendonian) in the Caliente Formation, Cuyama Valley Badlands, Ventura County, California. This tooth fragment with a cusp was referred to "Phyllostomatidae?" and designated as the anterior two-thirds of a right p2; it was likened to the same tooth in the large phyllostomids *Phyllostomus*, *Vampyrum*, *Chrotopterus* and

Notonycteris (James, 1963, p. 29). However, the few features preserved on the fragment do not precisely match this tooth in any of these genera. We believe its incompleteness precludes definitive assignment to a family (or even to tooth locus) and it may not even represent a bat. Hutchison and Lindsay (1974) referred two other specimens to "Phyllostomatidae incertae sedis." These are UCMP 82144, left cı, and UCMP 80324, edentulous right dentary in two pieces, both specimens from V6761, Vedder locality, Branch Canyon Formation (Hemingfordian), Santa Barbara County, California. The specimens were not associated and, as noted by Hutchison and Lindsay (1974, p. 7), "may represent more than one taxon." The cı is distinctive for a bat in that the main cusp is very short (low-crowned). It does not match exactly the cı of any known bat, but alone is insufficient to establish a diagnosis for a new taxon. In the dentary fragments, the posterior piece is not very informative, but the anterior piece shows alveoli for three lower premolars, a single-rooted p2, a very small single-rooted p3, displaced lingually from the toothrow, and a two-rooted p4. As already noted by Hutchison and Lindsay (1974, p. 8), this alveolar configuration and other characteristics of the bone resemble those of the mormoopid *Pteronotus* and the phyllostomids *Macrophyllum* and *Lonchorhina*. However, lacking teeth or more complete specimens, little meaningful information can be gleaned from the fragments.

Finally, the earliest known vampire bat, *Desmodus archaeodaptes* (Phyllostomidae, Desmodontinae) is from the Late Pliocene (Late Blancan) of Florida, USA; it is as highly derived as extant vampires (Morgan *et al.*, 1988; Morgan, 1991) and is similarly distinct from dentally more typical noctilionoids.

The family Noctilionidae are derived in dental morphology; one of the two extant species, *Noctilio albiventris*, appeared as a fossil in the Middle Miocene La Venta fauna of Colombia about 13–12 million years ago (Czaplewski, 1997; Czaplewski *et al.*, 2003b). An extinct species *Noctilio lacrimaelunaris* from the Late Miocene (Huayquerian) of the Río Acre in western Amazonia is dentally as derived as extant noctilionids (Czaplewski, 1996a) and is distinguished from the other two named species only by its small size.

Like the Noctilionidae, the fossil record of Thyropteridae extends back to the Middle Miocene La Venta fauna, Colombia, with two species, *Thyroptera* cf. *T. tricolor* and *T. lavali*, both of which still survive in the modern Neotropical fauna (Czaplewski, 1996b, 1997; Czaplewski *et al.*, 2003b).

The families Furipteridae in the Neotropics and Myzopodidae in the Afrotropics have no fossil record in the Paleogene or Neogene. Their records extend back only to the Quaternary period. Fossils of the extant furipterid *Amorphochilus schnablii* have been found in a Holocene or Late Pleistocene context in Peru, whereas the other extant furipterid *Furipterus horrens* occurred in Late Pleistocene or Holocene cave deposits in Brazil. Myzopodidae, in terms of

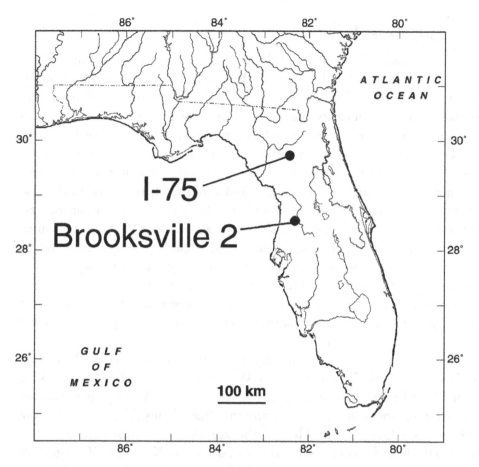

Figure 5.1 Map of Florida showing localities yielding bat fossils discussed herein.

molecular systematics among the most basal noctilionoids, is a small family of two extant species of *Myzopoda*. They presently persist only in Madagascar; however, Butler (1978) mentioned an Early Pleistocene fossil humerus of a *Myzopoda* from Olduvai, Tanzania, Africa that is larger than the extant species *Myzopoda aurita*. No Paleogene or Neogene fossils of myzopodids are yet known, and unfortunately even details of the African Pleistocene record are unpublished.

Given this record of mostly isolated teeth and fragments, it is understandable that the early evolutionary history of noctilionoids is weak. We hope to improve this situation somewhat through the description of two new species of bats from southeastern North America that hint at the basal radiation of Noctilionoidea. The fossils come from Oligocene sites in paleokarst (fissure fillings, sinkholes or deposits that represent filled paleocaves) of peninsular Florida that yielded the I-75 and Brooksville 2 local faunas (Figure 5.1).

5.2   Fossil localities

I-75 Local Fauna – The I-75 LF was discovered in 1965 during the construction of Interstate Highway 75 through Gainesville, Alachua County, in northern Florida. The I-75 site consisted of fossiliferous clays deposited in a small karst solution cavity about 5 m in diameter and 2 m deep, developed in marine Eocene limestone. Considering the small size of the fossil deposit, the I-75 site has a diverse vertebrate fauna containing about 45 species (Patton, 1969). The large mammal fauna is composed of two carnivores, including the amphicyonid *Daphoenus* and a mustelid, the three-toed horse *Miohippus*, a tayassuid, an oreodont and the small ruminant artiodactyl *Leptomeryx*. The leptomerycid is the most abundant species in the fauna. I-75 also has a rich microvertebrate fauna, including frogs, salamanders, lizards, snakes and small mammals (Patton, 1969; Holman, 1999). Small mammals include: the didelphid marsupial *Herpetotherium* cf. *H. merriami*; the large insectivore *Centetodon* cf. *C. wolffi*; six species of bats; the rabbit *Palaeolagus*; and at least four species of rodents, *Eutypomys* sp., a heteromyid and two eomyids.

Most previous workers (Patton, 1969; Emry *et al.*, 1987) placed the I-75 LF in the Whitneyan NALMA (32 to 30 Ma; Woodburne and Swisher, 1995), making it the oldest land vertebrate fauna known from Florida. Whitneyan faunas are otherwise unknown outside of the northern Great Plains (Tedford *et al.*, 1996; Prothero and Emry, 2004). Most of the age-diagnostic mammals from I-75 occur in the Orellan, Whitneyan and Early Arikareean NALMAs. The presence of *Centetodon*, *Eutypomys*, *Leptomeryx*, *Miohippus* and *Palaeolagus* establishes an Early Arikareean or older age (more than 24 Ma; Tedford *et al.*, 1987, 1996). A post-Orellan age is suggested by the presence of an advanced species of *Miohippus* near *M. intermedius*. The similarity of the bat faunas from I-75 and the Late Early Arikareean (28–24 Ma) Brooksville 2 LF from central Florida (Hayes, 2000) suggests that these two faunas are fairly close in age. Although the fauna contains quite a few taxa that make their last appearance in the Early Arikareean, this site does not contain any genera that are restricted to the Arikareean (e.g., *Nanotragulus*). Thus the I-75 site can be considered Late Whitneyan in age (Early Oligocene, about 30 Ma).

The I-75 site provides the earliest record of the taxonomic diversity and community structure of North American Middle Cenozoic chiropteran faunas. In addition to the two noctilionoids described here, five other species of bats are known from I-75, including a new genus and species of Mormoopidae, an undescribed genus and two new species of Emballonuridae, an indeterminate natalid and a large species of Vespertilionidae (Morgan and Czaplewski, 2003;

Figure 5.2 Photographs of Brooksville 2 locality and fissure fillings at time of collection (1970s). (A) Sublocality 1A; man holding onto rope is pointing at fissure filling. (B) and (C) Close-up views of same fissure filling as in (A). (Photos courtesy of Glynn Hayes.)

Morgan, unpublished data). Species belonging to families now restricted to the Neotropics are the most abundant bats in the I-75 fauna, including a large emballonurid and a small mormoopid.

Brooksville 2 – The Brooksville 2 site was discovered in 1994 in an abandoned limestone quarry about 10 km northeast of Brooksville in Hernando County in central Florida (Figures 5.2–5.3). The site consisted of clays and sands filling karst solution features in the marine lower Oligocene Suwannee Limestone (Hayes, 2000). FLMNH field crews visited the site on a number of occasions between April 1994 and February 1995. Vertebrate fossils were found scattered on the surface; however, because of the abundance of microvertebrates and the rarity of larger mammals, the fossils were collected primarily by screenwashing. A total of approximately 500 kg of sediments were collected and washed from the Brooksville 2 site, constituting a large portion of the fossiliferous sediments contained in the solution features (Hayes, 2000).

The vertebrate assemblage from Brooksville 2 is composed predominantly of small terrestrial vertebrates, including frogs, lizards, snakes and a diverse fauna of small mammals. Medium-sized mammals are represented primarily by

Figure 5.3 Photographs of Brooksville 2 locality at the time of collection (1970s). (A) Sublocality 1B in floor of quarry (near tools). (B) Sublocality 1C. (Photos courtesy of Glynn Hayes.)

isolated teeth, and large mammals such as rhinocerotids are absent. Hayes (2000) described the Brooksville insectivores, carnivores, lagomorphs and selected taxa of rodents, and briefly reviewed the ungulates. The ungulates and carnivores from this site include: the horse *Miohippus* sp.; an indeterminate oreodont; two small artiodactyls, *Nanotragulus loomisi* and the camelid *Notho-kemas waldropi*; and six species of carnivores, the tiny enigmatic *Palaeogale minuta*, the musteloids *Acheronicitis webbi* and *Arikarictis chapini*, and the canids *Enhydrocyon* cf. *E. pahinsintewakpa*, *Osbornodon wangi* and *Phlaocyon taylori*. In addition to bats, the small mammal fauna consists of the didelphid marsupial *Herpetotherium fugax*, the insectivores *Centetodon magnus* and the erinaceid *Parvericius montanus*, the lagomorph *Megalagus abaconis*, and numerous rodents, including several species of sciurids, several eomyids, two heteromyids, the castorid *Agnotocastor* sp., and an undescribed genus and species of entoptychine geomyid (Hayes, 2000, 2005).

Hayes (2000) determined the age of the Brooksville 2 LF through biochronologic comparisons with other Florida and Gulf Coast Arikareean faunas (Albright, 1998), and with Arikareean faunas from western North America. Preservation of the Brooksville fossils in an isolated karst feature precludes direct stratigraphic correlation with described Oligocene and Miocene stratigraphic units from northern Florida. The maximum age of Brooksville 2 is constrained by the presence of the erinaceid insectivore *Parvericius* and entoptychine rodents, both of which appear at the beginning of the Late Early Arikareean (~28 Ma; Tedford *et al.*, 1996). The minimum age of the site is limited by the occurrence of *Centetodon*, *Megalagus*, *Agnotocastor* and *Miohippus*, all of which became extinct at the end of the Early Arikareean (~24 Ma). On the basis of the first and last occurrences of these mammalian genera, Hayes (2000) placed the Brooksville 2 LF in the Late Early or "medial" Arikareean (Late Oligocene, between 24 and 28 Ma; Ar2 of Woodburne and Swisher, 1995, and Tedford *et al.*, 2004). Other Florida Arikareean faunas include Cow House Slough, Live Oak/SB 1A, White Springs and Buda (Frailey, 1978, 1979; Albright, 1998; Hayes, 2000; MacFadden and Morgan, 2003). The similarity between the chiropteran faunas of Brooksville 2 and I-75 indicates that Brooksville 2 probably belongs in the older portion of the Late Early Arikareean, between 28 and 26 Ma (Early Ar2).

The chiropteran sample from Brooksville 2 consists of about 200 fossils representing five species, including: a new genus and species of mormoopid; two new species, one large and one small, representing a new genus of emballonurid; a large genus and species of basal noctilionoid described herein; and an isolated tooth of a large molossid (Czaplewski *et al.*, 2003a; Morgan and Czaplewski, Chapter 4, this volume). Brooksville shares all of these species except the molossids with I-75. After Thomas Farm, Brooksville 2 has the second largest bat sample from any North American mid-Cenozoic site, and the third most diverse bat fauna following Thomas Farm and I-75. Both the Brooksville 2 and I-75 bat faunas are numerically dominated by two families currently restricted to the tropics, Mormoopidae and Emballonuridae.

5.3   Methods

Specimens were collected by screenwashing bulk samples of fissure fillings from the two localities. Some specimens received micropreparation with porcupine quills to enhance detail for molding, casting and photography. All are cataloged in the collections of the Florida Museum of Natural History, University of Florida (UF). Terminology for chiropteran cranial osteology mostly follows Giannini *et al.* (2006); dental terminology follows Menu and Sigé (1971), Legendre (1984) and Czaplewski *et al.* (2008). The abbreviations for tooth positions in mammals are standard, with upper case letters for upper teeth and lower case letters for lower teeth: I/i (upper/lower incisors), C/c (upper/ lower canines), P/p (upper/lower premolars) and M/m (upper/lower molars). The system of numbering the lower premolars as p2, p3 and p4 (rather than as p1, p3 and p4 or other system) is used here without any intention of indicating the proper homology of these teeth, because the homology is uncertain. Upper teeth are described as viewed with the tooth inverted (i.e., with the cusps manually oriented upward and roots downward) instead of in anatomical position. The terminology for chiropteran postcranial elements follows Vaughan (1959) and Smith (1972). Other abbreviations are: LF, Local Fauna; Ma, mega annum or millions of years before present; NALMA, North American Land Mammal Age; FLMNH and UF, Florida Museum of Natural History and University of Florida, Gainesville; USNM, United States National Museum of Natural History, Washington, DC; AMNH, American Museum of Natural History, New York; FMNH, Field Museum, Chicago; UCMP, University of California Museum of Paleontology, Berkeley. All measurements of fossils are in mm and were made using an Olympus SZX9 stereomicroscope with an eyepiece reticle. A camera lucida on the same microscope was used to draw selected specimens.

Recent molecular genetic data suggest that the Noctilionoidea are a monophyletic group consisting of the Phyllostomidae, Mormoopidae, Noctilionidae, Mystacinidae, Furipteridae, Thyropteridae and Myzopodidae (Kirsch *et al.*, 1998; Van Den Bussche and Hoofer, 2000, 2004; Hoofer *et al.*, 2003; Teeling *et al.*, 2003; Teeling *et al.*, 2005; Miller-Butterworth *et al.*, 2007). In our analysis of the Florida fossil noctilionoids, we included specimens from all of these families. We also scored members of the Emballonuridae as an outgroup.

We scored character states for 110 dental-osteological characters (listed in Appendix 5.1) in the following extinct and extant taxa (listed in Appendix 5.2): the two Florida fossil species (*Speonycteris aurantiadens* and *S. naturalis*); Noctilionidae: *Noctilio* (two spp.); Mormoopidae: *Mormoops* (two spp.), *Pteronotus* (six spp.) and an extinct Oligocene species (Morgan *et al.*, in press); Mystacinidae: *Mystacina* (two spp.) and extinct *Icarops* (one sp.); Phyllostominae (Phyllostomidae): most of the extant genera (21 spp. in 15 genera, including all currently recognized genera except *Tonatia*) plus two species of the extinct genus *Notonycteris*; Furipteridae: *Furipterus* (one sp.) and *Amorphochilus* (one sp.); Thyropteridae: *Thyroptera* (three spp.); Myzopodidae: *Myzopoda* (one sp.); Emballonuridae: *Peropteryx* (one sp.) and *Saccopteryx* (two spp.). Thus, our morphological analysis of the Florida fossils included virtually all known genera of potential noctilionoids except *Tonatia*. Very few fossil taxa from noctilionoid families are known by more than fragmentary specimens; we used the taxa with the best available samples. Originally we included several more species, especially of fossil bats, but they were later removed due to the low number of characters that could be scored. We attempted to maximize data completeness, which was high for all of the extant species and low for all of the fossil species (see data matrix in Appendix 5.3). Data completeness was low also in *Speonycteris naturalis* from I-75, represented by one tooth (25%). We excluded from the analysis this species and a few others for which data completeness was very low. Using computer-assisted phylogenetic analysis software (PAUP; Phylogenetic Analysis Using Parsimony) we did a heuristic search with all of the characters unordered and of equal weight.

5.4   Systematic paleontology

Mammalia Linnaeus, 1758
Chiroptera Blumenbach, 1779
Noctilionoidea Gray, 1821
Speonycteridae fam. nov.

> *Genotype* – *Speonycteris* gen. nov.
> *Diagnosis* – As for the only genus.
> *Stratigraphic distribution* – Oligocene, Whitneyan to Early Arikareean NALMAs.
> *Geographic distribution* – Southeasternmost North America, USA: northern peninsular Florida.

Speonycteris gen. nov.

> *Type species* – *Speonycteris aurantiadens*, sp. nov.
> *Included species* – *Speonycteris aurantiadens* and *Speonycteris naturalis* sp. nov., described below.
> *Diagnosis* – Noctilionoid bats with M2 wider than long, M2 having a large talon supported by extra roots lingually and posteriorly, M2 with low, indistinct hypocone at anterolingual end of talon cingulum and strongly offset lingually from the postprotocrista, postprotocrista connected to talon by a low, thin ridge curving posteriorly and lingually.

Lingual face of Cɪ with a single low central ridge (located between the prominent, sharp, anterolingual and posterolingual crests) running down from base to apex of main cusp. Cɪ with broad lingual cingular shelf that is very weakly bilobed. P4 with projecting anterior lobe that is as wide transversely as the main posterolabial crest is long, with indented labial margin in occlusal view, and with labial cingulum that lacks cuspules. Lower incisors probably one. Lower cheek teeth strongly exodaenodont. Three lower premolars present, a large one-rooted p2, a small, crowded (anteroposteriorly compressed), two-rooted p3 and a large two-rooted p4 with a metaconid. Lingual cingulids present on cɪ, p2, p3 and anterior part of p4, absent on molars. Labial cingulids strong on all lower teeth except at base of hypoconid of m3. Lower molars show nyctalodonty; carnassial-like notches absent from cristid obliquas and postcristids; m3 talonid is reduced with weak or absent distal cingulid, and indistinguishable to barely distinguishable entoconid and hypoconulid in unworn specimens. Proximal femur with greatly reduced trochanters.

Etymology – *Speos*, Greek, cave, grotto; and *nycteris*, Greek, bat.

Comparisons – *Speonycteris* differs from *Mormoops* and *Pteronotus* in having a P4 with anterior lobe wider than the posterolabial crest is long, in having an indented labial margin in occlusal view and in having a labial cingulum without a cuspule. Differs from Phyllostominae, Mystacinidae (*Mystacina* and *Icarops*), Myzopodidae (*Myzopoda*), Furipteridae (*Furipterus* and *Amorphochilus*) and Thyropteridae (*Thyroptera*) in having P4 with broader, squarish anterior lobe, lacking anterior sharp crest on main cusp. Differs from *Noctilio* in having P4 with broad, projecting anterior lobe and lacking an anterolabial main crest ("preparacrista").

Speonycteris differs from *Mormoops* in having M2 with a more expansive talon extending to or beyond the base of the protocone lingually, in having a weaker ridge connecting postprotocrista with hypocone and in having the lingual cingulum interrupted at the base of the protocone. Differs from *Pteronotus* in having M2 with a shorter postprotocrista, much lower hypocone, postprotocrista connected with the hypocone by small thin ridge and in having the lingual cingulum incomplete around the base of the protocone. Differs from many phyllostomines and from members of many other subfamilies of phyllostomids in the configuration of the postprotocrista and talon of M2 and in transversely wider than long upper molars. Differs from *Macrotus* in M2 proportions, in the presence of a paraloph and metaloph, and in possessing a strongly offset postprotocrista and hypocone. Differs from *Micronycteris* in having M2 with a paraloph, a relatively larger talon and a weak rather than strong ridge connecting postprotocrista with hypocone. Differs from *Lampronycteris* in having a low hypocone and in having a distinct terminus and posterior emargination between the labial end of the talon and the metacingulum. Differs from *Trinycteris* in having M2 with a low hypocone and in having a small curving ridge between postprotocrista and hypocone. Differs from *Glyphonycteris* in having M2 with a lower hypocone situated on the talon cingulum, in having an approximate right instead of obtuse angle in functional view between the preprotocrista and postprotocrista, in having a concave basined talon instead of convex or flat and in having a discontinuous lingual cingulum. Differs from *Lonchorhina* in having M2 with a small low hypocone and a thin weak ridge connecting at nearly right angles to the postprotocrista and hypocone-talon cingulum. Differs from *Vampyrum* and *Chrotopterus* in having M2 with an unreduced protocone-preprotocrista-postprotocrista and in having the mesostyle near the labial margin of the tooth. Differs from *Trachops* in having M2 with a low indistinct hypocone, in having a more expansive, basined talon and in having the commissures of the W-shaped ectoloph (preparacrista-postparacrista-premetacrista-postmetacrista) not slanted strongly backward from lingual to labial ends. Differs from *Macrophyllum* in having a low indistinct hypocone. Differs from *Tonatia* in having M2 with a greater offset between the postprotocrista and hypocone. Differs from *Lophostoma* in having M2 with a low indistinct hypocone and greater offset between the postprotocrista and hypocone. Differs from *Phyllostomus* in having M2 with a small thin ridge connecting postprotocrista with hypocone. Differs from *Phylloderma* in having M2 with a low indistinct hypocone situated on the talon cingulum, in having a strongly offset postprotocrista and hypocone and in having a small thin ridge connecting postprotocrista and hypocone. Differs from *Mimon* in having M2 talon with rounded rather than straight posterior margin. Differs from *Noctilio* in having M2 with trigon valley narrow at its labial end just lingual to the mesostyle, in having a much lower hypocone, in having a longer, weak, curving ridge connecting the postprotocrista with the hypocone, in lacking a paraconule and in lacking strong, high buttress-like paraloph and metaloph that close the trigon basin anteriorly and posteriorly. Differs from *Mystacina* and *Icarops* in having M2 with a larger, basined talon, in having a small ridge running from postprotocrista to hypocone and in having weaker paraloph and metaloph. Differs from *Myzopoda*, *Furipterus* and *Amorphochilus* in having M2 with a talon and hypocone. Differs from *Thyroptera* in having M2 with a much larger, basined talon.

Speonycteris aurantiadens sp. nov.
(Figures 5.4–5.11; Table 5.1)

Holotype – UF 240000, left mandible with p2–m3 and alveoli for anterior teeth.

Table 5.1 Measurements in mm of fossil Noctilionoidea from the Oligocene of Florida, USA. Abbreviations: a.p.l. = anteroposterior length; frag. = fragment; () = estimated measurement of broken specimen. *indicates specimens from I-75 local fauna.

Length of m2 to m3 in UF 240000 is 5.71 mm, in UF 157763, 5.65 mm.

Element	UF specimen number	Labial depth at anterior root of m2	Alveolar length c1–m3	Toothrow length p2–m3	Toothrow length p3–m3	Toothrow length m1–m3	Toothrow length m2–m3
Speonycteris aurantiadens							
Dentary	240000	3.70	14.92	12.86	11.50	8.89	5.71
	157663	3.80	–	–	–	–	5.65
	157664	3.95	14.10	–	11.40	8.50	5.81

Element	UF specimen number	a.p.l.	Labial a.p.l.	Lingual a.p.l.	Transverse width	Trigonid width	Talonid width
Speonycteris naturalis							
L M1 or M2	121717*	–	1.95	1.85	2.65	–	–
Speonycteris aurantiadens							
R C1 frag.	182797	2.95	–	–	–	–	–
L P4	182784	3.00	–	–	2.10	–	–
R M1 or M2	157760	–	–	2.90	4.05	–	–
R c1 frag.	182799	2.30	–	–	(2.40)	–	–
R p2	182849	2.05	–	–	1.65	–	–

L p2	240000	2.08	—	—	1.90	—	—
L p2	182805	2.20	—	—	1.95	—	—
L p3	240000	1.25	—	—	1.42	—	—
R p3	157764	1.25	—	—	1.30	—	—
L p4	240000	2.15	—	—	2.05	—	—
L p4	157761	2.25	—	—	2.00	—	—
R p4	157764	2.10	—	—	1.75	—	—
R p4	16733*	1.90	—	—	1.75	—	—
R m1	157759	3.15	—	—	—	1.90	2.05
R m1	157764	3.00	—	—	—	2.16	2.05
L m1	240000	3.40	—	—	—	2.25	2.35
L m1 or m2	179952	—	—	—	—	—	2.40
L m2	240000	3.35	—	—	—	2.20	2.12
R m2	157763	3.15	—	—	—	1.95	2.00
R m2	157764	3.05	—	—	—	1.90	2.00
L m3	240000	2.85	—	—	—	1.80	1.30
R m3	157763	2.70	—	—	—	1.90	1.30
R m3	157764	2.75	—	—	—	1.70	1.30

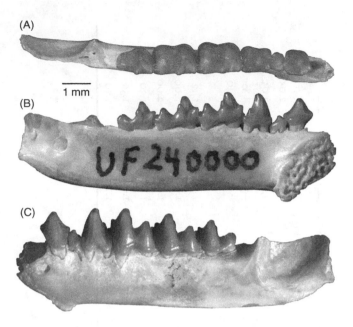

Figure 5.4 *Speonycteris aurantiadens*, gen. et sp. nov., holotype UF 240000, from Brooksville 2 locality. (A) occlusal, (B) lingual and (C) labial views.

Hypodigm – From I-75 locality: UF 16733, right p4. From Brooksville 2 locality: holotype plus UF 157764, right mandible with p3–m3 and alveoli for other lower teeth; UF 182799, right c1; UF 182849, right p2; UF 182805, left p2; UF 157761, left p4; UF 157759, right m1; UF 157763, right mandible fragment with m2–m3; UF 179952, left m1 or m2 fragment; UF 182797, right partial C1; UF 182784, left P4; UF 157760, right M2 (or possibly M1); UF 179978, proximal fragment of left femur.

Type locality and age – University of Florida locality Brooksville 2, fissure fillings within the Suwanee Limestone (Figures 5.2–5.3) near the town of Brooksville, Hernando County, Florida; Arikareean, late Oligocene (Hayes, 2000). Also known from the older I-75 locality, Alachua County, Florida; Early Oligocene, Whitneyan.

Diagnosis – A large species with M2 exceeding 4 mm in transverse width and alveolar length of c1-m3 exceeding 14 mm. M2 with relatively thin ridge running between offset postprotocrista and hypocone.

Etymology – *aurantium*, Latin, orange; *dens*, Latin, tooth. Named for the bright orange color of the teeth as preserved in fossils from the Brooksville 2 fissure fillings.

Description and comparisons – The hypodigm represents a large bat about the size of the living *Phyllostomus hastatus*, the second largest New

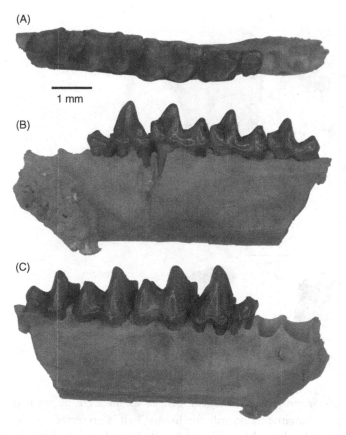

Figure 5.5 *Speonycteris aurantiadens*, gen. et sp. nov., UF 157764 from Brooksville 2 locality. (A) occlusal, (B) lingual and (C) labial views.

World bat. Using predictive equations for body mass based on the correlation between molar sizes and body mass of related living bats (NJC and Cindy L. Gordon, unpublished data), this bat weighed about 60 to 63 grams. The sample includes three partial lower jaws and several isolated teeth. The amount of wear on the teeth in the three jaws is identical. Given the same wear and the very similar preservation it was considered possible that the holotype left and one of the two right jaws could represent contralateral elements from an individual bat. By manipulating and artificially rearticulating the two specimens that preserve the mandibular symphysis (UF 240000 and 157764), these two specimens are not mates. The same test cannot be applied between UF 240000 and UF 157763 because 157763 lacks the symphysis. Nevertheless, the minimum number of individual bats represented in the sample based on all preserved elements is two.

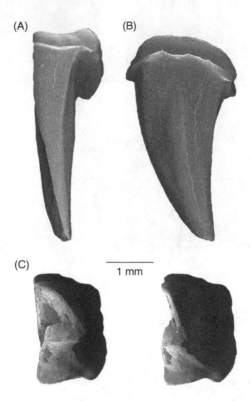

Figure 5.6 *Speonycteris aurantiadens*, gen. et sp. nov., UF 182797, upper canine (split longitudinally; only the lingual half is preserved) from Brooksville 2 locality. (A) anterior view, (B) lingual view and (C) occlusal views, stereo-pair.

The jaws and teeth of this large bat are robust and strongly built. Their morphology and overall similarity to large phyllostomines suggests that dietarily, *S. aurantiadens* was animalivorous, but generalized, capable of omnivory.

The horizontal ramus is uniformly deep from front to rear between the canine and m3 (Figure 5.4). Measurements of the dentary are provided in Table 5.1. The median synchondrosis surface of the mandibular symphysis is very rugose and the mandibles were clearly unfused. The symphysis extends from the anterior edge of the mandible to the level of the posterior root of the p3. In mesial view, the outline of the symphysis is an inclined oval with a rather strong, hook-like posteroventral projection. Between this projection and the mandibular body is a small, deep concavity, presumably for insertion of the genioglossus muscle. The left and right dentaries diverge caudally at an angle of about 24°. The body of the mandible is robust. At the level of the interalveolar septum between m1 and m2 (where broken in UF 157763), the mandibular body is gently curved in vertical cross section, with the labial side slightly concave and the lingual side convex.

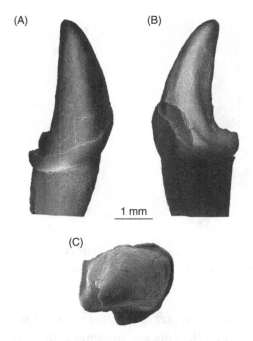

Figure 5.7 *Speonycteris aurantiadens*, gen. et sp. nov., UF 182799, lower canine from Brooksville 2 locality. (A) labial view, (B) lingual view and (C) occlusal view.

Posteriorly, most of the caudal ramus of the mandible is missing from all available specimens. The mandibular foramen opens low on the lingual side of the mandibular body at the junction of the mandibular body and caudal ramus. From there the mandibular canal runs forward near the ventral edge of the mandibular corpus beneath the molar roots. In both lower jaw specimens that preserve the mandibular symphysis, there is a relatively large, vertically oriented, anterior mental foramen that opens dorsally from a point directly ventral to the incisor alveolus. This anterior mental foramen communicates through several tiny perforations in the dentary bone with the intermandibular contact surface of the symphysis. A second, absolutely larger but relatively small posterior mental foramen opens high on the side of the dentary between the root of the lower canine and the root of p2. The opening of this foramen is directed anterodorsally. Caudally along the ventral margin where the horizontal body of the mandible meets the caudal ramus, there is a small curved upsweep to the ventral margin of the caudal ramus. The ventral margin of the caudal ramus then continues nearly straight backward an unknown distance, presumably to an angular process which is not preserved. Only the very base of the anterior edge of the coronoid process is preserved a short distance behind the m3; on its lingual side at the level of the alveolar row there is a small muscle insertion scar,

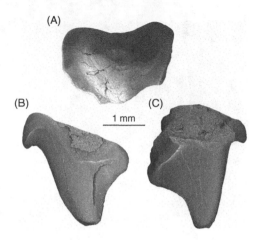

Figure 5.8 *Speonycteris aurantiadens*, gen. et sp. nov., UF 182784, right P4 from Brooksville 2 locality. (A) occlusal view, (B) labial view and (C) posterolingual view.

probably for part of the m. temporalis or m. pterygoideus. The lateral side of the ramus is dominated by a deep, smooth, concave masseteric fossa bordered anteriorly by the ventral continuation of the coronoid crest.

Although no incisors are preserved in any of the specimens, two of the mandibles (the holotype dentary UF 240000 and UF 157764) preserve the alveolus for a large mesial incisor. However, the incisor-bearing portion of the mandibles is somewhat damaged and we cannot rule out the possibility of a smaller distal incisor. We tentatively infer the presence of one lower incisor in *S. aurantiadens*.

The C1 is split longitudinally from apex of root to apex of the main cusp, exposing a long internal pulp cavity (Figure 5.6). The tooth is split approximately in half, with the medial (lingual) half preserved. Crown height of C1 from the dentino-enamel junction to the apex of the crown is 5.50 mm. The root is stouter and shorter in height than the crown. The crown is recurved backward, but not inward. The apex of the main cusp is barely worn, indicating a young adult animal. The basal cingulum is slightly arched on the lingual side; it is continuous with the anterior cingulum and bears a small cingular cusp at the posterolingual corner of the crown. A small gap separates the posterolingual cingular cusp from the posterolabial cingulum. The small portions of the labial cingulum that are preserved posteriorly and anteriorly indicate that the cingulum dropped lower on the missing labial side of the tooth. There is a series of small, thin, low styles or cuspules on the lingual cingulum, of which the central one is best developed. In occlusal view the lingual cingular shelf is barely bilobed, moderately wide and slightly basined. The main cusp is high, with a sharp crest running down the posterolingual height of the crown and another crest running down the anterolingual edge of the crown. The anterolingual crest is sharp on its proximal half, but is already slightly worn on its distal half to a rounded ridge; this wear is due to abrasion against the lower canine. In lingual view the lingual face of the main cusp is relatively flat except for a low, but prominent central longitudinal ridge.

One isolated ci is available, UF 182799, and it is intact except for breaks on the anterior and posterior portions of the basal cingulum and the tip of the root. Crown height of ci is 3.75 mm. The root is stout and grooved on the anterior, posterior and lingual faces; its labial face is convex. The crown also is stout and sits at a slight outward angle relative to the root axis. The summit of the primary cusp and its main crests are unworn in this specimen. As in many bats, the cingulum is lowest at the posterolabial corner of the crown. The cingulum curves upward anteriorly along the labial side and then rises more steeply to its highest level on the anterior side near the point where it would have contacted the lower incisor. The cingulum drops again to a lower level along the lingual side of the crown base. There is a small cuspid at the posterolingual corner of the cingulum. From base to summit, the main cusp has a moderate mediad curve. The labial and anterior faces of the main cusp are convex. There are strong longitudinal ridges running from base to summit on the posterolabial and posterolingual slopes, with the posterior face between them flat to slightly concave. A shallow groove runs adjacent and parallel to the posterolingual ridge immediately anterior to it on the lingual face. Anterior to the groove, the lingual face is convex, especially just above the cingulum.

Only one upper premolar, a partial P4, is available. It is missing the lingual portion due to breakage. The tooth is three-rooted, with roots supporting the anterior, lingual and posterolabial portions of the crown (although only the anterior root is preserved intact). In labial view, the base of the anterior lobe sweeps sharply upward from front to back so that the anterior and posterior portions are at different levels, indicating a premolar row higher than the molar-bearing portion of the toothrow. The P4 is dominated by a single main cusp, the protocone, which is recurved posteriorly. A relatively short, thick shearing crest curves downward from the apex of the protocone back to a cingular style at the posterolabial corner of the tooth. This crest is about as long as the anterior lobe of the tooth is wide. In occlusal view, the tooth has an indented labial margin and resembles the P4 of *Mormoops* and *Pteronotus* in having a broadened anterior lobe that is separated by an indentation from the missing lingual portion of the tooth. There is no labial cingular cuspule. A basal cingulum surrounds the entire preserved portion of the crown base. The cingulum is very narrow labially, but forms a fairly wide, guttered shelf around the anterior lobe. Two small styles sit astride the anterior cingulum.

As in many bats, the lower premolars show strong exodaenodonty (i.e., the labial portion of the crown is expanded labially and ventrally, such that the labial ventral margin of the crown is lower than the lingual ventral margin; term from Hürzeler, 1944).

There are three specimens of p2: one *in situ* in the holotype jaw and two isolated specimens, one of which is broken labially and missing most of its root. The p2 has a single long narrow root that is squarish with rounded corners in cross section, with longitudinal grooves running down the anterior, posterior and lingual faces. The cingulum dips low on the labial side where it is widest and deepest, and sweeps upward anteriorly and posteriorly. The cingulum disappears at the lower ends of the anterior and posterior crests of the main cusp. The lingual cingulum also is wide and deep, and dips downward, but not as low as on the labial side. Under the overhangs beneath the cingulum at the anterior and posterior projections of the crown are tiny wear facets caused by interdental contact with the c1 and p3, respectively. The main cusp of p2 is stout and conical – appearing approximately as long as high in lingual profile – and relatively large, about four-fifths or more of the height of the main cusp of the p4 and much larger than p3. Two strong crests descend the crown, one anteriorly and one posteriorly; both curve lingually towards the base and connect with the high points of the cingulum.

The p3 is represented by two specimens: it remains *in situ* in each of the most complete dentaries UF 157664 and UF 240000. The p3 is small and distinctly double-rooted. In occlusal view the crown is wider than long. The tooth has a very deep, narrow cingulum surrounding the crown. As in p2, the cingulum sweeps downward further on the labial than the lingual side. The main cusp is short and in lateral profile appears triangular, but much longer than high, with the apex of the main cusp situated over the anterior root. In occlusal view, the outline of the crown is irregular, rounded and convex lingually, straight anteriorly, and angular or indented labially and posteriorly. A crest descends anterolingually from the main cusp and curves to the anterolingual corner of the base. Posteriorly another crest extends from the summit of the main cusp to the rear of the tooth. As seen in lingual view in the lower toothrow, the posterior end of the p2 overhangs the anterior end of the p3, and the anterolingual corner of the p4 overhangs the posterior end or posterolingual corner of the p3.

Three p4s are available (two *in situ* in dentaries) from Brooksville 2 and one is available from the older I-75 locality. The Brooksville 2 specimens are barely worn, but the I-75 specimen is moderately worn on the apex of the main cusp (protoconid). The I-75 specimen is damaged in that the enamel is lost from much of the base and cingulum of the tooth; enamel is still present on the upper portion of the main cusp. The p4 is large and has two large roots, anterior and posterior. The roots are wide just beneath the base of the crown, but taper quickly and are slender at their apices. The main cusp is tall, about the same height as the protoconid of mɪ in labial view. In lingual view the p4 actually appears taller overall than the mɪ, but its base is situated lower than the base of mɪ in the dentary so that the two summits are about the same height. The main cusp is slightly recurved posterolingually. In occlusal view the p4 is longer than wide; it has a roughly rectangular occlusal outline except that it is wider posteriorly than anteriorly. A sharp longitudinal crest descends the main cusp anteriorly and near the base curves to meet a cingular cuspid at the anterolingual corner of the tooth. Another crest descends the posterolingual face of the main cusp to meet a moderately developed metaconid halfway down before continuing to a cingular cuspid at the posterolingual corner of the crown base. The basal cingulum is wide and deep on the anterior and posterior ends of the tooth. The cingulum is much narrower (and deep) on the labial and lingual sides. On the labial side of the base are two downswept portions of the crown and cingulum, one over each root, separated by a prominent median notch; these curves and the strong exodaenodonty give the tooth a very sinuous lower margin along the labial cingulum in labial view. This sinuosity is much less pronounced on the lingual side of the crown.

The only available upper molar is an M2 (or less likely, Mɪ) with the parastylar area broken away (Figure 5.9A, B). Its morphology is very similar to the M2 in several phyllostomines, especially *Lampronycteris, Micronycteris, Macrotus, Chrotopterus, Mimon, Phyllostomus* and *Tonatia*, and also in the mormoopids *Mormoops* and *Pteronotus*. The tooth is distinctly five-rooted with three large roots supporting the parastyle-paracone (root broken away), metastyle-metacone and posterior base of the protocone, with a smaller but substantial root beneath the lingual portion of the talon, and a still smaller rootlet supporting the posterolabial edge of the talon. The tooth bears the usual three main cusps protocone, paracone and metacone, plus an incipient hypocone at the anterolingual corner of the talon. The talon is large and broadly rounded posteriorly. A W-shaped ectoloph connects the stylar cusps parastyle (broken), mesostyle and metastyle with the paracone and metacone. The labial cingulum is very weak. A small, crested paraloph follows a sinuous route from the base of the paracone to the apex of the protocone. Metaloph is absent.

In *Speonycteris aurantiadens*, the hypocone of M2 is not developed as a raised cusp but occurs as a low, strong angle in the crests connecting the postproto-crista with the anterolingual end of the talon cingulum. The talon portion of the tooth seems to be evolutionarily highly labile in many bats. In this region some phyllostomines and mormoopids possess a cuspate hypocone that may or may not bear cristae that join the postprotocrista and/or talon cingulum,

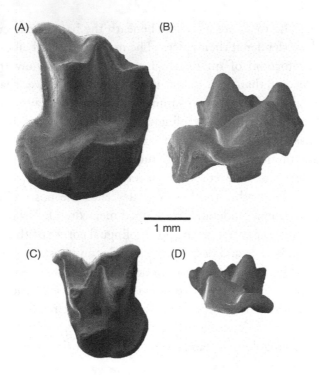

Figure 5.9 *Speonycteris* upper molars. *Speonycteris aurantiadens*, gen. et sp. nov., UF 157760 right M2 from Brooksville 2 locality in (A) occlusal view and (B) lingual view. *Speonycteris naturalis*, sp. nov., UF 121717 left M2 from I-75 locality in (C) occlusal view and (D) lingual view.

whereas in others the hypocone is not apparent, and in still others the hypoconal crests are emphasized as a sweeping crest continuous with the talon cingulum, but the hypocone itself is lacking. In other noctilionoids the hypocone and the talon itself are absent. In *S. aurantiadens* the preprotocrista is strong and continuous with a thin paracingulum along the anterior margin of the tooth. The postprotocrista is strong but short, ending in a metaconule-like tiny swelling. A pre-hypoconal crest extends labially from the hypoocone before curving anteriorly and upward to meet the postprotocrista at the metaconule-like swelling. A post-hypoconal crest continues at a right angle posteriorly from the hypocone and is continuous with the cingulum encircling the posterior margin of the talon. The talon encloses a large basin. The talon cingulum is continuous with a fairly wide, guttered metacingulum that extends labially to the metastyle.

The lingual cingulum in the M2 is low and weak. It consists of two short segments, one anterolingual and one posterolingual to the base of the protocone. The posterior segment in lingual view is at a lower level than the

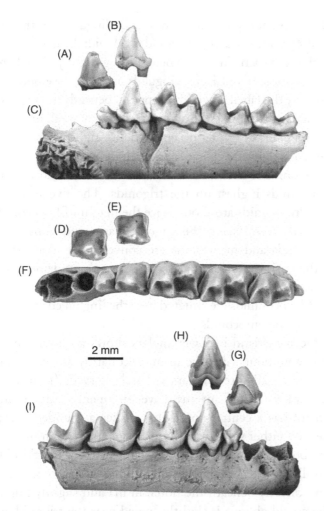

Figure 5.10 *Speonycteris aurantiadens*, dentary and comparison of p4s from I-75 and Brooksville 2 localities. (A), (D), (G) UF 16733, right p4 missing much enamel, from I-75, in lingual, occlusal and labial views. (B), (E), (H) UF 157761, left p4 (reversed for comparison) from Brooksville 2, in lingual, occlusal and labial views. (C), (F), (I) UF 157764, right dentary with p3–m3 from Brooksville 2, in lingual, occlusal and labial views. Photos of ammonium hydroxide-coated casts.

post-hypoconal crest/talon cingulum, and it encloses the foot of the lingual valley between the protocone and hypocone.

The lower molars, as in most bats, are two-rooted and exodaenodont (term explained above under Lower Premolars; Figure 5.10). The m1 and m2 are distinctly nyctalodont; m3 has a reduced talonid with no discernible hypoconulid, and the postcristid extends to the entoconid, giving the appearance of myotodonty in m3. Lingual cingulids are absent from the lower molars. In lingual view, the ventral margins of the molars drop downward from trigonid to talonid; the margin reaches its lowest point beneath the entocristid of each before rising slightly again beneath the entoconid. Anterior (mesial) and labial cingulids are well developed on all three lower molars; on m3 the labial cingulid disappears beneath the hypoconid. Posterior (distal) cingulids are strong on m1 and m2, but very weak on m3. The anterolingual end of each molar's anterior cingulid fits tightly into the area between the hypoconulid and postcingulid of the tooth in front of it.

In lingual view, paraconids of the lower molars are moderately low, metaconids higher and protoconids highest on the trigonids. The five main lower molar cusps are stout. Protoconids are about twice the height of hypoconids on m1 and m2. The lingual faces (facing the trigonid basin and talonid basin, respectively) of protoconids and hypoconids are convex with small, shallow, concave hollows on the basinward sides between the cusp's fundamental cone and the adjacent cristids. Entoconids on m1 and m2 reach about the same height as paraconids. Hypoconulids are situated near the lingual edge of m1 and m2, low and posterior to the entoconids.

In occlusal view, the paracristid in lower molars shows an obtuse bend or open curve in m1, a less pronounced curve in m2 and hardly any curve in m3, whereas the protocristid in all three molars is barely curved. In anterior or posterior view, the paracristids of m1–m3 have an openly curved profile, whereas the protocristid has a central notch that makes a rounded apex; the two halves of the crest on either side of the notch form an angle in m1 slightly less than 90°, in m2 about 90° and in m3 more than 90°. The cristid obliqua meets the posterior wall of the trigonid below and slightly labial to the notch in the protocristid in m1, directly beneath the notch in m2 and slightly lingual to the notch in m3. The cristid obliqua is slightly angled near the trigonid end in m1, less angled in m2 and straight in m3. The entocristid in m1 and m2 has a mild curve, concave on the lingual side. In m3 there is no discernible entoconid, with the cristid obliqua, postcristid and entocristid forming a continuous border to the trough-like talonid basin, with nearly straight to barely curved sides and angled corners. The m3 talonid is longer than wide, whereas the m1 and m2 talonids are wider than long.

Postcranially, speonycterids are as yet practically unknowable. The only postcranial element presently known is the proximal end of a left femur, UF 179978 (Figure 5.11), referred to *S. aurantiadens*. Measurements of this fragment are: shaft diameter distal to head, 2.35 mm; greatest proximal width across trochanters, 3.75 mm; diameter of head, 2.40 mm.

Despite the paucity of material, this element shows a derived morphology and most closely resembles the femur of phyllostomids and mormoopids.

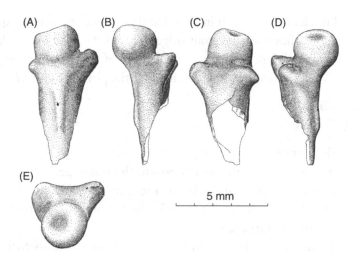

Figure 5.11 *Speonycteris aurantiadens*, UF 179978, proximal portion of left femur from Brooksville 2 locality. (A) anterior view; (B) dorsal view; (C) posterior view; (D) ventral view and (E) proximal end view.

The head is offset anteriorly from the longitudinal axis of the shaft. The fovea capitis is a circular, shallow and relatively large-diameter pit within a large flattened surface on the head. The flattened surface is tilted or offset in the direction of the greater trochanter, as in Noctilionidae and many Phyllostomidae.

The trochanters are situated close together on the posterior side of the femur; they are greatly reduced such that the femoral head is far larger than either trochanter. The trochanters are less reduced than the extreme reduction seen in mormoopids and furipterids. They are a little more reduced than in most other Western Hemisphere bat families, including more reduced than in noctilionids and many phyllostomids, but equivalently to certain other phyllostomids. The lesser trochanter is smaller in volume and diverges more from the axis of the shaft, but extends slightly farther headward than the greater trochanter. The greater trochanter is blunt, shorter and bulkier than the lesser. Both extend proximally only to the level of the base of the head.

Very little of the femur shaft is preserved. As far as one can tell, there was no curve or bend in the shaft near its proximal end. Enough of the shaft remains to show a small but stout longitudinal ridge for a muscle or tendinal insertion (iliacus group?) distal to the head on the anterior surface opposite the trochanters. This ridge is also present in phyllostomids, where it is strong in some (*Trachops, Desmodus, Glossophaga*) and weak in others (*Carollia, Artibeus*). It is also present in mormoopids, but low and indistinct. It is absent in *Noctilio*.

Greatly reduced trochanters are seen in bats that do not frequently crawl quad-
rupedally on the ground; however, the stout ridge distal to the head may indicate
relative agility in "walking" bipedally across the ceiling of a cave while hanging
by the hind feet, as Vaughan (1959, p. 99) described for the phyllostomid *Macrotus*.

Speonycteris naturalis sp. nov.
(Figure 5.9C, D; Table 5.1)

> *Holotype* – UF 121717, left M2.
>
> *Hypodigm* – Known only from a single tooth, the holotype.
>
> *Type locality and age* – University of Florida locality I-75, Alachua County,
> Florida; from deposits filling a fissure in the Crystal River Formation;
> Early Oligocene, Whitneyan.
>
> *Diagnosis* – Virtually identical in morphology and proportions to M2 of
> *S. aurantiadens*, but much smaller in size; the sole available M2 is about
> 65% the size of the M2 of *S. aurantiadens* in linear dimensions (Table
> 5.1). M2 with relatively heavy ridge running between offset postproto-
> crista and hypocone.
>
> *Etymology* – *naturalis*, Latin, natural.
>
> *Description* – The available specimen represents a bat about the size of the
> living *Mimon crenulatum* or *M. bennettii*. Using predictive equations for
> body mass based on the correlation between molar sizes and body mass
> of related living bats (NJC and Cindy L. Gordon, unpublished data),
> this bat weighed about 18 g. Although it is older (*c.* 30 Ma), the single
> upper molar representing *Speonycteris naturalis* is nearly indistinguish-
> able from the same tooth in *S. aurantiadens* (28–24 Ma), except for the
> significant size difference. Unlike the available M2 in *S. aurantiadens*,
> which has a broken preparacrista-parastylar area, the *S. naturalis* M2 is
> intact. The preparacrista is about the same length as the postmetacrista.
> It bears a strongly hooked parastyle. The mesial cingulum (paracingu-
> lum) extends as a continuation of the preprotocrista all the way to the
> parastyle; it forms a narrow shelf along the mesial wall of the paracone.
> The labial margin of the tooth is indented more deeply between the
> parastyle and mesostyle than it is between the mesostyle and metastyle.
> Along the labial margin of the M2 ahead of the mesostyle there is a
> short segment of labial cingulum that is oriented anterolabially, almost
> in parallel with the trend of the preparacrista, and which is abruptly
> terminated at its labial end. This segment of labial cingulum contrasts
> with the posterior segment between the mesostyle and metastyle,
> which is aligned anteroposteriorly and is rather straight. It differs from
> the posterior segment of labial cingulum in *S. aurantiadens*, which

follows the curved indentation of the labial margin behind the mesostyle, and is also more strongly developed than the same segment in *S. aurantiadens*. Roots are not preserved in the *S. naturalis* M2, but there appears to be a broken remnant of the edge of the base of a root supporting the lingual end of the talon as in *S. aurantiadens*. Unlike the morphology of the M2 in *S. aurantiadens*, a metaloph is present in M2 of *S. naturalis*, the small curving ridge that runs between the offset postprotocrista and hypocone is heavier in *S. naturalis* than in *S. aurantiadens* and there is a deep concavity within the trigon basin in *S. naturalis* that is absent in *S. aurantiadens*. These differences are possibly attributable to individual variation. The available M2 of *S. naturalis* is slightly worn, whereas that of *S. aurantiadens* is unworn.

Comparisons – The upper molar (UF 121717) of *Speonycteris naturalis* bears definite similarities in morphology to mormoopids and phyllostomids, but differs in parallel with the differences between speonycterids and other noctilionoid families already listed above. *Speonycteris naturalis* is closer in size than *S. aurantiadens* to the other known noctilionoids, yet is larger than any known mormoopid, mystacinid, thyropterid, furipterid or myzopodid. It is similar in size to M2s of the medium-sized noctilionid *Noctilio albiventris* and the phyllostomine *Lophostoma silvicola*.

5.5 Discussion

Both the small and large speonycterids are from the Late Early Oligocene, Late Whitneyan (ca. 30 mya) I-75 local fauna, northern Florida. The large species also occurs in the Early Late Oligocene, Early Arikareean (ca. 28–24 Ma; probably Ar2) Brooksville 2 local fauna. Despite no similarities in the mammals between these two local faunas other than the bats, the two occurrences of these speonycterids may only span about four million years from roughly 30 to 26 Ma.

5.6 Phylogenetic relationships

Recent phylogenetic analyses of bats suggest that the Noctilionoidea are a monophyletic group comprising six or seven families. Traditionally included in the Noctilionoidea are the families Noctilionidae, Mormoopidae and Phyllostomidae (e.g., Simmons, 1998, 2000; Simmons and Geisler, 1998). Additionally, several authors have found support for a relationship of the Mystacinidae with Noctilionoidea (Pierson *et al.*, 1986; Kirsch *et al.*, 1998; Kennedy *et al.*, 1999; Van Den Bussche and Hoofer, 2000). Molecular studies aimed at clarifying the relationship between *Mystacina* and other noctilionoids (summarized by Van Den Bussche and Hoofer, 2000) gave estimated divergence times as follows: *Mystacina* from the common ancestor of the mormoopid-phyllostomid clade, 32–41 Ma; and *Noctilio* from the common ancestor of the mystacinid-mormoopid-phyllostomid clade, 39–50 Ma. More recently, Hoofer *et al.* (2003) analyzed all families of vespertilionoid and noctilionoid bats. For the Noctilionoidea, their results indicated strong support for including the Furipteridae and Thyropteridae (as well as the other families mentioned above) in Noctilionoidea rather than in Nataloidea (Simmons, 1998; Simmons and Geisler, 1998) or Vespertilionoidea. Even more recently, molecular genetic work has suggested Myzopodidae should be included in Noctilionoidea (Teeling *et al.*, 2003).

We scored dental and morphological characters of putative phylogenetic importance (Appendix 5.1) on the fossils and on recent bat specimens (Appendix 5.2). A parsimony analysis of the morphological data matrix (Appendix 5.3) using PAUP yielded 194 equally parsimonious trees of 548 steps. In a 50% majority-rule consensus of the 194 trees (Figure 5.12) *Speonycteris* clusters outside a clade containing the Mormoopidae, Noctilionidae and Mystacinidae that is sister to a clade containing most of the Phyllostominae. Except for the untenable position of *Macrophyllum* outside the other phyllostomids, most relationships in our dental-morphological tree concur in general with previous authors' phylogenies. The position of *Macrophyllum* is probably the result of the limited character set in the teeth and jaws, and the likelihood of homoplasy in the characters. The branching sequence conflicts with the molecular tree of Hoofer *et al.* (2003) in finding a close relationship between mormoopids and noctilionids-mystacinids instead of mormoopids-phyllostomids with thyropterids and furipterids.

In several additional runs using subsets of the data and varying the settings and outgroups, *Speonycteris* always fell in between the core defined families of Noctilionoidea, sometimes closer to phyllostomines and sometimes to mormoopids. With or without the furipterids and thyropterids, the Florida fossils always nested within a group containing the Noctilionidae, Mormoopidae, Mystacinidae and Phyllostomidae. Together with the morphological intermediacy of the fossils' dental and femur characters, we interpret this computer-assisted phylogenetic tree as indicating an evolutionarily intermediate position for *Speonycteris* and conclude that *S. aurantiadon* and *S. naturalis* should be placed in a new family, Speonycteridae. The phylogenetic analysis suggests that *Speonycteris* was a stem noctilionoid based on the absence of derived characters shared by the other members of the Noctilionoidea. Aside from the Florida fossils, within the Phyllostominae the relative positions of the various genera varied considerably from run to run, except for the Vampyrini (*Notonycteris, Chrotopterus, Vampyrum, Trachops* and sometimes *Lophostoma*), which consistently formed a strongly supported, monophyletic clade.

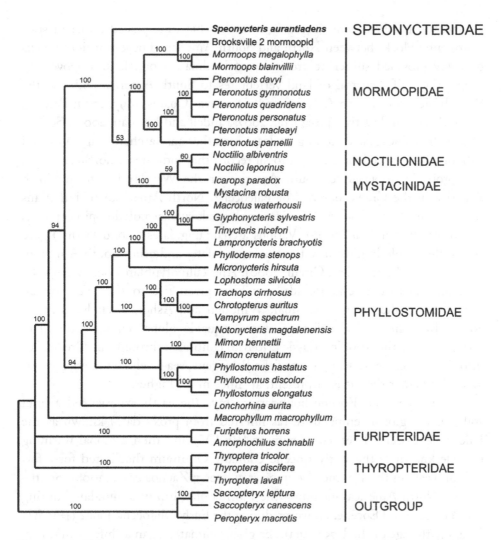

Figure 5.12 Fifty percent majority-rule consensus of 194 trees based on 110 dental morphological characters, 28 genera and 40 species of bats.

5.7 Biogeography

Bats probably originated from laurasiatheres in the Northern Hemisphere and underwent their early evolution there. The oldest fossils of bats come from Early Eocene localities in North America and Europe (Simmons *et al.*, 2008). Molecular genetic evidence suggests that they arose near the Cretaceous/Paleocene boundary, and underwent their fundamental radiations giving rise to extinct as well as the existing families in the Eocene (Teeling *et al.*, 2005; Simmons, 2005a, 2005b). In terms of Western Hemisphere geography, the region between North and South America was complex during the Late

Cretaceous and Paleocene. Tectonic events at that time placed transient sub-continental blocks between North and South America. These transient blocks probably provided sufficient proximity of landmasses or island stopovers to enable the waif dispersal of land vertebrates to South America and also the West Indies from North America and vice versa (Coates, 1997, 2003; Pascual, 2006, Pascual and Ortiz-Jaureguizar, 2007; Pindell and Kennan, 2009). Several mammalian lineages, including at least metatherians, archaic ungulates and eulipotyphlan insectivores have a fossil record that indicates a northern origin and probable dispersal to South America or the West Indies from North America in the Cretaceous and/or Paleocene. North American metatherians were the ancestors of South American marsupials such as polydolopimorphians and others (Goin *et al.*, 2006). Various archaic ungulates ("condylarths") gave rise to the didolodontids, litopterns, notoungulates and other South American ungulates (de Muizon and Cifelli, 2000). And eulipotyphlan insectivores such as geolabidids or apternodontids may have given rise to the West Indian solenodons (Lillegraven *et al.*, 1981; Whidden and Asher, 2001; Roca *et al.*, 2004). In terms of paleoclimate, warm, equable global climates that had prevailed since the Middle Triassic (230 Ma) continued through the Cretaceous into the Paleogene. During most of the Paleogene, sea level fluctuated, but was generally higher than it is at present, at times much higher.

At the Paleocene–Eocene boundary 55 mya the earth experienced a brief spike in the global temperature (as inferred from proxy data) known as the Paleocene–Eocene Thermal Maximum (PETM) and then a global warming episode known as the Early Eocene Climatic Optimum that lasted for a few million years in the Eocene (Lourens *et al.*, 2005; Zachos *et al.*, 2008). South-eastern North America, including the Florida platform was inundated during the Paleocene and Eocene. However, during the Early Oligocene about 34–30 Ma (roughly the age of the I-75 fauna), the global climate began a shift from warm, equable and ice-free (as it had been since the Triassic) to the cooler, temperate pattern of the modern world. The uplift of the Tibetan Plateau in Asia (Dupont-Nivet *et al.*, 2007) and the opening of the Drake Passage between South America and the Antarctic Peninsula allowed the circum-Antarctic current to isolate Antarctica and began a major glaciation that has kept ice there until the present day (Zachos *et al.*, 2001). Attendant with this cooling there was an abrupt and general sea-level decline that periodically exposed the northern part of the Florida platform as dry land through much of the rest of the Cenozoic. From the Early Oligocene through the Middle Miocene, fluctuations in sea level periodically exposed and then inundated the northern part of the Florida peninsula. From the Late Miocene onward, the northern portion of the peninsula has been continually above sea level.

In the Paleocene the near-connection between North and South America was lost. By the Late Eocene-Early Oligocene, South America lost its final connection with Antarctica and entered its famous period of "splendid isolation" that lasted until the Pliocene. In the Caribbean region, however, tectonic models suggest that the northeastern edge of the Caribbean tectonic plate shoved Cuba against the Florida platform on the North American plate by the Middle Eocene (Pindell and Kennan, 2009). The same plate movement possibly pushed up an ephemeral island chain or long, narrow peninsula "Gaarlandia," partly bridging the South America to North America gap (Iturralde-Vinent and MacPhee, 1999). These tectonic movements positioned portions of the landmasses of southern and southeastern North America, the West Indies, Middle America and South America within potential flying distance of one another. After their appearance in the Early Eocene or earlier in North America or Eurasia, bats might have dispersed to tropical proto-Middle America, overwater to the West Indies and even South America by the Middle or Late Eocene or later, although there are as yet no fossils to support this. Changing sea levels and landmass positions and connections in the region between North and South America in the Paleogene would have provided a complex, ever-changing playing field for the evolution and potential dispersal of bats and other organisms.

At least the northern Florida peninsula was probably above sea level by the Early Oligocene (30 Ma). This newly exposed Florida platform is composed of Eocene marine limestones; the earliest fossils of land vertebrates in Florida are from the I-75 local fauna (Florea, 2008; Morgan and Hulbert, 2008). The west-central and north-central Florida areas, including I-75 and Brooksville 2, are exceedingly prone to karstification today and have been since becoming dry land. As at Riversleigh, Australia, in the Quercy phosphorites in France, and in various parts of Africa, karst and paleocaves developed in regions of appropriate age, geographic setting and climate (Morgan and Czaplewski, Chapter 4, this volume). The paleocaves are capable of preserving fossils of bats and other vertebrates. As noted above, the Florida speonycterid fossils were found along with remains of several other kinds of bats, as well as other vertebrates in paleokarst deposits: fissure fillings, sinkholes or paleocaves. The karstic features were developed in marine limestones of the upper Eocene Crystal River Formation (at I-75) and lower Oligocene Suwannee Limestone (at Brooksville 2), relatively shortly before the occurrences of the respective local faunas. Most paleogeographic reconstructions of the region show Florida as completely submerged during the Late Oligocene (maximally 32 Ma to 24 Ma) when remains of the speonycterids were interred along with other members of the I-75 and Brooksville 2 local faunas (e.g., Scotese's Caribbean animation

100 Ma–present: www.scotese.com/caribanim.htm; Blakey's paleomap of the North Atlantic region at 27 Ma: http://jan.ucc.nau.edu/~rcb7/Miocene.html). Of course, the contradiction needs to be resolved and perhaps is simply due to the resolution of the currently available paleomaps; certainly at least a short northern peninsula or small portions of land periodically occurred above sea level in order for the land vertebrates to have existed there. At about the same time as the I-75 local fauna was deposited (32–30 Ma), the southern end of South America probably lost its last tenuous connection with Antarctica and appreciable continental glaciers formed in Antarctica (Zachos et al., 2001, 2008; Woodburne, 2004; Pascual, 2006). Along with the Antarctic glaciation, sea levels would have dropped and exposed the northern portion of the Florida peninsula. At its northern end, South America had already been separated from Middle America and North America in the Paleocene by a seaway, so that dispersal between these continents was restricted (Coates, 2003). In any case, the then-young limestones must have gone from submarine to subaerial exposure relatively quickly in the Early Oligocene, and then experienced karstification and fissure development at elevations barely above sea level, followed quickly by deposition of the fissure fillings. The paleocaves of Florida continued to provide roosting habitat for speonycterids and other bats at least until the Late Oligocene, when the Brooksville 2 local fauna was deposited. The two Speonycteridae described above that came from this karstic region were certainly cave-roosting species. As has probably occurred elsewhere around the world, the cave region in Florida (and potentially in carbonate rocks in the West Indies, too) probably acted as a center for bat speciation (Morgan and Czaplewski, Chapter 4, this volume).

Thus, the Speonycteridae occurred in southeastern North America at about the time when South America began its long Late-Paleogene–Late-Neogene isolation. According to Pascual's (2006) synthesis of South American land-mammal biogeography (excluding bats), most mammal groups that diversified in South America during that continent's Tertiary isolation were a combination of ancient (Mesozoic) mammal lineages left over from the continent's "Gondwanan Episode" (which became extinct there in the Paleogene) and North American invaders, mostly ungulate groups, that reached South America via Middle American islands early in its "South American Episode" in the Paleocene. If the speonycterids as a basal noctilionoid group reached South America early in its long Middle Cenozoic isolation, noctilionoids could have begun their radiation there relatively free from competition with other bats.

During the Late Paleogene and Neogene, tectonism continued to rearrange Central America and the Caribbean Basin. This allowed non-volant terrestrial mammals or waif dispersers to enter the Antilles as early as the Eocene (as evidenced by the perissodactyl *Hyrachyus* in the Eocene of Jamaica), possibly by island-hopping or even overland via the hypothesized Gaarlandia (Iturralde-Vinent and MacPhee, 1999; Coates, 2003; Pascual, 2006).

As noted above in the locality descriptions, bats in the I-75 and Brooksville 2 fossil faunas are dominated by taxa with Neotropical biogeographic affinities, including emballonurids, mormoopids, speonycterids and natalids, as well as the families Vespertilionidae and Molossidae. Among these, the Noctilionoidea are a clade of bat families that radiated in the southern continents and are still primarily restricted to Australia, Madagascar and South America. In the present day, they extend into Central America and the subtropical portions of North America. Clues in fossils as to their origins clearly should be sought in the formerly Gondwanan continents. Lim (2009) suggested an African origin for the Noctilionoidea and most other South American bat groups based on molecular evidence. Some molecular systematic evidence suggests that the Myzopodidae are basal to the Noctilionoidea among living bats (Miller-Butterworth *et al.*, 2007) but other molecular evidence puts Myzopodidae in a nataloid clade (Van Den Bussche and Hoofer, 2001; Van Den Bussche *et al.*, 2003). However, the occurrence of the stem-noctilionoid family Speonycteridae in the extreme southeastern part of North America in peninsular Florida hints at a greater complexity to the biogeography of their basal radiation.

5.8   Acknowledgments

Our thanks go to Glynn Hayes, Art Poyer, Ann Pratt and Erika Simons for help in fieldwork and screenwashing, to Tiffany Stephens for measuring specimens, to Brian Davis and Charles Baker for scanning electron microscopy (SEM) and to Roger Burkhalter for help in bellows photography. The following curators and collection managers allowed use of collections and made loans of specimens: Richard C. Hulbert, Jr. (who also acquired the type specimen of *S. aurantiadens* for FLMNH), Bruce J. MacFadden and S. David Webb, FLMNH (UF); Linda K. Gordon, David Schmidt and Robert Emry, USNM; Nancy Simmons and Bob Randall, AMNH; Bruce Patterson, FMNH; and Patricia Holroyd, UCMP. Drs. Webb and Hulbert also gave permission to mold and cast specimens for distribution and SEM photography. We thank the late Richard Tedford and L. Barry Albright for discussion of the ages of various Florida faunas. Glynn Hayes kindly provided us with his photographs of the Brooksville 2 locality and fissure fillings. Cheryl D. and Jessica L. Czaplewski provided moral and logistical support.

Appendix 5.1

Morphological characters used in assessing *Speonycteris*. Characters 21 and 27 were excluded from some of the computer-assisted phylogenetic analyses.

1. Hypocristid pattern on m1 and m2 nyctalodont (0); submyotodont (1); myotodont (2).
2. Talonid of m3 bears three cusps, hypoconid, entoconid and hypoconulid (0); two cusps, hypoconid and entoconid (1); two cusps, hypoconid and hypoconulid (2); one cusp, hypoconid (3); no distinct cusps, talonid an oblong loph (4).
3. Lower molar (m1 and m2) talonid width including cingula equal to trigonid width (0); narrower than trigonid (1); wider than trigonid (2).
4. Lower molars relative size m1 = m2 = m3 (0); m1 = m2 > m3 (1); m1 > m2 > m3 (2); m2 > m3 > m1 (3); m2 > m1 > m3 (4).
5. Lingual cuspids of lower molars aligned (0); metaconid lingual (1); metaconid slightly labial (2).
6. Position of m1 and m2 hypoconulid relatively labial (0); lingual (1).
7. m1 with paraconid tall (0); short (1) relative to metaconid.
8. Lingual cingulum at foot of trigonid valley of m1 strong (0); weak (1); absent (2).
9. Lingual cingulum at foot of trigonid valley of m2 strong (0); weak (1); absent (2).
10. Cristid obliqua on m1 meets posterior wall of trigonid centrally, below the notch in postcristid (0); labial to notch (1); lingual to notch (2); at metaconid (3); drops off before reaching posterior wall of trigonid (4).
11. Shape of entocristid of m1/m2 in occlusal view nearly straight (0); curved, concave lingually (1); angled, concave lingually (2); entocristid absent (3).
12. Notches in cristid obliqua and postcristid of m1 absent (0); weakly developed (1); strongly developed, carnassial-like (2).
13. Angle formed between paracristid and protocristid of m1 open (0); closed (1).
14. Angle formed between paracristid and protocristid of m2 open (0); closed (1).
15. Paracristid of m1 with acute angular notch (0); concave edge (1).
16. Labial cingulum of lower molars narrow (0); wide and regular in height (1); strong and irregular (2).
17. Number of lower premolars three (0); two (1); one (2).
18. p2 with two roots (0); one root (1).
19. p4 with two roots (0); one root (1); three roots (2).
20. Labial cingulum of p4 in lateral view without posteroventral expansion (0); with rounded posteroventral expansion (1); with two convexities (2).
21. Relative height of lower premolars p2 = p3 = p4 (0); p2 < p3 < p4 (1); p3 < p2 < p4 (2); p4 = p3 > p2 (3); p4 = p2 > p3 (4).
22. Lower premolars in occlusal view laterally compressed (0); not compressed (1); anteroposteriorly compressed (2).

23. Lower premolars aligned with axis of tooth row (o); oblique to axis (1).
24. Anterolingual cingular cusp of p4 present (o); absent (1).
25. Posterolingual cingular cusp of p4 present (o); absent (1).
26. p4 molariform, with talonid-like heel (o); talonid reduced or absent, vestigial metaconid present (1); talonid and metaconid absent (2).
27. Lower incisor alveoli aligned (o); staggered (1).
28. Number of lower incisors three (o); two (1); one (2).
29. Number of upper premolars three (o); two (1); one (2).
30. Middle upper premolar with three roots (o); two roots (1); one root (2).
31. M3 with four commissures (o); three commissures (1); two commissures (2); one commissure (3).
32. Upper molars (M1 and M2) hypocone absent (o); hypocone cuspate (1); hypocone cristate (2).
33. M1 and M2 with expanded talon (o); with talon weakly developed (1); without talon (2).
34. Hypoconal crest not confluent with cingulum of talon or absent (o); confluent (1); talon cingulum absent (2).
35. M1 and M2 with conule(s) (o); without conules (1).
36. M1 and M2 with paraloph and metaloph (o); paraloph only (1); metaloph only (2); neither paraloph nor metaloph (3).
37. Preprotocrista extends to parastyle (o); to base of paracone (1).
38. Postprotocrista does not reach metacone and lacks hypoconal swelling (o); extends to base of metacone (1); extends to metaconule (2); extends to hypocone (3); extends to metastyle (confluent with postcingulum), bypassing hypocone (if any) (4).
39. Postcingulum on M1 and M2 present (o); absent (1).
40. Lateral mental foramen positioned between roots of c1 and anterior-most premolar (o); between roots of p2 and next posterior premolar (1); between roots of the p4 (2).
41. Mandibular symphysis short (o); elongate (1).
42. Coronoid process tall (o); moderate in height (1); short (2).
43. Horizontal ramus of moderate thickness (about as deep as teeth are tall) (o); gracile (shallower than the teeth are tall) (1); robust (deeper than the teeth are tall) (2).
44. Angular process projects at or below level of occlusal plane of tooth row, well below coronoid process (o); or angular process projects above level of occlusal plane of tooth row, at same level as the coronoid process (1).
45. Greater tuberosity of humerus does not reach level of humeral head (o); reaches level of humeral head (1); extends proximally slightly past level of humeral head (2); extends well past humeral head (3).
46. Humeral head rounded (o); ovoid and canted toward greater tuberosity (1); ovoid and canted toward lesser tuberosity (2); ovoid, not canted (3).

47. Supraglenoid fossa in proximal end of humerus shallow (o); deep (1).

48. Deltopectoral ridge of humerus short (o); long (1).

49. Shaft of humerus straight (o); with slight sigmoid curve (1); strongly sigmoid (2).

50. Distal articular surface of humerus aligned with shaft (o); moderately offset (1); strongly offset (2).

51. Central surface of capitulum of humerus rounded (o); reduced (1); ridged and straight (2); ridged and tilted (3).

52. Epitrochlea of distal humerus narrow (o); broad (1).

53. Distal spinous process of humerus absent (o); shorter than trochlea (1); level with trochlea (2); long, extending beyond trochlea (3).

54. Flexor fossa of radius ventral (o); anterior (1).

55. On m1, distance between paraconid and metaconid is: (o) less than; (1) greater than; (2) equal to distance between metaconid and entoconid.

56. On m1, paraconid height is: (o) about half the height of the metaconid; (1) low, less than half height of metaconid; (2) essentially absent, forming a low flat shelf level with the trigonid valley.

57. p3 configuration: (o) large, double-rooted, elongated in occlusal outline; (1) small, single-rooted, circular or oval in occlusal outline; (2) elongated and single-rooted; (3) absent; (4) small, double-rooted and anteroposteriorly compressed.

58. p3 alveolus (or alveoli): (o) in line with tooth row; (1) alveolus displaced lingually from longitudinal axis of toothrow.

59. Lower molars crests: (o) thin and sharp-edged; (1) thicker and slightly inflated.

60. Mandibular angle (lower edge of dentary in lateral view): (o) little or no angle between horizontal ramus and ascending ramus; (1) moderate angle reflecting moderate cranial flexion; (2) strong bend reflecting exaggerated cranial flexion.

61. Humeral head: (o) separated from greater tuberosity by groove; (1) not well separated from greater tuberosity.

62. Lateral surface of greater tuberosity of humerus: (o) with no groove; (1) weakly grooved; (2) strongly grooved.

63. Humerus with flange of bone extending distally from lesser tuberosity along medial edge of shaft: (o) weak to absent; (1) represented by a tuberosity at distal extremity; (2) a continuous broad flange; (3) a continuous narrow flange.

64. Humerus groove separating medial and lateral ridges of capitulum: (o) present; (1) absent.

65. Groove between capitulum and trochlear groove of humerus: (o) narrow and shallow; (1) intermediate in width and depth; (2) wide and deep; (3) narrow and deep.

66. Humerus with medial edge of trochlea: (o) at same level as capitulum; (1) extending slightly distal to capitulum; (2) extending far distal to capitulum.

67. Distal spinous process or expanded medial epicondyle of humerus: (o) connected to trochlea; (1) separated from trochlea by distinct notch.

68. Distal spinous process of humerus in medial view: (0) oriented vertically (in line with shaft); (1) angled posteriorly; (2) absent.

69. Medial epicondyle of humerus: (0) narrow and lacks process or concavity; (1) moderately broad with concavity at proximal medial corner; (2) very broad with rounded process at proximal medial corner; (3) broad with no processes.

70. Distal shaft of humerus flange on posterior wall just proximal to distal articulation: (0) flange absent; (1) flange low with distinct convexity; (2) flange tall and rounded-triangular in outline.

71. Distal portion of humerus posterior surface of shaft: (0) rounded; (1) flattened.

72. m3 talonid distal cingulum: (0) well developed; (1) moderately developed; (2) absent.

73. Mandibular symphysis unfused (0); fused (1).

74. Mandibular symphysis with posteroventral projection absent (0); weak to moderate (1); strong (2).

75. p3 normal in size and position in toothrow (0); reduced in size and displaced labially from toothrow (1); reduced and displaced lingually (2); p3 absent (3); p3 with tall, laterally compressed, curving blades (4); p3 reduced and not displaced (5).

76. p3 with one root (0); with two roots obliquely positioned in jaw (1); having two roots in line with toothrow (2).

77. Lower teeth with normal cingula (0); cingula greatly reduced (1).

78. Lower premolar crowns not expanded downward on labial and lingual faces (0); expanded (1).

79. m1 hypoconulid normal in size (0); reduced (1).

80. p2 normal in size (0); reduced (1).

81. m1 trigonid valley present, normal (0); absent or much reduced (protoconid positioned about on midline of tooth (see Savage's (1951) obliquity and eccentricity) and paracristid reduced) (1).

82. m1 with labial cingulum normal (0); with strong cingular cuspid anterior to protoconid (1); labial cingulum rises on base of a very short protoconid with sharp notch in paracristid (2).

83. m1 with distal metacristid absent (0); present and separated by a notch from entocristid (1).

84. m1 in occlusal view having lingual wall of tooth between entoconid and metaconid concave (0); vertical (1); convex (2).

85. m1 with small cuspule at lingual end of cristid obliqua adjacent to metaconid absent (0); present (1).

86. M1 with hooked parastyle without posterolabially directed ridge (0); with distinct short ridge (1); with preparacrista curving posteriorly and not connected to parastyle (2); with long ectocingulum spanning the anterior "vee" of the ectoloph (3).

87. M1 and M2 preprotocrista broad all the way to parastyle (0); weak (thin) along anterior wall of paracone (1); absent along anterior wall of paracone (2).

88. M1 and M2 lingual margin of tooth in occlusal view with shallow or no indentation between base of protocone and talon hypocone (0); deep indentation (1).

89. Upper canine lingual cingulum entire (0); bilobed (1).

90. M1 metastyle simple and straight (in line with postmetacrista) or slightly curved (0); metastyle hooked (1).

91. P4 with lingual cingular cusp absent (0); weak (1); strong (2).

92. P4 anterior portion with small cingular cusp positioned anterior to the main cusp (0); with broad lobe positioned anterolingual to the main cusp (1); positioned posterior to the main cusp (2); absent (3); P4 expanded, shield-like with central broad longitudinal groove (4).

93. M1 and M2 lingual cingulum absent (0); weak and limited to posterior base of protocone (1); weak and connects to hypoconal crest/talon cingulum (2); weak and interrupted dorsal to protocone but present anterior to protocone (3); complete all the way from parastyle, around lingual side, around talon, to metastyle (4).

94. M1 and M2 protocone positioned far from paracone and metacone (0); close to paracone and metacone, the preprotocrista and postprotocrista shortened (1).

95. M1 and M2 mesostyle single (0); doubled (small notch between swollen cuspule at end of postparacrista and that at end of premetacrista) (1).

96. P4 with broadly rounded lingual half including large talon base (0); bilobed lingually with anterolingual flat shelf and posterolingual small basined talon (1); P4 transversely elongated with rounded lingual border and talon basin (2); not bilobed and with reduced talon (3); broad and flattened with longitudinal central basin (4).

97. M1 and M2 hypocone absent or formed as an indistinct low rise on the post-protocrista (0); hypocone represented by a low rise on the anterolingual corner of the talon, continuous with the talon cingulum (1); hypocone tall and cuspate on the anterolingual corner, and aligned with but not distinctly connected by sharp ridges to postprotocrista nor talon cingulum (2); hypocone low and cuspate and situated internally within talon (i.e., away from the lingual edge), its ridge not connected to talon cingulum (or lingual cingulum, if any) (3); hypocone well developed and cristate on the anterolingual corner of the talon, and connected by crests or ridges with postprotocrista and talon cingulum (4); hypocone low or absent on anterolingual corner of talon, connected to a strongly lingually curving postprotocrista and also continuous with talon cingulum (5).

98. p2 in lateral profile tall and pointed (0); a rounded or elongated, curved blade (1); p2 reduced in size (2).

99. p4 without posterolabial crest (0); with posterolabial crest and concavity between it and posterolingual crest (1).

100. M1 relative lengths of ectoloph crests (preparacrista, postparacrista, premetacrista and postmetacrista): preparacrista and postmetacrista normal, about same length

as postparacrista and premetacrista (0); preparacrista short and postmetacrista elongated (1).

101. Radius flexor fossa for insertion of M. biceps an open, triangular area bounded by bony ridges (0); a deep narrow groove (1).

102. Femur proximal end of shaft with long, longitudinal, flange-like third trochanter absent (0); weak (1); strong (2).

103. Femur distal end spline absent (0); present (1).

104. Femur proximal end greater and lesser trochanters normal (0); reduced (1).

105. Femur head central fovea small (0); large (1).

106. Femur head spheroidal (0); cylindrical (1).

107. Femur with additional flange-like longitudinal spline distal to head absent (0); weak or developed as tubercle only (1); strong (2).

108. Femur proximal shaft straight (0); bent medially towards greater trochanter (1).

109. M1 and M2 mesostyle position: falls on a line drawn between the parastyle and metastyle (0); indented lingually, situated well lingual of a line through the parastyle and metastyle (1).

110. M1 and M2 W-shaped ectoloph "vees": normal (lines drawn such that they bisect the angles between the preparacrista and postparacrista and between the preme-tacrista and postmetacrista will be approximately at right angles to the midline of the skull) (0); obliquely oriented (lines bisecting the vees will be oblique to the midline of the skull) (1).

111. m1 and m2 crown height relatively low, with robust crests (0); relatively high, with slender crests (1).

112. m1 and m2 talonid anteroposterior length relative to trigonid length: relatively long, talonid about as long as trigonid (0); talonid reduced, relatively short (1).

Appendix 5.2

Recent specimens examined in the completion of this study.
Emballonuridae:
Peropteryx macrotis, USNM 391022 female, Brazil: Minas Gerais; USNM 313146 male, Panama: Buena Vista; USNM 393000 male, Brazil: Pará.
Saccopteryx canescens, USNM 392996 male, Brazil: Pará.
Saccopteryx leptura, USNM 540657 female, Tobago: St. John Parish; USNM 392999 male, Brazil: Pará; USNM 513430 male, Ecuador: Zamora-Chinchipe.

Mormoopidae:
Pteronotus davyi, USNM 361894 male, Dominica: Grand Bay; USNM 583006 male, Belize: Stann Creek.
Pteronotus gymnonotus, USNM 431544 male, Colombia: Bolivar; USNM 305177 male, Panama: Chilibre.

Pteronotus macleayii, USNM 113771 male, Cuba: Baracoa; USNM 260730 male, Jamaica: Montego Bay.

Pteronotus parnellii, AMNH 189595 male, Mexico: Oaxaca; USNM 562734 male, Costa Rica: Guanacaste; USNM 511714 female, Mexico: Nayarit; USNM 545140 female, USNM 545144 female, Jamaica: Trelawny Parish; USNM 523047 female, Mexico: Nayarit.

Pteronotus personatus, USNM 430190 male, USNM 433531 female, Colombia: Bolivar.

Pteronotus quadridens, USNM 113786 female, Cuba: Baracoa; USNM 121061 sex unknown, Cuba: Baracoa.

Mormoops blainvillei, USNM 113766 male, Cuba: Baracoa; USNM 545146 male, Jamaica: Trelawny Parish.

Mormoops megalophylla, AMNH 190138 male, Mexico: Oaxaca; USNM 513432 male, Ecuador: Pichincha; USNM 417184 male, USNM 407207 male, Venezuela: Sucre; USNM 431641 female, Colombia: Bolivar.

Noctilionidae:

Noctilio leporinus, AMNH 254605 male, Mexico: Chiapas; USNM 281311 male, Colombia: Río Guaimaral.

Noctilio albiventris, AMNH 210594 female, Bolivia: El Beni; USNM 390586 male, Bolivia: Beni; USNM 483299 female, Colombia: Cauca.

Phyllostomidae:

Chrotopterus auritus, AMNH 261373 male, Bolivia: Santa Cruz; AMNH 267852 female, French Guiana: Paracou; USNM 530910 male, Peru: Madre de Dios.

Glyphonycteris daviesi, USNM 335105 female, Panama: San Blas.

Glyphonycteris sylvestris, USNM 388734 female, Venezuela: T. F. Amazonas.

Lampronycteris brachyotis, USNM 361501 male, Brazil: Pará.

Lonchorhina aurita, AMNH 230122 male, Peru: Pasco; USNM 499290 male, Colombia: Antioquia.

Lonchorhina orinocoensis, USNM 373422 male, Venezuela: Apure.

Lophostoma evotis, AMNH 267635 male, Guatemala: Izabal.

Lophostoma silvicola, AMNH 262425 male, Bolivia: Pando; AMNH 267923 male, French Guiana: Paracou; USNM 364278 female, Peru: Pasco; USNM 281053 male, Colombia: Marimondas.

Macrophyllum macrophyllum, AMNH 262424, Bolivia: Pando; USNM 554526 female, Paraguay: Parque Nac. Cerro.

Macrotus waterhousii, AMNH 2148 male, Mexico: Jalisco; USNM 508483 male, Mexico: Nayarit; UF 20816 male, Haiti.

Micronycteris hirsuta, AMNH 267096 male, French Guiana: Paracou.

Mimon bennetti, AMNH 265107 male, Guatemala: Izabal; USNM 123393 male, Brazil: Ypanema; USNM 172080 male, Mexico: Yucatan.

Mimon crenulatum, AMNH 267885 male, French Guiana: Paracou; USNM 393652 male, Brazil: Mato Grosso.

Neonycteris pusilla, AMNH 78830 (holotype) male, Brazil: Rio Waupes.

Phylloderma stenops, AMNH 124834 female, Honduras: Tegucigalpa; AMNH 205371 female, Trinidad: Arima; AMNH female French Guiana: Paracou; USNM 457937 male, Panama: Colon.

Phyllostomus discolor, AMNH 254612 female, Mexico: Chiapas; AMNH 267123 female, French Guiana: Paracou; USNM 531242 male, Peru: Piura.

Phyllostomus elongatus, AMNH 210674 female, Bolivia: Beni; AMNH 266058 female, French Guiana: Paracou; USNM 564305 male, Bolivia: Beni.

Phyllostomus hastatus, AMNH 209334 male, Brazil: Rondonia; USNM 361516 male, Brazil: Pará.

Trachops cirrhosus, AMNH 209349 male, Bolivia: Beni; AMNH 266084 male, French Guiana: Paracou; USNM 281037 male, Colombia: Río Guaimaral.

Trinycteris nicefori, AMNH 267878 female, French Guiana: Paracou; USNM 385422 male, Venezuela: Bolivar.

Vampyrum spectrum, AMNH 261379 male, Bolivia: Beni; AMNH 267446 female, French Guiana: Paracou; USNM 562766 female, Costa Rica: Heredia.

Furipteridae:

Amorphochilus schnablii, USNM 152261 sex unknown, Peru: Dintorni de Lima; USNM 269981 female, Peru: Arequipa.

Furipterus horrens, USNM 405785 male, Venezuela: T. F. Amazonas; USNM 315737 male, Panama: Changuinola.

Thyropteridae:

Thyroptera discifera, USNM 102922 male, Venezuela: San Julian; USNM 457965 female, Panama: Barro Colorado Island; USNM 105422 female, Venezuela: San Julian.

Thyroptera lavali, FMNH 89118 male, FMNH 89120 female, Peru: Loreto.

Thyroptera tricolor, USNM 564337 female, Bolivia: Beni; USNM 281200 male, Colombia: Colonia Agricola.

Mystacinidae:

Mystacina robusta, AMNH 160269 female, New Zealand: Stuart Island.

Mystacina tuberculata, USNM 120576 female, University of California Museum of Vertebrate Zoology174825 sex unknown, New Zealand.

Myzopodidae:

Myzopoda aurita, USNM 448931 male, Madagascar: Fianarantsoa Province.

Appendix 5.3

Input data matrix of character states for fossil and recent bats and morphological characters used in this study. Morphological characters are listed in Appendix 5.1; taxa are listed in Methods and Appendix 5.2. Character states for characters 21 and 27 are blank because they were excluded from the analysis. Character states for taxa *Speonycteris naturalis*, *Notonycteris sucharadeus* and *Neonycteris pusilla* were excluded from the analysis due to incompleteness of data.

An accompanying data matrix (MS Excel file) is available at www.cambridge.org/9780521768245.

5.9   REFERENCES

Albright, L. B., III. (1998). The Arikareean land mammal age in Texas and Florida: southern extension of Great Plains faunas and Gulf Coastal Plain endemism. In *Depositional Environments, Lithostratigraphy, and Biostratigraphy of the White River and Arikaree Groups (Late Eocene to Early Miocene, North America)*, ed. D. O. Terry, Jr., H. E. LaGarry and R. M. Hunt, Jr. Geological Society of America Special Paper, **325**, 167–183.

Butler, P. M. (1978). Insectivora and Chiroptera. In *Evolution of African Mammals*, ed. V. J. Maglio and H. B. S. Cooke. Cambridge, MA: Harvard University Press, pp. 56–68.

Coates, A. G. (ed.). (1997). *Central America: A Natural and Cultural History.* New Haven, CT: Yale University Press.

Coates, A. G. (2003). *Paseo Pantera: Una Historia de la Naturaleza y Cultura de Centroamérica.* Washington, DC: Smithsonian Books.

Czaplewski, N. J. (1996a). Opossums (Didelphidae) and bats (Noctilionidae and Molossidae) from the Late Miocene of the Amazon Basin. *Journal of Mammalogy*, **77**, 84–94.

Czaplewski, N. J. (1996b). *Thyroptera robusta* Czaplewski, 1996, is a junior synonym of *Thyroptera lavali* Pine, 1993 (Mammalia, Chiroptera). *Mammalia*, **60**, 153–156.

Czaplewski, N. J. (1997). Chiroptera. In *Vertebrate Paleontology in the Neotropics: The Miocene Fauna of La Venta, Colombia*, ed. R. F. Kay, R. H. Madden, R. L. Cifelli and J. J. Flynn. Washington, DC: Smithsonian Institution Press, pp. 410–431.

Czaplewski, N. J. (2010). Colhuehuapian bats (Mammalia: Chiroptera) from the Gran Barranca, Chubut province, Argentina. In *The Paleontology of Gran Barranca: Evolution and Environmental Change through the Middle Cenozoic of Patagonia*, ed. R. H. Madden, A. A. Carlini, M. G. Vucetich and R. F. Kay. Cambridge: Cambridge University Press, pp. 240–252.

Czaplewski, N. J. and Campbell, K. E. (2004). *A Possible Bat (Mammalia: Chiroptera) from the ?Eocene of Amazonian Perú.* Natural History Museum of Los Angeles County, Science Series, **4**, pp. 141–144.

Czaplewski, N. J., Morgan, G. S. and Naeher, T. (2003a). Molossid bats from the late Tertiary of Florida with a review of the Tertiary Molossidae of North America. *Acta Chiropterologica*, **5**, 61–74.

Czaplewski, N. J., Takai, M., Naeher, T. M., Shigehara, N. and Setoguchi, T. (2003b). Additional bats from the middle Miocene La Venta fauna of Colombia. *Revista de la Academia Colombiana de Ciencias Exactas, Físicas y Naturales*, **27**, 263–282.

Czaplewski, N. J. Morgan, G. S. and McLeod, S. A. (2008). Chapter 12, Chiroptera. In *Evolution of Tertiary Mammals of North America, Vol. 2: Small Mammals, Xenarthrans, and Marine Mammals*, ed. C. M. Janis, G. F. Gunnell and M. D. Uhen. Cambridge: Cambridge University Press, pp. 174–197.

de Muizon, C. and Cifelli, R. L. (2000). The "condylarths" (archaic Ungulata, Mammalia) from the early Palaeocene of Tiupampa, Bolivia: implications on the origin of the South American ungulates. *Geodiversitas*, **22**, 47–150.

Dupont-Nivet, G., Krijgsman, W., Langereis, C. G. *et al.* (2007). Tibetan plateau aridification linked to global cooling at the Eocene-Oligocene transition. *Nature*, **445**, 635–638.

Emry, R. J., Bjork, P. R. and Russell, L. S. (1987). The Chadronian, Orellan, and Whitneyan North American land mammal ages. In *Cenozoic Mammals of North America: Geochronology and Biostratigraphy*, ed. M. O. Woodburne. Berkeley, CA: University of California Press, pp. 118–152.

Florea, L. J. (2008). Geology and hydrology of karst in west-central and north-central Florida. In *Caves and Karst of Florida. A guidebook for the 2008 National Convention of the National Speleological Society*, ed. L. J. Florea. Huntsville, AL: National Speleological Society, pp. 225–239.

Frailey, C. D. (1978). An early Miocene (Arikareean) fauna from northcentral Florida (the SB-1A) local fauna. *Occasional Papers, Museum of Natural History, University of Kansas*, **75**, 1–20.

Frailey, C. D. (1979). The large mammals of the Buda local fauna (Arikareean: Alachua County, Florida). *Bulletin of the Florida State Museum, Biological Sciences*, **24**, 123–173.

Giannini, N. P., Wible, J. R. and Simmons, N. B. (2006). On the cranial osteology of Chiroptera. I. *Pteropus* (Megachiroptera: Pteropodidae). *Bulletin of the American Museum of Natural History*, **295**, 1–134.

Goin, F. J., Pascual, R., Tejedor, M. F. *et al.* (2006). The earliest Tertiary therian mammal from South America. *Journal of Vertebrate Paleontology*, **26**, 505–510.

Gray, J. E. (1821). On the natural arrangement of vertebrose animals. *London Medical Repository, Monthly Journal, and Review*, **15**, 296–311.

Gray, J. E. (1825). An attempt at a division of the family Vespertilionidae into groups. *Zoological Journal*, **2**, 242–243.

Gunnell, G. F. and Simmons, N. B. (2005). Fossil evidence and the origin of bats. *Journal of Mammalian Evolution*, **12**, 209–246.

Gunnell, G. F., Simons, E. L. and Seiffert, E. R. (2008). New bats (Mammalia: Chiroptera) from the late Eocene and early Oligocene, Fayum Depression, Egypt. *Journal of Vertebrate Paleontology*, **28**, 1–11.

Gunnell, G. F., Worsham, S. R., Seiffert, E. R. and Simons, E. L. (2009). *Vampyravus orientalis* Schlosser (Chiroptera) from the Early Oligocene (Rupelian), Fayum, Egypt – body mass, humeral morphology and affinities. *Acta Chiropterologica*, **11**, 271–278.

Hand, S. J., Murray, P. F., Megirian, D., Archer, M. and Godthelp, H. (1998). Mystacinid bats (Microchiroptera) from the Australian Tertiary. *Journal of Paleontology*, **72**, 538–545.

Hand, S., Archer, M. and Godthelp, H. (2001). New Miocene *Icarops* material (Microchiroptera: Mystacinidae) from Australia, with a revised diagnosis of the genus. *Association of Australasian Palaeontologists Memoir*, **25**, 139–146.

Hand, S., Archer, M. and Godthelp, H. (2005). Australian Oligo-Miocene mystacinids (Microchiroptera): upper dentition, new taxa, and divergence of New Zealand species. *Geobios*, **38**, 339–352.

Hand, S. J., Novacek, M. J., Godthelp, H. and Archer, M. (1998). First Eocene bat from Australia, *Journal of Vertebrate Paleontology*, **14**, 375–381.

Hand, S. J., Worthy, T., Beck, R. *et al.* (2007). New Zealand's first Tertiary bats and the evolution of bats in the southern hemisphere. Abstract, *Australasian Evolution Society 5th Conference, 12–15 June 2007*. Sydney: University of New South Wales.

Hayes, F. G. (2000). The Brooksville 2 local fauna (Arikareean, latest Oligocene): Hernando County, Florida. *Bulletin of the Florida Museum of Natural History*, **43**, 1–47.

Hayes, F. G. (2005). Arikareean (Oligocene-Miocene) *Herpetotherium* (Marsupialia, Didelphidae) from Nebraska and Florida. *Bulletin of the Florida Museum of Natural History*, **45**, 335–353.

Holman, J. A. (1999). Early Oligocene (Whitneyan) snakes from Florida (USA), the second oldest colubrid snakes in North America. *Acta Zoologica Cracoviensia*, **42**, 447–454.

Hoofer, S. R., Reeder, S. A., Hansen, E. W. and Van Den Bussche, R. A. (2003). Molecular phylogenetics and taxonomic review of noctilionoid and vespertilionoid bats (Chiroptera: Yangochiroptera). *Journal of Mammalogy*, **84**, 809–881.

Hürzeler, J. (1944). Beiträge zur Kenntnis der Dimylidae. *Schweizerischen Paläontologische Abhandlungen*, **65**, 1–44.

Hutcheon, J. M. and Kirsch, J. A. W. (2004). Camping in a different tree: results of molecular systematic studies of bats using DNA–DNA hybridization. *Journal of Mammalian Evolution*, **11**, 17–47.

Hutchison, J. H. and Lindsay, E. H. (1974). The Hemingfordian mammal fauna of the Vedder locality, Branch Canyon Formation, Santa Barbara County, California. Part 1: Insectivora, Chiroptera, Lagomorpha, and Rodentia (Sciuridae). *PaleoBios*, **15**, 1–19.

Iturralde-Vinent, M. A. and MacPhee, R. D. E. (1999). Paleogeography of the Caribbean region: implications for Cenozoic biogeography. *Bulletin of the American Museum of Natural History*, **238**, 1–95.

James, G. T. (1963). Paleontology and nonmarine stratigraphy of the Cuyama Valley Badlands, California. Part I. Geology, faunal interpretations, and systematic descriptions of Chiroptera, Insectivora, and Rodentia. *University of California Publications in Geological Sciences*, **45**, 1–154.

Kennedy, M., Patterson, A. M., Morales, J. C. *et al.* (1999). The long and short of it: branch lengths and the problem of placing the New Zealand short-tailed bat, *Mystacina. Molecular Phylogenetics and Evolution*, **13**, 405–416.

Kirsch, J. A. W., Hutcheon, J. M., Byrnes, G. P. and Lloyd, B. D. (1998). Affinities and historical zoogeography of the New Zealand short-tailed bat, *Mystacina tuberculata* Gray 1843, inferred from DNA-hybridization comparisons. *Journal of Mammalian Evolution*, **5**, 33–64.

Legendre, S. (1984). Étude odontologique des représentants actuels du groupe *Tadarida* (Chiroptera, Molossidae). Implications phylogéniques, systématiques et zoogéographiques. *Revue Suisse de Zoologie*, **91**, 399–442.

Lillegraven, J. A., McKenna, M. C. and Krishtalka, L. (1981). Evolutionary relationships of middle Eocene and younger species of *Centetodon* (Mammalia, Insectivora, Geolabididae) with a description of the dentition of *Ankylodon* (Adapisoricidae). *University of Wyoming Contributions to Geology*, **45**, 1–115.

Lim, B. K. (2009). Review of the origins and biogeography of bats in South America. *Chiroptera Neotropical*, **15**, 391–410.

Lourens, L. J., Sluijs, A., Kroon, D. *et al.* (2005). Astronomical pacing of late Palaeocene to early Eocene global warming events. *Nature*, **435**, 1083–1087.

MacFadden, B. J. and Morgan, G. S. (2003). New oreodont (Mammalia, Artiodactyla) from the late Oligocene (early Arikareean) of Florida. *Bulletin of the American Museum of Natural History*, **279**, 368–396.

Menu, H. and Sigé, B. (1971). Nyctalodontie et myotodontie, importants caractères de grades évolutifs chez les chiroptères entomophages. *Comptes Rendus de Séances de l'Académie des Sciences*, **272**, 1735–1738.

Miller-Butterworth, C. M., Murphy, W. J., O'Brien, S. J. *et al.* (2007). A family matter: conclusive resolution of the taxonomic position of the long-fingered bats, *Miniopterus*. *Molecular Biology and Evolution*, **24**, 1553–1561.

Morgan, G. S. (1991). Neotropical Chiroptera from the Pliocene and Pleistocene of Florida. *Bulletin of the American Museum of Natural History*, **206**, 176–213.

Morgan, G. S. and Czaplewski, N. J. (2003). A new bat (Chiroptera: Natalidae) from the Early Miocene of Florida, with comments on natalid phylogeny. *Journal of Mammalogy*, **84**, 729–752.

Morgan, G. S. and Hulbert, Jr., R. C. (2008). Cenozoic vertebrate fossils from paleokarst deposits in Florida. In *Caves and Karst of Florida. A Guidebook for the 2008 National Convention of the National Speleological Society*, ed. L. J. Florea. Huntsville, AL: National Speleological Society, pp. 248–271.

Morgan, G. S., Linares, O. J. and Ray, C. E. (1988). New species of fossil vampire bats (Mammalia: Chiroptera: Desmodontidae) from Florida and Venezuela. *Proceedings of the Biological Society of Washington*, **101**, 912–928.

Pascual, R. (2006). Evolution and geography: the biogeographic history of South American land mammals. *Annals of the Missouri Botanical Garden*, **93**, 209–230.

Pascual, R. and Ortiz-Jaureguizar, E. (2007). The Gondwanan and South American Episodes: two major and unrelated moments in the history of the South American mammals. *Journal of Mammalian Evolution*, **14**, 75–137.

Patton, T. H. (1969). An Oligocene land vertebrate fauna from Florida. *Journal of Paleontology*, **43**, 543–546.

Pierson, E. D., Sarich, V. M., Lowenstein, J. M., Daniel, M. J. and Rainey, W. E. (1986). A molecular link between the bats of New Zealand and South America. *Nature*, **324**, 60–63.

Pindell, J. L. and Kennan, L. (2009). Tectonic evolution of the Gulf of Mexico, Caribbean and northern South America in the mantle reference frame: an update. *Geological Society, London, Special Publications*, **328**, 1–55.

Prothero, D. R. and Emry, R. J. (2004). The Chadronian, Orellan, and Whitneyan North American land mammal ages. In *Late Cretaceous and Cenozoic Mammals of North America Biostratigraphy and Geochronology*, ed. M. O. Woodburne. New York: Columbia University Press, pp. 156–168.

Roca, A. L., Bar-Gal, G. K., Eizirik, E. *et al.* (2004). Mesozoic origin for West Indian insectivores. *Nature*, **429**, 649–651.

Savage, D. E. (1951). A Miocene phyllostomatid bat from Colombia, South America. *University of California Publications, Bulletin of the Department of Geological Sciences*, **28**, 357–366.

Schlosser, M. (1910). O. Über einige fossile Säugetiere aus dem Oligocän von Ägypten. *Zoologischer Anzeiger, Leipzig*, **35**, 500–508.

Schlosser, M. (1911). Beiträge zur Kenntnis der oligozänen Land-säugetiere aus dem Fayum: Ägypten. *Beiträge zur Paläontologie und Geologie Österreich-Ungarns Orients*, **24**, 51–167.

Sigé, B. (1985). Les chiroptères oligocènes du Fayum, Egypte. *Geologica et Palaeontologica*, **19**, 161–189.

Simmons, N. B. (1998). A reappraisal of interfamilial relationships of bats. In *Bat Biology and Conservation*, ed. T. H. Kunz and P. A. Racey. Washington, DC: Smithsonian Institution Press, pp. 3–26.

Simmons, N. B. (2000). Bat phylogeny: an evolutionary context for comparative studies. In *Ontogeny, Functional Ecology, and Evolution of Bats*, ed. R. A. Adams and S. C. Pedersen. Cambridge: Cambridge University Press, pp. 9–58.

Simmons, N. B. (2005a). Chiroptera. In *The Rise of Placental Mammals*, ed. K. D. Rose and J. D. Archibald. Baltimore, MD: Johns Hopkins University Press, pp. 159–174.

Simmons, N. B. (2005b). An Eocene big bang for bats. *Science*, **307**, 527–528.

Simmons, N. B. and Geisler, J. H. (1998). Phylogenetic relationships of *Icaronycteris*, *Archaeonycteris*, *Hassianycteris*, and *Palaeochiropteryx* to extant bat lineages, with comments on the evolution of echolocation and foraging strategies in Microchiroptera. *Bulletin of the American Museum of Natural History*, **235**, 1–182.

Simmons, N. B., Seymour, K. L., Habersetzer, J. and Gunnell, G. F. (2008). Primitive early Eocene bat from Wyoming and the evolution of flight and echolocation. *Nature*, **451**, 818–822.

Smith, J. D. (1972). Systematics of the chiropteran family Mormoopidae. *University of Kansas Museum of Natural History, Miscellaneous Publication*, **56**, 1–132.

Tedford, R. H., Galusha, T., Skinner, M. F. *et al.* (1987). Faunal succession and biochronology of the Arikareean through Hemphillian interval (late Oligocene through earliest Pliocene epochs) in North America. In *Cenozoic Mammals of North America: Geochronology and Biostratigraphy*, ed. M. O. Woodburne. Berkeley, CA: University of California Press, pp. 153–210.

Tedford, R. H., Swinehart, J. B., Swisher, III, C. C. *et al.* (1996). The Whitneyan-Arikareean transition in the High Plains. In *The Terrestrial Eocene-Oligocene Transition in North America*, ed. D. R. Prothero and R. J. Emry. Cambridge: Cambridge University Press, pp. 312–334.

Tedford, R. H., Albright III, L. B., Barnosky, A. D. *et al.* (2004). Mammalian biochronology of the Arikareean through Hemphillian interval (Late Oligocene through Early Pliocene epochs). In *Late Cretaceous and Cenozoic Mammals of North America Biostratigraphy and Geochronology*, ed. M. O. Woodburne. New York: Columbia University Press, pp. 169–231.

Teeling, E. C., Madsen, O., Murphy, W. J., Springer, M. S. and O'Brien, S. J. (2003). Nuclear gene sequences confirm an ancient link between New Zealand's short-tailed bat and South American noctilionoid bats. *Molecular Phylogenetics and Evolution*, **28**, 308–319.

Teeling, E. C., Springer, M. S., Madsen, O. *et al.* (2005). A molecular phylogeny for bats illuminates biogeography and the fossil record. *Science*, **307**, 580–584.

Tejedor, M. F., Czaplewski, N. J., Goin, F. J. and Aragón, E. (2005). The oldest record of South American bats. *Journal of Vertebrate Paleontology*, **25**, 990–993.

Tejedor, M. F., Goin, F. J., Gelfo, J. N. *et al.* (2009). New early Eocene mammalian fauna from western Patagonia, Argentina. *American Museum Novitates*, **3638**, 1–43.

Van Den Bussche, R. A. and Hoofer, S. R. (2000). Further evidence for inclusion of the New Zealand short-tailed bat (*Mystacina tuberculata*) within Noctilionoidea. *Journal of Mammalogy*, **81**, 865–874.

Van Den Bussche, R. A. and Hoofer, S. R. (2001). Evaluating monophyly of Nataloidea (Chiroptera) with mitochondrial DNA sequences. *Journal of Mammalogy*, **82**, 320–327.

Van Den Bussche, R. A. and Hoofer, S. R. (2004). Phylogenetic relationships among Recent chiropteran families and the importance of choosing appropriate out-group taxa. *Journal of Mammalogy*, **85**, 321–330.

Van Den Bussche, R. A., Reeder, S. A., Hansen, E. W. and Hoofer, S. R. (2003). Utility of the dentin matrix protein 1 (*DMP1*) gene for resolving mammalian intraordinal phylogenetic relationships. *Molecular Phylogenetics and Evolution*, **26**, 89–101.

Vaughan, T. A. (1959). Functional morphology of three bats: *Eumops*, *Myotis*, and *Macrotus*. *University of Kansas Publications, Museum of Natural History*, **12**, 1–153.

Whidden, H. P. and Asher, R. J. (2001). The origin of the Greater Antillean insectivorans. In *Biogeography of the West Indies: Patterns and Perspectives*, ed. C. A. Woods and F. E. Sergile. Boca Raton, FL: CRC Press, pp. 237–252.

Woodburne, M. O. (2004). Global events and the North American mammalian biochronology. In *Late Cretaceous and Cenozoic Mammals of North America: Biostratigraphy and Geochronology*, ed. M. O. Woodburne. New York: Columbia University Press, pp. 315–343.

Woodburne, M. O. and Swisher, III, C. C. (1995). Land mammal high-resolution geochronology, intercontinental overland dispersals, sea level, climate, and vicariance. Geochronology, time scales, and global stratigraphic correlation. *SEPM Special Publication*, **54**, 336–364.

Zachos, J. C., Pagani, M., Sloan, L., Thomas, E. and Billups, K. (2001). Trends, rhythms, and aberrations in global climate 65 Ma to present. *Science*, **292**, 686–693.

Zachos, J. C., Dickens, G. R. and Zeebe, R. E. (2008). An early Cenozoic perspective on greenhouse warming and carbon-cycle dynamics. *Nature*, **451**, 279–283.

Necromantis Weithofer, 1887, large carnivorous Middle and Late Eocene bats from the French Quercy Phosphorites: new data and unresolved relationships

SUZANNE HAND, BERNARD SIGÉ AND ELODIE MAITRE

6.1 Introduction

In 1887, Weithofer described the fossil bat *Necromantis adichaster* on the basis of fragmentary material from the Paleogene Quercy phosphorite fillings of southwestern France. The Phosphorites are composed mostly of phosphate-rich clays, including fossil materials, that fill the caves and fissures riddling the karstic landscape in the large regional Quercy area (including parts of four departments, but mostly the Lot and Tarn-et-Garonne) and extending west to the hills of the Massif Central (e.g., Thévenin, 1903; Gèze, 1938). The fillings were deposited over a period of about 30 million years from the late Early Eocene to the late Early Miocene (e.g., Legendre *et al.*, 1997), but also as recently as the Late Pliocene and Quaternary (Crochet *et al.*, 2006; Aguilar *et al.*, 2007).

From 1870 to 1907, many mines were established in the Quercy region to exploit the naturally occurring phosphate, a widely used fertilizer, with much of the Quercy ore being exported (see Durand-Delga, 2006 for a detailed history of the Quercy mining operations). Geologists were rarely allowed to visit the localities during the mining period, with fortunate exceptions, including Trutat, who took photographs (see Duranthon and Ripoll, 2006), and Thévenin, who produced a learned report (Thévenin, 1903). During the *c.* 40 years of intensive mining, the Quercy Phosphorites became renowned for producing many fossil bones and teeth, in good and even exceptional condition (Daubrée, 1871; Delfortrie, 1872). Today the Quercy Phosphorites are globally recognized for the thousands of vertebrate fossils they have produced, including many specimens of bats. For a recently updated record of Quercy micromammals see

Evolutionary History of Bats: Fossils, Molecules and Morphology, ed. G. F. Gunnell and N. B. Simmons. Published by Cambridge University Press. © Cambridge University Press 2012.

Sigé and Hugueney (2006) and for a catalog of classic Quercy mammalian taxa (i.e., before more recent excavations) see Sigé *et al.* (1979).

In the late nineteenth century, the Quercy vertebrate fossils, derived from various, mostly unspecified localities, were sold to French and foreign museums, universities and private collectors. Today these collections are referred to as the "Anciennes Collections du Quercy" or "old Quercy collections." Despite lacking exact locality data, the Anciennes Collections were nevertheless the focus of several paleontological studies, including those by Filhol (e.g., 1872, 1876), Schlosser (1887), Weithofer (1887) and Milne-Edwards (1892). For the first half of the twentieth century, following progressive cessation of Quercy mining operations before World War I, European paleontologists continued to study these collections. Notable studies of the fossils derived from the mining operations include those by Teilhard de Chardin on carnivores (1914–1915) and primates (1921), Revilliod on bats (1917, 1920, 1922) and more recently Koenigswald *et al.* (2009). Only in the early 1960s did teams of French paleontologists from Paris's Muséum National d'Histoire Naturelle and then Montpellier University commence new joint excavations of precisely recognized localities in the Quercy Phosphorites (see papers in *Palaeovertebrata*, 6, 1974), a scientific program that has continued for the last 45 years (see papers in *Strata*, series 1, 13, 2006).

In 1887, when Weithofer described Quercy's largest known bat, *Necromantis adichaster*, on the basis of limited material from the mined Paleogene Phosphorites, he speculated that this bat had some affinities with South America's phyllostomids. Winge (1893) questioned Weithofer's attribution of *Necromantis* to Phyllostomidae, as did Leche (1911). Later, Revilliod (1920) revised *Necromantis* on the basis of more complete Quercy material, including a skull (but still of unknown provenance and exact age). In his revision, Revilliod (1920) attributed *Necromantis* to the paleotropical bat family Megadermatidae as its oldest member, a position it has held without formal challenge since then.

Additional fossil material referable to *Necromantis* from the Anciennes Collections du Quercy has since become available from various institutions (Paris and Montauban museums, Lyon and Montpellier universities; see below), and modern fieldwork in the Quercy pits has provided further specimens from four well-located and dated localities (Maitre, 2008). These fossils now document the presence of *Necromantis* in the Quercy Phosphorites from at least the Middle to Late Eocene (~36 to 44 Ma), with three distinct species represented (Maitre, 2008). They also invite reappraisal of the affinities of *Necromantis* to other bats, including megadermatids, as well as the likely paleoecological role of *Necromantis* in the Eocene of Western Europe.

6.2   Material and methods

Here, new fossil material referable to the western European bat genus *Necromantis* is described, and material originally reported by Weithofer (1887) and Revilliod (1920) revisited and analyzed. The fossil material we examine here includes three well-preserved skulls, several dentaries and maxillae containing teeth, as well as many isolated teeth, and two distal humerus fragments.

Specimens from the "Anciennes Collections du Quercy" are from unknown or imprecise localities of indeterminate age in the Quercy Phosphorites, southwestern France. Specimens discussed in this chapter are held by the following institutions: Naturhistorisches Museum, Basel (specimen prefix QP, QS, QW, QH); Museum National d'Histoire Naturelle, Paris (QU); Museum d'Histoire Naturelle, Montauban (MnCh); Université Montpellier 2 (AQM/QU); University Lyon 1 (FSL) and Natural History Museum of Vienna (NHMV). Appendix 6.1 provides an itemized list of available known *Necromantis* specimens. Additionally, there may be other unrecognized specimens collected in the nineteenth century (or even recently extracted, exchanged or sold) from Quercy held by various provincial museums, or in foreign museums or universities, and private collections. In such cases, locality names are generally confused, inaccurate or incomplete, and hence lack meaningful chronological information.

Modern collecting by federally funded French scientists has produced *Necromantis* specimens from Perrière near Bernadelle hamlet and Rosières-5 near Escamps (both reference level MP 17b of the European Paleogene mammal scale), La Bouffie (MP 17a) near Sindou and Cuzal (MP 13) near Lamandine, Quercy Phosphorites, southwestern France. Specimens from the new collections are denoted: Perrière (PRR), Rosières-5 (ROS-5), La Bouffie (BFI) and Cuzal (CUZ). For detailed discussions of Quercy mammalian faunas and biochronology, see Remy *et al.* (1987), Marandat *et al.* (1993), Sigé and Hugueney (2006) and Maitre (2008).

Cranial, skeletal and dental terminology follows Miller (1907), Revilliod (1922), Van Valen (1979), Szalay (1969), Sigé (1971, 1976, 1985) and Hand (1985, 1993).

6.3   Systematic paleontology

Superfamily incertae sedis
Family incertae sedis
Genus *Necromantis* Weithofer, 1887
Necromantis adichaster Weithofer, 1887
Figures 6.1–6.10

Synonymy:

1893: *Necromantis adichaster in* Winge, note 39 p. 60

1898: *Necromantis adichaster in* Trouessart, p. 154

1904: *Necromantis adichaster in* Trouessart, p. 111

1911: *Necromantis adichaster in* Leche, p. 557

1920: *Necromantis adichaster in* Revilliod, pp. 69–79, pl. 2, figs. 1–6

1920: *Necromantis grandis in* Revilliod, p. 81, fig. 20

1920: *Necromantis planifrons in* Revilliod, p. 81, fig. 21

1967: *Necromantis adichaster in* Miguet, pp. 104–106

1979: *Necromantis adichaster, grandis, planifrons in* Sigé *et al.*, p. 48

1981: *Necromantis adichaster in* Crochet *et al.*, tab. 2–2 (Perrière, Rosières-5)

1983: *Necromantis adichaster in* Sigé and Legendre, N.R.C. (Perrière level)

1987: *Necromantis adichaster in* Remy *et al.*, tab. 1a (Perrière, Rosières-5)

1993: *Necromantis* cf. *adichaster in* Marandat *et al.*, p. 619, fig. 6 (Cuzal)

2006: *Necromantis adichaster, grandis, planifrons, sp., in* Sigé and Hugueney, tab. 2c, p. 220

Dental formula – I o/o, C 1/1, P 3/3, M 3/3

Emended diagnosis – (Revilliod, 1920; translated from the French): cranium long, very broad, flattened, its very tilted axis forming an angle of about 115° with that of the facial region; low sagittal crest; postorbital bones very strong and frontal-maxillary inflations delimiting the long facial region, close to pentagonal in shape and strongly titled forward; nasal bones distinct anteriorly, their suture visible for a few millimeters; supraorbital foramina present, opening into two grooves; premaxillae probably absent, remnants of maxillary branches possibly present; well-developed exoccipital overlying mastoid; carotid groove excavated in basisphenoid; two basal pits, elongated and shallow; presphenoid region narrow with median groove; alisphenoid canal and optic foramen present; infraorbital foramen situated above P4, but without infraorbital canal; horizontal ramus very deep, its labial surface subtly concave; maximum depth below m2; alveoli for canines contiguous; angular apophysis/process short, in axis of mandible; no upper or lower incisors; maxilla, like mandible, with two premolars anterior to P4 with posterior more reduced and lingually displaced; P4 with square outline, large posterior heel and well-developed anterior cingulum bearing two small cusps; M1–2 wider than long; without mesostyle cuspid at the junction of the postpara- and premetacristae; internal branchs of ectoloph reduced; heel large and semilunar; M3 with three labial crests only (lacking postmetacrista); protocone well developed; paraconid and metaconid on lower molars of similar height and volume; talonid very short and narrow; entoconid poorly developed; hypoconulid median or submedian; talonid of m3 rectangular with rudimentary cusps.

Type – Weithofer (1887) did not designate a type specimen and the figured specimen, a fragmentary lower jaw (Weithofer, 1887, figs. 18–21), had no catalog number. Subsequently, Revilliod (1920) gave the number QW627 to this same lower jaw from the Geological Museum of the University of Vienna (Revilliod, 1920, fig. 19), regarding it as the type specimen of the genus *Necromantis*.

Type locality – Anciennes Collections du Quercy (unknown locality, indeterminate age), southwestern France.

Other localities – Perrière and Rosières-5 (MP 17b) and other indeterminate Quercy localities (e.g., Collection MNHN Paris, Collection MHN Montauban, Collection Montpellier 2 University, Collection Lyon 1 University).

Material – Three skulls, six mandibles, two distal humerus fragments and 38 teeth isolated or in maxillae or mandibles.

Measurements – Table 6.1.

6.4 Description

The description of this species by Revilliod (1920) was based on one skull (QP630; Revilliod, 1920, plate II, 1–6) and a lower jaw (QW627; Revilliod, 1920, fig. 19). Here, other specimens from known localities as well as from the Anciennes Collections du Quercy are figured, measured and described. Their description complements those given previously in the literature and facilitates examination of intrageneric variation.

6.4.1 Skull description

General outline and proportions – (Based on Revilliod, 1920, and specimens QP630, QU16367, AQM7 and QS799; Figures 6.1–6.3). The skull is broad and robust. The rostrum is broad and short, in length approximately two-thirds the length of the braincase, in width half the mastoid width and just less than twice the interorbital width. Because the frontal shield extends posteriorly well beyond the interorbital constriction, the rostrum is difficult to define. In dorsal view, it is almost half the length of the skull, but from point of maximal interorbital constriction it is just over one-third skull length and approximately two-thirds braincase length. The braincase is broadest across the mastoids at the level of the post-tympanic processes. Maximum zygomatic width is

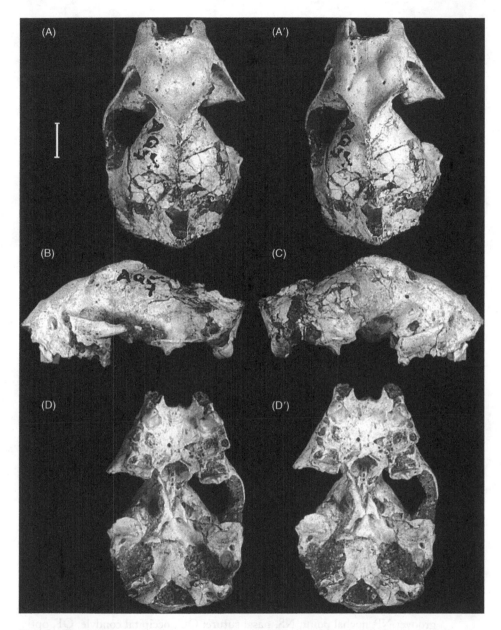

Figure 6.1 *Necromantis adichaster*, AQM7, skull with left and right P4 and M3.
(A)–(A′) dorsal view (stereopair); (B) left lateral view; (C) right lateral view; (D)–(D′)
ventral view (stereopair). Scale bar: 5 mm.

unknown, but its remnants suggest that it is broader than mastoid width. The
highest point of the skull is at the anterior extremity of the sagittal crest, related
to the marked angle in the basicranial axis through this point. In lateral view,
the angle of dorsoventral tilting of the head on the basicranial axis (as defined

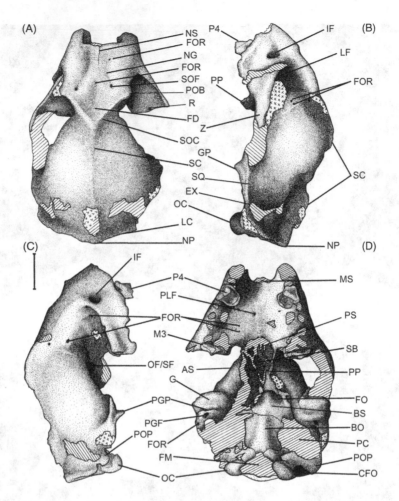

Figure 6.2 Terminology. *Necromantis adichaster*, AQM7, skull (see also Figure 6.1). (A) dorsal; (B) left lateral; (C) right lateral; (D) ventral. Scale bar: 5 mm. Abbreviations: AS, alisphenoid; BO, basioccipital; BS, basisphenoid; CFO, condyloid fossa; E, exoccipital; FD, frontal depression; FM, foramen magnum; FO, foramen ovale; FOR, foramen; G, glenoid; GP, glenoid process; IF, infraorbital foramen; LC, lambdoidal crest; LF, lacrimal foramen; M3, posterior upper molar; MF, maxillary fossa; MS, median spine; NG, nasal groove; NP, nuchal point; NS, nasal suture; OC, occipital condyle; OF, optic foramen; P4, posterior upper premolar; PC, cavity for periotic; PGF, postglenoid foramen; PGP, postglenoid process; PLF, palatine foramen; POB, postorbital bar; POP, paroccipital process; PS, presphenoid; PT, pterygoid; PTP, post-tympanic process; R, posterior root M3; SB, sphenorbital bridge; SC, sagittal crest; SOC, supraorbital crest; SF, sphenorbital fissure; SO, supraoccipital; SQ, squamosal; Z, zygoma.

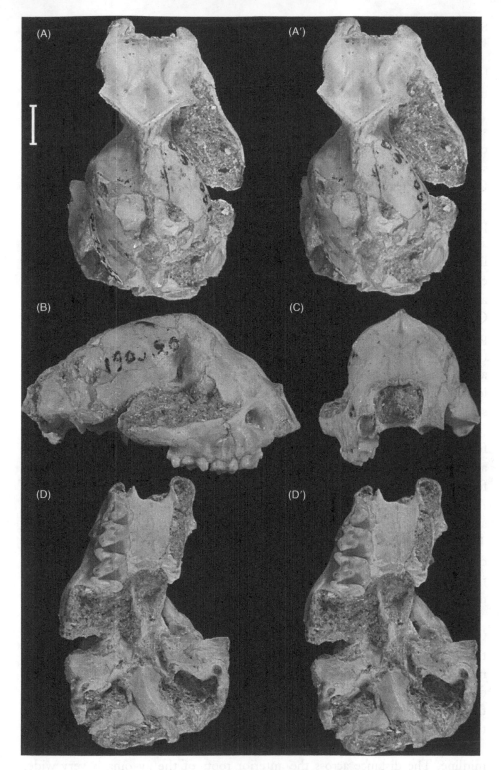

Figure 6.3 *Necromantis adichaster*, QU16367, skull with right P4-M3 (see also Figure 6.7). (A)–(A′) dorsal view (stereopair); (B) right lateral view; (C) anterior view; (D)–(D′), ventral view (stereopair). Scale bar: 5 mm.

by Freeman, 1984) or rostral rotation (Pedersen, 1993, 1995) is extremely negative, perhaps as great as −65°.

Rostrum − Dorsally, the rostrum is dominated by the pentagonal frontal shield defined by supraorbital crests ending in postorbital processes and broadly concave frontals with two foramina and grooves. The supraorbital crests are strongly but variably developed in the four cranial specimens − marked in QP630 and QU16367, less so in AQM7 and QS799. The point of junction of the ridges is slightly (QU16367) or significantly (QP630, AQM7, QS799) posterior to maximum interorbital constriction. The crests form between them an angle of 80° (QP630, AQM7, QS799) or 90° (QU16367) and are subrectilinear (AQM7), slightly concave laterally (QP630), slightly asymmetrically concave right to convex left (QU16367) or weakly sinuous (QS799). At their anterior extremity, they extend as postorbital processes that are more (QU16367, QP630) or less (QS799, AQM7) developed. The anterior frontal fossa at the point of bifurcation of the supraorbital ridges appears to be relatively deep (QP630) or only slightly depressed (QS799, AQM7). The longitudinal median axis of this depression may be marked, in the form of a short groove more or less clear (notably QU16367). A deeper fossa appears to be associated with stronger supraorbital crests.

In QP630, as noted by Revilliod, between and anterior to paired supraorbital foramina, is a small single foramen. This foramen is absent in the other three skulls and appears to be a variable feature in *Necromantis*. In contrast, paired supraorbital foramina are present in all four specimens, although slightly asymmetrical in AQM7. The anteriorly directed grooves that issue from these foramina are also always present, being more (QP630, QU16367) or less (QS799, AQM7) accentuated. There is some rostral inflation, with the nasal grooves showing a slight medially convex curve because of it. Long lateral inflations in the rostrum correspond to the region occupied by the alveoli of the very robust and long upper canines. A single median nasal foramen is associated with a very low median swelling.

At the rostrum's anterior margin, the indentation is square, particularly in QP630, QU16367 and AQM7, yet slightly more rounded in QS799 (in which this region is poorly exposed). In all four specimens, the nasal bones do not meet at their most anterior point and a suture is distinct posteriorly for a distance of *c*.3 mm. The anterior border of the nasals is also more or less thickened, relieved by some scalloping (notably on QU16367). In anterior view, the nasal aperture is almost as deep as wide.

Dorsally, the zygomatic arch forms a very obtuse angle with the nasal midline. The distance across the anterior roots of the zygoma is very wide, and hence the area for attachment of cheek muscles extensive. The zygomatic arch extends onto the face, joining the lateral wall of the maxilla at the level of

P4. The zygomatic plate is relatively tall at its anterior origin (QP630, QU16367), correlated with the great relative height of the rostrum and decreases gradually in height posteriorly (AQ7).

The position and importance of the foramina in the orbital region, in all specimens, corresponds well to the original observations of Revilliod. However, the infraorbital region invites further description: the canal is very short, reduced to a wide, oval foramen occuring above P4, covered by a small bony bridge of variable width (narrow in QP630, AQM7, QU16367, wider in QS799). Clearly related is another, important, trait not noted by Revilliod. The extreme anterior border of the orbit is conspicuously raised above the orbit's floor and therefore relatively distant from the labial face of the teeth. The maxilla forms a high plate, thin, acute at its upper edge, as an anterior extension of the zygomatic arch. This plate is crossed anteriorly by the large infraorbital foramen and its upper edge continuous with the weak anterior supraorbital crest.

The lacrimal occurs within the orbit, at its most anterior point, large and round (half to one-third the size of the infraorbital foramen) separated from the infraorbital foramen by the maxillary plate. Ventral to the lacrimal, the floor of the anterior recess of the orbit forms a very broad, deep fossa that is pockmarked anteriorly, but otherwise covers the maxillary bone, except for the posterior root of M3. Posterior and medial the postpalatal foramen appears to have been very large.

Palate – The palate is relatively short and broad, and the toothrows converge anteriorly. The anterior edge of the palate is level with the posterior edge of the C1 alveolus but it is not a straight edge as described by Revilliod; on all specimens, including QP630, it consists of two indentations on either side of a small median spine. The palate's posterior border has two deep lateral indentations that extend anteriorly to the level of the M2 metacone. The posterior extension of the midline is level with the anterior face of M2. The maximum length of the palate extends to the level of M3. The palate is concave in anteroposterior profile. Palatal foramina occur medial to the anterior face of M1 and M2–3. The posterior margin of the palate is very round.

The co-occurrence on skull QP630 of a narrow facet in the medial aspect of the maxilla on each side of the nasal aperture, as well as a subvertical sulcus immediately posterior to this facet, was suggested by Revilliod (1920) to indicate the presence of a premaxilla, albeit reduced to a maxillary branch. In Revilliod's interpretation, the facet and furrow served for the insertion of ligamentous attachments of the premaxilla. In fact, these features are poorly and variably expressed, and we are not convinced that premaxillae were present in *Necromantis*. The "facet" described by Revilliod occurs in QP630 and QU16367, but not in AQM7, while the "deep groove" appears at most as the limit of a swelling in the inner wall of the nasal cavity, observed in QP630 and QU16367, but not in AQM7. Compared with recent bats which lack premaxillae and upper incisors (i.e., megadermatids), *Necromantis* is characterized by a strong tendency to reduce the range of teeth even more effectively than that characterizing recent megadermatids (see below); the nasal notch of *Necromantis* is less deep than in megadermatids, which are themselves devoid of the maxillary branch of premaxilla; in *Necromantis*, as in recent megadermatids lacking upper incisors, the upper canine bears an antero-internal cingular cuspule occupying the position, and presumably the function, of a small upper lateral incisor.

Interorbital and pterygoid region – Between the orbital fossae, the braincase narrows. The point at which the two supraorbital crests unite (i.e., the origin of the sagittal crest) is clearly posterior to the point of narrowest constriction. The pterygoid wing slopes ventrally and laterally. QP630 best preserves this part of the skull, the right side better than the left. The posterior edge of the pterygoid wing curves ventrally and posteriorly and slightly medially. There is no evidence of a second flange. If hammular processes were present, they are now missing. The tallest point of the pterygoids occurs at a distance posterior to M3 that is equal to the length of M2–3.

The internal wall of the orbit is excavated midway by a structure, of rounded contour, more or less marked, at the outlet of the extension of the wide sphenorbital fissure between the base of the cerebral skull and the upper surface of the pterygoid plate. This structure is relatively faint in AQM7, but clearer in QP630, and in QU16367 and QS799 it is very accentuated and clearly delimited by a small crest anteriorly and superiorly.

The sphenorbital fissure and optic foramen are concealed on the left by crushing and calcite infill. On the right side, the ventrally opening sphenorbital fissure is at least visible, although the exact shape is unclear. The optic foramen is close to the sphenorbital fissure and posterior to maximum interorbital constriction. The sphenorbital fissure's posterior edge is posterior to the level of the tallest point of the pterygoid. The sphenorbital bridge is probably slightly constricted posterior to the tallest point of the pterygoid. The vomer extends posteriorly beyond the palate for a distance equal to M1 length. Any canal existing in the presphenoid-basisphenoid does not appear to be deep.

Zygomatic arch – This is preserved best on the left hand side of QP630, low anteriorly, unknown height posteriorly. Zygomatic width appears to have been greater than mastoid width. The zygoma are probably parallel for much of their length.

Cranial vault – Of the three subcomplete skulls known (QP630, QU16367, AQM7), the braincase is widest at the post-tympanic process, narrowest in the interorbital region, highest at the junction with the supraorbital ridges. At this junction, the sagittal crest is well developed, but variable in form – in QP630 it forms a distinct peak, but this is not the case in the other two specimens. Regardless, in all three skulls the sagittal crest is present but low, extending posteriorly to well-developed lambdoidal crests. Although there is no nuchal crest there is a nuchal process. The cranium is almost tubular in lateral view, but flattened rather than inflated. The profile of its roof is not absolutely straight, but rather weakly convex (QP630, QU16367). Parietal/interparietal sinuses and foramina are not preserved (due to crushing), with no information about their trajectory. In skulls QU16367 and AQM7, damage to the parietal region makes it impossible to comment on venous sinuses in QP630 described by Revilliod (1920).

Glenoid – The glenoid surface is slightly concave, and wider than long. It is not possible to determine whether it ends medial to the zygomatic process or not, but probably the latter. It is best preserved on the right side of QP630, where only the zygomatic process is missing. The postglenoid process is not tall, approximately one-quarter the length of the glenoid surface, and not especially recurved. The large, elliptical postglenoid foramen opens vertically on the posterior face of the glenoid. The posterior, medial and anterior faces of the glenoid are all gently sloping. Evidently, from QP630, the foramen ovale opens in the alisphenoid on the side of a bony ridge or prominence. The position of the foramen ovale is partially obscured in other specimens and because of calcite and breaks in this region we cannot confirm or refute an anterior foramen association with it.

Temporal – Posterior to the glenoid and immediately anterior to the mastoid region, the post-tympanic process is small. In lateral view, there is no indentation for the tympanic ring. The bone is continuous in the posterolateral surface of the skull, and there is only a very small area where the squamosal/exoccipital does not completely cover the mastoid. The lambdoidal crest is continuous from the exoccipital (nuchal point) to the post-tympanic process.

Basicranium – The medial sphenoid series (basisphenoid and presphenoid) are broadly fused with the palate, pterygoid, alisphenoid and basioccipital. The anterior margin of the basicranial area is the sphenorbital bridge, which is probably slightly constricted posterior to the pterygoid wing (tips). The sphenorbital fissure is widest at this point. In QP630 there are several breaks in this region, but the basisphenoid appears to have had relatively deep pits, probably deepest anteriorly (perhaps representing pockets or invaginations), which gradually disappear posteriorly. There are no second pits posteriorly. A bony ridge apparently separates the alisphenoid from the basisphenoid (clearest on left side) with the foramen ovale developed in the side of, or under, this ridge. There appears to have been a lateral extension of the basisphenoid into the petrosal. Anterolaterally to this is a foramen for a blood vessel. The periotic was evidently not greatly enlarged, the width of the basioccipital being almost as wide as the cavity for the periotic. The basioccipital, in all three specimens, has a longitudinal median swelling widening towards the rear at the foramen magnum.

Occipital – This region as a whole is best preserved in QP630, but AQM7 and QU16367 enable verification of some of features noted by Revilliod. The posterior-most part of the skull is the junction of the parietals (or possibly interparietals) and supraoccipitals. It is not particularly convex, and in lateral view is almost hidden by lambdoidal crests of the squamosal/occipital. In ventral view, the posterior skull contour of the supraoccipital at the level of the lambdoidal crest is pointed. In posterior view (QP630, AQM7), the contact of the occipital and the parietals makes a sharp angle of about 90°, and is accentuated by the strong development of the lambdoidal crests. The obtuse angle formed by the latter forms a peak of comparable importance in the two specimens that sufficiently preserve this region (QP630, AQM7). The lambdoidal crest is particularly well developed and thickened in the exoccipital region, where it forms a strip approximately 1.50 mm in height. In the three skulls (QP630, QU16367, AQM7), the crest connects to a small horizontal ridge forming an otic shelf, itself the rectilinear extension of the zygomatic process of the squamosal. In QU16367 (right and left sides), as in QP630 (right side), the exoccipital is complete, forming a slight swelling in front of the lambdoid crest and above the post-tympanic process.

The foramen magnum is directed ventrally or posteroventrally. It is oval, twice as wide as deep and very flattened, making it almost rectangular in shape. In posterior view, the anterior margin of the foramen magnum is almost invisible, because of the compression of the foramen magnum. Its posterior edge is very straight and not thickened. The lambdoidal crest is very well developed, and runs probably unbroken from the nuchal point (the most posteriorly directed point on the skull) to the zygomatic process of the squamosal. The occipital condyle projects strongly ventrally and posteriorly, its articular surface developed as a semicircle of 2 mm (AQM7) to 4 mm (QU16367) diameter. Lateral to the occipital condyle is a transversely elongated occipital foramen. Between the occipital condyle and postoccipital process is the condyloid fossa. The postoccipital process is a flat, circular to triangular piece of bone, covering the mastoid. The condyloid foramen in the anterolateral face of the condyle is concealed by damage.

Ear region – The periotics, ectotympanics and auditory ossicles are missing. The periotic was evidently only weakly articulated with the surrounding basicranial elements with its attachment probably via lateral extension of the basisphenoid, and postoccipital process covering the mastoid.

6.4.2   Mandible and teeth

The dentary (e.g., QW 627, QH427, QU16369, QH 16370, QH 16371, MnCh 1, AQM8; Figure 6.4) is massive, the horizontal ramus particularly deep. The latter ascends steeply directly posterior to m3 to become subvertical. The coronoid process is robust, weakly extending labially; the coronoid process is relatively short with respect to the deep horizontal ramus; the quadrate angular process is short. Two specimens (MnCh 1, AQM8) preserve the complete symphysis. In both specimens, there is no indication of the existence of incisors: the exposed bone surface is porous and there is no evidence of alveoli. Very close to the symphysis border, both specimens show a foramen opening on the lateral side.

Lower dentition – The c1 is massive, with three faces (distal and lingual faces subplanar, mesolabial face convex) and a cingulid encircling the tooth (Figures 6.4 and 6.5). p2 is as large as p4, single-rooted and consists of a single cusp surrounded by a simple sublingual cingulid. p3 is relictual, and so far known only

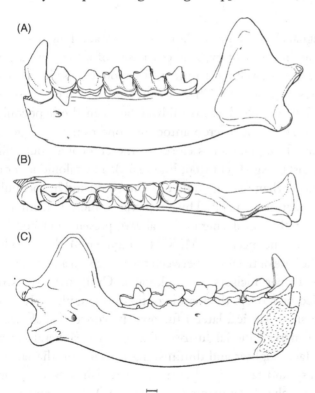

(A)

(B)

(C)

Figure 6.4 *Necromantis adichaster*, QU16369, complete left mandible with c1–m3 and p3 alveolus. (A) labial view; (B) occlusal view; (C) lingual view. Scale bar: 1 mm.

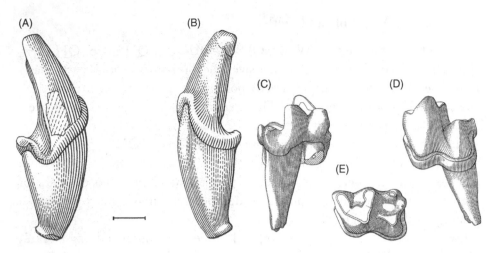

Figure 6.5 *Necromantis adichaster.* A–B, QU16369, right c1. (A) labial view; (B) lingual view. (C)–(E) Ros5–51, left m1 or 2. (C) lingual view; (D) occlusal view; (E) labial view. Scale bar: 1 mm.

from a very small lingual alveolus. p4 is relatively narrow and long, its two roots oriented longitudinally in the dentary; it is composed of a median cusp and a small anterior and posterior swelling. In m1–3, the more or less inflated hypoconulid (e.g., Rosières-5–51 and MnCh 1; Figures 6.5 and 6.6) is somewhat variable in position, but typically located midway between the hypoconid and entoconid (thus illustrating the necromantodont condition; see Sigé *et al.*, Chapter 13, this volume), or sometimes closer to the entoconid, but still well separated and clearly distal (e.g., QU 16369; Figure 6.5). The talonid is short and wide, and markedly lower than the trigonid. The entoconid and hypoconulid are equally distinct and sometimes separated by a transverse sinus. The entoconid is variably developed, but the labial cingulid is always present and continuous, although rather slight. In one specimen (MNCH 1; Figure 6.6), there is a lingual rounded swelling at the base of the notch between the entoconid and hypoconulid.

Upper dentition – Upper incisors are unknown. C1 is massive, strongly recurved and encircled by a deep, thick and continuous cingulum. Its internal face is flat, with two slight vertical lateral furrows. Its convex labial face has a faint posterior and a marked mesial furrow. There is a peripheral cingulum, thick on the lingual face, slighter and diminishing mesially on the labial face, with mesial lower cusp and taller distal cusp. P2–3 remain unknown but the space, organization and alveoli remnants (e.g., skull AQM7; Figure 6.1) indicate the existence of a very tiny vestigial lingual premolar (P3) and a slightly larger, but nevertheless small, labial premolar (P2), as suggested for skull QP630 by Revilliod (1920, p. 77). P4 is large with an extended lingual heel,

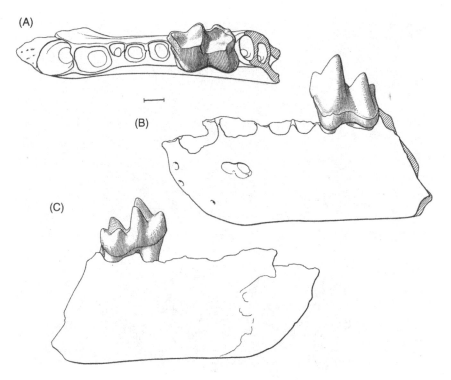

Figure 6.6 *Necromantis adichaster*. MNCH 1, partial left mandible bearing m1 and alveoli for c1–p4 and m2. (A) occlusal view; (B) labial view; (C) lingual view. Scale bar: 1 mm.

somewhat square in shape, and with parastyle more or less labially located. The structure of M1–2 is simple and massive with a relatively small protofossa, large heel expanding backwards (slightly lingually on M1), with rounded and non-cuspate cingular border, mesostyle without cusp and lingually displaced, marked submedian ectoflexus and moderately expanded parastyle. M3 is strongly reduced, without pre- and postmetacristae, paracone present, low protocone, with rudimentary heel preserved (Figure 6.7).

6.4.3   Humerus

Humerus – (ROS-5-55, ROS-5-56; Figures 6.8–6.9). Two specimens preserve the distal epiphysis and distal shaft of the right humerus referable to Quercy's very large *Necromantis adichaster*. The following description accords with the general orientation of the forelimb in bats in which there has been a 90° dorsal rotation with respect to that of most mammals, such that in the distal humerus epiphysis (with the arm resting along the body) the most

(A)

(B)

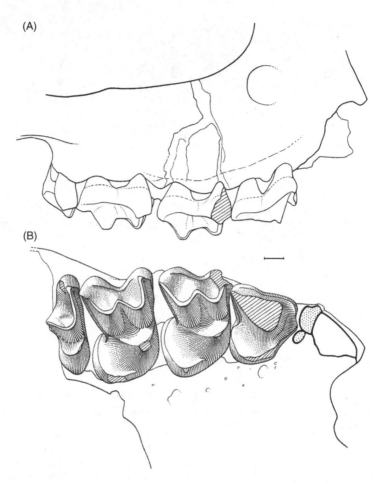

Figure 6.7 *Necromantis adichaster*, QU16367 (see also Figure 6.3), toothrow with P4–M3 and alveoli for C1–P3. (A) labial view; (B) occlusal view. Scale bar: 1 mm.

articular face is lateral, the opposite surface medial, the epitrochlea anterior and the epicondyle posterior.

The shaft is slightly wider than transversely deep, and distally is slightly flattened and gently curved outwards. The anteroposterior length of the epiphysis is 8.5 mm and that of the shaft 3.5 mm. The articular surface (trochlea, condyle and epicondyle) is offset with respect to the shaft, so that the epicondyle is fully outside the line of the shaft and condyle partly so. The whole articular axis is perpendicular to the long axis of the shaft. The epitrochlea projects forward from upper mid-height of the mid-medial trochlea surface and has a small distal knob rather than a true styloid process. As a whole, the rather limited trochlea represents only approximately one-quarter the width of the entire distal epiphysis, while the spherical condyle occupies conspicuously more. Two other

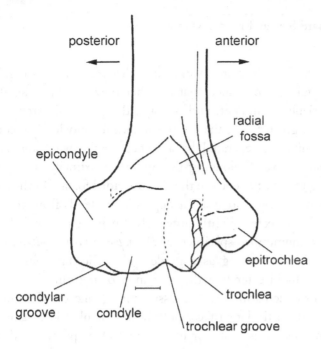

Figure 6.8 *Necromantis adichaster*, Ros-5–51, right distal humerus fragment; lateral view; terminology. Scale bar: 1 mm.

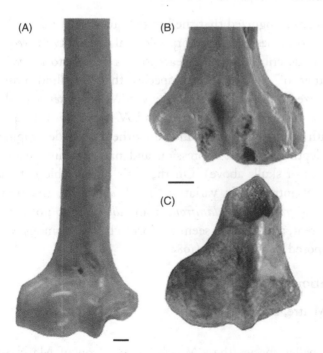

Figure 6.9 *Necromantis adichaster* Weithofer, 1887, right distal humeral fragments. (A)–(B), Ros-5–55; (A) lateral view; (B) medial view. (C) Ros-5–56; medial view. Scale bars: 1 mm.

noteworthy features are: first, the condyle and epicondyle are feebly distinguished by a very shallow or faint groove, such that together they form a wide, almost spherical surface, variably prominent; and, second, the proximal extent of the epicondyle is distinctly greater than that of the spherical condyle. The condyle and the pretrochlea hull are separated by a rather gentle, wide and almost symmetrical trochlear groove. Prolonging this groove proximally, a low ridge variably extends along the shaft to enclose the shallow radial fossa. On the medial face of both specimens, a deep and straight depression extends from the distal trochlear groove, and is directed slightly anteriorly. Such a structure, like that observed in a modern mummified specimen of *Rhinopoma microphyllum*, relates to extension of the forearm, acting as a guide and anchoring point for the presumed separate sesamoid extending the olecranon of the ulna. This would strengthen the fully extended arm as well as assist during take-off. The anterior and posterior faces of the articular surfaces both bear a circular depression. The anterior, shallower depression is posterior to the epitrochlea process and opens lateral to it; the deeper, posterior depression has an oval opening and rounded rim.

6.4.4   Remarks *Necromantis adichaster*

Revilliod (1920) recognized that the limited quantity of material available to him did not allow assessment of morphological variability in *Necromantis*. The new material described above appears to be referable to *N. adichaster* rather than to either of the additional species that Revilliod proposed: *N. grandis* (characterized by identical morphology to *N. adichaster* with slightly wider talonid on m2, but similarly sized m3) and *N. planifrons* (same size as *N. adichaster* but with cranial differences, including the relative development of postorbital crests, depth of frontal depression and nasal opening width and depth; see description of skulls above). On the basis of available data, and a better appreciation of intraspecific variation in *N. adichaster*, the proposed features distinguishing *grandis* and *planifrons* from *adichaster* do not appear to be significantly different, and the taxa seem a priori to be synonymous with the type species, as proposed also by Hand (1985).

6.5   Additional species

Necromantis gezei Maitre, 2008
Figure 6.10 F–H

> *Diagnosis* – Smaller species than *N. adichaster*; trigon of M2 closed by postprotocrista.
> *Derivatio nominis* – In homage to Professor Bernard Gèze, who, at the beginning of modern research in the Quercy Phosphorites, demonstrated

Figure 6.10 (A)–(D), *Necromantis adichaster*. (A)–(B), PRR-71, right C1. (A) lingual view; (B) labial view. (C) ROS-5_NaA.1.1, left P4. (D) PRR-66, left M1. (E) PRR-70, left M2. (F)–(H) *Necromantis marandati* Maitre, 2008. (F) BFI_NmA.1.4, right P4 (partial). (G) BFI_NmA.1.5, holotype, right M2. (H) BFI_NmA.1.1, right m1. (I)–(K) *Necromantis gezei* Maitre, 2008. (I)–(J) CUZ_NgA.1.1, left C1. (I) labial view; (J) lingual view. (K) CUZ-383, holotype, right M2. lingual view. Scale bar: 1 mm.

that a given karstic filling is homogeneous in its faunal contents and geological age, and potentially significantly different from another, even adjacent filling, as exemplified by the Aubrelong pits that he quarried, studied and published before World War II (Gèze, 1938).

Holotype – BFI_NgA.1.5, right M2, figured pl. 1 fig. i (Maitre, 2008), University Montpellier 2 Collections, France.

Type locality – La Bouffie (MP 17a), Quercy Phosphorites, Lot department, southwestern France.

Material – Five teeth.

Measurements – Table 6.1.

Description – m1–2 structure as typically found in *Necromantis* with necromantodont lower molar morphology (see Sigé *et al.*, this volume), strong cingulid, trigonid tall with respect to talonid. P4 with strong, sharp paracone, and deep parastyle with respect to labial border. M2 with cusped parastyle; uncusped and deep mesostyle with respect to the labial notch; trigon closed by link between postprotocrista and

Table 6.1 Measurements (mm) of teeth of *Necromantis adichaster*, *N. gezei* and *N. marandati*. Upper case denotes upper dentition, lower case denotes lower dentition (e.g., M1 is upper first molar, M2 is lower first molar). r, right; l, left; L, maximum length; W, maximum width; Wtri, trigonid width; Wtal, talonid width.

Collection	No.	Type	r/l	L	W	Wtri	Wtal
Nineteenth-century Quercy Collections							
Necromantis adichaster							
MNHN	QU16369	c1	l.	2.28	2.57		
MNHN	QU16369	p2	l.	2.43	2.04		
C3G Lyon	FSLNa1	p4	l.	2.48	1.47		
MNHN	QU16370	p4	l.	2.6	1.44		
MNHN	QU16371	p4	r.	2.56	1.51		
MNHN	QU16369	p4	l.	2.4	1.74		
UM2	QUNa1	p4	l.	2.18	1.26		
	Average	p4		2.44	1.48		
		Nb		5	5		
		SM		0.074	0.077		
		SD		0.166	0.172		
		V		7.148	12.179		
MHN Mont.	MnCh1	m1	l.	3.03	2.21	1.87	2.19
C3G Lyon	FSLNa1	m1	l.	2.91	1.92	1.76	1.92
MNHN	QU16370	m1	l.	3	1.97	1.87	1.94
MNHN	QU16371	m1	r.	3.08	2.13	1.93	2.11
MNHN	QU16369	m1	l.	3.02	2.12	1.92	2.08
UM2	QUNa1	m1	l.	2.68	1.8	1.76	1.85
	Average	m1		2.95	2.03	1.85	2.02
		Nb		6	6	6	6
		SM		0.059	0.063	0.031	0.053
		SD		0.145	0.154	0.075	0.131
		V		5.114	7.934	4.231	6.770
C3G Lyon	FSLNa1	m2	l.	2.93	2.05	2.05	1.9
MNHN	QU16370	m2	l.	2.92	2.24	2.24	2.06
MNHN	QU16371	m2	r.	3.06	2.16	2.14	2.12
MNHN	QU16369	m2	l.	3.06	2.3	2.3	2.13
UM2	QUNa1	m2	l.	2.64	2.03	2.03	1.82
	Average	m2		2.92	2.16	2.15	2.01
		Nb		6	6	6	6
		SM		0.070	0.048	0.048	0.057
		SD		0.172	0.117	0.117	0.139
		V		6.115	5.661	5.680	7.210
C3G Lyon	FSLNa1	m3	l.	2.64	1.92	1.92	1.37

Table 6.1 (*cont.*)

Collection	No.	Type	r/l	L	W	Wtri	Wtal
MNHN	QU16370	m3	l.	2.74	2	2	1.44
MNHN	QU16371	m3	r.	2.74	1.95	1.95	1.47
MNHN	QU16369	m3	l.	2.81	2.17	2.17	1.6
UM2	QUNa1	m3	l.	2.37	1.9	1.9	1.26
	Average	m3		2.66	1.988	1.988	1.428
		Nb		5	5	5	5
		SM		0.077	0.049	0.049	0.056
		SD		0.173	0.108	0.108	0.126
		V		6.831	5.730	5.730	9.234
MNHN	QU16367	P4	r.	2.94	2.92		
MNHN	QU16367	M1	r.	3.26	4.29		
MNHN	QU16367	M2	r.	3.26	4.41		
MNHN	QU16367	M3	r.	1.51	3.97		
Total number of specimens					27		

Cuzal (CUZ)
Necromantis marandati n. sp.

UM2	NmA.1.1	C1	l.	2.39	1.81		
UM2	NmA.1.2	M2	r.	3.24	4.41		
Total number of specimens					2		

La Bouffie (BFI)
Necromantis gezei n. sp.

UM2	NgA.1.1	m1	r.	2.75	1.93	1.93	1.89
UM2	NgA.1.2	m2	r.	2.62	1.91	1.88	1.79
UM2	NgA.1.3	m2	l.	2.62	1.89	1.89	1.75
UM2	NgA.1.1	m2		2.61	1.93	1.86	1.79
	average	m2		2.62	1.91	1.88	1.78
		Nb		3	3	3	3
		SM		0.003	0.012	0.009	0.013
		SD		0.006	0.020	0.015	0.023
		V		0.239	1.134	0.880	1.406
UM2	NgA.1.5	M2	r.	3.5	3.95		
Total number of specimens					5		

Perrière (PRR)
Necromantis adichaster

UM2	69	m2	r.	2.84	2.09	1.92	
UM2	71	C1	r.	3.12	2.52		

Table 6.1 (*cont.*)

Collection	No.	Type	r/l	L	W	Wtri	Wtal
UM2	66	M1	l.	3.25	4.29		
UM2	67	M2	broken				
UM2	68	M2	broken				
UM2	70	M2	l.	3.33	4.21		
Total number of specimens					6		
Rosières-5 (ROS5)							
Necromantis adichaster							
UM2	51	m1	l.	3.12	2.1	1.92	2.05
UM2	NaA.1.2	C1	l.	3.15	2.46		
UM2	NaA.1.3	C1	l.	3.08	2.41		
UM2	50	C1	l.	3.27	2.6		
	Average	C1		3.17	2.49		
		Nb		3	3		
		SM		0.055	0.057		
		SD		0.096	0.098		
		V		3.287	4.285		
UM2	NaA.1.1	P4	l.	2.93	2.82		
Total number of specimens					5		

postcingulum; large, posteriorly developed heel with thick cingulum, with some variation in the state of its differentiation.

Remarks – The smaller size of *N. gezei* nov. sp. with respect to *N. adichaster* and *N. marandati* nov. sp. is the most distinctive feature of this new taxon, this size difference being too marked to be the result of sexual dimorphism (Maitre, 2008). Further, although the lower molars in *N. gezei* are morphological mirrors of those in *N. adichaster*, P4 and M2 morphology is quite distinct, the latter clearly distinguished by complete closing of the trigon by the postprotocrista, as also found in *N. marandati* (see below).

Necromantis marandati Maitre, 2008
Figure 6.10 I–K

Diagnosis – Same size as *N. adichaster*, but M2 with longer inner ectoloph branches and trigon closed by postprotocrista.

Derivatio nominis – In homage to Mr. Bernard Marandat for his significant and lasting contribution to Quercy fieldwork, fossil preparation and conservation, as well as his own research, including Early Eocene bats from southern France (Marandat *et al.*, 1993).

Holotype – Right M2, CUZ_383, figured pl. 25 fig. k (Maitre, 2008), University Montpellier 2 Collections, France.

Type locality – Cuzal (MP 13), Quercy Phosphorites, France.

Material – Two teeth.

Measurements – Table 6.1.

Description – M2 markedly wider than long; uncusped but deep mesostyle with respect to labial margin (lingually displaced); strong labial notch between paracone and mesostyle; large parastyle; trigon closed by link between postprotocrista and postcingulum; weak paraloph; strong cingulum on heel. CUZ_NgA.1.1 is referred to this taxon by Maitre (2008). It is a left C1 with strong cingulum encircling tooth, thicker lingually, with very well-developed posterior cuspid; labial face convex and lingual face very flat.

Remarks – In *N. marandati* nov. sp. C1 is proportionately smaller than M2 and markedly smaller than C1 in *N. adichaster*, perhaps the consequence of sexual dimorphism. Moreover, this tooth has a narrower and thinner cingulum. In the upper molars, some differences can be observed in the M2 trigon which in *N. marandati* is sharply closed (where the postprotocrista joins the postcingulum as in *N. gezei* nov. sp.), while in *N. adichaster* specimens the postprotocrista stops before joining the posterior border of the tooth. In *N. marandati*, the labial cingulum is less developed, the ectoloph inner branches (postparacrista and premetacrista) are longer, the uncusped mesostyle is closer to the labial edge, the heel is slightly narrower, its outline is rounded and the anterolingual cingulum extremity bears a small cusp.

6.6   Comparison of *Necromantis* with other bats

Revilliod (1920) referred *Necromantis adichaster* to the family Megadermatidae as its earliest member. Here we compare *Necromantis* with megadermatids, as well as other extant and extinct bat lineages.

As noted by Weithofer (1887) and Revilliod (1920) among others, *Necromantis adichaster* is unique among extinct and extant bats in the striking robustness of its craniomandibular morphology. Other characteristic features of *Necromantis* include: an extremely pronounced angle in basicranial axis; a flattened, tube-like cranium; loss or near-loss of bony premaxilla; probable loss of all upper and lower incisors; partially unfused anterior nasal suture; conspicuous grooves developed in pentagonal nasofrontal region; dorsal arching of palate; retention of minute P3 and p3; M1–2 transversely developed and with lingually displaced mesostyle; heels of P4, M1, M2 developed posterolabially; markedly median ectoflexus in upper molars; large and median hypoconulid in lower molars; lambdoidal crest continuous from exoccipital (nuchal point) to post-tympanic process; oval to nearly rectangular foramen magnum; and squamosal almost completely covering mastoid.

Many, but not all, of these features are shared with megadermatids, and also to various degrees with other bats, as follows: extremely pronounced angle in basicranial axis (shared with nycterids, rhinolophids); a flattened, tube-like cranium (archaic emballonurid *Vespertiliavus*); partially unfused anterior nasal suture (*Vespertiliavus*, emballonurids); conspicuous grooves developed in pentagonal nasofrontal region (*Vespertiliavus*, emballonurids); pronounced supraorbital crest development (megadermatids, nycterids, emballonurids, pteropodids); dorsal arching of palate (emballonurids); loss or near-loss of bony premaxilla (megadermatids, emballonurids, nycterids); loss of all upper incisors (megadermatids); retention of p3 (archaeonycterids, rhinolophids, early hipposiderids); M1–2 transversely developed and with lingually displaced mesostyle (archaeonycterids, some hipposiderids, some phyllostomids, the Eocene *Ageina, Honrovits, Palaeophyllophora*); heels of P4, M1, M2 developed posterolabially (emballonurids, nycterids, phyllostomids); two basisphenoid pits (many taxa including megadermatids, phyllostomids, emballonurids, nycterids); markedly median ectoflexus (*Ageina, Honrovits, Palaeophyllophora*, archaeonycterids); large, median necromantodont hypoconulid (*Ageina, Honrovits, Palaeophyllophora*); lambdoidal crest continuous from exoccipital (nuchal point) to post-tympanic process (megadermatids, emballonurids); and squamosal almost completely covering mastoid (megadermatids, emballonurids).

Other conspicuous cranial differences between *Necromantis* and megadermatids include the following. In *Necromantis*, the infraorbital canal is very short, reduced to a wide, oval foramen covered by a small bony bridge, whereas all recent megadermatids possess a true infraorbital canal, very long and narrow, that opens in the orbit at the level of M2, and anteriorly, by one or more orifices, above P4. The zygomatic arch is much more robust than in recent megadermatids, and the palate relatively much shorter and wider. Compared to megadermatids, in *Necromantis* the nasal indentation is angular rather than round, much less deep and in anterior view almost square rather than semicircular. Maximum cranial height occurs at the level of minimum interorbital width rather than far more posteriorly as in megadermatids in which it is generally at the level of the glenoid. In *Necromantis*, occipital-parietal contact follows a sharp angle of *c*. 90°, while the same area has a rounded profile in recent megadermatids; the exoccipital is better ossified than in megadermatids and most rhinolophoids in which the mastoid participates more or less in the skull wall. In *Necromantis* an indentation for the tympanic ring reaches only the level of the glenoid surface and does not affect the horizontality of the otic shelf, whereas in recent megadermatids the indentation is more pronounced, extending dorsally to above the upper level of the zygomatic process, with the otic shelf either undeveloped or slightly curved.

As noted by Revilliod (1920), Sigé (1976) and Hand (1985), dental features shared by *Necromantis* and megadermatids include the following: reduction in number of lower incisors; anteroposterior compression of m1 such that it is shorter than or equal in length to m2; anteroposterior compression of the talonids of m1–3; marked reduction of M3; reduction in the number of pre-molars to a maximum of two in the upper and lower toothrows (*N. adichaster* retains a tiny P3 and p3); and posterior extension of the palate lingual to M3 (see below). Dental features shared by *Necromantis* and the largest living megadermatid, Australia's *Macroderma gigas*, are described by Maitre (2008) and include the following: M1–2 with lingually displaced mesostyle, wide labial cingulum rising on the occlusal surface (notably at parastyle level), paracone near metacone, narrow trigon basin partly closed by postprotocrista, reduced before joining the posterior margin, well-developed posteriorly directed heel; very reduced M3, transversely stretched and composed of paracone, protocone and only two branches of the ectoloph; nearly identical p4; lower molars lacking crest from metaconid to entoconid, with thick and undulating cingulid, reduced talonid basin; very reduced m3. Dental differences between the two taxa include the following. In the upper dentition: in *Macroderma*, upper teeth proportionately longer; loss of P2; more elongate P4; M1–2 with larger para-style, taller mesostyle, absence of postcingulum; M1–2 with more labial meso-style, postprotocrista reduced before posterior margin, protofossa reduced and slightly open, more elongate stylar crests, wide labial notch, longer heel; and in *Necromantis*, posterior margin of M3 with little discontinuity at level of paracone, sinus between paracone and protocone and well-developed parastyle, pre- and postprotocrista. Differences in the lower dentition include: c1 of *Macroderma* with narrower cingulum notably on lingual side, less inclined on the labial side; loss of p3, p2 more reduced; lower molars rather nyctalodont, with more open trigonid, notably in m1 and more reduced talonid.

The distal humerus of *Necromantis* (Figures 6.8–6.9) is relatively underived and hence shares features with many bat families, including megadermatids, most noctilionoids, emballonurids, rhinopomatids, hipposiderids, pteropodids and archaeonycteridids. In these groups, the condyle is spherical and articula-tion is offset laterally with respect to the shaft (Smith, 1972, figs. 3–5; Felten *et al.*, 1973, figs. 7–12). The epitrochlea in *Necromantis* is wide with respect to the condyle, as it is also in archaeonycteridids, palaeochiropterygids, rhinolophoids, emballonurids, rhinopomatids, natalids, most noctilionoids and pteropodids. The styloid (distal) process is poorly developed, as in archaeonycteridids, archaic emballonurids (*Vespertilavus*), archaic molossids (*Cuvierimops*) and unlike (e.g.) hipposiderids, rhinolophids, megadermatids, molossids and mystacinids, in which it is clearly well developed. The radial fossa is relatively shallow and, as in most bats, the distal part of the shaft is not especially flattened. The trochlea is not especially prominent, as in most bats except molossids and mystacinids. The proximally extended epicondyle, developed beyond the proximal surface of the rounded condyle, and an only indistinct groove between the condyle and epicondyle, are features shared with emballonurids such as *Taphozous* species. Rather than the more common gentle groove medially prolonging that on the distal epiphysis, both Rosières-5 specimens have a deep, narrow and short depression, which presumably helps lock, by partly lodging, the small sesamoid, thereby reinforcing the joint during full arm extension, as well as assisting during take-off (see above). Rather comparable although less-pronounced structures are observed among vespertilionids, emballonuroids and palaeochiropterygids. The distal humerus morphology of *Necromantis* differs from that of megadermatids specifically as follows: trochlea length less than, not greater than, lateral epicondyle length; trochlear groove shallower and condylar groove deeper; depression between trochlea and distal styloid process broader; epitrochlea narrower (with respect to articular surface); and radial fossa deeper and more centrally located.

6.7   Functional morphology

Cranial and limited skeletal material of *Necromantis* species is so far known only for *N. adichaster*, and the following discussion on functional morphology relates to that species.

With a skull length of 32 mm (Revilliod, 1920) and estimated weight of 47 g (Maitre, 2008), *Necromantis adichaster* is one of the largest known Quercy bats (and Middle to Late Eocene bats in general) along with some species of *Pseudorhinolophus* and *Vespertiliavus*. *Necromantis* was not as large as the living megadermatid *Macroderma gigas* (skull length 40 mm, weight 105 g; Richards *et al.*, 2008) or phyllostomid *Vampyrum spectrum* (51 mm, 190 g; Navarro and Wilson, 1982). However, it is unique among extinct and extant bats in the striking robustness of its craniomandibular morphology. Weithofer (1887), in his original description of this bat, noted the strength of the *Necromantis* masticatory apparatus as reflected by a very deep and thick mandible with deep fossa for insertion of large masseter muscles. The association of these mandibular features with *Necromantis*'s large, low and crowded teeth suggested to

Weithofer a strong capacity to grind or crush, like some carnivorans. The name *Necromantis* ("death-eater") alludes to the diet that Weithofer assumed for this bat – carrion, but probably also small, live vertebrates, including mammals such as rodents and other bats.

As noted above, *Necromantis* has a relatively short, broad skull, its wide zygoma corroborating Weithofer's deduction (based on only the lower jaw and dentition) of great masseter muscle volume. In mammals, including bats, greater masseter volume is associated with a strong, deeply piercing bite, and typically the inclusion of hard prey items in the diet (Maynard Smith and Savage, 1959). Freeman (1984) categorized animalivorous bats as wide-faced or long-faced bats (according to the ratio of zygomatic width to condylocanine length). In this scheme, *N. adichaster* (with ratio of *c*.72%) would be classed as a wide-faced bat (ratios of 70–80%), most of which represent specialists on hard prey such as armored beetles, but in some cases bone. Freeman (1984) compared the craniomandibular morphology and bite force of short-faced bats to that of carnivorans such as *Hyaena* (ratio of 71%), *Felis* (79%) and *Ailuropoda* (82%), which include flesh and bone, and in the latter tough vegetation, in their diets.

The sagittal crest in *Necromantis* is conspicuous and complete, but appears to have been relatively low compared to some bats, such as, for example, *Vampyrum spectrum*, *Macroderma gigas*, *Hipposideros commersoni*, *Cheiromeles parvus* and *Scotophilus gigas*. Sagittal crest development in mammals reflects temporal muscle mass volume, suggesting temporal muscles were relatively less developed than masseters in *Necromantis*. In *Macroderma*, the mandibular condyle is relatively low and the coronoid process high, a condition typical in carnivores whereby the temporalis (the primary prey-seizing muscle) is developed at the expense of the masseter (Maynard Smith and Savage, 1959). This allows greater gape, vertical movement and slicing ability, but less transverse activity, a function that Freeman (1984) found to be important in wide-faced bats processing hard prey. In *Necromantis*, the angular process of the dentary is very small, indicating that crushing force was much more important than width of gape. The height of the condyle above the toothrow is relatively tall (more than twice molar height), allowing for longer moment arm and greater attachment area for masseter and pterygoid muscles (Maynard Smith and Savage, 1959).

In carnivorous bats, metastylar crest (postmetacrista) length in $M1-2$ is greater than preparacrista length, the mesostyle is lingually displaced, para- and metacristae more longitudinally oriented, trigonids more open particularly on $m1$ and sectorial crests lengthened on all teeth. These classically carnivorous specializations are clearly present in the phyllostomid *Vampyrum spectrum*, and, among megadermatids, in *Megaderma spasma* and *M. lyra*. With respect to megadermatids, these features are curiously less developed in the larger *Macroderma gigas* (Hand, 1985), which nevertheless includes significantly more vertebrate prey in its diet (Boles, 1999). This appears to reflect the fact that *M. gigas* is a less specialized, more opportunistic predator with a very broad diet (Tidemann *et al.*, 1985).

Necromantis exhibits many of these classically carnivorous dental features (Figures 6.4–6.7, 6.10). Other striking features include the relatively large tooth area with respect to cranial dimensions, including greatly expanded heel on M_{I-2} and broad stylar shelf, which are indicative of a crushing rather than purely slicing function. Further, marked wear of the teeth in several *Necromantis adichaster* specimens suggest it was eating very hard or abrasive prey rather than purely slicing flesh and/or arthropod cuticle (seen also in *Cheiromeles* specimens, among living bats, and the Eocene species of *Palaeophyllophora*; Maitre and Sigé, 2006). Striking dental features, such as a markedly median ectoflexus, lingually displaced mesostyle and large and median hypoconulid, as also seen in Early Eocene bats such as *Ageina* and *Honrovits* and retained in *Palaeophyllophora*, further suggest that *Necromantis* was a relict form specialized in carnivory and bone-crushing.

As noted above, some features shared by *Necromantis* and *Macroderma* may be due to convergence related to brachycephaly, and possibly correlated with a powerful bite by bringing the canines closer to the fulcrum of the jaw. Hand (1996) suggested megadermatids are distinguished from other bats by the following combination of cranial and dental characters associated with a shortening of the face and probably increased biteforce: almost complete loss of premaxilla; loss of upper incisors; reduction of lower incisors; anteroposterior compression of m_I such that it is shorter than or equal in length to m_2; anteroposterior compression of m_{I-3} talonids; marked reduction of M_3; reduction in the number of premolars to maximum of two in upper and lower toothrows (*N. adichaster* retains a vestigial P_3 and p_3); and posterior extension of the palate lingual to M_3 presumably for compensatory support during mastication.

One of the most distinctive cranial features of *Necromantis* is the extreme negative tilting of the head on the basicranial axis. Although suggested to be most comparable to nycteridids (Revilliod, 1920), in *Necromantis* this is developed to a much greater degree than in any other bat. Negative tilting of the head on the basicranial axis is thought to be controlled at least in part by the functional requirements of echolocation: orally emitting echolocators (e.g., emballonurids, noctilionids, vespertilionids, molossids) have positively tilted heads with respect to basicranial axis, while nasal emitters (hipposiderids, rhinolophids, phyllostomids, mormoopids, nycteridids, megadermatids and possibly rhinopomatids) have negatively tilted heads (Freeman, 1984, fig. 6; Pedersen, 1993). Rostral rotation is one of the most conspicuous correlates of nasal emission (Freeman, 1984; Pedersen, 1993). In nasal-emitting bats, the rostrum is rotated ventrally to align the nasopharynx with the direction of flight, the foramen magnum is moved ventrally and the inner ear is rotated posteriorly to compensate for the general rotation of the skull ventrally about the cranio-cervical axis (Pedersen, 1993). Like known nasal emitters, *Necromantis* exhibits pronounced rostral rotation, the angle between the planes of the hard palate and basicranial axis being approximately 60° and hence greater than that observed in any other bat (Freeman, 1984; Pedersen 1993, table 2). Rostral fontanels, found by Pedersen (1995) between the nasal, maxillary and frontal bones in Old World nasal emitters, are also well developed in *Necromantis*.

Most nasal-emitting bats have conspicuous noseleaves, but others do not; for example, rhinopomatids have a small nosepad and mormoopids have chin flaps and other fleshy facial protruberances. Hence, it is not clear whether *Necromantis* might have had a noseleaf or not. Hand (1998) found that rostrum shape and size is not directly correlated with, nor an accurate predictor of, noseleaf size and complexity in at least some bat lineages (e.g., hipposiderids). Further, structure/function relationships of noseleaves are not completely understood, although it has been hypothesized that they beam ultrasonic calls and thus contribute to the directionality of nasally emitted sound energy (e.g., Pye, 1988) and recent studies demonstrate that noseleaf furrows act as resonance cavities shaping the sonar beam in nasal emitters (Zhuang and Muller, 2006).

The periotic of *Necromantis* is not yet known, but the petrosal was evidently not greatly enlarged (nor the number of turns in the cochlea increased) as judged by the relatively unexpanded cavity for this bone (basioccipital width being almost as wide), suggesting that its echolocation call was not of particularly high frequency (see below). Revilliod (1920) remarked that the structure of the orbital region in *Necromantis* suggests that its eyes were well developed. Its very constricted interorbital region suggests that its olfactory lobe was probably not large (although there is no information about the cribriform plate itself) (Bhatnagar and Kallen, 1974), and the vomeronasal organ was probably not well developed (Cooper and Bhatnagar, 1976).

The only postcranial element yet referred to *Necromantis* is the distal portion of the humerus. In bats, the morphology of the distal humerus is well recognized as a rich source of taxonomic information, but has also been widely used to infer flight mode in extinct and extant taxa (e.g., Miller, 1907; Revilliod, 1917; Vaughan, 1959, 1970; Sigé, 1971, 1985; Smith, 1972; Habersetzer and Storch, 1987; Hand et al., 2009). In *Necromantis*, the humeral condyle (Figures 6.8–6.9) is mostly spherical, its articular surface is clearly offset with respect to the shaft, the epitrochlea is broad, and its wide separation from the trochlea and deep scars for muscle attachment, and locking of the radial patella or sesamoid suggest a relatively large muscle mass. This arrangement results in the ability to rotate the forearm around the humeral axis and is associated with relatively slow, maneuvrable flight (Vaughan, 1959, 1970; Smith, 1972) that is more agile and responsive to constraints imposed by a cluttered environment such as dense forest. It would also facilitate ground-based predation, and take-off with a heavy load.

Like extant carnivorous bats with large eyes and unexpanded petrosals, such as megadermatids and phyllostomids, the hunting style of *Necromantis* may have involved relatively low-frequency echolocation calls as well as passive listening. Such bats inspect their surroundings from a perch, locate their prey visually and by sound, before ambushing and killing it usually with one swift bite to the back of the head (Kulzer *et al.*, 1984). Smaller prey (*c.* 10 g) may be completely crushed and eaten (bones included), whereas parts of larger prey (remnants of skull, legs, digestive tract and tail) are sometimes dropped to the floor from a feeding station (Douglas, 1967; Vehrencamp *et al.*, 1977; Guppy and Coles, 1983; Kulzer *et al.*, 1984; Nowak, 1999). *Necromantis adichaster* probably also hunted in this way and, as one of the largest evidently carnivorous species in the Quercy bat faunas, could probably carry, crush and devour heavy prey, as required by its presumed necromantic habit.

6.8   Evolutionary relationships

The small quantity of material and evidently limited diversity known for *Necromantis* restrict phyletic interpretations. However, it seems that two lineages may be identified within this genus, now known from the Middle Eocene (reference level MP 13 of the European Paleogene mammal scale) to Late Eocene (reference level MP 17b). On the basis of available data, derivation of *N. adichaster* from *N. marandati* seems probable (Maitre, 2008). The size and the general structural pattern of the dentition remain the same, but with the protofossa on the upper molars opening posteriorly in *adichaster*. It is noteworthy that this same kind of morphological evolution is also observed in the mixopterygid *Carcinipteryx* lineage (Maitre *et al.*, 2008) and appears to represent a further example of mosaic morphological evolution in bats.

The species *N. gezei* Maitre, 2008 of reference level MP 17a represents a second lineage slightly smaller in size, but with similar primary morphological characters as those of *N. marandati* Maitre, 2008. As in the genus *Carcinipteryx*, this morphological evolution during anagenesis appears to be comparable within the two lineages (Maitre, 2008).

In terms of familial relationships, Weithofer (1887) suggested affinities of *Necromantis* to the Phyllostomidae on the basis of morphological similarities in a mandible preserving alveoli and the last two molars. In particular, he noted the presence of a small intermediate premolar between larger anterior and posterior premolars, reduced talonids on m2–3 and what he interpreted to be reduction of the lower incisors to a single pair. He considered this dental pattern to be similar to that found in living phyllostomids such as *Chrotopterus auritus* and *Trachops fuliginosus*, and referred *Necromantis* to that family.

Winge (1893) criticized Weithofer's conclusions, suggesting that the dentary was not sufficiently preserved to refer it to any known family. He remarked that many of the features noted by Weithofer as being shared with phyllostomids were also found among molossids, vespertilionids and megadermatids, and he considered the presence of the small intermediate premolar to be of little importance. This was also the opinion of Leche (1911).

On the basis of more complete material including a skull (QP630), rostrum (QS799) and two dentaries (QW627, QH427), Revilliod (1920) referred *Necromantis* to the family Megadermatidae where it has since remained (but see Sigé *et al.*, 1976). Hand (1985) included *Necromantis adichaster* in a phylogenetic analysis of extinct and extant megadermatids, upholding its position as the oldest and most plesiomorphic member of that family. In her analysis, polarity of (mostly dental) character states was facilitated by recognition of a Megadermatidae-Nycterididae sister-group relationship within the Rhinolophoidea (following Miller, 1907), but this latter relationship is now challenged by molecular and other data (e.g., Hutcheon *et al.*, 1998; Eick *et al.*, 2005; Teeling *et al.*, 2005; Simmons and Gunnell, 2005; Miller-Butterworth *et al.*, 2007; Simmons *et al.* 2008). Griffiths *et al.* (1992) conducted a phylogenetic analysis of living megadermatids on the basis of hyoid morphology, and included emballonurids, as well as nycteridids as outgroups. Contrary to Hand's (1985) interpretation, Griffiths *et al.* (1992) found *Macroderma gigas* to be the most plesiomorphic of living megadermatids and *Lavia frons* the most derived, a result probably better supported by current understanding of interfamilial relationships and therefore outgroup recognition and character-state polarity.

Collectively, the current evidence does not support the referral of *Necromantis* to Megadermatidae. The new data also indicate that some similarities between *Necromantis* and recent paleotropical bats such as megadermatids may be convergent. In dated molecular phylogenies, *Necromantis* is commonly used as a calibration point for the minimum age of the split between megadermatids and other rhinolophoids. Data presented here suggest that this calibration is probably invalid, and its removal may lead to revisions in molecular estimates for divergences within bats.

On the basis of available data, it seems prudent to avoid refering *Necromantis* to any extant bat family. Probably diverging early from other lineages such as emballonurids, hipposiderids, megadermatids, vespertilionoids and noctilio-noids, *Necromantis* might be independently derived from an older and more primitive "archaeonycterid" radiation in the Early Eocene, but it is also quite likely that its true origins lie within the early emballonuroid radiation. *Necromantis* probably soon began to specialize in hunting and consuming armored and bony prey, an evolutionary strategy favoring larger body size, increased crushing and chewing masticatory power, and morphological adaptations in flight mechanics, and take-off and carrying capacity. Its diet almost certainly included hard items such as armored arthropods and small vertebrates, mammals among them and probably bats. Such a dietary niche possibly represents the most derived known among carnivorous bats; the recent *Macroderma* and *Vampyrum* may have an overlapping diet but their more blade-like molar morphology and relatively gracile craniomandibular apparatus would favor more slicing and less blunt-force crushing abilities than in French Eocene *Necromantis*.

Current and future excavations of targeted Quercy karstic deposits are expected to recover new specimens of the paleokarstic bat *Necromantis adichaster* that will provide new data about its postcranium, particularly the forearm and pectoral girdle, and hence new information about this peculiar bat's biology, ecology and evolutionary relationships.

6.9   Paleoenvironment

The species of *Necromantis* lived in Europe during at least the Middle to Late Eocene (*c.* 44–36 Ma). In the Early Eocene, from about 53 Ma, Europe north of the proto-Atlantic Ocean had become isolated from North America to the west and Asia to the south, and a period of marked endemic faunal evolution began (Legendre and Hartenberger, 1992). This period was characterized by the coexistence of relictual mammalian taxa and new species arising from the new regional endemism. Towards the end of the Late Eocene, several archaic mammal groups declined in diversity, while many new species made their appearance. The extinctions, probably related to environmental change linked to lower temperatures, drying and opening of habitats, were accompanied by significant immigration. This first episode of the so-called "Grande Coupure" (Stehlin, 1909), as clearly shown by rodents, was followed by a phase of marked faunal turnover at the beginning of the Oligocene lasting a few million years. This was the consequence of new continental crossings, especially in northeastern Europe, with the closure of the Turgai Strait, and with faunal exchange resuming between Europe, Asia and America.

The paleoenvironment of the Eocene Quercy faunas is recognized as being that of warm and wet tropical forest which covered the Quercy limestone plateau of regional southwestern France (Legendre and Hartenberger, 1992). Within these forests, well-developed karst structures provided subterranean roosts for the region's bat fauna. In the modern world, wet tropical forests are widely recognized as the most favorable for high bat diversity, with plentiful arboreal roosts particularly attractive for bats (Brosset, 1966; Charles-Dominique *et al.*, 2001). Quercy Paleogene bat diversity has been examined by Maitre (2008) who records 7 bat families, 10 genera and 57 species from karst deposits spanning the Middle Eocene (*c.* 44 Ma) to late Early Oligocene (*c.* 30 Ma), an interval representing approximately the first two-thirds of the period of karstic filling in the Quercy. The fossil bats have been recovered from, so far, 90 separate localities in the Quercy region, with as many as 22 (evidently syntopic) bat species being recorded from individual MP reference levels (e.g., MP17a; see Maitre, 2008). Nevertheless, in terms of truly reflecting overall regional bat diversity in the Paleogene, the Quercy deposits offer an incomplete picture, there being a significant lack of similar data about regional non-cave bats (which, with rare exceptions, generally represent different familial and generic taxa).

Of the seven bat families represented in the Paleogene Quercy deposits, four are modern (families Vespertilionidae, Molossidae, Hipposideridae, Emballonuridae), with just two of these (Vespertilionidae and Molossidae) still represented in Europe today. The Hipposideridae and Emballonuroidea (i.e., including Rhinopomatidae) are now restricted to Old World tropical and subtropical regions, having disappeared from a cooling Europe by the Middle Miocene (Sigé, 1968; Legendre, 1980). Members of the extinct families Palaeochiropterygidae and Mixopterygidae made their last appearance in the Quercy in the Late Oligocene (Maitre, 2008, appendix 1.2). As yet it is not clear to which if any known family *Necromantis* might be referred, but it does not appear to be a member, or ancestor, of any group still surviving in Europe.

6.10   Summary and conclusions

Necromantis adichaster Weithofer, 1887 is among the largest bats recorded from southwestern France's Quercy phosphorite fillings. It was described on the basis of fragmentary material of uncertain age from the Anciennes Collections du Quercy. From the same collections, Revilliod (1920) revised *Necromantis*, describing the first cranial material and erecting two additional species, *N. planifrons* and *N. grandis*, which we regard as synonyms of *adichaster*. Revilliod attributed *Necromantis* to the paleotropical family Megadermatidae as its oldest member. Additional material of *Necromantis* from the Anciennes Collections du Quercy and further specimens from four well-dated Quercy localities now document the presence of *Necromantis* in western Europe from at least the Middle to Late Eocene (~36–44 Ma). Dental, cranial and postcranial morphology indicates that *Necromantis adichaster* was a large, robust bat adapted to predation on hard prey, including small vertebrates and possibly also carrion (as suggested by Weithofer's name for the taxon). The new data indicate that some similarities between *Necromantis* and recent paleotropical bats such as megadermatids may be convergent. Collectively, the current evidence does not support the referral of *Necromantis* to Megadermatidae. In dated molecular phylogenies, *Necromantis* is commonly used as a calibration point for minimum age of the split between megadermatids and other rhinolophoids. Our research suggests that this calibration is invalid, and its removal may lead to revisions in molecular estimates for divergences within bats.

6.11   Acknowledgments

We thank the following institutions for access to and loan of specimens in their care: Museum of Geology, Vienna University, Vienna; Muséum National d'Histoire Naturelle, Paris; Université Claude Bernard, Lyon 1, Lyon; Muséum d'Histoire Naturelle, Montauban; Naturhistorisches Museum, Basel; and the Université de Montpellier, Montpellier. We also thank: the late Roger Remy, C. Pondeville and J. Muirhead for illustrations; K. Black and J.-P. Aguilar for photographs; M. Archer for additional preparation of *Necromantis* skull specimens; and K. Black and G. Gunnell for constructive criticism of an earlier draft of this manuscript.

Appendix 6.1

Known available *Necromantis* material

Abbreviations and symbols: 19°, material obtained from nineteenth-century mines in the Quercy paleokarstic fillings, including specimens preserved in various collections without precise known origin in the Quercy Phosphorites region; 20°, material obtained by paleontological teams since the 1960s, from precisely located, previous Quercy phosphatic mines; d, described and figured by previous authors; n, new material; r, restudied material; s/f, newly studied and/or figured; #, presently available; ##, presently lost; ###, cast available. For dentition, upper case denotes upper tooth, lower case denotes lower tooth (e.g., M1 is upper first molar, M1 is lower first molar).

Institutions:

(A) *Museum of Geology, Vienna University, Vienna*

Necromantis adichaster Weithofer, 1887 (19°, d, #): QW627: full horizontal ramus of a large and deep jaw; alveoli for c, p2, lingual p3, p4, m1; m2–3 present; posterior jaw lacking; no incisor alveoli in precanine area; necromantodont m2; origin of the specimen near Escamps village; holotype of *adichaster* species, type species of *Necromantis* genus.

(B) *Muséum National d'Histoire Naturelle, Paris*

1 – QU16367 (19°, s/f, #, ###), *Necromantis adichaster*: subcomplete skull, preserving most of cranial roof, palate and basicranium; left alveoli for C to P3 and M1; *in situ* P4 and M2–3; right alveoli for C to P3; *in situ* P4 to M3.

2 – QU16369 (19°, s/f, ##, ###), *N. adichaster*: complete left jaw preserving c to m3, low p4; necromantodont m1 and 2; complete ascending ramus; Maitre, 2008, fig. II. 5. 1).

3 – QU16370 (19°, s, ##, ###), *N. adichaster*: left jaw fragment preserving horizontal ramus, alveoli for c to p3; *in situ* p4 to m3; tall p4, necromantodont m1 and 2.

4 – QU16371 (19°, s, ##, ###), *N. adichaster*: right jaw fragment preserving horizontal ramus; alveoli for c to p3; *in situ* p4 to m3; lingually shifted necromantodont m1 and 2.

(C) *Collections de Paléontologie, Université Claude Bernard, Lyon 1, Lyon*

FSL8130 (19°, d, #, ###), *N. adichaster*: left jaw fragment preserving horizontal ramus; alveoli for p3; *in situ* p4 to m3; lingually shifted necromantodont m1 and 2; Miguet, 1967, pp. 104–106, fig. 1.

(D) *Collections de Paléontologie, Muséum d'Histoire Naturelle, Montauban*

MNCH 1 (19°, s, #), *N. adichaster*: anterior left jaw fragment; precanine area without alveoli; alveoli of c, p2, p3, bi-rooted p4 and m1; necromantodont.

(E) *Paläontologische Abteilung, Basel Naturhistorisches Museum, Basel*

1 – QP639 (19°, d, #), origin near Lamandine; *N. adichaster*: almost complete skull, left and right alveoli for C, P2, lingual P3, M1; *in situ* P4, M2, M3; lacking bullae, zygomatic and periotic parts; Revilliod, 1920, pl. 2, figs. 1–6.

2 – QW627 (19°, d, #), *N. adichaster*: posterior left jaw fragment, *in situ* m1 to 3, necromantodont; Revilliod, 1920, p. 70, fig. 19.

3 – QH427 (19°, d, #), *N. adichaster* (*grandis* as recognized synonym): left posterior jaw fragment, *in situ* m2–3, necromantodont; Revilliod, 1920, p. 71, fig. 20.

4 – QS799 (19°, d, #), *N. adichaster* (*planifrons* as recognized synonym): facial skull fragment; Revilliod, 1920, p. 80, fig. 21.

(F) *Collections de Paléontologie, Université de Montpellier (UM), Montpellier*

1 – AQM7 (19°, s, #, ###), *N. adichaster*: subcomplete skull; posterior cranium cracked; condyles, occipital area, palate and basal posterior area preserved; bullae lost; alveoli for C to P3; *in situ* P4 to M3 both sides.

2 – AQM8 (19°, s, #, ###), *N. adichaster*: left jaw fragment preserving horizontal ramus; short precanine area without alveoli; anterior lateral foramen present; alveoli for c, large p2, lingual p3; *in situ* p4 to m3; necromantodont m1 and 2.

3 – Perrière locality, Tarn-et-Garonne dept.; *in situ* clay filling. *N. adichaster*: (UM) PRR66 to 71 (20°, s/f, #): 66, complete left M1; PRR67: right M1 or 2, lingual part; 68, idem; 69, idem; 70, left M2 on maxillary fragment; 71, complete right C.

4 – Rosières-5 locality, Lot dept.; outside old mine chimney rubble. *N. adichaster*: (UM) ROS-5 50 to 56 (20°, s/f, #): 50, complete left C; 51, worn, lingually shifted necromantodont left m1; 52, left, worn P4; 53, complete, worn right c; 54, complete, worn right c; 55, partial distal right humerus diaphysis with epiphysis; 56, distal right humerus worn epiphysis.

5 – La Bouffie locality, Lot dept., *in situ* chimney filling. *Necromantis gezei* Maitre, 2008: (UM) BFI NmA1.1 to 5 (20°, s/f, #): 1, complete unworn necromantodont right m2 (nec 1); 2, complete unworn necromantodont right m2; 3, complete worn necromantodont right m2; 4, unworn P4 labial half; 5, worn right M1 with broken metacone (= holotype).

6 – Cuzal locality, Lot dept., *in situ* lateral filling within long extended unroofed cave (Marandat *et al.*, 1993). *Necromantis marandati* Maitre, 2008: (UM) CUZ (20°, s/f, #): 383, right unworn M2 (= holotype); NgA.1.1, left unworn C.

6.13   REFERENCES

Aguilar, J.-P., Michaux, J., Pélissié, T. and Sigé, B. (2007). Early late Pliocene paleokarstic fillings predating the major Plio-Pléistocene erosion of the Quercy table, SW-France. *Acta Carsologica*, **36**, 469–473.

Astruc, J.-G., Hugueney, M., Escarguel *et al.* (2003). Puycelci, nouveau site à vertébrés de la série molassique d'Aquitaine. Densité et continuité biochronologique dans la zone Quercy et bassins périphériques au Paléogène. *Geobios*, **36**, 629–649.

Bhatnagar, K. P. and Kallen, F. C. (1974). Morphology of the nasal cavities and associated structures in **Artibeus jamaicensis** and **Myotis lucifugus**. *American Journal of Anatomy*, **139**, 167–190.

Boles, W. E. (1999). Avian prey of the Australian Ghost Bat *Macroderma gigas* (Microchiroptera: Megadermatidae): prey characteristics and damage from predation. *Australian Zoologist*, **31**, 82–91.

Brosset, A. (1966). *La biologie des chiroptères*. Paris: Masson & Cie.

Charles-Dominique, P., Brosset, A. and Jouard, S. (2001). Atlas des chauve-souris de Guyane. *Patrimoines Naturels*, **49**, 1–172.

Cooper, J. G. and Bhatnagar, K. P. (1976). Comparative anatomy of the vomeronasal complex in bats. *Journal of Anatomy*, **122**, 571–601.

Crochet, J.-L., Hartenberger, J.-L., Rage, J.-C. *et al.* (1981). Les nouvelles faunes de vertébrés antérieures à la «Grande Coupure» découvertes dans les phosphorites du Quercy. *Bulletin du Museum National d'Histoire Naturelle, Paris*, **3**, 245–266.

Crochet, J.-Y., Aguilar, J.-P., Astruc, J. G. *et al.* (2006). Reprises plio-quaternaires du paléokarst du Quercy. *Strata*, **13**, 85–95.

Daubrée, A. (1871). Sur le gisement dans lequel la chaux phosphatée a été récemment découverte dans les départements de Tarn-et-Garonne et du Lot. *Comptes rendus de l'Académie des Sciences de Paris*, **43**, 1029–1036.

Delfortrie, M. (1872). Les Gîtes de chaux phosphatée dans le Département du Lot. *Actes de la Société linéenne de Bordeaux*, **28**, 1871–1873.

Douglas, A. M. (1967). The natural history of the Ghost Bat, *Macroderma gigas* (Microchiroptera, Megadermatidae), in Western Australia. *Western Australian Naturalist*, **10**, 125–138.

Durand-Delga, M. (2006). De la découverte des phosphorites du Quercy au renouveau de leur étude avec Bernard Gèze. *Strata*, **13**, 25–36.

Duranthon, F. and Ripoll, F. (2006). Documents photographiques inédits d'Eugène Trutat sur l'exploitation des phosphorites du Quercy. *Strata*, **13**, 37–49.

Eick, G. N., Jacobs, D. S. and Matthee, C. A. (2005). A nuclear DNA phylogenetic perspective on the evolution of echolocation and historical biogeography of extant bats (Chiroptera). *Molecular Biology and Evolution*, **22**, 1869–1886.

Felten, H., Helfricht, A. and Storch, G. (1973). Die Bestimmung der europäschen Fledermäus nach der distalen Epiphyse des Humerus. *Senckenbergiana biologica*, **54**, 291–297.

Filhol, H. (1872). Recherches sur les mammifères fossiles des dépôts de phosphate de chaux dans les départements du Lot, du Tarn et de Tarn-et-Garonne. *Annales des Sciences de Géologie*, **3**, 1–31.

Filhol, H. (1876). Recherche sur les Phosphorites du Quercy. *Annales de la Société Géologique*, **7**, 44–48.

Filhol, H. (1877). Recherches sur les Phosphorites du Quercy. Etude des fossiles qu'on y rencontre et spécialement des mammifères. *Annales des sciences géologiques*, **8**, 1–340.

Freeman, P. W. (1984). Functional cranial analysis of large animalivorous bats (Microchiroptera). *Biological Journal of the Linnean Society*, **21**, 387–408.

Gèze, B. (1938). Contribution à la connaissance des phosphorites du Quercy. *Bulletin de la Société Géologique de France*, **5** (VIII), 123–146.

Griffiths, T. A., Truckenbrod, A. and Sponholtz, P. J. (1992). Systematics of megadermatid bats (Chiroptera, Megadermatidae), based on hyoid morphology. *American Museum Novitates*, **3041**, 1–21.

Gunnell, G. F. and Simmons, N. (2005). Fossil evidence and the origin of bats. *Journal of Mammalian Evolution*, **12**, 209–246.

Guppy, A. and Coles, R. B. (1983). Feeding behaviour of the Australian Ghost Bat, *Macroderma gigas* (Chiroptera: Megadermatidae) in captivity. *Australian Mammalogy*, **6**, 97–99.

Habersetzer, J. and Storch, G. (1987). Klassifikation und funktionelle Flügelmorphologie paläogener Fledermäuse (Mammalia, Chiroptera). *Courier Forschungsinstitut Senckenberg*, **91**, 117–150.

Hand, S. J. (1985). New Miocene megadermatids (Chiroptera: Megadermatidae) from Australia with comments on megadermatid phylogenetics. *Australian Mammalogy*, **8**, 5–43.

Hand, S. J. (1993). First skull of a species of *Hipposideros* (*Brachipposideros*) (Microchiroptera: Hipposideridae), from Australian Miocene sediments. *Memoirs of the Queensland Museum*, **31**, 179–192.

Hand, S. J. (1996). New Miocene and Pliocene megadermatids (Microchiroptera: Megadermatidae) from Australia, with broader comments on megadermatid evolution. *Geobios*, **29**, 365–377.

Hand, S. J. (1998). *Xenorhinos*, a new genus of Old World leaf-nosed bats (Microchiroptera: Hipposideridae) from the Australian Miocene. *Journal of Vertebrate Paleontology*, **18**, 430–439.

Hand, S. J., Weisbecker, V., Beck, R. M. D. *et al.* (2009). Bats that walk: a new evolutionary hypothesis for the terrestrial behaviour of New Zealand's endemic mystacinids. *BMC Evolutionary Biology*, **9**, 167.

Hutcheon, J. M., Kirsch, J. A. W. and Pettigrew, J. D. (1998). Base-compositional biases and the bat problem. III. The question of microchiropteran monophyly. *Philosophical Transactions of the Royal Society of London B*, **353**, 607–617.

Kœnigswald, W. v., Rup, I. and Gingerich, P. D. (2009). Cranial morphology of a new apatemyid, *Carcinella sigei* n. gen. n.sp. (Mammalia, Apatotheria) from the late Eocene of southern France. *Palaeontographica, Abt. A*, **288**, 53–91.

Kulzer, E., Nelson, J. E., McKean, J. L. and Maehres, F. P. (1984). Prey-catching behaviour and echolocation in the Australian Ghost Bat, *Macroderma gigas* (Microchiroptera, Megadermatidae). *Australian Mammalogy*, **7**, 37–50.

Leche, W. (1911). Einige Dauertypen aus der Klasse der Saugetiere. *ZoologischerAnzeiger*, **38**, 551–559.

Legendre, S. (1980). Un chiroptère emballonuride dans le Néogene d'Europe occidentale; considérations paléobiogéographiques. *Géobios*, **13**, 839–847.

Legendre, S. and Hartenberger, J.-L. (1992). Evolution of mammalian faunas in Europe during the Eocene and Oligocene In *Eocene-Oligocene Climatic and Biotic Evolution*, ed. D. Prothero. New York: Princeton University Press, pp. 512–552.

Legendre, S., Sigé, B., Astruc, J. G. *et al.* (1997). Les phosphorites du Quercy: 30 ans de recherche. Bilan et perspectives. *Geobios, Suppl. 1*, **30**, 331–345.

Maitre, E. (2008). Les Chiroptères paléokarstiques d'Europe occidentale, de l'Eocène moyen à l'Oligocène inférieur, d'après les nouveaux matériaux du Quercy (SW France): systématique, phylogénie, paléobiologie. Unpublished Ph.D. thesis, Université Claude Bernard-Lyon 1.

Maitre, E. and Sigé, B. (2006). A fossil bat with extremely worn teeth and a question for tropical bat specialists. *Bat Research News*, **47**, 62.

Maitre, E., Sigé, B. and Escarguel, G. (2008). A new family of bats in the Paleogene of Europe: systematics and implications of emballonurids and rhinolophoids. *Neues Jahrbuch Geologisches und Palaeontologisches Abteilung*, **250**, 199–216.

Marandat, B., Crochet, J.-Y., Godinot, M. *et al.* (1993). Une nouvelle faune d'âge éocène moyen (Lutétien supérieur) dans les phosphorites du Quercy. *Geobios*, **26**, 617–623.

Maynard Smith, J. and Savage, R. J. G. (1959). The mechanics of mammalian jaws. *School Science Review*, **40**, 289–301.

Miguet, R. (1967). Observations nouvelles sur les chiroptères des phosphorites du Quercy. *Travaux des Laboratoires de Géologie de la Faculté des Sciences de Lyon*, **14**, 103–114.

Miller, G. S. (1907). The families and genera of bats. *Bulletin of the United States National Museum*, **57**, 1–282.

Miller-Butterworth, C. M., Murphy, W. J., O'Brien, S. J., Jacobs, D. S., Springer, M. S. and Teeling, E. C. (2007). A family matter: conclusive resolution of the taxonomic position of the long-fingered bats, *Miniopterus*. *Molecular and Biological Evolution*, **24**, 1553–1561.

Milne-Edwards, A. (1892). Sur les oiseaux fossils des dépots éocènes de phosphate de chaux du Sud de la France. *Comptes Rendus du Second Congrès Ornithologique International*, 60–80.

Navarro L. D. and Wilson, D. E. (1982). *Vampyrum spectrum*. *Mammalian Species*, **184**, 1–4.

Nowak, R. (1999). *Walker's Mammals of the World*, vol. 1, 6th edn. Baltimore, MD: Johns Hopkins University Press.

Pedersen, S. C. (1993). Cephalometric correlates of echolocation in the Chiroptera. *Journal of Morphology*, **218**, 85–98.

Pedersen, S. C. (1995). Cephalometric correlates of echolocation in the Chiroptera: II. Fetal development. *Journal of Morphology*, **225**, 107–123.

Pye, J. (1988). Noseleaves and bat pulses. In *Animal Sonar: Processes and Performance. Proceedings NATO Advanced Study Institute on Animal Sonar Systems*, ed. P. Nachtigall and P. Moore. New York: Plenum Press, pp. 791–796.

Remy, J. A., Crochet, J.-Y., Sigé, B. *et al.* (1987). Biochronologie des phosphorites du Quercy: mise à jour des listes fauniques et nouveaux gisements de mammifères fossiles. *Münchner Geowissenschaftliche Abhandlungen*, **10**, 169–188.

Revilliod, P. (1917). Contribution à l'étude de chiroptères des terrains tertiaires. Première partie. *Mémoires de la Société Paléontologique suisse*, **43**, 1–58.

Revilliod, P. (1920). Contribution à l'étude des chiroptères des terrains tertiaires. Deuxième partie. *Mémoires de la Société Paléontologique suisse*, **44** (1919), 63–129.

Revilliod, P. (1922). Contribution à l'étude des chiroptères des terrains tertiaires. Troisième partie et fin. *Mémoires de la Société Paléontologique suisse*, **45**, 131–195.

Richards, G. C., Hand, S. J., Armstrong, K. and Hall, L. S. (2008). Ghost Bat. In *Mammals of Australia*, ed. S. Van Dyck and R. Strahan, 2nd edn. Sydney: New Holland, pp. 449–450.

Schlosser, M. (1887). Die affen, Lemuren, Chiropteren, Insectivoren, Marsupialier, Creodonten und Carnivoren des europäischen Tertiärs und deren Beziehungen zu ihren lebenden und fossilen aussereuropäischen Verwandten. *Beiträge zur paläontologie Österreich-Ungarns und des Orients*, **6**, 1–224.

Sigé, B. (1968). Les chiroptères du Miocène inférieur de Bouzigues, I, Etude systématique. *Palaeovertebrata*, **1**, 65–133.

Sigé, B. (1971). Anatomie du membre antérieur chez un chiroptère molossidé (*Tadarida* sp.) du Stampiae de Céreste (Alpes-de-Haute-Provence). *Palaeovertebrata*, **4**, 1–38.

Sigé, B. (1976). Les Megadermatidae (Chiroptera, Mammalia) miocènes de Beni Mellal, Maroc. *Annale de l'Université de Provence, Géologie méditerranéenne*, **3**, 71–86.

Sigé, B. (1985). Les chiroptères oligocènes du Fayum, Egypte. *Geologica et Palaeontologica*, **19**, 161–189.

Sigé, B. and Hugueney, M. (2006). Les micromammifères des gisements à phosphate du Quercy (SW France). *Strata*, **13**, 207–227.

Sigé, B. and Legendre, S. (1983). L'histoire des peuplements de chiroptères du bassin méditerranéen: l'apport comparé des remplissages karstiques et des dépôts fluvio-lacustres. *Mémoires de Biospéologie*, **10**, 209–225.

Sigé, B., Crochet, J.-Y., Hartenberger, J.-L., Remy, J. A., Sudre, J. and Vianey-Liaud, M. (1979). Catalogue des Mammifères du Quercy. In *Fossilium Catalogus I: Animalia 126*, ed. F. Westphal. The Hague: Junk, pp. 1–99.

Simmons, N., Seymour, K. L., Habersetzer, J. and Gunnell, G. F. (2008). Primitive early Eocene bat from Wyoming and the evolution of flight and echolocation. *Nature*, **451**, 818–821.

Smith, J. D. (1972). Systematics of the chiropteran family Mormoopidae. *University Kansas Museum National History Miscellaneous Publication*, **56**, 1–132.

Stehlin, H. G. (1909). Remarques sur les faunules de mammifères des couches éocènes et oligocènes du Bassin de Paris. *Bulletin de la société géologique de France*, **9**, 488–620.

Szalay, F. S. (1969). Mixodectidae, Microsyopidae, and the insectivore-primate transition. *Bulletin, American Museum of Natural History*, **140**, 163–330.

Teeling, E. C., Springer, M. S., Madsen, O., Bates, P., O'Brien, S. J. and Murphy, W. J. (2005). A molecular phylogeny for bats illuminates biogeography and the fossil record. *Science*, **307**, 580–584.

Teilhard de Chardin, P. (1914–1915). Les carnassiers des Phosphorites du Quercy. *Annales de Paléontologie*, **9**, 101–192.

Teilhard de Chardin, P. (1921). Sur quelques primates des Phosphorites du Quercy. *Annales de Paléontologie*, **10**, 1–20.

Thévenin, A. (1903). Etude géologique de la bordure Sud-Ouest du Massif Central. *Bulletin Service de la Carte géologique de la France 1*, **14**, 353–554.

Tidemann, C. R., Priddel, D. M., Nelson, J. E. and Pettigrew, J. D. (1985). Foraging behaviour of the Australian Ghost Bat, *Macroderma gigas* (Microchiroptera: Mega-dermatidae). *Australian Journal of Zoology*, **33**, 705–713.

Trouessart, E.-L. (1898–1899). *Catalogus mammalium tam viventium quam fossilium. I, Primates, prosimiae, chiroptera, insectivora, carnivora, rodentia, pinnipedia.* Berolini: R. Friedländer und Sohn.

Trouessart, E.-L. (1904–1905). *Catalogus mammalium tam viventium quam fossilium. I, Primates, prosimiae, chiroptera, insectivora, carnivora, rodentia, pinnipedia.* Berolini: R. Friedländer und Sohn.

Van Valen, L. (1979). The evolution of bats. *Evolutionary Theory,* **4,** 103–121.

Vaughan, T. A. (1959). Functional morphology of three bats: *Eumops, Myotis, Macrotus. University of Kansas Publications Museum of Natural History,* **12,** 1–153.

Vaughan, T. A. (1970). The skeletal system. In *Biology of Bats,* vol. 1, ed. W. A. Wimsatt. New York: Academic Press, pp. 97–138.

Vehrencamp, S., Stiles, F. and Bradbury, J. (1977). Observations on the foraging behavior and avian prey of the neotropical carnivorous bat, *Vampyrum spectrum. Journal of Mammalogy,* **58,** 469–477.

Weithofer, A. (1887). Zur kenntniss der fossilen cheiropteren der französischen Phosphorite. *Mathematisch-naturwissenschaftlich Classe,* **96,** 341–360.

Winge, H. (1893). Jordfundne og nulevende Flagermus fra Lagoa Santa Minas Geraes, Brasilien: med udsigt over flagermusenes indbyrdes slægtskab. *Museo lundii,* **2,** 92 pp.

Zhuang, Q. and Muller, R. (2006). Noseleaf furrows in a horseshoe bat act as resonance cavities shaping the biosonar beam. *Physical Review Letters,* **97,** 218701, 1–4.

7

African Vespertilionoidea (Chiroptera) and the antiquity of Myotinae

GREGG F. GUNNELL, THOMAS P. EITING AND
ELWYN L. SIMONS

7.1 Introduction

Vesper and evening bats (Family Verspertilionidae) are a diverse group (about 350 living species) that has a nearly global distribution (being absent only in polar regions, on some oceanic islands and in harsher desert climates). Vespertilionids often have been included with molossids, mystacinids, myzopodids, thyropterids, furipterids and natalids in the superfamily Vespertilionoidea (Koopman, 1994), but many other variations of the superfamily exist (e.g., Simmons, 1998; Jones *et al.*, 2002; Hoofer and Van Den Bussche, 2003; Hoofer *et al.*, 2003; Horáček *et al.*, 2006; Miller-Butterworth *et al.*, 2007). Our prime focus in this chapter is on two subfamilies of the Vespertilionidae, Vespertilioninae and Myotinae, as defined by Simmons (2005).

Osteologically, the basic dichotomy between myotines and vespertilionines can be typified by differing patterns of dental morphology. All myotines share myotodont lower molar morphology, in which the postcristid extends to the entoconid and isolates the hypoconulid (as opposed to nyctalodonty where the postcristid extends to the hypoconulid and does not reach the entoconid). Myotines also share the presence of three premolars, with the middle premolar being reduced. Some vespertilionines have myotodont lower molars, but only a few exhibit both myotodonty and the retention of three premolars (e.g., *Plecotus* and *Idionycteris*). No vespertilionines have the middle premolar reduced. Recent phylogenetic analyses suggest that all vespertilionines are far removed from myotines (e.g., Hoofer and Van Den Bussche, 2003), implying that any shared morphological similarities are likely to be convergences.

Within Vespertilionidae, *Myotis* is one of earliest occurring genera in the fossil record, and it contains nearly one-third of modern vespertilionid specific diversity (Gunnell and Simmons, 2005; Simmons 2005; Eiting and Gunnell, 2009). A better understanding of the biogeographic history of *Myotis* is crucial for

Evolutionary History of Bats: Fossils, Molecules and Morphology, ed. G. F. Gunnell and N. B. Simmons. Published by Cambridge University Press. © Cambridge University Press 2012.

elucidating the origin and evolutionary radiation of Vespertilionidae. In this chapter we examine the evolutionary history of *Myotis* and allied forms and suggest that the geographic origin of this now diverse and widespread subfamily (Myotinae) may well have been in Africa. Even if the geographic point of origin was not African, it is becoming clearer that Africa and other Gondwanan land-masses played key roles as centers of diversification for bats as well as other mammals (Seiffert *et al.*, 2003, 2009; Miller *et al.*, 2005; Seiffert, 2007; Gunnell, 2010).

Abbreviations include: BALch, Baraval Chiropteran specimen, Université de Montpellier II (Montpellier, France); CGM, Cairo Geological Museum (Cairo, Egypt); IRSNB M, Institut Royal des Sciences Naturelles de Belgique (Brussels, Belgium); UMMZ, University of Michigan Museum of Zoology (Ann Arbor, MI, USA).

7.2   *Myotis* today

Myotis is a nearly ubiquitous and widespread genus perhaps only second to humans in breadth of global distribution among mammals (Nowak, 1994; Simmons, 2005). It is abundant and diverse (100+ species; see Figure 7.1B) and lives in a variety of habitats, including both dry and wet forests, open woodlands and along moist coastal planes. Most species either roost in caves or in trees, and higher latitude species hibernate during the winter months (Nowak, 1994). *Myotis* species range in body mass from around 2.5 g (*M. siligorensis*) to about 40 g (*M. myotis*), although the vast majority of species are 10 grams or less. With the exception of only a very few species (e.g., the fish- and crustacean-eating *M. vivesi*), all species of *Myotis* are insectivorous with most being aerial hawkers (Nowak, 1994).

Teeth of *Myotis* are often viewed as relatively primitive but, in fact, exhibit several apparently derived character states. These include reduction or loss of upper molar hypocones (hypocones are present in the primitive North American bats *Onychonycteris* and *Icaronycteris*, at least on M1 – see Smith *et al.*, Chapter 2, this volume) and reduction of upper and lower middle premolars (?P3/p3) with the lower middle premolar having only a single root (see Figure 7.5C). Thus the dentition of *Myotis* can be viewed as a combination of primitive and derived morphologies, with retention of three premolars being the most apparent primitive feature.

7.3   *Myotis* relationships and divergence times

Myotis has traditionally been divided into four or more subgenera based on differential morphological correlates with feeding strategies (Findley, 1972;

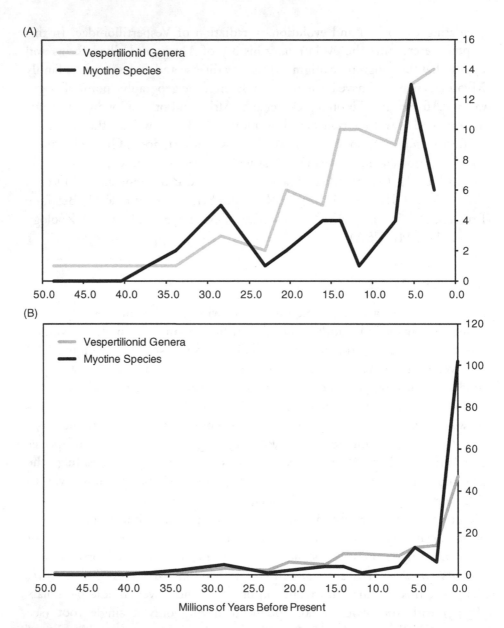

Figure 7.1 Number of vespertilionid genera (light gray line) plotted against number of myotine species (black line) through time: (A) plotted without modern taxa, (B) with modern taxa included. Note fairly stable diversity of myotines from the Late Eocene until the Pliocene (1–5 species) where diversity increases dramatically to 13 species. Note also the rather steady increase in vespertilionid generic diversity beginning in the Early Miocene. In the Recent, *Myotis* species diversity is more than twice vespertilionid generic diversity. The decrease in diversity for Late Pliocene (2.5 Ma) myotines is almost certainly artificial resulting from poor sampling through this interval.

Koopman, 1994). However, phylogenetic analysis of genetic information has failed to substantiate these subgeneric groups. Instead biogeographic groups are favored, suggesting that similar feeding strategies likely developed convergently several times within the *Myotis* clade (Fenton and Bogdanowicz, 2002; Hoofer and Van Den Bussche, 2003; Kawai *et al.*, 2003; Stadelmann *et al.*, 2004, 2007). Stadelmann *et al.* (2004, 2007) examined the relationships within *Myotis* based on both mitochondrial and nuclear genes, and were able to assign the majority of the *Myotis* species examined to one of five different biogeographic clades – a New World clade (a mixture of Nearctic and Neotropical species), an Ethiopian clade, two Palaearctic clades (essentially Europe and non-tropical Asia) and an Oriental clade (essentially Peninsular and Oceanic Asia). These results have recently been corroborated by a large-scale study based on nuclear DNA (Lack *et al.*, 2010).

The *Myotis* clades found by Stadelmann *et al.* (2004, 2007) and Lack *et al.* (2010) form a monophyletic clade that is the sister group to other vespertilionids (Figure 7.2). Estimated divergence times for the whole *Myotis* clade range from around 16 to 14 Ma (during the Middle Miocene). Figure 7.3 shows the estimated molecular divergence times of various extant global *Myotis* clades and the ages of the earliest known fossil specimens that may represent myotines (those taxa that could represent basal myotines that would form successive sister clades between extant myotines and other vespertilionids; see Figure 7.2).

In South America, Asia and Australia, molecular divergence times predate the earliest known fossil records of myotines by as much as seven million years (see Qui *et al.*, 1985 and Morgan and Czaplewski, Chapter 4 this volume, for discussions of the first appearances of myotines in Asia and South America, respectively), almost certainly the result of either a lack of appropriately aged localities in these areas or a lack of taphonomically appropriate samples.

The earliest records of potential myotines in North America predate the estimated divergence time of *Myotis* by at least five million years and perhaps much longer if Orellan-aged (32 Ma) *Oligomyotis* truly represents a myotine (Morgan and Czaplewski, Chapter 4, this volume). The same is true in Africa and Europe, where potential members of the myotine clade are present in fossil deposits long before estimated divergence times of the respective *Myotis* clades (Horáček, 2001, Eiting and Gunnell, 2009). The presence of what appear to be 30+ million-year-old fossil myotines from Europe (e.g., Horáček, 2001) and Africa (e.g., Gunnell *et al.*, 2008) suggest that the antiquity of the myotine clade can be traced much farther into the past than the Middle Miocene (as suggested by molecular evidence). Alternatively, the dental characters used to define the myotine clade may not be particularly informative, either because they are primitive retentions or because they may have arisen multiple times during the evolution of chiropterans (Horáček, 2001).

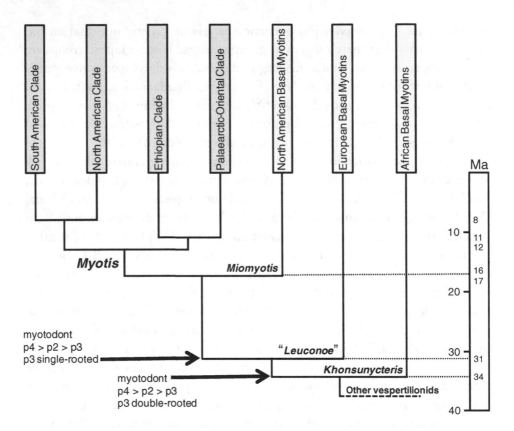

Figure 7.2 Phylogenetic tree depicting relationships of the four extant *Myotis* clades as defined by Stadelmann *et al.* (2004, 2007) with age of occurrence shown by time scale at right. Successive sister groups to extant *Myotis* clade are defined by first appearance of basal-most member of each clade and are arranged in order of age of first appearance. The North American basal clade (defined initially by *Miomyotis*) first appears at approximately 16 Ma, the European basal clade (initially represented by "*Leuconoe*") appears at 31 Ma, and the African basal clade (originally represented by *Khonsunycteris*) appears at 34 Ma and is thus the earliest known evidence of the existence of the myotine clade.

7.4   Fossil Myotines

Five dental features found in combination typify living *Myotis*: myotodont lower molars; the presence of three premolars; the middle premolar reduced; the fourth premolar larger than the second; and the middle premolar single-rooted. It is difficult to be certain of the polarity of these characteristics. Myotodonty has traditionally been viewed as a derived chiropteran character state (see Menu and Sigé, 1971; Sigé *et al.*, Chapter 13, this volume) and it has clearly evolved several times independently within chiropterans. The presence

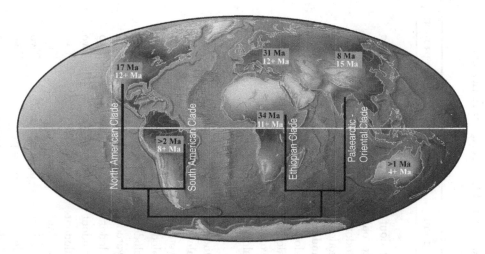

Figure 7.3 Extant *Myotis* clades superimposed on global map with molecular divergence times shown in white and ages of first known fossil evidence for each shown in black. For example, the North American clade has a basal divergence time of 12+ Ma based on molecular evidence (Stadelmann *et al.*, 2007) while the earliest known fossil specimens attributed to myotines are dated at 17 Ma, a discrepancy of approximately +5 million years. Based on molecular evidence, the basal divergence of the extant myotines clade is approximately 16 Ma while the oldest fossils that may be attributable to this clade are dated at 34 Ma, a discrepancy of +18 million years.

of three premolars is, as far as is known, primitive for bats. No bat, fossil or living, has four premolars and all of the earliest known primitive bats retain three premolars. It is also unclear what the homology of bat premolars is – we employ standard dental terminology in this chapter and refer to the retained myotine premolars as p2–4 (Miller, 1907), but other possibilities exist (Giannini and Simmons, 2007). A reduced p3 would seem to be a derived condition, but one of the most primitive known bats, *Onychonycteris finneyi*, has p3 reduced (Simmons *et al.*, 2008; Smith *et al.*, Chapter 2, this volume). A single-rooted p3 could be derived – even though *O. finneyi* has a reduced p3 it clearly had two roots. However, no other bats besides myotines have this combination of dental features. As noted above, in addition to these lower tooth features, myotines have apparently lost the hypocone on upper molars.

Table 7.1 summarizes relevant lower dentition characteristics for myotine and potential myotine sister taxa from the Eocene and Oligocene of Europe and Africa. The genus *Stehlinia*, known from the Middle Eocene through Early Oligocene (42.7 through 28.2 Ma) in Europe, has often been viewed as a primitive vespertilionid (McKenna and Bell, 1997) or as a member of an extinct family primarily known from the Middle Eocene of Europe, the

Table 7.1 Summary of dental morphology of relevant vespertilionid taxa from the Eocene and Oligocene of Africa and Europe.

Genus	Species	Epoch	Age	Date (Ma)	Molar form	Premolars	p3 roots
Oligocene							
Myotis	*lavocati*	L Oligocene	MP-26	27	Myotodont	p4>p2>p3	One
Myotis	*minor*	L Oligocene	MP-26	27	Myotodont	p4>p2>p3	One
Myotis	*intermedius*	L Oligocene	MP-26	27	Myotodont	p4>p2>p3	One
Myotis	*major*	L Oligocene	MP-26	27	Myotodont	p4>p2>p3	One
Leuconoe	**sp. indet. A**	**E Oligocene**	**MP-22**	**31**	**Myotodont**	**p4>p2>p3**	**One**
Quinetia	*missonnei*	E Oligocene	MP-21	32.4	Nyctalodont	p4>p3≥p2	One
Cuviermops	*legendrei*	E Oligocene	MP-22-23	30-31	Nyctalodont, Myotodont	p4>p2, p3 absent	Absent
Steblinia	*bonisi*	E Oligocene	MP-21	28.2-32.5	Nyctalodont	p4 = p3>p2	Two
Eocene							
Khonsunycteris	***aegypticus***	**L Eocene**	**MP-19**	**34**	**Myotodont**	**p4>p2>p3**	**Two**
Cuviermops	*parisiensis*	L Eocene	MP-19	34.3-34.7	Nyctalodont, Myotodont	p4>p2, p3 absent	Absent
Cuviermops	*priscus*	L Eocene	MP-17a	37.3-38	Nyctalodont	p4>p2, p3 absent	Absent
Cuviermops	*intermedius*	L Eocene	MP-17b-18	35.5-36.5	Nyctalodont, Myotodont	p4>p2, p3 absent	Absent
Steblinia	*gracilis*	L Eocene	MP17a-b	36.5-38	Nyctalodont	p4 = p3>p2	Two
Steblinia	*minor*	L Eocene	MP-17a-20	33.2-37.6	Nyctalodont	p4 = p3>p2	Two
Steblinia	*quercyi*	L Eocene	MP-17a-b	36.7-37.0	Nyctalodont	p3>p4>p2	Two
Steblinia	*pusilla*	M Eocene	MP-13-14	41.7-43	Nyctalodont	p4 = p3>p2	Two
Steblinia	*rutimeyeri*	M Eocene	MP-14	41.7	Nyctalodont	p4 = p3>p2	Two
Steblinia	*revilliodi*	M Eocene	MP-13	42.7	Nyctalodont	p4 = p3>p2	Two
Steblinia	*alia*	M Eocene	MP-13	42.7	Nyctalodont	p4 = p3>p2	Two

Palaeochiropterygidae (Maitre, 2008; Smith *et al.*, Chapter 2, this volume), which have often been viewed as potentially ancestral to vespertilionoids. Most species of *Stehlinia* have nyctalodont lower molars (some species occasionally show incipient myotodonty), p3 and p4 of the same size with both being larger than p2, and p3 always double-rooted. No species of *Stehlinia* shares the combination of dental characters found in *Myotis*.

Late Eocene through Early Oligocene (38 through 30 Ma) *Cuviermops* has often been included in the family Molossidae, one of the most closely related sister groups to Vespertilionidae (Hoofer and Van Den Bussche, 2003). *Cuviermops* is of some interest here because several of its included species show some sign of incipient myotodonty. However, in most other relevant dental features *Cuviermops* differs substantially from the pattern seen in *Myotis*. Importantly, all species of *Cuviermops* have only two lower premolars (by convention recognized as p2 and p4; see Miller, 1907) so any other comparisons with *Myotis* are moot.

The Early Oligocene taxon *Quinetia misonnei* (Figure 7.4C, 7.5B) from Hoogbutsel in Belgium was originally described as *Myotis misonnei* by Quinet (1965). Horáček (2001) correctly pointed out that "*Myotis*" *misonnei* had nyctalodont molars and also differed from typical *Myotis* in having broad and shallow molar talonids, well-differentiated lingual cingulids and lacking molar entoconid crests. He proposed the new genus *Quinetia* to accommodate *M. misonnei* and moved the taxon into Vespertilioninae and suggested that it was likely closely related to extant *Plecotus*.

The earliest European record of a fossil bat that shares all of the lower dental features of extant *Myotis* can be found at the Quercy locality of Baraval in France (Maitre, 2008). This locality is dated to the Early Oligocene (approximately 31 Ma, Rupelian, European reference level MP-22). The single specimen (Figure 7.4B) was assigned to the genus *Leuconoe* as an indeterminate species by Maitre (2008). Molecular analysis has failed to sustain *Leuconoe* as a genus (or subgenus) distinct from *Myotis* (Hoofer and Van Den Bussche, 2003; Lack *et al.*, 2010) so we tentatively recognize the Baraval bat as the first occurrence of *Myotis* in the fossil record (see also Horáček, 2001 for discussion of other potential Late Oligocene and earliest Miocene myotines). It is possible, with more complete specimens, that the genus *Leuconoe* could be confirmed as a separate fossil taxon, but such an action awaits further discoveries (see discussion in Ziegler, 2000). An occurrence of *Myotis* in the Early Oligocene in France may not be too surprising, given the relatively diverse radiation of *Myotis* species recognized from the Late Oligocene (Ziegler, 2000).

Until quite recently there were only two fossil bats known from the Late Eocene and Early Oligocene of Africa, the enigmatic *Vampyravus orientalis* Schlosser (1910, 1911; also see Gunnell *et al.*, 2009), from an uncertain level

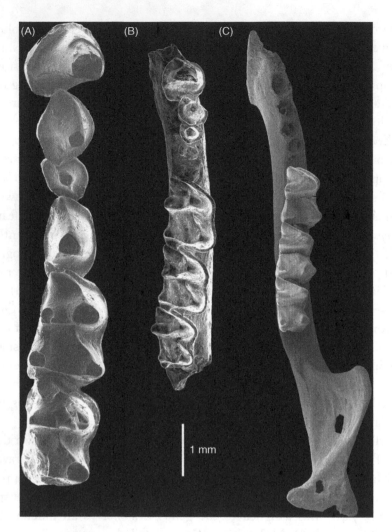

Figure 7.4 Representative specimens of some early myotines in occlusal view.
(A) Holotype of Late Eocene *Khonsunycteris aegyptiacus* (CGM 83673),
left dentary (reversed) c1-m2, from Egypt. (B) Early Oligocene "*Leuconoe*" sp.
(BALch-01), right dentary, c1-p3, m1–3 (Maitre, 2008) from France. (C) Holotype
of Early Oligocene *Quientia misonnei* (IRSNB M 1189), right dentary m1–3 and
alveoli of anterior teeth, from Belgium.

(but probably Early Oligocene) in the Fayum, Egypt, and *Philisis sphingis*, a
member of the African endemic fossil bat family Philisidae (Sigé, 1985). Other
limited records from Tunisia (Sigé, 1991) and Oman (Sigé *et al.*, 1994) came to
light mostly representing further occurrences of philisids, but with some
tantalizing, but poorly documented, evidence of more modern groups as well
(Sigé, 1991; Sigé *et al.*, 1994; Gunnell *et al.*, 2003; Gunnell, 2010).

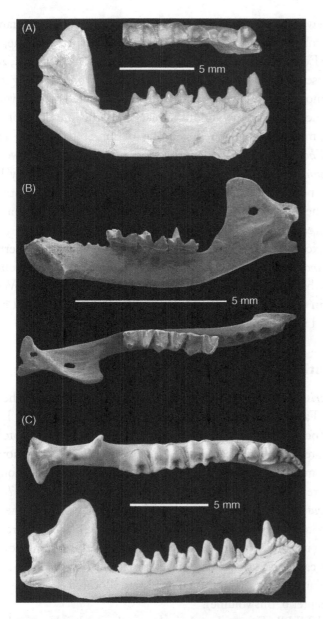

Figure 7.5 Representative specimens of some fossil and extant myotines showing tooth and dentary structures. (A) Holotype of *Khonsunycteris aegyptiacus* (CGM 83673), left dentary c1–m2 in occlusal (top) and lingual (bottom) views. (B) Holotype of *Quientia misonnei* (IRSNB M 1189), right dentary m1–3 and alveoli of anterior teeth in lingual (top) and occlusal (bottom) views. (C) Extant *Myotis myotis* (UMMZ 109335), left dentary i1–m3 in occlusal (top) and lingual (bottom) views.

In 2008, several new bat taxa, including good records of the modern bat families Rhinopomatidae, Emballonuridae, Megadermatidae and Vespertilionidae, were described from Late Eocene deposits in the Fayum of Egypt (Gunnell *et al.*, 2008). Included among these new bats was the vespertilionid *Khonsunycteris aegypticus*. *Khonsunycteris* has most of the lower dental features associated with myotines, including the presence of fully myotodont molars and the presence of three premolars with the middle premolar reduced and p4 larger than p2. The one difference between *Khonsunycteris* and extant myotines is the retention of two roots on the p3 of *Khonsunycteris*. Additionally, *Khonsunycteris* has a relatively high-crowned lower canine, which, along with the double-rooted p3, serves to distinguish it from the African endemic myotine *Cistugo* (Ziegler, 2000), a taxon traditionally recognized as a subgenus of *Myotis*, but recently raised to generic level (Lack *et al.*, 2010). Except for the presence of a double-rooted p3 (almost certainly primitive for bats), *Khonsunycteris* possesses all of the other dental features found in *Myotis* and suggests that *K. aegypticus* should not only be included within Vespertilionidae, but within the subfamily Myotinae as well. As such, it would be the earliest occurring and most primitive known member of that subfamily.

7.5   Potential implications

Khonsunycteris is one of three bat genera found at Quarry L-41 in the Jebel Qatrani Formation, Fayum, Egypt (Gunnell *et al.*, 2008; Gunnell, 2010). L-41 is placed in the Late Eocene (Priabonian) and is dated at 34 Ma (Seiffert, 2010). All three L-41 genera represent modern bat families (*Khonsunycteris* – Vespertilionidae, *Dhofarella* – Emballonuridae, *Saharaderma* – Megadermatidae) and are among the earliest known occurrences of these families. Among emballonurids, only *Tachypteron* and *Vespertiliavus* occur earlier (both from the Lutetian of Europe). The Early Eocene (Ypresian) potential emballonurid *Eppsinycteris* Hooker (1996) has recently been shown to belong in a different family (see Smith *et al.*, Chapter 2, this volume). The only potentially earlier occurrence of a megadermatid is *Necromantis* from the Lutetian and Bartonian of Europe, but this taxon is now allocated elsewhere as well (see Hand *et al.*, Chapter 6, this volume).

Several Early and Middle Eocene bats have been allocated to Vespertilionoidea (including Philisidae, see Sigé, 1991; Sigé *et al.*, 1994), but *Stehlinia* from the Lutetian and Bartonian of Europe remains the only earlier occurrence of a possible vespertilionid. However, based on the presence of nyctalodont lower molars, a reduced p2 and a double-rooted p3 equivalent in size to p4, it seems that *Stehlinia* is, at best, distantly related to myotines and more likely belongs to the extinct family Palaeochiropterygidae. This leaves *Khonsunycteris* as both the oldest and the most primitive member of Myotinae.

Southern (Gondwanan) origins for some bat groups have been discussed in the past (Hershkovitz, 1972; Sigé, 1991; Hand and Kirsch, 1998; Teeling *et al.*, 2005; Gunnell *et al.*, 2008). The presence of *Khonsunycteris* in the Fayum in the Late Eocene suggests that, even if myotines didn't originate there, North Africa was an important area for development and diversification of the group.

An additional aspect of the African bat assemblage may support the notion of a myotine origination on that continent. Other than the extant families from the Fayum, most other known African Eocene bats are philisids, including Early Eocene *Dizzya* (Sigé, 1991) and Late Eocene *Witwatia* (Gunnell *et al.*, 2008), known from earlier in the Priabonian than the L-41 bats. *Witwatia* and the Oligocene (*Philisis*) and Miocene (*Scotophilisis*) philisids have myotodont lower molars. Early Eocene *Dizzya* (from Chambi, Tunisia) does not have myotodont lower molars. However, *Dizzya* is represented only by two isolated specimens, an upper molar that is very philisid-like (distinct protofossa extending to the labial margin, distinct para- and metastylar fovea, mesostyle small, connected to the premetastylar crista and placed on the tooth margin, distinct and strong para-stylar hook) and a broken left dentary with two broken molars that are less philisid-like (nyctalodont, cusps more robust and less anteroposteriorly compressed, hypoflexid less deep). Either primitive philisids were nyctalodont, or *Dizzya* actually represents two separate taxa, one philisid and another of less certain affinities. The latter interpretation is possible since Sigé (1991) also described a rhinolophoid tooth from Chambi, and Ravel *et al.* (2010) have recently documented primitive archaeonycterid-like bats with submyotodont lower molars from the Early Eocene of Algeria (El Kohol). These other discoveries, although poorly represented, indicate that Early Eocene African bats were already diverse with at least one philisid already present.

If *Dizzya* turns out to be myotodont instead of nyctalodont, then myotodonty was established by the Early Eocene in Africa, and myotodonty may represent the primitive condition of Vespertilionidae. The evidence now available favors an African origin for Myotinae. More complete material of *Khonsunycteris*, including upper dentitions and cranial elements, would go a long way towards confirming the myotine affinities of this taxon and the presence of a primitive *Myotis*-like radiation in the Late Eocene of Africa – a radiation that predates the spread of *Myotis* across the Old and New Worlds.

7.6   Acknowledgments

We thank Thierry Smith and Eric De Bast (Royal Belgian Institute of Natural Sciences) and Elodie Rigaud for supplying images of *Quinetia* and *Leuconoe*, respectively. We thank Philip Myers and Steve Hinshaw (University of Michigan Museum of Zoology) for access to comparative specimens. Nancy Simmons (American Museum of Natural History) and Ivan Horáček (Charles University) provided guidance and sage advice. We also thank H. Hamouda, A. Abd el-Raouf, Z. el-Alfy, K. Soliman and the late Y. Attia for facilitating fieldwork in Fayum, and the staff of the Egyptian Mineral Resources Authority and Egyptian Geological Museum for their assistance. P. Chatrath has been instrumental in all aspects of fieldwork in Egypt and we thank him for his expertise.

7.7   REFERENCES

Eiting, T. P. and Gunnell, G. F. (2009). Global completeness of the bat fossil record. *Journal of Mammalian Evolution*, **16**, 151–173.

Fenton, M. B. and Bogdanowicz, W. (2002). Relationships between external morphology and foraging behaviour: bats in the genus *Myotis*. *Canadian Journal of Zoology*, **80**, 1004–1013.

Findley, J. S. (1972). Phenetic relationships among bats of the genus *Myotis*. *Systematic Zoology*, **21**, 31–52.

Giannini, N. P. and Simmons, N. B. (2007). Element homology and the evolution of dental formulae in megachiropteran bats. *American Museum Novitates*, **3559**, 1–27.

Gunnell, G. F. (2010). Chiroptera. In *Cenozoic Mammals of Africa*, ed. L. Werdelin and W. J. Sanders. Berkeley, CA: University of California Press, pp. 581–597.

Gunnell, G. F. and Simmons, N. B. (2005). Fossil evidence and the origin of bats. *Journal of Mammalian Evolution*, **12**, 209–246.

Gunnell, G. F., Jacobs, B. F., Herendeen, P. S. *et al.* (2003). Oldest placental mammal from sub-Saharan Africa: Eocene microbat from Tanzania – evidence for early evolution of sophisticated echolocation. *Palaeontologia Electronica*, **5**, 10.

Gunnell, G. F., Simons, E. L. and Seiffert, E. R. (2008). New bats (Mammalia: Chiroptera) from the late Eocene and early Oligocene, Fayum Depression, Egypt. *Journal of Vertebrate Paleontology*, **28**, 1–11.

Gunnell, G. F., Worsham, S. R., Seiffert, E. R. and Simons, E. L. (2009). *Vampyravus orientalis* Schlosser (Chiroptera) from the early Oligocene (Rupelian), Fayum, Egypt – body mass, humeral morphology and affinities. *Acta Chiropterologica*, **11**, 271–278.

Hand, S. J. and Kirsch, J. A. W. (1998). A southern origin for the Hipposideridae (Microchiroptera)? Evidence from the Australian fossil record. In *Bat Biology and Conservation*, ed. T. H. Kunz and P. A. Racey. Washington, DC: Smithsonian Institution Press, pp. 72–90.

Hershkovitz, P. (1972). The recent mammals of the Neotropical region: a zoogeographic and ecological review. In *Evolution, Mammals, and Southern Continents*, ed. A. Keast, F. C. Erk and B. Glass. Albany, NY: State University of New York Press, pp. 311–432.

Hoofer, S. R. and Van Den Bussche, R. A. (2003). Molecular phylogenetics of the chiropteran family Vespertilionidae. *Acta Chiropterologica*, **5** (Suppl.), 1–63.

Hoofer, S. R., Reeder, S. A., Hansen, E. W. and Van Den Bussche, R. A. (2003). Molecular phylogenetic and taxonomic review of noctilionoid and vespertilionoid bats (Chiroptera: Yangochiroptera). *Journal of Mammalogy*, **84**, 809–821.

Hooker, J. J. (1996). A primitive emballonurid bat (Chiroptera, Mammalia) from the earliest Eocene of England. *Palaeovertebrata*, **25**, 287–300.

Horáček, I. (2001). On the early history of vespertilionid bats in Europe: the lower Miocene record from the Bohemian Massif. *Lynx (Praha)*, **32**, 123–154.

Horáček, I., Fejfar, O. and Hulva, P. (2006). A new genus of vespertilionid bat from Early Miocene of Jebel Zelten, Libya, with comments on *Scotophilus* and early history of vespertilionid bats (Chiroptera). *Lynx (Praha)*, **37**, 131–150.

Jones, K. E., Purvis, A., MacLarnon, A., Bininda-Emonds, O. R. P. and Simmons, N. B. (2002). A phylogenetic supertree of the bats (Mammalia: Chiroptera). *Biological Reviews*, **77**, 223–259.

Kawai, K., Nikaido, M., Harada, M. *et al.* (2003). The status of the Japanese and East Asian bats of the genus *Myotis* (Vespertilionidae) based on mitochondrial sequences. *Molecular Phylogenetics and Evolution*, **28**, 297–307.

Koopman, K. F. (1994). Chiroptera: systematics. *Handbook of Zoology*. vol. **8**, pt. 60, *Mammalia*. Berlin: Walter de Gruyter.

Lack, J. B., Roehrs, Z. P., Stanley Jr., C. E., Ruedi, M. and Van Den Bussche, R. A. (2010). Molecular phylogenetics of *Myotis* indicate familial-level divergence for the genus *Cistugo*. *Journal of Mammalogy*, **91**, 976–992.

Maitre, E. (2008). Les Chiropteres paléokarstiques d'Europe Occidentale, de l'Eocene Moyen a l'Oligocene Inferieur, d'Apres les nouveaux materiaux du Quercy (SW France): Systematique, Phylogenie, Paleobiologie. Unpublished Ph.D. thesis, Universite Claude Bernard – Lyon 1.

McKenna, M. C. and Bell, S. K. (1997). *Classification of Mammals Above the Species Level*. New York: Columbia University Press.

Menu, H. and Sigé, B. (1971). Nyctalodontie et myotodontie, importants caractères de grades évolutifs chez les chiroptères entomophages. *Comptes Rendus de l'Académie des Sciences, Paris*, **272**, 1735–1738.

Miller E. R., Gunnell, G. F. and Martin, R. D. (2005). Deep time and the search for anthropoid origins. *Yearbook of Physical Anthropology*, **48**, 60–95.

Miller, G. S. (1907). The families and genera of bats. *Bulletin of the United States National Museum*, **57**, 1–282.

Miller-Butterworth, C. M., Murphy, W. J., O'Brien, S. J. *et al.* (2007). A family matter: conclusive resolution of the taxonomic position of the Long-fingered Bats, *Miniopterus*. *Molecular Biology and Evolution*, **24**, 1553–1561.

Nowak, R. M. (1994). *Walker's Bats of the World*. Baltimore, MD: Johns Hopkins University Press.

Qui, Z., Han, D., Qi, G. and Yufen, L. (1985). A preliminary report on a micromammalian assemblage from the hominoid locality of Lufeng Co. Yunnan Province. *Acta Anthropologica Sinica*, **4**, 13–32 (In Chinese, translated by Will Downs).

Quinet, G. E. (1965). *Myotis misonnei* n. sp. – Chiroptère de l'Oligocène de Hoogbutsel. *Bulletin de l'Institut royal des Sciences naturelles de Belgique*, **41**, 1–11.

Ravel, A., Marivaux, L., Tabuce, R. and Mahboubi, M. (2010). Oldest bat (Chiroptera, Eochiroptera) from Africa: Early Eocene from El Kohol (Algeria). *Journal of Vertebrate Paleontology, SVP Program and Abstracts Book*, **2010**, 149A.

Schlosser, M. (1910). Über einige fossile Säugetiere aus dem Oligocän von Ägypten. *Zoologischen Anzeiger*, **35**, 500–508.

Schlosser, M. (1911). Beiträge zur Kenntnis der oligozänen Landsäugetiere aud dem Fayum: Ägypten. *Beiträge Zur Paläontologie und Geologie Österr-Ungarns Orients*, **24**, 51–167.

Seiffert, R. E. (2007). A new estimate of afrotherian phylogeny based on simultaneous analysis of genomic, morphological, and fossil evidence. *BMC Evolutionary Biology*, **7**, 224.

Seiffert, E. R. (2010). Chronology of Paleogene mammal localities. In *Cenozoic Mammals of Africa*, ed. L. Werdelin and W. J. Sanders. Berkeley, CA: University of California Press, pp. 19–26.

Seiffert, E. R., Simons, E. L. and Attia, Y. (2003). Fossil evidence for an ancient divergence of lorises and galagos. *Nature*, **422**, 421–424.

Seiffert, E. R., Perry, J. M. G., Simons, E. L. and Boyer, D. M. (2009). Convergent evolution of anthropoid-like adaptations in Eocene adapiform primates. *Nature*, **461**, 1118–1121.

Sigé, B. (1985). Les Chiroptères Oligocènes du Fayum, Egypte. *Geologica et Palaeontologica*, **19**, 161–189.

Sigé, B. (1991). Rhinolophoidea et Vespertilionoidea (Chiroptera) du Chambi (Eocène inférieur de Tunisie). Aspects biostratigraphique, biogéographique et paléoécologique de l'origine des chiroptères modernes. *Neues Jahrbuch für Geologie und Paläontologie, Abhandlungen*, **182**, 355–376.

Sigé, B., Thomas, H., Sen, S. *et al.* (1994). Les Chiroptères de Taqah (Oligocène Inférieur, Sultanat D'Oman). Premier inventaire systématique. *Münchner Geowissenschaftliche Abhandlungen*, **26**, 35–48.

Simmons, N. B. (1998). A reappraisal of interfamilial relationships of bats. In *Bat Biology and Conservation*, ed. T. H. Kunz and P. A. Racey. Washington, DC: Smithsonian Institution Press, pp. 3–26.

Simmons, N. B. (2005). Order Chiroptera. In *Mammal Species of the World*, ed. D. E. Wilson and D. M. Reeder. Baltimore, MD: Johns Hopkins University Press, pp. 312–529.

Simmons, N. B., Seymour, K. L., Habersetzer, J. and Gunnell, G. F. (2008). Primitive Early Eocene bat from Wyoming and the evolution of flight and echolocation. *Nature*, **451**, 818–821.

Stadelmann, B., Jacobs, D. S., Schoeman, C. and Ruedi, M. (2004). Phylogeny of African *Myotis* bats (Chiroptera, Vespertilionidae) inferred from cytochrome *b* sequences. *Acta Chiropterologica*, **6**, 177–192.

Stadelmann, B., Lin, L.-K., Kunz, T. H. and Ruedi, M. (2007). Molecular phylogeny of New World *Myotis* (Chiroptera, Vespertilionidae) inferred from mitochondrial and nuclear DNA genes. *Molecular Phylogenetics and Evolution*, **43**, 32–48.

Teeling, E. C., Springer, M. S., Madsen, O. *et al.* (2005). A molecular phylogeny for bats illuminates biogeography and the fossil record. *Science*, **307**, 580–584.

Ziegler, R. (2000). The bats (Chiroptera, Mammalia) from the late Oligocene fissure fillings Herrlingen 8 and Herrlingen 9 near Ulm (Baden-Württemberg). *Senckenbergiana Lethaea*, **80**, 647–683.

Evolutionary and ecological correlates of population genetic structure in bats

KEVIN J. OLIVAL

8.1 Introduction

Studies of intraspecific genetic structure are of central importance to evolutionary biology, as population differentiation can lead to distinct evolutionary trajectories and may be viewed as the first step in the process of speciation (Wright, 1932). The distribution of genetic variation among populations is determined by a complex interaction of historical and contemporary evolutionary forces – including genetic drift, mutation, selection and gene flow – which are influenced by the specific life history, behavior and ecology of a species (Slatkin, 1987; Goodnight, 2006). Historical events such as vicariance, colonization events or glacial refugia can also dramatically shape contemporary patterns of genetic subdivision. Gene flow is the primary force that leads to genetic homogeneity among populations, and species may be subject to geographic, behavioral or ecological barriers to gene flow. The task of teasing out the relative importance of different life-history, behavioral and ecological characteristics in determining genetic structure, each taking place across a range of temporal and spatial scales, continues to challenge population geneticists (Hey and Machado, 2003; Estoup et al., 2004; Duminil et al., 2007).

Bats (order Chiroptera) which comprise nearly 1200 species, exhibit a wide range of life-history and ecological characteristics that make them an ideal group to test hypotheses regarding the correlates of population structure in animals (Nowak, 1999; Simmons, 2005). Being nocturnal and difficult to study by traditional ecological methods (including direct observation, telemetry and capture–mark–recapture), a growing number of studies have used molecular markers to examine the population biology of bats (Burland and Worthington-Wilmer, 2001). In general, population genetic studies of volant animals are of particular interest because of their potential for long-distance dispersal and high levels of gene flow over terrestrial and oceanic landscapes.

Evolutionary History of Bats: Fossils, Molecules and Morphology, ed. G. F. Gunnell and
N. B. Simmons. Published by Cambridge University Press. © Cambridge University Press 2012.

A number of recent studies in bats (Sinclair *et al.*, 1996; Webb and Tidemann, 1996; McCracken and Gassel, 1997; Ditchfield, 2000; Russell *et al.*, 2005), birds (Wennerberg *et al.*, 2002; Dallimer *et al.*, 2003; Sonsthagen *et al.*, 2004) and butterflies (Shephard *et al.*, 2002) have supported the assumption that volant taxa will have low levels of population structure across very broad geographic ranges. However, a capacity for flight alone does not necessitate genetic homogeneity among populations, and recent studies have shown that bats may be subject to biogeographic (Hisheh *et al.*, 1998; Heaney *et al.*, 2005; Weyandt *et al.*, 2005; Roberts, 2006), ecological (Miller-Butterworth *et al.*, 2003; Ruedi and Castella, 2003; Campbell *et al.*, 2006) or behavioral (Entwistle *et al.*, 2000; Furmankiewicz and Altringham, 2007) dispersal barriers, as are other vertebrates. Most bat population genetic studies to date have focused on one species at a time, although a handful have compared patterns of population structure across several sympatric bat species (Ditchfield, 2000; Campbell *et al.*, 2006; Roberts, 2006; Campbell *et al.*, 2007; Garcia-Mudarra *et al.*, 2009).

In order to better understand the processes that determine population structure in animals, or bats in particular, broad-scale comparative approaches are needed. Previous comparative studies have examined the correlation of life-history and ecological traits with population genetic structure in plants (Aguinagalde *et al.*, 2005; Moyle, 2006; Duminil *et al.*, 2007), but similar studies in animals are lacking (Waples, 1987). Comparative approaches and meta-analyses have also been used to understand broad-scale patterns in bats, including correlates of extinction risk (Jones *et al.*, 2003), basal metabolic rate (Cruz-Neto and Jones, 2006) and vulnerability to habitat fragmentation (Meyer *et al.*, 2008). However, no meta-analysis published to date has tested for correlates of population genetic structure across bats, or any other order of mammals.

Here I present a quantitative meta-analysis of population genetic studies for 61 bat species (from ten families) to better understand the intraspecific patterns of genetic subdivision in Chiroptera. I use F_{ST} and its analog for sequence data, Φ_{ST} (Excoffier *et al.*, 1992) as dependent variables for population genetic differentiation. F_{ST} and Φ_{ST} are the most commonly used metrics of genetic subdivision and are expressed as a ratio between population-level differentiation and total diversity (Wright, 1951); F_{ST} actually describes the probability that two alleles drawn at random from a subpopulation are identical by decent. As ratios, these statistics are largely independent from the mutation process, so that data from a range of studies using different molecular markers can be compared (Pannell and Charlesworth, 2000). Values of F_{ST} vary from 0 to 1, with zero being equal to no geographic structure (i.e., panmixia) and one equal to complete genetic subdivision among populations. I selected a number of categorical and continuous life-history variables to explore their significance as predictors of F_{ST} and test a number of morphological, ecological and behavioral hypotheses using both a traditional single and multiple regression approach, and a method that controls for phylogenetic dependence.

Population structure is a hierarchical measure and depends, to some degree, on the geographic scale of sampling (Wright, 1951) or the number of genetic loci examined (Beerli, 2006). Methodological approaches could potentially bias the levels of population structure that are detected in a given study (Epperson, 2005; Goodnight, 2006). Therefore I also examined variables related to sampling and study design for correlation with measures of F_{ST}. Significant interactions would suggest potential sampling biases that should be controlled for in further analyses.

A priori hypotheses and predicted effects on population genetic structure for each variable are summarized in Table 8.1 and in the sections below. The variables examined here are not exhaustive, but provide a good representation of the potential evolutionary and ecological determinants of population genetic structure in bats.

8.2   Wing morphology

Wing morphology in both bats and birds is known to be associated with dispersal ability, foraging and migratory behavior (Norberg and Rayner, 1987; Norberg, 1995; Iriarte-Diaz et al., 2002). Four measures of the wing that relate to aerodynamic efficiency and maneuverability (wingspan, wing loading, aspect ratio and tip shape index) were predicted to have an effect on dispersal ability and possibly population structure.

Wing aspect ratio (AR) is measured as the square of the wingspan divided by wing area and relates to flight efficiency. I hypothesize that AR will be negatively correlated with population genetic structure (F_{ST}). Bats with low AR are better suited for maneuverable flight and will exhibit behaviors such as foraging in cluttered habitats of the forest understory. Bat species with higher AR values lack the ability for acrobatic flight and are more capable of dispersing long distances with a lower energetic cost (Norberg and Rayner, 1987). Migratory species on average have higher AR than non-migratory species for birds (Norberg, 1995; Lockwood et al., 1998) and bats (Norberg and Rayner, 1987; Fleming and Eby, 2003), but no previous study has tested for correlation with measures of gene flow, or population genetic structure.

Wing loading is measured as mass × g (gravitational acceleration) divided by wing area and is positively related to flight speed and negatively to maneuverability (Norberg and Rayner, 1987). I hypothesize that wing loading is negatively correlated with genetic structure in bats, although current theory is equivocal.

Table 8.1 A priori hypotheses that certain morphology and life-history variables will affect population genetic structure in bats, and predicted effects on F_{ST}. Predictions are further justified with references in the Introduction.

	HYPOTHESIS	PREDICTIONS		Key references
		Low F_{ST} (panmictic)	High F_{ST} (structured)	
Morphology related to dispersal	Wing Loading	High WL	Low WL	Norberg and Rayner, 1987; Norberg, 1995
	Wing Aspect Ratio	High AR	Low AR	Norberg and Rayner, 1987; Norberg, 1995
	Wingspan	Long wings	Short wings	Norberg and Rayner, 1987
	Wing Tip Shape Index	Low value, pointed	High value, rounded	Norberg and Rayner, 1987
Morphology related to promiscuity	Body Mass	High Mass	Low Mass	Swartz et al., 2003; Willig et al., 2003
	Baculum length	Long	Short	Hosken et al., 2001; Hosken and Stockley, 2004
	Testis Size	Large	Small	Pitnick et al., 2006
	Brain Size/Neocortex Vol.	Small	Large	Pitnick et al., 2006
Ecology/Behavior	Food Resource	Fruit/Nectar	Insect/Carnivore	Dumont, 2003; Hodgkison et al., 2004
	Roost Type	Tree/Foliage	Cave/Building	Lewis, 1995
	Population Size	Large	Small	Russell et al., 2005
	Geographic Distribution	Continental	Insular	Schmitt et al., 1995; Hisheh et al., 1998
	Seasonal Migration	Migratory	Non-migratory	Fleming and Eby, 2003
	Mating System	Multimale groups	Harems	Storz, 1999; Racey and Entwistle, 2000
	Mating Seasonality	Seasonal/swarming	Year-round	Kerth et al., 2003

Lower wing loadings may translate to less energy expenditure during flight that may be good for long-distance migration and soaring behavior, but low wing loads also mean slower overall flight speeds and shorter distances traveled per unit of time (Norberg *et al.*, 2000; Iriarte-Diaz *et al.*, 2002). Conversely, species with high wing loading have higher flight speeds for long-distance movement.

Wingspan should covary with both AR and wing loading, as it is an important component of both variables. All else being equal with wing shape and weight, I predict that longer wingspans will be found in more mobile species and be negatively associated with population genetic structure.

Wing tip shape index measures the degree of roundedness of the distal portion of the wing and is independent of wing size (Norberg and Rayner, 1987). The influence of wing tip shape on bird and bat dispersal is not clear (Norberg and Rayner, 1987; Lockwood *et al.*, 1998). Larger values describe more rounded and maneuverable wing tips and smaller values for more pointed tip shapes (Norberg and Rayner, 1987). Pointed wing tips are likely to be associated with migratory behavior and long-distance dispersal (Lockwood *et al.*, 1998), therefore I predict that tip shape index values will positively correlate with F_{ST} across species.

8.3   Other morphology

Body mass – New World bats in the family Vespertilionidae show a positive correlation between body mass and geographic range size (Willig *et al.*, 2003). Species with larger geographic ranges are expected to have more genetic cohesiveness across the landscape (and potentially lower overall genetic structure). Higher body mass should also directly covary with higher wing loading, which translates to less maneuverable flight and higher flight speeds during dispersal. Body mass should therefore negatively correlate with F_{ST}, but may also be strongly influenced by phylogeny (see Giannini *et al.*, Chapter 16, this volume).

Brain size and neocortex volume – In bats, species with more promiscuous mating behaviors among females have relatively smaller brains than species that form monogamous or multimale groups (Pitnick *et al.*, 2006). Thus, as a proxy for promiscuity and potentially higher levels of gene flow, brain size (controlled for body size) should positively correlate with genetic structure. Alternatively, the social brain hypothesis states that larger brains, relative to body size, will be associated with more social behavior – and potentially promiscuity. Dunbar demonstrated a relationship between relative neocortex size and group size across primate species (Dunbar, 1992, 1998). Relative brain size has been correlated with increased sociality in three orders of mammals (Perez-Barberia *et al.*, 2007).

Testis size – Relative testes mass is positively correlated with sperm competition in a number of taxonomic groups, and is also positively associated with female promiscuity in bats (Pitnick *et al.*, 2006). Species with relatively larger testes potentially have greater numbers of mates and should have higher levels of gene flow and lower levels of genetic structure (Harcourt *et al.*, 1981).

Baculum length – The baculum, or os penis, is morphologically variable across bat species although not present in all families (Hosken *et al.*, 2001). Differences in genital morphology correlate with male fertilization success; polyandrous species tend to have longer bacula (Hosken *et al.*, 2001; Hosken and Stockley, 2004). Thus, species with longer bacula can also dramatically shape contemporary genetic patterns, for example, again a proxy for more promiscuous mating systems, are expected to have lower overall genetic structure.

8.4   Behavioral characteristics

Mating Behavior – Bats have a wide range of mating systems, including multimale, harem and monogamy, with most species (~90%) being polygynous (McCracken and Wilkinson, 2000). The genetic consequences of mammal and bat social structure have been previously reviewed (Chepko-Sade and Halpin, 1987; McCracken, 1987; vanStaaden, 1995; Storz, 1999). Breeding groups that are isolated because of selective mating behavior (e.g., harem formation) may differentiate more quickly over time due to the combined effects of inbreeding and genetic drift, leading to greater variance in population genetic structure (Storz, 1999). Species may also differ in the seasonality of mating associations, including year-round groups or seasonal, or swarming, aggregations (Rossiter *et al.*, 2000; Veith *et al.*, 2004). I predicted that species with seasonal mating aggregations will have lower levels of population structure than bats that form year-round mating colonies because of the potential for extensive gene flow between colonies during swarming (Kerth *et al.*, 2003).

Population Size – Bats are known to form the largest aggregations of any mammal, with upwards of eight million individuals in a single roost for some species (e.g., *Tadarida brasiliensis* and *Eidolon helvum*). Larger populations will retain genetic diversity for longer periods in the absence of gene flow because of incomplete lineage sorting and lower rates of genetic drift; and alleles will reach fixation in a population more quickly when population sizes are small (Hartl and Clark, 1997). Thus, larger populations will retain higher levels of genetic diversity when separated and these species should have lower overall genetic differentiation between populations (Russell *et al.*, 2005).

Migratory behavior and dispersal ability – The ability to disperse is negatively correlated with genetic population structure for a wide array of taxonomic groups (Bohonak, 1999). Seasonal migration is defined as the regular, seasonal movement of bats from summer to wintering colonies and is most common among Microchiroptera that live in temperate regions (Fleming and Eby, 2003). Previous studies have noted that genetic structure was universally low among migratory bat species (McCracken and Gassel, 1997; Burland and Worthington-Wilmer, 2001; Russell *et al.*, 2005). However, other studies have challenged this view, and suggest that morphological or ecological factors may be responsible for maintaining population structure even in highly migratory species (Miller-Butterworth *et al.*, 2003).

8.5 Ecological characteristics

8.5.1 Food resources

The availability of food resources may be an important factor affecting dispersal and thus shaping genetic structure of bat populations. Higher levels of genetic subdivision may be seen in bat species with stable and localized food resources (i.e., insects) and lower levels of population structure in bats that need to fly long distances to obtain their sources of food (i.e., nectar and fruit). Many frugivorous bats, in the Old and New World, move across very large geographic distances in response to the availability of food resources (Sinclair *et al.*, 1996; Webb and Tidemann, 1996; Wilkinson and Fleming, 1996; Moreno-Valdez *et al.*, 2000; Hodgkison *et al.*, 2004; Moreno-Valdez *et al.*, 2004; Tidemann and Nelson, 2004), and fruit bats on Pacific islands may even track food resources between islands (McConkey and Drake, 2007). Fruiting and flowering for many tropical forest species occur over broad areas in masting events that are variable and asynchronous, and may drive bat movement over long distances (Maycock *et al.*, 2005; Sakai *et al.*, 2006). Insect abundance may also affect bat movement, but, in general, insects represent a locally stable resource (Wickramasinghe *et al.*, 2004; Bontadina *et al.*, 2008) although there is evidence that Brazilian free-tail bats may track migratory insect movement (McCracken *et al.*, 2008).

8.5.2 Roosting type

Roost type can be used as a proxy for roost fidelity in bats (Lewis, 1995). Generally, bats that roost in trees or foliage tend to shift roosts more frequently than bats that roost in more solid structures such as caves or buildings (Lewis, 1995). Low roost fidelity should result in more frequent dispersal of individuals between populations, thereby increasing gene flow and lowering overall levels of population structure (Raghuram *et al.*, 2006). An alternate hypothesis would be that cave-roosting bats are more adapted to flying long distances as these species, which form dense aggregations of individuals, are subject to stronger intraspecific competition for local food resources.

8.5.3   Oceanic barrier to gene flow

Several studies have suggested that oceanic distances are a significant barrier to movement for many bat species and may be an important force causing population subdivision (Peterson and Heaney, 1993; Schmitt *et al.*, 1995; Hisheh *et al.*, 1998; Juste *et al.*, 2000; Maharadatunkamsi *et al.*, 2003; Olival, 2008). Data from other species suggest some level of recent or ongoing gene flow across purported oceanic barriers (Pumo *et al.*, 1996; Juste *et al.*, 2003; Roberts, 2006; Olival, 2008). The extent that oceanic distance represents a barrier to gene flow in bats is an open question and is likely species specific (Garcia-Mudarra *et al.*, 2009). I hypothesize that island distributed bats will have higher levels of genetic structure compared to continentally distributed species because of their current isolation. An equally valid alternate hypothesis would be that bat species that have successfully colonized oceanic islands have an evolutionary adaptation to flying long distances and may thus have lower levels of population structure because of their inherited dispersal ability.

8.5.4   IUCN status

Conservation status alone is not expected to be a determinant of population genetic structure, but a correlation with endangered status and population genetic structure may be expected. Bat species with limited mobility, and thus more structured populations, are more vulnerable to local threats such as forest fragmentation (Meyer *et al.*, 2008). Similarly, endangered species, by definition, often have reduced numbers of individuals and exist in isolated, fragmented populations – both of which would potentially increase the level of population structure over time.

8.6   Methods

8.6.1   Data collection, population genetic structure

Population genetic studies were identified by searching ISI Science Citation Index and Zoological Records for the years 1985–2009, and additional unindexed papers were identified from literature cited. Only studies that examined original data for ≥ 3 populations for a given species and reported F_{ST} or Φ_{ST} values were included in the analysis. Global F_{ST} was calculated as the mean of all pairwise F_{ST} values in each study, and used as the dependent variable. Separate global F_{ST} values were calculated for mitochondrial (Φ_{ST}, hereafter F_{ST_mt}) and nuclear (F_{ST_nuc}) DNA markers; a third dependent variable, F_{ST_all}, included global F_{ST_mt} and F_{ST_nuc} values pooled together or an average of the two when values for both markers were available ($n = 18$). As F_{ST} or Φ_{ST} evolve at different rates (mtDNA has one-quarter the effective population size as nuclear markers), most analyses were performed on the data sets independently. Three species had negative F_{ST_mt} values (*P. vampyrus*, *T. brasiliensis* and *T. rufus*), indicating that they were not significantly different from zero, and were coded as 0.001. All F_{ST} values were log-transformed to meet assumptions of normality.

8.6.2   Experimental and sampling variables

Variables related to study design and experimental methods were obtained from each population genetic study including: number of loci examined, total and average number of populations and individuals sampled, and geographic distance between populations (Table 8.2). Average geographic distance (km) was calculated as a mean of pairwise distances between sampled populations or a simple average between minimum and maximum geographic distance if pairwise values were unavailable. Geographic distance was estimated using locality information and Google Earth v. 4.0.2 if distances were not specified in the study.

8.6.3   Species-specific variables

I used morphological and life-history values as reported in the paper from which F_{ST} data were obtained if they were given, additional sources of data were as follows. Web of Science and IUCN Red List (www.iucnredlist.org) were used to identify additional references for morphological/life-history data for each species (Table 8.3). A majority of wing morphology and body mass variables were obtained from Norberg and Rayner's (1987) data set of 257 species; with additional AR values found in (Bullen and McKenzie, 2002; Jones *et al.*, 2003). For *P. hypomelanus, T. rufus, T. furculus, P. jagori* and *H. fischeri* values of wing area and wingspan were calculated from field-collected wing tracings using the program ImageJ (http://rsb.info.nih.gov/ij/). Species were coded as either threatened (CR, EN, VU) or non-threatened (LC, LR) for analyses based on 2009 IUCN designations. Bat species were assigned as being migratory or non-migratory based on definitions and data presented in Flemming and Eby (2003), and additional migratory status from the IUCN database or other published reviews (McCracken, 1987; Burland and Worthington-Wilmer,

Table 8.2 Experimental variables, global FST values and primary references for 61 bat species.

Family	Species	# of alloz	# of μsats (nuc)	# pop. (nuc)	# pop. (mt)	Total N (nuc)	Total N (mt)	Ave. N/pop. (nuc)	Ave. N/pop. (mt)	Ave. dist, km (nuc)	Ave. dist, km (mt)	F_{ST}_mt	F_{ST}_nuc	F_{ST}_all	References
Emballonuridae	*Saccopteryx bilineata*	3	4					10.3		90			0.059	0.059	McCracken, 1987
Hipposideridae	*Rhinonicteris aurantia*				8		52		6.5		1000	0.335		0.335	Armstrong, 2006
Hipposideridae	*Triaenops furculus*				8		43		5.4		281	0.089		0.089	Russell et al., 2007; pers. comm.
Hipposideridae	*Triaenops rufus*				9		54		6.0		434	0.00*		0.00*	Russell et al., 2007; pers. comm.
Megadermatidae	*Macroderma gigas*	6	9	9	9	217	217	24.0	24.0	1520	1520	0.804	0.337	0.571	Wilmer et al., 1999
Molossidae	*Otomops martiensseni*				16		50		3.1		3000	0.090		0.090	Lamb et al., 2008
Molossidae	*Tadarida brasiliensis*	5	6		11	274	94	45.6	8.5	1650	1637	0.00*	0.158	0.080	Russell et al., 2005; McCracken and Gassel, 1997
Mystacinidae	*Mystacina tuberculata*				13		241		18.5		631	0.338	0.338	0.338	Lloyd, 2003a, 2003b
Myzopodidae	*Myzopoda aurita*				5		12		2.4		502	0.686		0.686	Russell et al., 2008
Phyllostomidae	*Ardops nichollsi*				3		14		4.7		65	0.005		0.005	Carstens et al., 2004
Phyllostomidae	*Artibeus jamaicensis*	14	2		6	84	49	42.0	8.2	11	65	0.018	0.008	0.013	Carstens et al., 2004; Ortega et al., 2003
Phyllostomidae	*Brachyphylla cavernarum*				5		32		6.4		65	0.005		0.005	Carstens et al., 2004

Family	Species													Reference
Phyllostomidae	Carollia perspicillata			11		81		7.4		10	0.060		0.060	Meyer et al., 2009
Phyllostomidae	Desmodus rotundus	7	6	18	52	50	8.7	2.8	250	1625	0.860	0.075	0.468	Martins et al., 2007; Wilkinson, 1985
Phyllostomidae	Glossophaga longirostris			11		41		3.7		375	0.725		0.725	Newton et al., 2003
Phyllostomidae	Leptonycteris curasoae			12		47		4.0		1062	0.109		0.109	Wilkinson and Fleming, 1996; Newton et al., 2003
Phyllostomidae	Phyllostomus hastatus			9				17.2				0.031	0.031	McCracken and Bradbury, 1981
Phyllostomidae	Uroderma bilobatum			13		151		11.6		10	0.010		0.010	Meyer et al., 2009
Pteropodidae	Cynopterus brachyotis		10	6	55	99	5.5	16.5	650	607	0.093	0.055	0.074	Heaney et al., 2005; Campbell et al., 2006
Pteropodidae	Cynopterus horsfieldi		8	8	47	47	5.8	5.8	650	650	0.038	0.071	0.054	Campbell et al., 2006
Pteropodidae	Cynopterus nusatenggara	31	11						405			0.171	0.171	Schmitt et al., 1995
Pteropodidae	Cynopterus sphinx		27	7	431	47	15.9	6.7		650	0.247	0.082	0.164	Storz et al., 2001a, 2001b; Campbell et al., 2006
Pteropodidae	Eidolon helvum	15	5		170		34.0		191			0.153	0.153	Juste et al., 2000
Pteropodidae	Eonycteris spelaea	17	15		253		16.9		2000			0.120	0.120	Maharadatunkamasi et al., 2003

Table 8.2 (*cont.*)

Family	Species	# of alloz	# of μsats	# pop. (nuc)	# pop. (mt)	Total N (nuc)	Total N (mt)	Ave. N/pop. (nuc)	Ave. N/pop. (mt)	Ave. dist., km (nuc)	Ave. dist., km (mt)	F_{ST_mt}	F_{ST_nuc}	F_{ST_all}	References
Pteropodidae	*Haplonycteris fischeri*	9	32	12	6	123	46	9.5	7.7	625	610	0.606	0.373	0.490	Roberts, 2006; Peterson and Heaney, 1993
Pteropodidae	*Macroglossus minimus*		32	11						450			0.105	0.105	Heaney *et al.*, 2005
Pteropodidae	*Ptenochirus jagori*		32	11						450			0.081	0.081	Heaney *et al.*, 2005
Pteropodidae	*Ptenochirus minor*		32	11						450			0.041	0.041	Heaney *et al.*, 2005
Pteropodidae	*Pteropus alecto*				6	35		5.8		1497			0.023	0.023	Webb and Tidemann, 1996
Pteropodidae	*Pteropus hypomelanus*				5		50		10.0		1025	0.882	0.882	0.882	Olival, 2008
Pteropodidae	*Pteropus poliocephalus*	9		6							377		0.014	0.014	Webb and Tidemann, 1996
Pteropodidae	*Pteropus rodricensis*		9	2		94		47.0					0.026	0.026	O'Brien, 2005
Pteropodidae	*Pteropus scapulatus*	4	6	6		117		19.5		1362			0.028	0.028	Sinclair *et al.*, 1996
Pteropodidae	*Pteropus vampyrus*				10		129		12.9		1700	0.00*	0.00*	0.00*	Olival, 2008
Pteropodidae	*Rousettus aegyptiacus*	14		4		60		12.0		630			0.493	0.493	Juste *et al.*, 1996
Pteropodidae	*Rousettus amplexicaudatus*		32	11						650			0.100	0.100	Heaney *et al.*, 2005
Pteropodidae	*Thoopterus nigrescens*	4	4	2	2	34	34	17.0	17.0	150	150	0.730	0.480	0.605	Campbell *et al.*, 2007

Table (rotated 90° on page; column headers not printed).

Family	Species													Reference
Rhinolophidae	*Hipposideros ruber*	5	10	16	215	216	13.5	21.5	251	420	0.678	0.142	0.410	Pires DB, 2007
Rhinolophidae	*Hipposideros speoris*	5	5			40		8.0		102		0.210	0.210	Rajan and Marimuthu, 2000
Rhinolophidae	*Hipposideros turpis*	6	7		266		38.0		75			0.041	0.041	Echenique-Diaz et al., 2009
Rhinolophidae	*Rhinolophus comutus*	6	11		288	288	26.2	26.2	60	60	0.158	0.017	0.088	Yoshino et al., 2008
Rhinolophidae	*Rhinolophus ferrumequinum*	6	8		161		20.1		850			0.117	0.117	Rossiter et al., 2000
Rhinolophidae	*Rhinolophus monoceros*	6	18		455	455	25.3	25.3	177.5	177.5	0.303	0.009	0.156	Chen et al., 2006, 2008
Vespertilionidae	*Antrozous pallidus*		36			194	5.4	5.4	—	650	0.244		0.244	Weyandt and Van Den Bussche, 2007
Vespertilionidae	*Corynorhinus townsendii*	6	12		167	163	13.9	13.6	1225	1225	0.605	0.057	0.331	Piaggio et al., 2009
Vespertilionidae	*Eptesicus fuscus*	5	5		88	88	17.6	17.6	2500	2500	0.820	0.107	0.464	Turmelle, 2002; pers. comm.
Vespertilionidae	*Miniopterus schreibersii*	10			78	78	7.8	7.8		725	0.620		0.620	Miller-Butterworth et al., 2003
Vespertilionidae	*Myotis bechsteinii*	7	70		36		15.5	15.5	1200	1200	0.510	0.020	0.265	Kerth et al., 2008
Vespertilionidae	*Myotis capaccinii*	7	14		36	36	2.6	2.6	825	825	0.974	0.000	0.487	Bilgin et al., 2007
Vespertilionidae	*Myotis daubentonii*	8	27		671		24.8		125			0.017	0.017	Ngampasertwong et al., 2008
Vespertilionidae	*Myotis muricola*	32	10		218	218	21.8		600			0.360	0.360	Hisheh et al., 2004
Vespertilionidae	*Myotis myotis*	15	13		260	260	20.0	20.0	148	148	0.540	0.022	0.281	Castella et al., 2001
Vespertilionidae	*Myotis nattereri*	8	13		803		61.8		68.2			0.017	0.017	Rivers et al., 2005

Table 8.2 (*cont.*)

Family	Species	# of alloz (nuc)	# of µsats	# pop. (nuc)	# pop. (mt)	Total N (nuc)	Total N (mt)	Ave. N/pop. (nuc)	Ave. N/pop. (mt)	Ave. dist., km (nuc)	Ave. dist., km (mt)	F_{ST}_mt	F_{ST}_nuc	F_{ST}_all	References
Vespertilionidae	*Myotis septentrionalis*		5	22		155		7.0		80			0.002	0.002	Arnold, 2007
Vespertilionidae	*Nyctalus azoreum*				14		159		11.4		155	0.211		0.211	Salgueiro et al., 2004
Vespertilionidae	*Nyctalus noctula*		8	13	18	264	364	20.3	20.2	2022	2022	0.230	0.020	0.125	Petit and Mayer, 1999
Vespertilionidae	*Pipistrellus pipistrellus*		6	32		407		12.7		1035			0.044	0.044	Racey et al., 2007
Vespertilionidae	*Pipistrellus pygmaeus*		6	50		398		17.9		1415			0.024	0.024	Racey et al., 2007
Vespertilionidae	*Plecotus auritus*		5	20		101		5.0		50			0.019	0.019	Burland et al., 1999
Vespertilionidae	*Scotophilus kuhlii*	32	10			189		18.9		600			0.070	0.070	Hisheh et al., 2004
Vespertilionidae	*Vespertilio murinus*				5		247		49.4		390	0.071		0.071	Safi et al., 2007

* F_{ST} value negative, not significantly different from zero.

\# of alloz = Number of allozyme markers used in study.

\# of µsats = Number of microsatellite markers used in study.

\# pop = Total number of populations examined, as determined a priori by researcher.

N = Sample size, number of bats examined in Total and Average per population.

2001). If a species had populations that were both migratory and non-migratory, e.g., *Tadarida brasiliensis* (McCracken and Gassel, 1997), then the species was coded as migratory. Average colony size was taken from direct census or exit count data as published or in the IUCN database. Species' distribution was coded as being either "continental" if populations sampled were from a continent or contained within one island (e.g., Great Britain), or "island" if sampled from multiple islands. Roost types were compiled into two categories: temporary roosts (trees and foliage) vs. permanent roosts (caves and buildings). Mating seasonality and harem status were additionally taken from McCracken and Wilkinson (2000) and Racey and Entwistle (2000).

8.6.4   Statistical analysis, single-predictor

All continuous data were log-transformed to meet assumptions of normality. Dependent variables, F_{ST}_all, F_{ST}_mt and F_{ST}_nuc, were first tested singly against experimental variables. Significant covariation with experimental variables would suggest that methodology and study design affect the amount of genetic structure detected and these confounding variables should be controlled for in subsequent analyses. Average geographic sampling distance significantly correlated with population genetic structure across all species (see results), and was therefore controlled for by saving the residuals from the global F_{ST}/average geographic distance regression as the new dependent variables for all subsequent analyses. I then initially tested species-specific life-history and morphological variables against the new measures of population genetic. I used single-predictor Pearson correlations (Zar, 1999) to identify significant predictor variables, and scatter plots and linear regressions were run for each significant correlation to visually assess data and the influence of outliers using PASW Statistics 18.0.

8.6.5   Statistical, phylogenetic independent contrasts

Species should not be treated as independent data points in comparative analysis, because traits may be more similar among related taxa due to a shared common ancestry rather than convergent adaptation to selection. Variables identified as significant in standard single-predictor analysis were then tested using a least squares regression (through the origin), a method of phylogenetic independent contrasts (PIC) (Felsenstein, 1985). Phylogenetic relationships for taxa used in this analysis were based primarily on the bat supertree topology (Jones *et al.*, 2002). A few changes to the phylogeny of Jones *et al.* (2002) were made to reflect new information regarding the position of

Table 8.3 Morphological and ecological variables examined for 61 bat species.

Family	Species	Baculum (mm)	Brain mass (mg)	Neocortex (mm3)	Testes (g)	Colony size	Mass	Wingspan
Emballonuridae	Saccopteryx bilineata	0.57	228	43.9	0.011	6	0.0075	0.275
Hipposideridae	Rhinonicteris aurantia					2150	0.0086	0.308
Hipposideridae	Triaenops furculus							0.262
Hipposideridae	Triaenops rufus							0.305
Megadermatidae	Macroderma gigas		1704	510.5	0.168	1000	0.123	0.66
Molossidae	Otomops martiensseni		756	153.6	0.17	1000	0.0355	0.467
Molossidae	Tadarida brasiliensis	0.55			0.091	500 000	0.0125	0.295
Mystacinidae	Mystacina tuberculata					150	0.0135	0.274
Myzopodidae	Myzopoda aurita						0.009	
Phyllostomidae	Ardops nichollsi			148.6	0.099	7	0.028	0.38
Phyllostomidae	Artibeus jamaicensis		1016	281.1	0.051	10	0.047	0.42
Phyllostomidae	Brachyphylla cavemarum		1196	322.9	0.064	2750	0.049	0.45
Phyllostomidae	Carollia perspicillata	0	546	129.1	0.125	100	0.0191	0.316
Phyllostomidae	Desmodus rotundus		999	312.5	0.083	50	0.0285	0.366
Phyllostomidae	Glossophaga longirostris		435	92.9		30	0.0128	0.212
Phyllostomidae	Leptonycteris curasoae		610	140.3	0.263	17 085	0.0245	
Phyllostomidae	Phyllostomus hastatus		1517	438.9	0.308	55	0.093	
Phyllostomidae	Uroderma bilobatum		612	155.8	0.098	30	0.0154	0.307
Pteropodidae	Cynopterus brachyotis	1.83	981	287.6		8	0.0265	0.39
Pteropodidae	Cynopterus horsfieldi		1373	432.3		5	0.0529	0.489
Pteropodidae	Cynopterus nusatenggara							
Pteropodidae	Cynopterus sphinx	2.08	1061	352.2	0.36	35	0.0415	0.415
Pteropodidae	Eidolon helvum	4.4	4290	1406	5.5	500 000	0.274	0.777
Pteropodidae	Eonycteris spelaea	3.69	1310	384	0.798	5000	0.0428	0.434
Pteropodidae	Haplonycteris fischeri				0.81		0.0181	0.356
Pteropodidae	Macroglossus minimus	2.77	561	155.7	0.287	2	0.0135	0.264
Pteropodidae	Ptenochirus jagori						0.0868	0.576
Pteropodidae	Ptenochirus minor							
Pteropodidae	Pteropus alecto	9.26	7039	2716	5.1	5000	0.78	1.19
Pteropodidae	Pteropus hypomelanus	6.76	5302	1906		250	0.35	0.76
Pteropodidae	Pteropus poliocephalus	6.5	7230	2710	6.75	25 000	0.7	1.33
Pteropodidae	Pteropus rodricensis					1000		
Pteropodidae	Pteropus scapulatus	10.98	5360	2078	5.3	200 000	0.375	0.904

Wing area	WAR	Wing load	Tip shape	Breeding	Harem	IUCN	Roost	Food	Migratory	Distribution
0.0125	6.1	5.9	1.53		Yes	LC	o	i	N	C
0.01507	6.29	5.6		S	No	LC	I	i	N	C
0.0113	6.12					LC	I	i		C
0.0132	7.04					LC	I	i		C
0.0717	6.1	16.8	0.83	S	No	EN	I	i	N	C
0.0234	9.3	14.9	1.33	S	Yes	EN	I	i	Y	C
0.0106	8.2	11.5	1.48	S	No	LC	I	i	Y	C
0.0108	7	12.3	1.43	Y	No	EN	o	i	N	I
						LC	o	i		C
						LC	o	f	N	I
0.0277	6.4	16.6		Y	Yes	LC	o	f	N	I
	6.36				No	LC	I	f	N	I
0.0165	6.1	11.4	2.22	S	Yes	LC	o	f	N	I
0.02	6.7	14	1.38	Y	No	LC	I		N	C
	5.71			S	No		I	f	N	I
	5.92			S	No	EN	I	f	Y	C
	7.6	25.2		Y	Yes	LC	I	i	N	C
0.015	6.3	10.1			No	LC	o	f	N	C
0.0198	7.7	13.1	1.33		Yes	LC	o	f	N	I
0.0302	7.9	17.2			Yes	LC	I	f	N	C
						LC		f	N	I
0.0258	6.7	15.6	1.34	Y	Yes	LC	o	f	N	C
0.0879	6.9	30.6	1.2		No	EN	o	f	Y	I
0.0218	8.6	19.3		Y	No	LC	I	f	N	I
0.0224	5.68	7.93				LC	o	f	N	I
0.0108	6.5	12.3	1.49	Y	No	LC	o	f	N	I
0.061	5.4	13.96				LC	o	f	N	I
						LC	o	f	N	I
	6.81	22.95	1.13	Y	No	LC	o	f	Y	C
0.1025	5.52	33.5		Y	No	LC	o	f		I
0.2582	7.41	24.5		Y	No	EN	o	f	Y	C
				Y	No	EN	o	f	N	C
0.112	7.3	32.8		Y	No	LC	o	f	Y	C

Table 8.3 (*cont.*)

Family	Species	Baculum (mm)	Brain mass (mg)	Neocortex (mm3)	Testes (g)	Colony size	Mass	Wingspan
Pteropodidae	*Pteropus vampyrus*	7.05	9121	3610		5000	1.179	1.3
Pteropodidae	*Rousettus aegyptiacus*	1.2				5000	0.14	0.572
Pteropodidae	*Rousettus amplexicaudatus*	2.76	1352	433	0.74	2000	0.075	0.529
Pteropodidae	*Thoopterus nigrescens*						0.066	0.494
Rhinolophidae	*Hipposideros ruber*		265			200 000	0.0113	0.3
Rhinolophidae	*Hipposideros speoris*	0.47	284	57.6	0.08	150	0.0109	0.293
Rhinolophidae	*Hipposideros turpis*		599	128.9		2307	0.025	
Rhinolophidae	*Rhinolophus cornutus*		194	37.5		500	0.0066	
Rhinolophidae	*Rhinolophus ferrumequinum*	4.32	350			125	0.0226	0.332
Rhinolophidae	*Rhinolophus monoceros*					1501	0.0048	
Vespertilionidae	*Antrozous pallidus*	1.04			0.362	60	0.021	0.353
Vespertilionidae	*Corynorhinus townsendii*					50	0.011	
Vespertilionidae	*Eptesicus fuscus*	0.8	238	43.3		50	0.0165	
Vespertilionidae	*Miniopterus schreibersii*	0	266.3	43.8	0.08	5000	0.0142	0.309
Vespertilionidae	*Myotis bechsteinii*	0.69	265	36.3		20	0.0101	0.256
Vespertilionidae	*Myotis capaccinii*	0.83				40	0.006	
Vespertilionidae	*Myotis daubentonii*	0.69	230			100	0.007	0.248
Vespertilionidae	*Myotis muricola*		124			2	0.0045	
Vespertilionidae	*Myotis myotis*	0.98	485	78.3		4	0.0265	0.383
Vespertilionidae	*Myotis nattereri*	0.8	220	34.8		275	0.007	0.268
Vespertilionidae	*Myotis septentrionalis*					60	0.007	0.241
Vespertilionidae	*Nyctalus azoreum*							
Vespertilionidae	*Nyctalus noctula*	5.76	364	54.9	0.54	500	0.0265	0.344
Vespertilionidae	*Pipistrellus pipistrellus*	1.66			0.225	50	0.0052	0.108
Vespertilionidae	*Pipistrellus pygmaeus*	1.5				250	0.0052	0.109
Vespertilionidae	*Plecotus auritus*	1.06	238		0.044	50	0.009	0.267
Vespertilionidae	*Scotophilus kuhlii*		348.5	66.8		500	0.0194	0.339
Vespertilionidae	*Vespertilio murinus*	1.72	445			100	0.0115	0.278

Wing area	WAR	Wing load	Tip shape	Breeding	Harem	IUCN	Roost	Food	Migratory	Distribution
0.2	8.4	57.8	1.24	Y		No	EN	o	f	Y	I
0.0558	5.9	24.6	1.45			No	LC	1	f	N	I
	6.3	17.8				No	LC	1	f	N	I
0.0367	6.7	17.6	1				LC		f	N	C
	6.55	7.96					LC	1	i	N	I
0.0116	7.4	9.2	8.7	Y		No	LC	1	i		C
	5.6		1.5			No	EN	1	i	N	I
	5.23						EN	1	i	N	I
0.0182	6.1	12.2	2.13	S		Yes	LC	1	i	N	I
	5.45			S		No		1	i	N	C
0.0194	6.55	10.6		S		No	LC	1	i	Y	C
0.013	6.61	7.54		S		No	LC	1	i	N	C
	6.4	9.4		S		No	LC	1	i	N	C
0.0137	7	10.2	1.03	S		No	EN	1	i	Y	C
0.011	6	9	4.37	S		No	EN	o	i	N	C
0.0056		10.5		S		No	EN	1	i		C
0.0098	6.3	7	2.05			No	LC	1	i		C
	6.8	5.5					LC	o	i	N	I
0.0233	6.3	11.2	1.89	S		Yes	LC	1	i	Y	C
0.0113	6.4	6.1	1.38	S		No	LC	1	i	N	C
0.0101	5.8		2.24	S		No	LC	o	i	N	C
				S		No	EN	1	i	N	I
0.0161	7.4	16.1	0.99	S		No	LC	o	i	Y	C
0.085	5.46		2.75	S		No	LC	1	i	Y	I
0.082	5.82		2.98				LC	1	i	Y	I
0.0124	5.7	7.1	1.43	Y		No	LC	1	i	N	C
						No	LC	o	i	N	I
0.0111	7	10.2	0.95	S		No	LC	1	i	Y	C

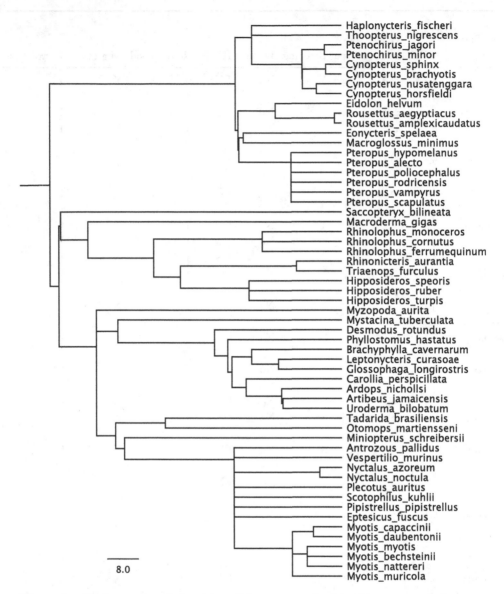

Figure 8.1 Phylogenetic relationships of 61 taxa examined in this meta-analysis. Topology based on Jones *et al.*, 2002 supertree, with modifications per Teeling *et al.*, 2005; Hoofer and Van den Bussche, 2003; and Giannini and Simmons, 2005. "Pruned" trees that only included a subset of taxa analyzed for each variable were used in Independent Contrast analysis in Mesquite v.2.01; branch lengths set to 1.

rhinolophoids and paraphyly of the suborder Microchiroptera (Teeling *et al.*, 2005), and to add resolution to the Vespertilionidae (Hoofer and Van den Bussche, 2003) and Pteropodidae (Giannini and Simmons, 2005) (Figure 8.1).

All branch lengths were given a value of 1 and treated as equal, a method that is unlikely to bias results (Ackerly, 2000). PIC were also examined using a phylogeny with transformed branch lengths per Pagel's (1992) with comparable results. PIC were calculated using the PDAP package implemented in Mesquite v.2.01 (Midford *et al.*, 2005; Maddison and Maddison, 2006). Each PIC analysis was performed using a "pruned" phylogenetic tree, including only those taxa with complete data sets for each variable.

8.6.6 Statistical analysis, ANOVA

Species were pooled into ecological or behavioral categories to test for significant differences in F_{ST} values between groups (per hypotheses in Table 8.1). Comparisons included: migratory ability (migratory vs. non-migratory), distribution (continental vs. island), diet (fruit/nectar vs. insects), mating aggregations (year-round vs. seasonal), harem mating (single male vs. multimale groups), roost fidelity (temporary vs. permanent roost types) and IUCN status (threatened vs. non-threatened). Groups were compared using one-way ANOVAs in PASW 18.0.

8.6.7 Statistical analysis, multivariate

Incomplete data for several variables precluded full multivariate analysis of all predictor variables across species. Variables available for a sufficient number of taxa included: harem mating, migratory ability, roost type, and diet as binomial categorical predictors and colony size, wing AR, wingspan and wing loading as covariates. Variables were screened and selected for the model using a mixed selection procedure with a cutoff for variables to enter or leave at $\alpha = 0.20$. Akaike's information criterion (AIC) was used to select the best model from subsets of selected variables for F_{ST}_all and F_{ST}_mt separately. A generalized linear framework was used to test models using a normal distribution and identity link function, and a multiple least squares regression was used to obtain r^2 and p values for each model. Multivariate analyses run in JMP v.7.0.1 (SAS Institute, Inc.).

8.7 Results

8.7.1 Population genetic structure and experimental variables

Population genetic structure data were compiled for 61 bat species (Table 8.2). Overall, the central tendency of F_{ST}_all values was skewed towards zero (median = 0.100, mean = 0.194, $n = 61$). Four families that together

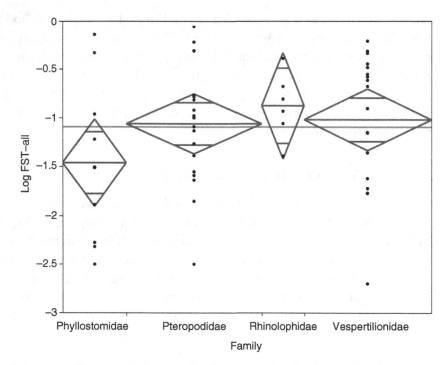

Figure 8.2 One-way ANOVA of log F_{ST} across bat families with a majority of population genetic structure data. Data show a wide range of F_{ST} values for species within each family, and no significant differences in means between them. This suggests that F_{ST} itself does not have a strong phylogenetic bias. The line across each diamond shows mean global F_{ST}, vertical span represents 95% confidence interval.

represented 85% of the species included: Phyllostomidae ($n = 9$ spp.), Pteropodidae ($n = 19$), Rhinolophidae ($n = 6$) and Vespertilionidae ($n = 18$). There was a wide range of global F_{ST} values among species for each family (Figure 8.2). Phyllostomids had the lowest mean (0.158) and median (0.03) F_{ST} values compared with the other families but there were no significant differences between families in a one-way ANOVA of log-transformed F_{ST} (Figure 8.2).

Average geographic sampling distance (km) was the only experimental/sampling variable that was significantly correlated with population structure (F_{ST}_all, $r^2 = 0.164$, $p = 0.001$, $n = 59$ species) (Figure 8.3). Thus, geographic distance was controlled for in subsequent analyses using residuals of the regression of F_{ST} and average geographic distance (per methods). Other experimental variables, including the number of loci, populations or individuals examined were not significantly correlated with global F_{ST}.

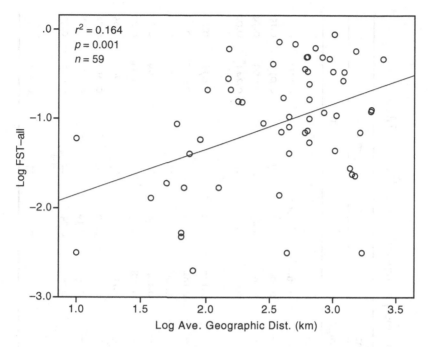

Figure 8.3 Significant correlation between average sampling distance (km) between populations and global F_{ST} for each species, log scale. Geographic sampling bias was controlled for in subsequent analyses by using the regression residuals as the new dependent variable.

8.7.2   Morphological variables

In standard single-predictor analyses, wing AR, wing loading, body mass, brain mass (uncorrected), neocortex volume (uncorrected) and baculum length (uncorrected and corrected for body mass) were each significantly and negatively related to F_{ST}_all at the $p < 0.05$ level (Table 8.4, Figures 8.4–8.5). Wing AR, wing loading, body mass and brain mass (uncorrected) were also significant and negatively associated with F_{ST}_mt data, with r^2 values higher than in regressions of F_{ST}_all (Table 8.4, Figures 8.4–8.5). Wing AR was the strongest predictor variable and explained 36% of shared variance in the global F_{ST}_mt data (Figure 8.4B), but only 8.7% for F_{ST}_all (Figure 8.4A). Within Pteropodidae only, AR was a better predictor (F_{ST}_all: $r^2 = 0.40$, $p = 0.008$, $n = 16$, Figure 8.6; F_{ST}_mt: $r^2 = 0.64$, $p = 0.03$, $n = 7$), and body mass ($r^2 = 0.28$) and wingspan ($r^2 = 0.34$) were also significant predictors of F_{ST}_all ($p < 0.05$, data not shown). For all bats, in analysis of F_{ST}_nuc data alone, tip shape was also found to be a significant predictor (Table 8.4) and body-mass-corrected neocortex volume was positively associated with F_{ST}_nuc, but not significant ($r^2 = 0.171$, $p = 0.07$).

Table 8.4 Single-predictor regressions between global FST values and morphological/behavioral trait variables.

Variable	F_{ST-all}				F_{ST-mt}				F_{ST-nuc}			
	N	Sign	r^2	p	N	Sign	r^2	p	N	Sign	r^2	p
Wing Morphology												
Wing Aspect Ratio	52	−	**0.086***	0.03	31	−	**0.360****	<0.001	34	+	0.011	0.56
Wing Span	47	−	0.014	0.42	27	−	**0.104†**	0.10	29	+	0.077	0.15
Wing Loading	42	−	**0.130***	0.02	26	−	**0.181***	0.03	29	+	0.003	0.79
Tip Shape	29	−	0.000	0.97	15	+	0.0034	0.84	21	−	**0.244***	0.02
Other Morphology												
Body Mass	54	−	**0.073***	0.05	33	−	**0.147***	0.03	36	+	0.078	0.10
Brain Mass	38	−	**0.118***	0.03	22	−	**0.182***	0.05	25	−	0.000	0.99
Brain Mass (corrected)	38	−	0.007	0.62	22	−	0.048	0.33	25	+	0.010	0.63
Neocortex Vol.	33	−	**0.144***	0.03	21	−	0.119	0.13	20	+	0.069	0.26
Neocortex Vol. (corrected)	33	−	0.000	0.98	21	−	0.018	0.56	20	+	**0.171†**	0.07
Testes Mass	26	−	0.011	0.60	15	+	0.132	0.18	15	−	0.003	0.85
Testes Mass (corrected)	26	+	0.013	0.58	15	+	0.084	0.30	15	+	0.010	0.72
Baculum Length	28	−	**0.108†**	0.09	12	−	0.057	0.46	21	−	0.001	0.88
Baculum Length (corrected)	26	−	**0.164***	0.04	11	−	0.012	0.75	19	−	0.048	0.37
Behavioral												
Ave. Colony Size	50	−	0.017	0.37	30	−	0.088	0.11	33	+	0.032	0.32

**$p<0.01$; *$p\leq0.05$; † $=0.05<p<0.1$.

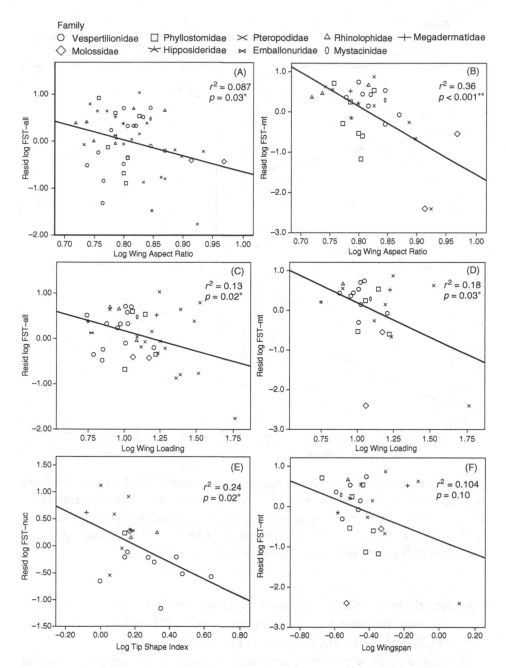

Figure 8.4 Correlations between wing morphology variables and F_{ST} (after controlling for geographic sampling distance using residuals). Data grouped by bat family. (A) Aspect ratio and F_{ST}_all; (B) aspect ratio and F_{ST}_mt; (C) wing loading and F_{ST}_all; (D) wing loading and F_{ST}_mt; (E) tip shape index and F_{ST}_nuc; (F) wingspan and F_{ST}_mt. Note that (F) is not significant, $p = 0.10$, and the significance of (D) is driven by one outlier species, *P. vampyrus*.

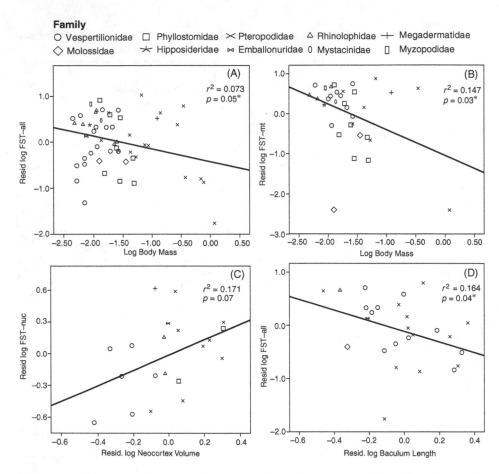

Figure 8.5 Correlation between other morphological variables and F_{ST}. Note that body mass against F_{ST}_all (A) is driven by large-bodied species in the family Pteropodidae. Baculum length (D, after controlling for body mass) was negatively correlated with F_{ST} ($p = 0.04$) and neocortex volume (C, controlled for body mass) was positively associated with genetic structure (although not significant, $p = 0.07$).

However, the only variable that remained significant after controlling for phylogeny in single-predictor, least squares regressions on PIC was wing AR against F_{ST}_mt ($N_c = 30$, $r^2 = 0.217$, $p = 0.008$) (Table 8.5, Figure 8.7).

As expected, there was some degree of covariance among morphological predictor variables. For example, wingspan and body mass were highly correlated ($r = 0.951$, $p < 0.001$) as were morphological variables calculated with a shared metric (e.g., wing loading and body mass; and wingspan, wing loading and AR). Total mass and colony size also were positively correlated ($r = 0.426$, $p = 0.002$, $n = 51$), a relationship that was primarily driven by large-bodied

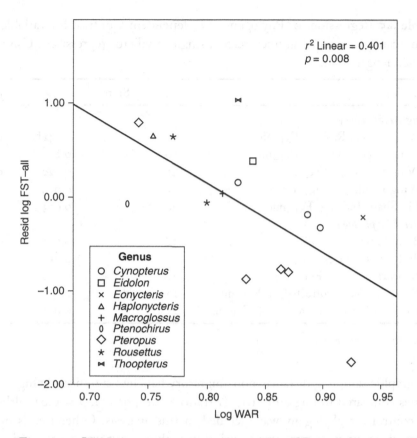

r^2 Linear = 0.401
$p = 0.008$

Genus
○ Cynopterus
□ Eidolon
× Eonycteris
△ Haplonycteris
+ Macroglossus
0 Ptenochirus
◇ Pteropus
* Rousettus
⋈ Thoopterus

Figure 8.6 Linear regression, wing aspect ratio and F_{ST}_all for the family Pteropodidae only.

pteropodids that roost in gregarious colonies of several thousand individuals. Variables of wing morphology were also found to be significantly higher in migratory vs. non-migratory bat species in one-way ANOVAs: wing AR (migratory $\mu = 7.05$, $n = 15$ vs. non-migratory $\mu = 6.39$, $n = 33$, F-ratio $= 6.37$, $p = 0.015$); wing loading (migratory $\mu = 21.11$, $n = 12$ vs. non-migratory $\mu = 12.67$, $n = 27$, F-ratio $= 6.83$, $p = 0.013$) and wing area (migratory $\mu = 0.073\text{m}^2$, $n = 13$, vs. non-migratory $\mu = 0.024\text{m}^2$, $n = 23$).

8.7.3 Ecology and behavior, ANOVAs

Results from one-way ANOVAs for roost type, seasonal migration, mating systems, diet, geographic distribution and IUCN threatened status are presented in Table 8.6 and Figure 8.8. All comparisons of F_{ST} means between these groups were not significant (Figure 8.8). Except, after controlling for

Table 8.5 Regressions of Phylogenetic Independent Contrast for variables significantly associated in uncorrected, single-predictor regressions. Using branch lengths $= 1$.

Variables	N_c	Sign	r^2	p
Wing Morphology				
Wing Aspect Ratio + F_{ST}_all	51	−	**0.052†**	0.100
Wing Aspect Ratio + F_{ST}_mt	30	−	**0.217****	0.008
Wing Loading + F_{ST}_all	41	+	0.020	0.373
Wing Loading + F_{ST}_mt	25	−	0.004	0.755
Tip Shape Index + F_{ST}_nuc	20	−	0.093	0.180
Other Morphology				
Body Mass + F_{ST}_all	53	−	0.017	0.343
Body Mass + F_{ST}_mt	32	−	0.013	0.522
Neocortex Vol. + F_{ST}_all	32	−	0.020	0.432
Neocortex Vol. (corrected) + F_{ST}_nuc	19	+	0.133	0.144
Baculum Length (corrected) + F_{ST}_all	25	−	0.017	0.527

**$p<0.01$; † $= 0.05<p<0.1$.

geographic sampling distance, non-migratory bats had significantly higher F_{ST} values compared to migratory bats (F-ratio $= 5.61$, $df = 1$, $p = 0.02$) although no control for phylogeny was included in this analysis. Other trends in categorical comparisons supported a priori hypotheses, but were not significant at the $p<0.05$ level. For example, bats that roost in permanent structures had higher mean F_{ST} values than bats that use temporary roosts and insectivorous bats had higher mean F_{ST} values compared to frugivorous bats. Groups with higher mean F_{ST} values are italicized in Table 8.6.

8.7.4   Multivariate analysis

Multivariate model selection and testing were done separately for F_{ST}_all and F_{ST}_mt data; analysis of F_{ST}_nuc was excluded due to low sample size. The stepwise variable selection procedure for F_{ST}_all identified wing AR, wing loading, mass and roost type as significant factors. AIC values for these models and their subsets are summarized in Table 8.7. The best model (lowest AIC score) for F_{ST}_all included two morphological variables as predictors, wing AR and wing loading, and explained 25% of the variance in population genetic structure among species (r^2, least squares regression). However, these two factors were found to significantly covary across species (Pearson correlation, $r = 0.33$, $p = 0.03$, $n = 42$), as they are both derived from a formula

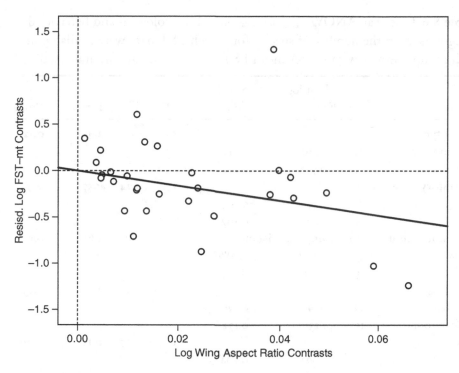

Figure 8.7 Standard least squares regression through origin on phylogenetic independent contrasts for wing aspect ratio and F_{ST_mt}. Relationship between F_{ST} and wing shape is still significant after controlling for phylogeny, but weaker than in non-phylogenetic regressions.

including wing area. The next best model that included only one variable of wing morphology was AR and roost type (AIC $= -47.49$) that explained 11% of variation in F_{ST_all}.

The stepwise variable selection procedure for F_{ST_mt} identified wing AR, colony size, harem mating and diet as significant predictors. No significant interactions between these variables were observed. The best model comprised of wing AR, average colony size and harem mating, and explained 54% of variation in population genetic structure (measured using mitochondrial DNA) across species (Table 8.7). A complete model that included all significant predictor variables from stepwise selection: colony size, AR, harem mating and diet was significant and explained a full 72% of the variance in mtDNA F_{ST} data (Figure 8.9), but this latter model ranked lowest with the highest AIC score (Table 8.7). Overall, the six models for F_{ST_mt} data had twice the predictive power as the six models for the complete data set, F_{ST_all}, which included nuclear data (average $r^2 = 0.53$ vs. 0.26, Table 8.7).

Table 8.6 One-way ANOVAs, species grouped in ecological and behavioral categories. N = the number of species for which FST data were available in each group. Groups with higher mean FST (more structure) are italicized.

Source	N	Mean log FST	Std.E FST	DF	SS	F-ratio	Prob.
Roost type				I	1.11	2.49	0.12
Temporary	25	−1.249	0.133				
Permanent	34	*−0.972*	*0.114*				
Migratory				I	0.24	0.57	0.45
No	39	−1.046	0.104				
Yes	15	−1.195	0.167				
Migratory (controlled for sampling distance)				I	1.80	5.61	0.02
No	37	*0.998*	*0.086*				
Yes	15	−0.311	0.173				
Breeding				I	0.94	2.12	0.15
Seasonal	23	*−0.910*	*0.139*				
Year-round	15	−1.231	0.172				
Harem				I	0.07	0.15	0.69
No	38	*−1.063*	*0.108*				
Yes	10	−1.156	0.210				
Food				I	0.76	1.70	0.20
Fruit/Nectar	26	−1.210	0.131				
Insects/other	34	*−0.983*	*0.115*				
Distribution				I	0.01	0.01	0.91
Contintental	33	−1.079	0.132				
Island	27	*−1.059*	*0.119*				
IUCN Status				I	0.23	0.49	0.49
Non-Endangered	45	−1.124	0.100				
Endangered	14	*−0.979*	*0.181*				

8.8   Discussion

In this chapter, I used a comparative and quantitative approach to examine taxon-specific ecological, behavioral and morphological traits that could influence population genetic structure in bats. This is the first comparative analysis of its kind across an order of animals, made possible by a growing body of molecular studies in bats. Previous reviews of bat population genetic studies were largely narrative (McCracken, 1987; Burland and Worthington-Wilmer, 2001) and over 20+ population genetic studies in bats have been

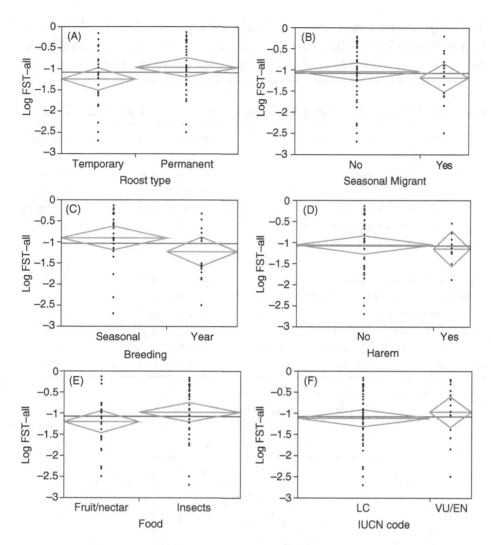

Figure 8.8 One-way ANOVAs using $F_{ST}\_$all for different categorical variables related to a species behavior, ecology and conservation status. All comparisons between groups were non-significant, except migratory category after controlling for sampling distance was significant (F-ratio = 5.61, p = 0.02, figure not shown). Trends support predictions (see Table 8.1), except breeding seasonality and harem mating thst are contrary to a priori predictions. The line across each diamond shows group mean for global F_{ST}, vertical span represents 95% confidence interval.

published since 2001. Standardized literature reviews based on meta-analyses (per this chapter) are relatively rare and are valuable as they minimize conclusion biases by focusing on statistical comparisons (Arnqvist and Wooster, 1995). Only by comparing data across a large number of species can we begin to shed

Table 8.7 Results from multivariate model selection sorted by Akaike's information criterion values. Standard least squares regressions (r^2) on selected models and subsets for F_{ST}_all and F_{ST}_mt data analyzed separately.

F_{ST}_mt Generalized Linear Model (all + intercept)	AIC	r^2, least squares	p value
colony size, aspect ratio, harem	−13.5145	0.54	0.0006**
aspect ratio, harem	−13.7379	0.42	0.0018**
colony size, aspect ratio	−15.5143	0.45	0.0005**
aspect ratio, diet	−16.1142	0.42	0.0006**
colony size, aspect ratio, diet	−24.0904	0.62	0.0001**
colony size, aspect ratio, diet, harem	−25.1722	0.72	0.0001**
F_{ST}_all			
	AIC	r^2, least squares	p value
aspect ratio, wing loading	−46.8532	0.25#	0.004*
wing loading, aspect ratio, roost type	−47.1785	0.32#	0.003*
aspect ratio, roost type	−47.4898	0.11	0.054†
mass, wing loading, aspect ratio	−48.5385	0.32#	0.002*
mass, wing loading, aspect ratio, roost type	−49.3783	0.39#	0.001*
mass, aspect ratio, roost type	−49.6809	0.14	0.073†

**$p<0.001$; *$p<0.01$; † = $0.05<p<0.1$.
note that these models include two wing measurements that covary.

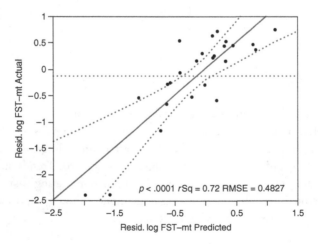

Figure 8.9 Actual by predicted plot of standard least squares fit for model with highest predictive ability for population genetic structure (explaining 72% of variance in F_{ST}_mt). Dashed lines represent 95% confidence limits. Model includes colony size, wing aspect ratio, diet and harem mating. See also color plate section.

light on the common processes affecting the evolution of bat populations and, more generally, drivers of speciation in mammals.

I found a number of variables to be significant predictors of F_{ST} in this analysis, including: wing AR, colony size, harem mating and diet when examining data from mitochondrial DNA markers alone; and AR, wing loading, mass and roost type for the complete data set (F_{ST}_all). Additionally, after controlling for sampling distance, migratory bats were shown to have populations with significantly less genetic structure than non-migratory species.

8.8.1 Sampling distance and experimental variables

Geographic sampling distance was the only experimental variable significantly correlated with F_{ST} across 59 bat species. This suggests some overall level of isolation by distance (IBD) across species. Increasing the geographic extent of sampling may also bias levels of population genetic structure detected by sampling across a greater number of vicariance events, thus confusing contemporary patterns of gene flow with the historical legacy of population subdivision (e.g., Flanders et al., 2009). Other experimental variables such as number of individuals or populations were not significant and may not be as important to control for. Similarly, Crispo and Hendry (2005) did not find any significant effect of marker type, divergence metric type, total number of individuals or number of loci examined on IBD in their meta-analysis. IBD results from preferential dispersal and mating between individuals in close proximity which leads to a positive correlation of genetic and geographic distances (Wright, 1943; Rousset, 1997). The significant association between F_{ST} and geographic distance also suggests that IBD is an important force driving the diversification of bat populations and that genetic drift may have a relatively stronger influence than gene flow in cases of limited dispersal (Hutchison and Templeton, 1999). Bat species with low dispersal ability and continuous distributions may best be described by an IBD model (e.g., Ruedi and Castella, 2003; Salgueiro et al., 2004), and about half of the bat studies in this analysis that tested for IBD ($n = 27$) found it to be significant (data not shown).

8.8.2 Wing morphology

The most unique feature that distinguishes bats from other mammals is their capacity for self-powered flight. Interestingly, a morphological measure of the wing itself (i.e., AR) was the strongest single predictor of population genetic structure among bats, even after controlling for phylogeny.

This single variable explained 36% of shared variance across all species, and 40% in Pteropodidae alone, for F_{ST}_mt. This was surprising considering all the other potential covariates and factors influencing genetic structure, but nicely fits with predictions from aerodynamic theory and confirms the importance of dispersal ability in genetically homogenizing populations. Bat species with higher ARs are capable of more energetically efficient, long-distance flight and combined with high wing loading, tend to be straight and fast fliers (Norberg and Rayner, 1987). On the other hand, bats with lower ARs have wings that are more maneuverable, and preferentially forage in cluttered habitat (Hodgkison *et al.*, 2004). Thus it follows that species with higher wing ARs will disperse more easily between populations and spread their genes over greater distances, thus having lower overall levels of genetic structure (lower F_{ST}).

This idea is not entirely new, but has never been tested in comparative analyses across the order. Garcia-Mudarra *et al.* (2009) suggested the relationship, but found that AR did not correlate with mean pairwise cytB divergence (%) for 18 bat species sampled on both sides of the Straits of Gibraltar, and no general pattern of the strait limiting bat gene flow was observed. Previous empirical investigations have also shown that population genetic structure *within* bat species may be correlated with wing morphology (Entwistle *et al.*, 2000; Miller-Butterworth *et al.*, 2003). Interestingly, wing AR was also found to be a significant predictor of global extinction risk in bats (Jones *et al.*, 2003). This morphological measurement, a ratio nonetheless, appears to be an important surrogate for dispersal ability, population-level processes and threat susceptibility in bats.

Wing loading was only found to be a significant predictor in the multivariate model for F_{ST}_all. As hypothesized, species with high wing loads had lower levels of population structure. These species may expend more energy, but are able to fly faster than more maneuverable species with lower wing loadings (Norberg, 1995), potentially leading to increased gene flow over longer distances. Covariation with AR in multivariate models confounds this significance of wing loading alone; AR appears to be more important and had a stronger effect in each model.

8.8.3   Migration and dispersal

After controlling for geographic sampling distance, bat species that undergo seasonal migration were shown to have significantly lower F_{ST} values than non-migratory species. These results support previous theories on the effect of migratory behavior on genetic structure (Fleming and Eby, 2003).

However, the spatial extent of seasonal migration can vary widely among species and it would be of interest to compare accurate estimates of dispersal distances against F_{ST}. As banding returns are typically very low (Ellison, 2008), better data may be acquired in the future via miniaturization of satellite and GPS transmitters (Olival and Higuchi, 2006). Satellite telemetry has recently been used on large fruit bats, *Pteropus vampyrus*, to record extensive international movement (>800 km) between Sumatra, Peninsular Malaysia and Thailand (Epstein *et al.*, 2009). Roost type, a proxy for roost fidelity, was also significant as a predictor in multivariate models of genetic structure. This supports the prediction that bats roosting in more permanent structures have higher roost site fidelity, which results in less interpopulation movement and higher levels of population subdivision. Similar to migration, short-term movement between roost locations has a number of evolutionary costs and benefits that should not be overlooked (Lewis, 1995).

8.8.4 Mating behavior

Harem mating was a significant predictor in the F_{ST_mt} model, although in a direction opposite to a priori predictions. Extra-harem copulations from subordinate males are common in harem groups and may partially explain this trend (Heckel *et al.*, 1999). Interlineage polygyny has been shown to increase levels of F_{ST} between matrilineal groups (Rossiter *et al.*, 2005). Seasonal mating aggregation was not found to be a significant categorical predictor in this analysis. Nonetheless, there is clear evidence that population genetic structure is reduced as individuals from multiple colonies and nursery sites aggregate and mate during seasonal swarming events (Kerth *et al.*, 2003; Veith *et al.*, 2004; Rivers *et al.*, 2005; Furmankiewicz and Altringham, 2007). This genetic exchange is often male-biased (Veith *et al.*, 2004), and female philopatry in a species may have diminished the effect of seasonal mating in this analysis, especially when comparing studies using mtDNA.

8.8.5 Colony size

Colony size was a significant predictor in the multivariate model for F_{ST_mt}, fitting the general prediction of greater retention of genetic polymorphisms (diversity) in large populations (Hartl and Clark, 1997). Colony size was a rather weak predictor of F_{ST}; this is likely due to difficulties in determining what an "average" population size is for a given species. Population sizes will vary by location depending on local resource abundance, available roosting habitat and seasonal behavior, and one average value per species does not capture this diversity. It would be interesting to test if local colony sizes correlate with intraspecific F_{ST} values for different bat species.

8.8.6   Historical and ecological barriers to gene flow

In comparing insular vs. continental species, geographic distribution was not found to be an important factor affecting broad patters of intraspecfic evolution in bats. The role of oceanic distances as a barrier to gene flow and the importance of vicariance in shaping population structure probably should continue to be examined on a case-by-case basis in bats. Several studies have shown that historical habitat and distribution changes have shaped past demography and contemporary patterns of population structure for several bat species (Petit *et al.*, 1999; Wilmer *et al.*, 1999; Lloyd, 2003; Ruedi and Castella, 2003; Piaggio and Perkins, 2005; Roberts, 2006; Olival, 2008). More recent ecological changes in the distribution of available habitat and food resources may also be an important determinant of population genetic subdivision in other bat species (Kalko, 1998; Estrada and Coates-Estrada, 2001; Lumsden *et al.*, 2002; Storz and Beaumont, 2002; Hisheh *et al.*, 2004; Campbell *et al.*, 2006). Including these factors was beyond the scope of this analysis, however with use of GIS tools and historical climate models a more quantitative analysis of landscape/ecological change and patterns of spatial genetic structure across Chiroptera could be examined in the future. Although, it is not certain that F_{ST} and other molecular measurements would be very informative for processes occurring at an ecological and anthropogenic time scales (e.g., <250 years).

8.8.7   Mitochondrial vs. nuclear DNA

Multivariate models of F_{ST_mt} alone had twice as much predictive power (r^2) as those that used F_{ST_all}, and the relationship between genetic structure and AR was much weaker in analysis of F_{ST_all} vs. F_{ST_mt}. This discrepancy is due to two possible factors: (1) an effective population size for mtDNA that is one-quarter that for nuclear markers which allows new mutations to reach fixation in fewer generations (Hartl and Clark, 1997) and (2) female philopatry in bats (Prugnolle and de Meeus, 2002). Most species of mammals show a bias towards male dispersal and female philopatry (Greenwood, 1980), and this has been shown for several bat species (Weyandt *et al.*, 2005; Kerth, 2006; Rossiter *et al.*, 2006; Safi *et al.*, 2007). However, many bats are biased towards female dispersal (McCracken, 1987; Fleming and Eby, 2003; Salgueiro *et al.*, 2004), thus the stronger relationship detected in analysis of F_{ST}_mt is likely due to a combination of faster fixation rate of mtDNA and female philopatry for some taxa.

8.8.8   Caveats and limitations of this analysis

First, there is *intra*specific variation for each of the predictor variables examined and I used a single averaged value per species. This focus on *inter*specific comparisons necessarily suffers from a loss of information and statistical power. Certain continuous traits (e.g., AR) may vary significantly *within* species and affect levels of genetic structure among populations (Miller-Butterworth *et al.*, 2003). Second, bat aerodynamics are much more complicated than the simple aerodynamic theory developed for fixed-wing aircraft (Swartz *et al.*, 2003). The elasticity of the wing membrane, 3-D structure of the camber and wingbeat frequency will profoundly influence the aerodynamics and efficiency of bat flight (Stockwell, 2001; Bullen and McKenzie, 2002; Iriarte-Diaz *et al.*, 2002), and these should also be taken into account along with values of AR, wing loading, etc. (Swartz *et al.*, 2006). Third, although a linear fit was used in the models presented, it is likely that the interaction of F_{ST} with life-history and morphological variables follows non-linear dynamics. Fourth, a control for phylogeny in multivariate analyses is needed. Lastly, F_{ST} itself – by nature of it being a summary statistic that relies on some biologically untenable assumptions (migration-drift equilibrium and constant population sizes) – has limitations (Bossart and Prowell, 1998; Hey and Machado, 2003). By using an average F_{ST} value per species, a significant amount of power and information is lost in the analysis; but this seemed the best approach to apply here. F_{ST} may also be biased by other factors, such as unequal migration rates between different populations (Wilkinson-Herbots and Ettridge, 2004).

8.8.9   Consequences of bat population structure

Although valuable from an evolutionary perspective, shedding light on the patterns and processes of population genetic structuring in bats has two important practical consequences – biological conservation and disease ecology. Bats perform several invaluable ecosystem services, including insect control (Leelapaibul *et al.*, 2005) and seed dispersal and pollination for over 300 Old World tropical forest species (Fujita, 1988; Hodgkison *et al.*, 2003). Unfortunately, about a quarter of bat species are considered globally threatened by the IUCN (Mickleburgh *et al.*, 2002). A better understanding of population connectivity and intraspecific genetic diversity for these threatened species would be of great value to implement management plans and facilitate their conservation (Mickelburg *et al.*, 1992; Hutson *et al.*, 2001). For example, a regional conservation strategy is needed to conserve the panmictic and highly vagile species, *Pteropus vampyrus*, in Southeast Asia (Olival, 2008; Epstein *et al.*, 2009).

Bats also are increasingly being recognized as natural reservoir hosts of important emerging zoonotic diseases (Calisher *et al.*, 2006), including SARS (Li *et al.*, 2005), Ebola (Leroy *et al.*, 2005) and Henipaviruses (Mackenzie *et al.*, 2003). Many of these viruses emerge in human populations because of a change in host ecology, e.g., dispersal patterns leading to interspecific contact, most often precipitated by anthropogenic factors (Daszak and Cunningham, 2003). Better understanding of host population genetic structure may be critical to modeling the dynamics of viral circulation and horizontal transmission of infectious diseases (Gonzalez Ittig and Grardenal, 2004; Davis *et al.*, 2005). Furthermore, structuring of host populations may influence the evolution of pathogen virulence (Boots *et al.*, 2004) and host population structure (global F_{ST}) was recently shown to be an important predictor of viral diversity in bats (Turmelle and Olival, 2009).

8.9   Conclusions

Many families and genera were completely unrepresented in the small subset (61/1100+ species) of chiropteran diversity examined here. There are relatively few bat species in the world that have been studied thoroughly, over the long term, using both genetic and ecological methods (e.g., *Rhinolophus ferrumequinum* (Rossiter *et al.*, 2006)). I gathered here the most thorough data set of population genetic studies in bats to date, although several additional species were excluded, as F_{ST} values were not reported (e.g., Ditchfield, 2000) and others had few variables available for ecological or life-history traits. The paucity of taxa represented here points to the need for: (1) more consistent reporting of population genetic summary statistics and the use of standardized mitochondrial and nuclear DNA markers so that results may be compared among species; (2) an increase in molecular studies for under-represented taxa; (3) better collection of ecological, behavioral and life-history information in concert with field sampling for genetic studies.

Despite certain limitation and caveats, a number of important results have emerged from this meta-analysis – expanding our knowledge of chiropteran evolution at the population level. Most notably, wing morphology was a significant predictor of population genetic structure across all bats, even after controlling for phylogenetic influence. Other a priori hypotheses of ecological and behavioral drivers of population differentiation were supported, including the importance of population size and roosting, migratory and mating behaviors. Expanding the data set of morphological, ecological and behavioral predictor variables to create a more robust, multivariate analysis with controls for phylogeny may give us the power to (roughly) estimate values of population structure in the 1100+ species that have not yet been examined with DNA. These estimates, although coarse-grained, could have important consequences for bat conservation and emerging disease ecology.

8.10 Acknowledgments

I'm indebted to Amy Turmelle, Amy Russell, Toni Piaggio, Larry Heaney, Paul Heideman, Julie Ranivo, Steve Goodman, Shiang-Fan Chen and Stephen Rossiter for sharing unpublished data used in this chapter. I thank George Amato, Kate Jones, Susan Perkins, Nancy Simmons, Don Melnick, Juan Carlos Morales and Rob Desalle for valuable discussions and comments that improved earlier versions of this chapter; and an anonymous reviewer for their comments. KJO is currently supported as a Global Health Fellow by an ARRA award (3R01TW005869–06S1) from the National Institutes of Health's Fogarty International Center.

8.11 REFERENCES

Ackerly, D. D. (2000). Taxon sampling, correlated evolution, and independent contrasts. *Evolution*, **54**, 1480–1492.

Aguinagalde, I., Hampe, A., Mohanty, A. *et al.* (2005). Effects of life-history traits and species distribution on genetic structure at maternally inherited markers in European trees and shrubs. *Journal of Biogeography*, **32**, 329–339.

Armstrong, K. N. (2006). Phylogeographic structure in *Rhinonicteris aurantia* (Chiroptera: Hipposideridae): implications for conservation. *Acta Chiropterologica*, **8**, 63–81.

Arnold, B. D. (2007). Population structure and sex-biased dispersal in the forest dwelling vespertilionid bat, *Myotis septentrionalis*. *American Midland Naturalist*, **157**, 374–384.

Arnqvist, G. and Wooster, D. (1995). Meta-analysis: synthesizing research findings in ecology and evolution. *Trends in Ecology and Evolution*, **10**, 236–240.

Beerli, P. (2006). Comparison of Bayesian and maximum-likelihood inference of population genetic parameters. *Bioinformatics*, **22**, 341–345.

Bilgin, R., Karatas, A., Coraman, E. and Morales, J. C. (2008). The mitochondrial and nuclear genetic structure of *Myotis capaccinii* (Chiroptera: Vespertilionidae) in the Eurasian transition, and its taxonomic implications. *Zoologica Scripta*, **37**, 253–262.

Bohonak, A. J. (1999). Dispersal, gene flow, and population structure. *Quarterly Review of Biology*, **74**, 21–45.

Bontadina, F., Schmied, S. F., Beck, A. and Arlettaz, R. (2008). Changes in prey abundance unlikely to explain the demography of a critically endangered Central European bat. *Journal of Applied Ecology*, **45**, 641–648.

Boots, M., Hudson, P. J. and Sasaki, A. (2004). Large shifts in pathogen virulence relate to host population structure. *Science*, **303**, 842–844.

Bossart, J. L. and Prowell, D. P. (1998). Genetic estimates of population structure and gene flow: limitations, lessons and new directions. *Trends in Ecology and Evolution*, **13**, 202–206.

Bullen, R. D. and McKenzie, N. L. (2002). Scaling bat wingbeat frequency and amplitude. *Journal of Experimental Biology*, **205**, 2615–2626.

Bullen, R. D. and McKenzie, N. L. (2007). Bat wing airfoil and planform structures relating to aerodynamic cleanliness. *Australian Journal of Zoology*, **55**, 237–247.

Burland, T. and Worthington-Wilmer, J. (2001). Seeing in the dark: molecular approaches to the study of bat populations. *Biological Review*, **76**, 389–409.

Burland, T. M., Barratt, E. M., Beaumont, M. A. and Racey, P. A. (1999). Population genetic structure and gene flow in a gleaning bat, *Plecotus auritus*. *Proceedings of the Royal Society of London Series B - Biological Sciences*, **266**, 975–980.

Caceres, C. and Barclay, R. M. R. (2000). *Myotis septentrionalis*. *Mammalian Species*, **634**, 1–4.

Calisher, C. H., Childs, J. E., Field, H. E., Holmes, K. V. and Schountz, T. (2006). Bats: important reservoir hosts of emerging viruses. *Clinical Microbiology Reviews*, **19**, 531–545.

Campbell, P., Akbar, Z., Adnan, A. M. and Kunz, T. H. (2006). Resource distribution and social structure in harem-forming Old World fruit bats: variations on a polygynous theme. *Animal Behaviour*, **72**, 687–698.

Campbell, P., Schneider, C. J., Adnan, A. M., Zubaid, A. and Kunz, T. H. (2006). Comparative population structure of *Cynopterus* fruit bats in peninsular Malaysia and southern Thailand. *Molecular Ecology*, **15**, 29–47.

Campbell, P., Putnam, A. S., Bonney, C. *et al.* (2007). Contrasting patterns of genetic differentiation between endemic and widespread species of fruit bats (Chiroptera: Pteropodidae) in Sulawesi, Indonesia. *Molecular Phylogenetics and Evolution*, **44**, 474–482.

Carstens, B. C., Sullivan, J., Dávalos, L. M., Larsen, P. A. and Pedersen, S. C. (2004). Exploring population genetic structure in three species of Lesser Antillean bats. *Molecular Ecology*, **13**, 2557–2566.

Castella, V., Ruedi, M., Excoffier, L. *et al.* (2000). Is the Gibraltar Strait a barrier to gene flow for the bat *Myotis myotis* (Chiroptera: Vespertilionidae)? *Molecular Ecology*, **9**, 1761–1772.

Castella, V., Ruedi, M. and Excoffier, L. (2001). Contrasted patterns of mitochondrial and nuclear structure among nursery colonies of the bat *Myotis myotis*. *Journal of Evolutionary Biology*, **14**, 708–720.

Chen, S. F., Rossiter, S. J., Faulkes, C. G. and Jones, G. (2006). Population genetic structure and demographic history of the endemic Formosan lesser horseshoe bat (*Rhinolophus monoceros*). *Molecular Ecology*, **15**, 1643–1656.

Chen, S. F., Jones, G. and Rossiter, S. J. (2008). Sex-biased gene flow and colonization in the Formosan lesser horseshoe bat: inference from nuclear and mitochondrial markers. *Journal of Zoology (London)*, **274**, 207–215.

Chepko-Sade, D. and Halpin, Z. T. (eds.). (1987). *Mammalian Dispersal Patterns: The Effects of Social Structure on Population Genetics*. Chicago, IL: University of Chicago Press.

Crispo, E. and Hendry, A. P. (2005). Does time since colonization influence isolation by distance? A meta-analysis. *Conservation Genetics*, **6**, 665–682.

Cruz-Neto, A. P. and Jones, K. E. (2006). Exploring the evolution of the basal metabolic rate in bats. In *Functional and Evolutionary Ecology of Bats*, ed. Z. Akbar, G. F. McCracken and T. H. Kunz. New York: Oxford University Press, pp. 56–89.

Dallimer, M., Jones, P. J., Pemberton, J. M. and Cheke, R. A. (2003). Lack of genetic and plumage differentiation in the red-billed quelea *Quelea quelea* across a migratory divide in southern Africa. *Molecular Ecology*, **12**, 345–353.

Daszak, P. and Cunningham, A. A. (2003). Anthropogenic change, biodiversity loss and a new agenda for emerging diseases. *Journal of Parasitology*, **89**, S37–S41.

Davis, P. L., Holmes, E. C., Larrous, F. *et al.* (2005). Phylogeography, population dynamics, and molecular evolution of European bat lyssaviruses. *Journal of Virology*, **79**, 10487–10497.

Ditchfield, A. D. (2000). The comparative phylogeography of Neotropical mammals: patterns of intraspecific mitochondrial DNA variation among bats contrasted to non-volant small mammals. *Molecular Ecology*, **9**, 1307–1318.

Duminil, J., Fineschi, S., Hampe, A. *et al.* (2007). Can population genetic structure be predicted from life-history traits? *American Naturalist*, **169**, 662–672.

Dumont, E. R. (2003). Bats and fruit: an ecomorphological approach. In *Bat Ecology*, ed. T. H. Kunz and B. Fenton. Chicago, IL: University of Chicago Press, pp. 398–429.

Dunbar, R. I. M. (1992). Neocortex size as a constraint on group-size in primates. *Journal of Human Evolution*, **22**, 469–493.

Dunbar, R. I. M. (1998). The social brain hypothesis. *Evolutionary Anthropology*, **6**, 178–190.

Echenique-Diaz, L., Yokoyama, J., Takahashi, O. and Kawata, M. (2009). Genetic structure of island populations of the endangered bat *Hipposideros turpis turpis:* implications for conservation. *Population Ecology*, **51**, 153–160.

Ellison, L. (2008). Summary and analysis of the U.S. Government Bat Banding Program. Open-File Report 2008–1363, US Geological Survey, **117**.

Entwistle, A. C., Racey, P. A. and Speakman, J. R. (2000). Social and population structure of a gleaning bat, *Plecotus auritus. Journal of Zoology*, **252**, 11–17.

Epperson, B. K. (2005). Mutation at high rates reduces spatial structure within populations. *Molecular Ecology*, **14**, 703–710.

Epstein, J. H., Olival, K. J., Pulliam, J. R. C. *et al.* (2009). *Pteropus vampyrus*, a hunted migratory species with a multinational home-range and a need for regional management. *Journal of Applied Ecology*, **46**, 991–1002.

Estoup, A., Beaumont, M., Sennedot, F., Moritz, C. and Cornuet, J. M. (2004). Genetic analysis of complex demographic scenarios: spatially expanding populations of the cane toad, *Bufo marinus. Evolution*, **58**, 2021–2036.

Estrada, A. and Coates-Estrada, R. (2001). Bats in continuous forest fragments and in an agricultural mosaic habitat-island at Los Tuxtlas, Mexico. *Biological Conservation*, **103**, 237–245.

Excoffier, L., Smouse, P. E. and Quattro, J. M. (1992). Analysis of molecular variance inferred from Mmtric distances among DNA haplotypes – application to human mitochondrial-DNA restriction data. *Genetics*, **131**, 479–491.

Felsenstein, J. (1985). Phylogenies and the comparative method. *American Naturalist*, **125**, 1–15.

Flanders, J., Jones, G., Benda, P. *et al.* (2009). Phylogeography of the greater horseshoe bat, *Rhinolophus ferrumequinum*: contrasting results from mitochondrial and microsatellite data. *Molecular Ecology*, **18**, 306–318.

Fleming, T. H. and Eby, P. (2003). Ecology of bat migration. In *Bat Ecology*, ed. T. H. Kunz and B. Fenton. Chicago, IL: University of Chicago Press, pp. 156–208.

Fujita, M. (1988). Flying foxes and economics. *Bats*, **6**, 4–9.

Furmankiewicz, J. and Altringham, J. (2007). Genetic structure in a swarming brown long-eared bat (*Plecotus auritus*) population: evidence for mating at swarming sites. *Conservation Genetics*, **8**, 913–923.

Garcia-Mudarra, J. L., Ibanez, C. and Juste, J. (2009). The Straits of Gibraltar: barrier or bridge to Ibero-Moroccan bat diversity? *Biological Journal of the Linnean Society*, **96**, 434–450.

Giannini, N. P. and Simmons, N. B. (2005). Conflict and congruence in a combined DNA-morphology analysis of megachiropteran bat relationships (Mammalia: Chiroptera: Pteropodidae). *Cladistics*, **21**, 411–437.

Gonzalez Ittig, R. and Grardenal, C. (2004). Recent range expansion and low levels of contemporary gene flow in *Calomys musculinus*: its relationship with the emergence and spread of Argentine haemorrhagic fever. *Heredity*, **93**, 535–541.

Goodnight, C. J. (2006). Genetics and evolution in structured populations. In *Evolutionary Genetics: Concepts and Case Studies*, ed. C. W. Fox and J. B. Wolf. New York: Oxford University Press, pp. 80–100.

Greenwood, P. J. (1980). Mating systems, philopatry and dispersal in birds and mammals. *Animal Behaviour*, **28**, 1140–1162.

Harcourt, A. H., Harvey, P. H., Larson, S. G. and Short, R. V. (1981). Testis weight, body weight and breeding systems in primates. *Nature*, **293**, 55–57.

Hartl, D. L. and Clark, A. G. (1997). *Principles of Population Genetics*. Sunderland, MA: Sinauer Associates, Inc.

Heaney, L. R., Walsh, J. S. and Peterson, A. T. (2005). The roles of geological history and colonization abilities in genetic differentiation between mammalian populations in the Philippine archipelago. *Journal of Biogeography*, **32**, 229–247.

Heckel, G., Voight, C. C., Mayer, F. and von Helversen, O. (1999). Extra-harem paternity in the white-lined bat *Saccopteryx bilineata*. *Behaviour*, **136**, 1173–1185.

Hey, J. and Machado, C. A. (2003). The study of structured populations – new hope for a difficult and divided science. *Nature Reviews Genetics*, **4**, 535–543.

Hisheh, S., Westerman, M. and Schmitt, L. (1998). Biogeography of the Indonesian archipelago: mitochondrial DNA variation in the fruit bat, *Eonycteris spelaea*. *Biological Journal of the Linnean Society*, **65**, 329–345.

Hisheh, S., How, R. A., Suyanto, A. and Schmitt, L. H. (2004). Implications of contrasting patterns of genetic variability in two vespertilionid bats from the Indonesian archipelago. *Biological Journal of the Linnean Society*, **83**, 421–431.

Hodgkison, R., Balding, S. T., Zubaid, A. and Kunz, T. H. (2003). Fruit bats (Chiroptera: Pteropodidae) as seed dispersers and pollinators in a lowland Malaysian rain forest. *Biotropica*, **35**, 491–502.

Hodgkison, R., Balding, S. T., Zubaid, A. and Kunz, T. H. (2004). Habitat structure, wing morphology, and the vertical stratification of Malaysian fruit bats (Megachiroptera: Pteropodidae). *Journal of Tropical Ecology*, **20**, 667–673.

Hodgkison, R., Balding, S. T., Zubaid, A. and Kunz, T. H. (2004). Temporal variation in the relative abundance of fruit bats (Megachiroptera: Pteropodidae) in relation to the availability of food in a lowland Malaysian rain forest. *Biotropica*, **36**, 522–533.

Hoofer, S. R. and Van den Bussche, R. A. (2003). Molecular phylogenetics of the chiropteran family Vespertilionidae. *Acta Chiropterologica*, **5**, 1–59.

Hosken, D. J. and Stockley, P. (2004). Sexual selection and genital evolution. *Trends in Ecology and Evolution*, **19**, 87–93.

Hosken, D. J., Jones, K. E., Chipperfield, K. and Dixson, A. (2001). Is the bat os penis sexually selected? *Behavioral Ecology and Sociobiology*, **50**, 450–460.

Hutchison, D. W. and Templeton, A. R. (1999). Correlation of pairwise genetic and geographic distance measures: inferring the relative influences of gene flow and drift on the distribution of genetic variability. *Evolution*, **53**, 1898–1914.

Hutson, A. M., Mickleburgh, S. P. and Racey, P. A. (2001). *Microchiropteran Bats: Global Status Survey and Conservation Action Plan. IUCN Chiroptera Specialist Group. IUCN.* Cambridge, United Kingdom and Gland, Switzerland.

Iriarte-Diaz, J., Novoa, F. F. and Canals, M. (2002). Biomechanic consequences of differences in wing morphology between *Tadarida brasiliensis* and *Myotis chiloensis*. *Acta Theriologica*, **4**, 193–200.

Jones, G. and Vanparijs, S. M. (1993). Bimodal echolocation in pipistrelle bats – are cryptic species present. *Proceedings of the Royal Society of London Series B - Biological Sciences*, **251**, 119–125.

Jones, K. E., Purvis, A., MacLarnon, A., Bininda-Emonds, O. R. P. and Simmons, N. B. (2002). A phylogenetic supertree of the bats (Mammalia: Chiroptera). *Biological Reviews*, **77**, 223–259.

Jones, K. E., Purvis, A. and Gittleman, J. L. (2003). Biological correlates of extinction risk in bats. *American Naturalist*, **161**, 601–614.

Juste, J. B., Machordom, A. and Ibanez, C. (1996). Allozyme variation of the Egyptian rousette (*Rousettus aegyptiacus*) in the Gulf of Guinea. *Biochemical Systematics and Ecology*, **24**, 499–508.

Juste, J., Ibanez, C. and Machordom, A. (2000). Morphological and allozyme variation of *Eidolon helvum* (Mammalia: Megachiroptera) in the islands of the Gulf of Guinea. *Biological Journal of the Linnean Society*, **71**, 359–378.

Juste, J., Ibanez, C., Trujillo, D., Munoz, J. and Ruedi, M. (2003). Phylogeography of barbastelle bats (*Barbastella barbastellus*) in the western Mediterranean and the Canary Islands. *Acta Chiropterologica*, **5**, 165–175.

Kalko, E. K. V. (1998). Organization and diversity of tropical bat communities through space and time. *Zoology: Analysis of Complex Systems*, **101**, 281–297.

Kerth, G. (2006). Relatedness, life history, and social behavior in the Long-Lived Bechstein's bat. In *Functional and Evolutionary Ecology of Bats*, ed. A. Zubaid, G. F. McCracken and T. H. Kunz. Oxford: Oxford University Press, pp. 199–212.

Kerth, G., Kiefer, A., Trappmann, C. and Weishaar, M. (2003). High gene diversity at swarming sites suggest hot spots for gene flow in the endangered Bechstein's bat. *Conservation Genetics*, **4**, 491–499.

Kerth, G., Petrov, B., Conti, A. *et al.* (2008). Communally breeding Bechstein's bats have a stable social system that is independent from the postglacial history and location of the populations. *Molecular Ecology*, **17**, 2368–2381.

Lamb, J. M., Abdel-Rahman, E. H., Ralph, T. *et al.* (2006). Phylogeography of southern and northeastern African populations of *Otomops martiensseni* (Chiroptera: Molossidae). *Durban Museum Novitates*, **31**, 42–53.

Leelapaibul, W., Bumrungsri, S. and Pattanawiboon, A. (2005). Diet of wrinkle-lipped free-tailed bat (*Tadarida plicata* Buchannan, 1800) in central Thailand: insectivorous bats potentially act as biological pest control agents. *Acta Chiropterologica*, **7**, 111–119.

Leroy, E. M., Kumulungui, B., Pourrut, X. *et al.* (2005). Fruit bats as reservoirs of Ebola virus. *Nature*, **438**, 575–576.

Lewis, S. E. (1995). Roost fidelity of bats – a review. *Journal of Mammalogy*, **76**, 481–496.

Li, W. D., Shi, Z. L., Yu, M. *et al.* (2005). Bats are natural reservoirs of SARS-like coronaviruses. *Science*, **310**, 676–679.

Lloyd, B. D. (2003). Intraspecific phylogeny of the New Zealand short-tailed bat, *Mystacina tuberculata*, inferred from multiple mitochondrial gene sequences. *Systematic Biology*, **52**, 460–476.

Lloyd, B. D. (2003). The demographic history of the New Zealand short-tailed bat *Mystacina tuberculata* inferred from modified control region sequences. *Molecular Ecology*, **12**, 1895–1911.

Lockwood, R., Swaddle, J. P. and Rayner, J. M. V. (1998). Avian wingtip shape reconsidered: wingtip shape indices and morphological adaptations to migration. *Journal of Avian Biology*, **29**, 273–292.

Lumsden, L. F., Bennett, A. F. and Silins, J. E. (2002). Location of roosts of the lesser long-eared bat *Nyctophilis geoffroyi* and Gould's wattled bat *Chalinolobus gouldii* in a fragmented landscape in south-eastern Australia. *Biological Conservation*, **106**, 237–249.

Mackenzie, J. S., Field, H. E and Guyatt, K. J. (2003). Managing emerging diseases borne by fruit bats (flying foxes), with particular reference to henipaviruses and Australian bat lyssavirus. *Journal of Applied Microbiology*, **94**, 59S–69S.

Maddison, W. P. and Maddison, D. R. (2006). Mesquite: a modular system for evolutionary analysis. http://mesquiteproject.org/mesquite1.o/mesquite/download/MesquiteManual.pdf.

Maharadatunkamsi, H. S., Kitchener, D and Schmitt, L. (2003). Relationships between morphology, genetics, and geography in the cave fruit bat *Eonycteris spelaea* (Dobson, 1871) from Indonesia. *Biological Journal of the Linnean Society*, **79**, 511–522.

Martins, F. M., Ditchfield, A. D., Meyer, D. and Morgante, J. S. (2007). Mitochondrial DNA phylogeography reveals marked population structure in the common vampire bat, *Desmodus rotundus* (Phyllostomidae). *Journal of Zoological Systematics and Evolutionary Research*, **45**, 372–378.

Maycock, C. R., Thewlis, R. N., Ghazoul, J., Nilus, R. and Burslem, D. (2005). Reproduction of dipterocarps during low intensity masting events in a Bornean rain forest. *Journal of Vegetation Science*, **16**, 635–646.

McConkey, K. R. and Drake, D. R. (2007). Indirect evidence that flying foxes track food resources among islands in a Pacific Archipelago. *Biotropica*, **39**, 436–440.

McCracken, G. F. (1987). Genetic structure of bat social groups. In *Recent Advances in the Study of Bats*, ed. B. Fenton, P. A. Racey and J. M. V. Rayner. Cambridge: Cambridge University Press, pp. 281–298.

McCracken, G. F. and Bradbury, J. W. (1981). Social organization and kinship in the polygynous bat *Phyllostomus hastatus*. *Behavioral Ecology and Sociobiology*, **8**, 11–34.

McCracken, G. F. and Gassel, M. F. (1997). Genetic structure in migratory and nonmigratory populations of Brazilian free-tailed bats. *Journal of Mammalogy*, **78**, 348–357.

McCracken, G. F. and Wilkinson, G. S. (2000). Bat mating systems. In *Reproductive Biology of Bats*, ed. E. G. Crichton and P. H. Krutzsch. San Diego, CA: Academic Press, pp. 321–362.

McCracken, G. F., Gillam, E. H., Westbrook, J. K. *et al.* (2008). Brazilian free-tailed bats (*Tadarida brasiliensis*: Molossidae, Chiroptera) at high altitude: links to migratory insect populations. *Integrative and Comparative Biology*, **48**, 107–118.

Meyer, C. F. J., Frund, J., Lizano, W. P. and Kalko, E. K. V. (2008). Ecological correlates of vulnerability to fragmentation in Neotropical bats. *Journal of Applied Ecology*, **45**, 381–391.

Meyer, C. F. J., Kalko, E. K. V. and Kerth, G. (2009). Small-scale fragmentation effects on local genetic diversity in two phyllostomid bats with different dispersal abilities in Panama. *Biotropica*, **41**, 95–102.

Mickelburg, S., Hutson, A. M. and Racey, P. A. (1992). *Old World Fruit Bats: An Action Plan for their Conservation*. Gland, Switzerland: International Union for Conservation of Nature.

Mickelburg, S., Hutson, A. M. and Racey, P. A. (2002). A review of the global conservation status of bats. *Oryx*, **36**, 18–34.

Midford, P. E., Garland, T. and Maddison, W. P. (2005). PDAP Package of Mesquite. Version 1.07.

Miller-Butterworth, C., Jacobs, D. and Harley, E. (2003). Strong population substructure is correlated with morphology and ecology in a migratory bat. *Nature*, **424**, 187–191.

Moreno-Valdez, A., Grant, W. E. and Honeycutt, R. L. (2000). A simulation model of Mexican long-nosed bat (*Leptonycteris nivalis*) migration. *Ecological Modelling*, **134**, 117–127.

Moreno-Valdez, A., Honeycutt, R. L. and Grant, W. E. (2004). Colony dynamics of *Leptonycteris nivalis* (Mexican long-nosed bat) related to flowering agave in Northern Mexico. *Journal of Mammalogy*, **85**, 453–459.

Moyle, L. C. (2006). Correlates of genetic differentiation and isolation by distance in 17 congeneric *Silene* species. *Molecular Ecology*, **15**, 1067–1081.

Newton, L. R., Nassar, J. M. and Fleming, T. H. (2003). Genetic population structure and mobility of two nectar-feeding bats from Venezuelan deserts: inferences from mitochondrial DNA. *Molecular Ecology*, **12**, 3191–3198.

Ngamprasertwong, T., Mackie, I. J., Racey, P. A. and Piertney, S. B. (2008). Spatial distribution of mitochondrial and microsatellite DNA variation in Daubenton's bat within Scotland. *Molecular Ecology*, **17**, 3243–3258.

Norberg, U. M. (1995). Wing design and migratory flight. *Israel Journal of Zoology*, **41**, 297–305.

Norberg, U. M. and Rayner, J. M. V. (1987). Ecological morphology and flight in bats (Mammalia, Chiroptera) – wing adaptations, flight performance, foraging strategy and echolocation. *Philosophical Transactions of the Royal Society of London Series B - Biological Sciences*, **316**, 337–419.

Norberg, U. M. L., Brooke, A. P. and Trewhella, W. J. (2000). Soaring and non-soaring bats of the family Pteropodidae (flying foxes, *Pteropus* spp.): wing morphology and flight performance. *Journal of Experimental Biology*, **203**, 651–664.

Nowak, R. M. (1999). *Walker's Mammals of the World*. Baltimore, MD: Johns Hopkins University Press.

O'Brien, J. (2005). Phylogeography and conservation genetics of the fruit bat genus Pteropus (Megachiroptera) in the western Indian Ocean. Unpublished Ph.D. thesis, National University of Ireland.

Olival, K. J. (2008). Population genetic structure and phylogeography of Southeast Asian flying foxes: implications for conservation and disease ecology. Unpublished Ph.D. thesis, Columbia University.

Olival, K. J. and Higuchi, H. (2006). Monitoring the long-distance movement of wildlife in Asia using satellite telemetry. In *Conservation Biology in Asia*, ed. J. McNeely, T. McCarthy, A. Smith, L. Whittaker and E. Wikramnayake. Kathmandu, Nepal: Society for Conservation Biology Asia Section and Resources Himalaya Foundation, pp. 319–339.

Ortega, J., Maldonado, J. E., Wilkinson, G. S., Arita, H. T. and Fleischerr, R. C. (2003). Male dominance, paternity, and relatedness in the Jamaican fruit-eating bat (*Artibeus jamaicensis*). *Molecular Ecology*, **12**, 2409–2415.

Pagel, M. D. (1992). A method for the analysis of comparative data. *Journal of Theoretical Biology*, **156**, 431–442.

Pannell, J. R. and Charlesworth, B. (2000). Effects of metapopulation processes on measures of genetic diversity. *Philosophical Transactions of the Royal Society of London Series B - Biological Sciences*, **355**, 1851–1864.

Perez-Barberia, F. J., Shultz, S. and Dunbar, R. I. M. (2007). Evidence for coevolution of sociality and relative brain size in three orders of mammals. *Evolution*, **61**, 2811–2821.

Peterson, A. T. and Heaney, L. R. (1993). Genetic differentiation in Philippine bats of the genera *Cynopterus* and *Haplonycteris*. *Biological Journal of the Linnean Society*, **49**, 203–218.

Petit, E. and Mayer, F. (1999). Male dispersal in the noctule bat (*Nyctalus noctula*): where are the limits? *Proceedings of the Royal Society of London Series B - Biological Sciences*, **266**, 1717–1722.

Petit, E., Excoffier, L. and Mayer, F. (1999). No evidence of bottleneck in the postglacial recolonization of Europe by the noctule bat (*Nyctalus noctula*). *Evolution*, **53**, 1247–1258.

Piaggio, A. J. and Perkins, S. L. (2005). Molecular phylogeny of North American long-eared bats (Vespertilionidae: *Corynorhinus*); inter- and intraspecific relationships inferred from mitochondrial and nuclear DNA sequences. *Molecular Phylogenetics and Evolution*, **37**, 762–775.

Piaggio, A. J., Navo, K. W. and Stihler, C. W. (2009). Intraspecific comparison of population structure, genetic diversity, and dispersal among three subspecies of Townsend's big-eared bats, *Corynorhinus townsendii townsendii*, *C-t. pallescens*, and the endangered *C-t. virginianus*. *Conservation Genetics*, **10**, 143–159.

Pires, D. B. (2007). Patterns of genetic, morphological, and behavioral differentiation in the African bat, *Hipposideros ruber*. Unpublished Ph.D. thesis, University of California, Los Angeles.

Pitnick, S., Jones, K. E. and Wilkinson, G. S. (2006). Mating system and brain size in bats. *Proceedings of the Royal Society of London Series B - Biological Sciences*, **273**, 719–724.

Prugnolle, F. and de Meeus, T. (2002). Inferring sex-biased dispersal from population genetic tools: a review. *Heredity*, **88**, 161–165.

Pumo, D. E., Kim, I., Remsen, J., Phillips, C. J and Genoways, H. H. (1996). Molecular systematics of the fruit bat, *Artibeus jamaicensis*: origin of an unusual island population. *Journal of Mammalogy*, **77**, 491–503.

Racey, P. A. and Entwistle, A. C. (2000). Life-history and reproductive strategies in bats. In *Reproductive Biology of Bats*, ed. E. G. Crichton and P. H. Krutzsch. San Diego, CA: Academic Press, pp. 363–414.

Racey, P. A., Barratt, E. M., Burland, T. M. *et al.* (2007). Microsatellite DNA polymorphism confirms reproductive isolation and reveals differences in population genetic structure of cryptic pipistrelle bat species. *Biological Journal of the Linnean Society*, **90**, 539–550.

Raghuram, H., Chattopadhyay, B., Nathan, P. T. and Sripathi, K. (2006). Sex ratio, population structure and roost fidelity in a free-ranging colony of Indian false vampire bat, *Megaderma lyra*. *Current Science*, **91**, 965–968.

Rajan, K. E. and Marimuthu, G. (2000). Genetic diversity within and among populations of the microchiropteran bat *Hipposideros speoris* based on a RAPD analysis. *Zeitschrift fur Saugetierkunde*, **65**, 301–306.

Rivers, N. M., Butlin, R. K. and Altringham J. D. (2005). Genetic population structure of Natterer's bats explained by mating at swarming sites and philopatry. *Molecular Ecology*, **14**, 4299–4312.

Roberts, T. E. (2006). History, ocean channels, and distance determine phylogeographic patterns in three widespread Philippine fruit bats (Pteropodidae). *Molecular Ecology*, **15**, 2183–2199.

Rossiter, S. J., Jones, G., Ransome, R. D. and Barratt, E. M. (2000). Genetic variation and population structure in the endangered greater horseshoe bat *Rhinolophus ferrumenquinum*. *Molecular Ecology*, **9**, 1131–1135.

Rossiter, S. J., Ransome, R. D., Faulkes, C. G., Le Comber, S. C. and Jones, G. (2005). Mate fidelity and intra-lineage polygyny in greater horseshoe bats. *Nature*, **437**, 408.

Rossiter, S. J., Jones, G., Ransome, R. D. and Barratt, E. M. (2006). Causes and consequences of genetic structure in the greater horseshoe bat, *Rhinolophus ferrumequinum*. In *Functional and Evolutionary Ecology of Bats*, ed. A. Zubaid, G. F. McCracken and T. H. Kunz. Oxford: Oxford University Press, pp. 213–226.

Rossiter, S. J., Ransome, R. D., Faulkes, C. G., Dawson, D. A. and Jones, G. (2006). Long-term paternity skew and the opportunity for selection in a mammal with reversed sexual size dimorphism. *Molecular Ecology*, **15**, 3035–3043.

Rousset, F. (1997). Genetic differentiation and estimation of gene flow from F-statistics under isolation by distance. *Genetics*, **145**, 1219–1228.

Ruedi, M. and Castella, V. (2003). Genetic consequences of the ice ages on nurseries of the bat *Myotis myotis*: a mitochondrial and nuclear survey. *Molecular Ecology*, **12**, 1527–1540.

Russell, A. L., Medellin, R. A. and McCracken, G. F. (2005). Genetic variation and migration in the Mexican free-tailed bat (*Tadarida brasiliensis mexicana*). *Molecular Ecology*, **14**, 2207–2222.

Russell, A. L., Ranivo, J., Palkovacs, S., Goodman, S. M. and Yoder, A. D. (2007). Working at the interface of phylogenetics and population genetics: a biogeographical analysis of *Triaenops* spp. (Chiroptera: Hipposideridae). *Molecular Ecology*, **16**, 839–851.

Russell, A. L., Goodman, S. M., Fiorentino, I, and Yoder, A. D. (2008). Population genetic analysis of *Myzopoda* (Chiroptera: Myzopodidae) in Madagascar. *Journal of Mammalogy*, **89**, 209–221.

Safi, K., Konig, B. and Kerth, G. (2007). Sex differences in population genetics, home range size and habitat use of the parti-colored bat (*Vespertilio murinus*, Linnaeus 1758) in Switzerland and their consequences for conservation. *Biological Conservation*, **137**, 28–36.

Sakai, S., Harrison, R. D., Momose, K. *et al.* (2006). Irregular droughts trigger mass flowering in aseasonal tropical forests in Asia. *American Journal of Botany*, **93**, 1134–1139.

Salgueiro, P., Coelho, M. M., Palmeirim, J. M. and Ruedi, M. (2004). Mitochondrial DNA variation and population structure of the island endemic Azorean bat (*Nyctalus azoreum*). *Molecular Ecology*, **13**, 3357–3366.

Schmitt, L. H., Kitchener, D. J. and How, R. A. (1995). A genetic perspective of mammalian variation and evolution in the Indonesian archipelago: biogeographic correlates in the fruit bat genus *Cynopterus*. *Evolution*, **49**, 399–412.

Shephard, J. M., Hughes, J. M. and Zalucki, M. P. (2002). Genetic differentiation between Australian and North American populations of the monarch butterfly *Danaus plexippus* (L.) (Lepidoptera: Nymphalidae): an exploration using allozyme electrophoresis. *Biological Journal of the Linnean Society*, **75**, 437–452.

Simmons, N. B. (2005). Order Chiroptera. In *Mammal Species of the World: A Taxonomic and Geographic Reference*, ed. D. E. Wilson and D. M. Reeder. Baltimore, MD: Johns Hopkins University Press, pp. 312–529.

Sinclair, E. A., Webb, N. J., Marchant, A. D. and Tidemann, C. R. (1996). Genetic variation in the little red flying-fox *Pteropus scapulatus* (Chiroptera: Pteropodidae): implications for management. *Biological Conservation*, **76**, 45–50.

Slatkin, M. (1987). Gene flow and the geographic structure of natural populations. *Science*, **236**, 787–792.

Sonsthagen, S. A., Talbot, S. L. and White, C. M. (2004). Gene flow and genetic characterization of Northern goshawks breeding in Utah. *Condor*, **106**, 826–836.

Stockwell, E. F. (2001). Morphology and flight manoeuvrability in New World leaf-nosed bats (Chiroptera: Phyllostomidae). *Journal of Zoology*, **254**, 505–514.

Storz, J. F. (1999). Genetic consequences of mammalian social structure. *Journal of Mammalogy*, **80**, 553–569.

Storz, J. F. and Beaumont, M. A. (2002). Testing for genetic evidence of population expansion and contraction: an empirical analysis of microsatellite DNA variation using a hierarchial bayesian model. *Evolution*, **56**, 154–166.

Storz, J. F., Bhat, H. R. and Kunz, T. H. (2001). Genetic consequences of polygyny and social structure in an Indian fruit bat, *Cynopterus sphinx*. I. Inbreeding, outbreeding, and population subdivision. *Evolution*, **55**, 1215–1223.

Storz, J. F., Bhat, H. R. and Kunz, T. H. (2001). Genetic consequences of polygyny and social structure in an Indian fruit bat, *Cynopterus sphinx*. II. Variance in male mating success and effective population size. *Evolution*, **55**, 1224–1232.

Swartz, S. M., Freeman, P. W. and Stockwell, E. F. (2003). Ecomorphology of bats: comparative and experimental approaches relating structural design to ecology. In *Bat Ecology*, ed. T. H. Kunz and B. Fenton. Chicago, IL: University of Chicago Press, pp. 257–300.

Swartz, S. M., Bishop, K. and Aguirre, M.-F. I. (2006). Dynamic complexity of wing form in bats: implications for flight performance. In *Functional and Evolutionary Ecology of Bats*, ed. A. Zubaid, G. F. McCracken and T. H. Kunz. New York: Oxford University Press, pp. 110–130.

Teeling, E. C., Springer, M. S., Madsen, O. *et al.* (2005). A molecular phylogeny for bats illuminates biogeography and the fossil record. *Science*, **307**, 580–584.

Tidemann, C. R. and Nelson, J. E. (2004). Long-distance movements of the grey-headed flying fox (*Pteropus poliocephalus*). *Journal of Zoology*, **263**, 141–146.

Turmelle, A. S. (2002). Phylogeography and population structure of the big brown bat *(Eptesicus fuscus)*. Unpublished Honors thesis, Boston University.

Turmelle, A. S. and Olival, K. J. (2009). Correlates of viral richness in bats (Order Chiroptera). *EcoHealth*, **6**, 522–539.

van Staaden, M. J. (1995). Breeding tactics, social structure and genetic variation in mammals: problems and prospects. *Acta Theriologica*, **Suppl.3**, 165–182.

Veith, M., Beer, N., Kiefer, A., Johannesen, J. and Seitz, A. (2004). The role of swarming sites for maintaining gene flow in the brown long-eared bat (*Plecotus auritus*). *Heredity*, **93**, 342–349.

Waples, R. S. (1987). A multispecies approach to the analysis of gene flow in marine shore fishes. *Evolution*, **41**, 385–400.

Webb, N. J. and Tidemann, C. R. (1996). Mobility of Australian flying-foxes, *Pteropus* spp. (Megachiroptera): evidence from genetic variation. *Proceedings of the Royal Society of London Series B - Biological Sciences*, **263**, 497–502.

Wennerberg, L., Klaassen, M. and Lindstrom, A. (2002). Geographical variation and population structure in the White-rumped Sandpiper *Calidris fuscicollis* as shown by morphology, mitochondrial DNA and carbon isotope ratios. *Oecologia*, **131**, 380–390.

Weyandt, S. E. and Van Den Bussche, R. A. (2007). Phylogeographic structuring and volant mammals: the case of the pallid bat (*Antrozous pallidus*). *Journal of Biogeography*, **34**, 1233–1245.

Weyandt, S. E., Van Den Bussche, R. A., Hamilton, M. J. and Leslie, Jr., D. M. (2005). Unraveling the effects of sex and dispersal: Ozark big-eared bat (*Corynorhinus townsendii ingens*) conservation genetics. *Journal of Mammalogy*, **86**, 1136–1143.

Wickramasinghe, L. P., Harris, S., Jones, G. and Jennings, N.V. (2004). Abundance and species richness of nocturnal insects on organic and conventional farms: effects of agricultural intensification on bat foraging. *Conservation Biology*, **18**, 1283–1292.

Wilkinson, G. S. (1985). The social organization of the common vampire bat. 2. Mating system, genetic structure, and relatedness. *Behavioral Ecology and Sociobiology*, **17**, 123–134.

Wilkinson, G. S. and Fleming, T. H. (1996). Migration and evolution of lesser long-nosed bats *Leptonycteris curasoae*, inferred from mitochondrial DNA. *Molecular Ecology*, **5**, 329–339.

Wilkinson-Herbots, H. M. and Ettridge, R. (2004). The effect of unequal migration rates on Fst. *Theoretical Population Biology*, **66**, 185–197.

Willig, M. R., Patterson, B. D. and Stevens, R. D. (2003). Patterns of range size, richness, and body size in Chiroptera. In *Bat Ecology*, ed. T. H. Kunz and B. Fenton. Chicago, IL: University of Chicago, pp. 580–621.

Wilmer, J. W., Hall, L., Barratt, E. and Moritz, C. (1999). Genetic structure and male-mediated gene flow in the ghost bat (*Macroderma gigas*). *Evolution*, **53**, 1582–1591.

Wright, S. (1932). The roles of mutation, inbreeding, crossbreeding, and selection in evolution. *Proceedings of the Sixth International Genetics Congress*, **1**, 356–366.

Wright, S. (1943). Isolation by distance. *Genetics*, **28**, 114–138.

Wright, S. (1951). The genetical structure of populations. *Annals of Eugenics*, **15**, 323–354.

Yoshino, H., Armstrong, K. N., Izawa, M., Yokoyama, J. and Kawata, M. (2008). Genetic and acoustic population structuring in the Okinawa least horseshoe bat: are intercolony acoustic differences maintained by vertical maternal transmission? *Molecular Ecology*, **17**, 4978–4991.

Zar, J. H. (1999). *Biostatistical Analysis*. Upper Saddle River, NJ: Prentice Hall.

A bird? A plane? No, it's a bat: an introduction to the biomechanics of bat flight

SHARON M. SWARTZ, JOSÉ IRIARTE-DÍAZ, DANIEL K. RISKIN AND KENNETH S. BREUER

9.1 Introduction

Bats are unique among mammals for their ability to fly. A substantial body of research has focused on understanding how they do so, and in 1990, Norberg's landmark volume provided an up-to-date understanding of diverse aspects of bat flight (Norberg, 1990). Building on work accomplished before 1990, our understanding of bat flight has changed significantly in the last two decades, and warrants an updated review. For example, many hypotheses about how bats fly were based either on aircraft aerodynamics or on studies of birds. In some respects, these predictions did fit bats well. However, recent advances in the study of bat flight have also revealed important differences between winged mammals and other fliers. Although we have, of course, always known that a bat is neither a bird nor a plane, the significance of the differences among bats and all other flyers are only now becoming clear.

In this chapter, we provide an overview of the morphology of bats from the perspective of their unique capacity for powered flight. Throughout the chapter, we provide references to classic literature concerning animal flight and the bat flight apparatus, and direct readers to sources of additional information where possible. We focus on relatively newer work that over the last 20 years has begun to change the ways in which we understand how bats carry out their remarkable flight behavior, and that has altered the way we understand the structural underpinnings of bat flight.

This chapter is organized to provide a review of several topics relevant to bat flight, and we hope that readers will understand each section better for having read them all. First, we explain the basic principles of aerodynamics necessary to understand bat flight. These include Reynolds number, lift and drag forces,

Evolutionary History of Bats: Fossils, Molecules and Morphology, ed. G. F. Gunnell and N. B. Simmons. Published by Cambridge University Press. © Cambridge University Press 2012.

unsteady effects and Strouhal number. Next, we review the morphological characters of bats relevant to flight, which include the compliant skin and bones of the wings, the overall geometry of the wings and their bones, the distribution of sensory hairs across the wings and the physiology of the musculature that drives the wings. Finally, we review whole-bat flight performance, from forward flight to hovering flight, maneuvering and landing. We believe that it is only through study of all these disparate topics – fluid mechanics, anatomy and behavior – that one can have a truly integrative understanding of bat flight.

9.2   Aerodynamic principles of flight

The aerodynamics of flapping flight is a complex subject, and we will not attempt to convey a detailed summary of the aerodynamic underpinnings of the flapping flight of bats here. For more detailed discussions, we refer the reader to excellent sources on general aerodynamics (e.g., Anderson, 2005) or animal flight (Norberg, 1990; Azuma, 2006). Our much more limited objective is to introduce the reader to fundamental concepts in aerodynamics that are necessary to appreciate the flight performance of bats.

To understand how an animal flies, one must first identify the requirements of flight. In simple terms, a bat must move the air with its wings in such a way as to produce aerodynamic force. The component of the aerodynamic force that moves the bat forward is thrust, and the component that keeps the bat from falling and moves it vertically is lift. These are opposed by drag and gravity, respectively. In comparison with bats, airplanes are simple: engines provide constant thrust, and the resulting movement of air over fixed wings also constantly produces lift. Bat flight aerodynamics are more complicated because neither thrust nor lift are constant; both are produced in a cyclic manner because the wings are flapping.

One fundamental concept necessary to understanding flapping flight is the Reynolds number, a non-dimensional number that characterizes the relative magnitude of inertial and viscous forces, and hence the overall character of a fluid flow around or within a solid object (see also Purcell, 1977; Vogel, 1981 for more on the Reynolds number in biological systems). The Reynolds number, *Re*, is defined as:

$$Re = \frac{\rho U c}{\mu} \tag{9.1}$$

where U is flight speed, c is a typical length scale, usually the average wing chord, ρ is the fluid density, approximately 1.21 kg m^{-3} for air at standard atmospheric conditions and μ is the fluid viscosity, approximately 1.7×10^{-5} kg m^{-1} s^{-1} for air at room temperature. The way a fluid moves over a wing is entirely dependent on Reynolds number, so it is impossible to understand how bats fly without considering it.

At low Reynolds numbers, such as those relevant for insect flight, for example ($Re < 1000$), viscous forces dominate, while at higher Reynolds numbers ($Re > 10^5$), as in the case of air moving across a fast-flying giant albatross, inertial forces dominate. Bats span a wide range of sizes and flight speeds, where Re ranges from approximately 10^3 to 10^5; this range does not overlap with that of human-engineered aircraft. Indeed, bat flight occurs in a very complex regime for aerodynamic analysis, where the onset of critical flow phenomena, such as laminar separation and the transition from laminar to turbulent flow, are extremely difficult to predict reliably (Shyy *et al.*, 1999; Torres and Müller, 2004; Song *et al.*, 2008). This, combined with the thin wing geometries typical of bats, indicates that conventional airplane aerodynamics are of limited help in interpreting bat flight aerodynamics.

When inertial forces are important, as they are at the Re of bat flight, thrust and lift arise from fluid momentum generated by motions of the wings. In flight, a bat can add downward and rearward momentum to the air, and that imparts a net force on the body that permits flight. In this case, the aerodynamic force is proportional to the flight speed, U, multiplied by the air momentum generated by the wing: $\rho U A$, where ρ is air density and A is the wing area. We can then write the specific aerodynamic forces, lift, L, and drag, D, as:

$$L = C_L \frac{\rho U^2}{2} A \tag{9.2}$$

and

$$D = C_D \frac{\rho U^2}{2} A \tag{9.3}$$

where C_L and C_D are the coefficients of lift and drag, respectively. These coefficients are non-dimensional constants with values that typically range from 0.1 to 3.0; the exact value of these aerodynamic coefficients is determined by the shape and motion of the wing. For example, a highly streamlined wing would have a high lift coefficient and low drag coefficient; a wing that is less streamlined would have a lower lift coefficient and higher drag coefficient. One important complexity of bat flight is exemplified here; because the three-dimensional conformation of bat wings changes continuously as they flap, so the lift and drag coefficients of bat wings change continuously during the wingbeat cycle. This also, however, illustrates an avenue by which bats have the potential to actively control flight dynamics (see also below).

A *wing*, in aerodynamic terminology, is a three-dimensional lifting surface. The simplest analysis of the generation of lift comes from the examination of the local shape of an airfoil, the two-dimensional cross-sectional shape of a wing. Lift is generated when air moves over the top surface of the airfoil at higher speed than it moves over the bottom surface. The difference in airspeed between the top and bottom wing surfaces can be accomplished in several ways, such as a curvature in the airfoil surface, giving it *camber*, or an inclination of the foil relative to the oncoming air, producing a positive *angle of attack* (Figure 9.1). When we consider the shapes of bat wings in an aerodynamic context, then, any features that influence camber or angle of attack are important for performance, even without the additional effects of flapping. Examples of such features might include the length of the fifth digit and the position of the metacarpophalangeal and interphalangeal joints of this digit, the ability of the muscles of the wing to control angle of attack, or the stiffness of the wing membrane skin and its resultant state of billowing, and hence camber, when it experiences pressure differences between the wing's top and bottom surfaces.

Bats in flight, of course, do not employ fixed, static wings, but instead flap them in characteristic and complex ways. When we consider lift in relation to local flow at the wing surface, it is immediately clear that lift changes dynamically over the course of every wingbeat cycle. In general, during the downstroke, the wing has a positive angle of attack and hence generates positive lift, but during the upstroke, the effective angle of attack is lower, and may even be negative (see also below, Flight performance). This overall pattern can be modulated in a number of ways, such as by pronating and supinating the wing. Furthermore, bats do not simply flap the wing up and down, but sweep the wings through some angle other than strictly vertical, with forward or cranial motion during the downstroke and backward or caudal motion during the upstroke (Figure 9.1). The degree to which these various motions occur appears to vary with speed, for specific flight behaviors and among species, and has yet to be well described. The result of the wing posture and motion during the flapping motions of bat flight is that bat flight is characterized by a *stroke plane angle* that is not vertical (Figure 9.1). This stroke plane angle has an important influence on the relative speed and angle of attack experienced by the flapping wing: as the stroke plane angle becomes more horizontal, the speed of the wind with respect to the wing surface increases during the downstroke and decreases during the upstroke. Moreover, a wing can undergo twisting about its long axis at the same time that it undergoes flapping, and the magnitude of the twist may change along the span of the wing, and with the timing of the wingbeat cycle. This additional complexity is yet another way that the angle of attack of the wing may come to vary locally depending on the precise location within the wing, and dynamically, depending on the timing within the wingbeat cycle.

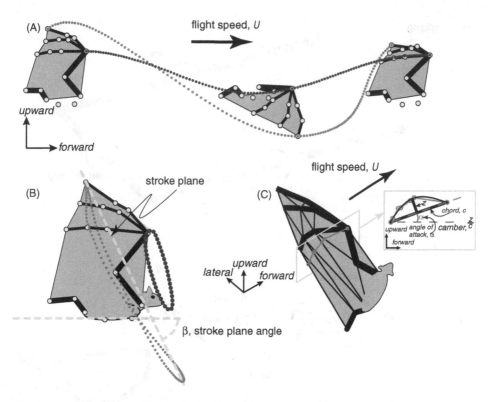

Figure 9.1 Schematic of bat in flight illustrating aerodynamic terms and concepts. (A) Lateral view of a bat wing moving through a wingbeat cycle, tracing out the motion of wingtip and carpus in frame of reference of external world. (B) In the frame of reference of the bat's body, motion of landmarks on the bat's wing can be seen as cyclical, tracing out a trajectory similar to a flattened, tilted ellipse. The movement of the wingtip defines, from its uppermost to lowermost positions, a stroke plane, which can be defined by β, the stroke plane angle, the angle between the line connecting these two points and the horizontal plane. (C) To define angle of attack, α, and camber, consider a parasagittal section though the wing, as outlined, and then shown on the right of schematic. The wing chord is the line connecting the frontmost, or leading, edge and rearmost, or trailing, edges of the wing in parasagittal section of interest. In flight, bat wings are typically curved in an upwardly convex fashion; the circles indicate the locations where an imaginary parasagittal cutting plane intersects the wing skin, and the lines connect those points, estimating the length of the wing in the cutting plane. Camber is then computed as the maximum height of the wing in the plane divided by the wing chord. See also color plate section.

9.2.1 Wake flows and trailing vortices

Although it is the wing motion that is directly responsible for the generation of lift and thrust, we can gain considerable insight into the mechanisms of aerodynamic force production by looking at fluid motion in the wake

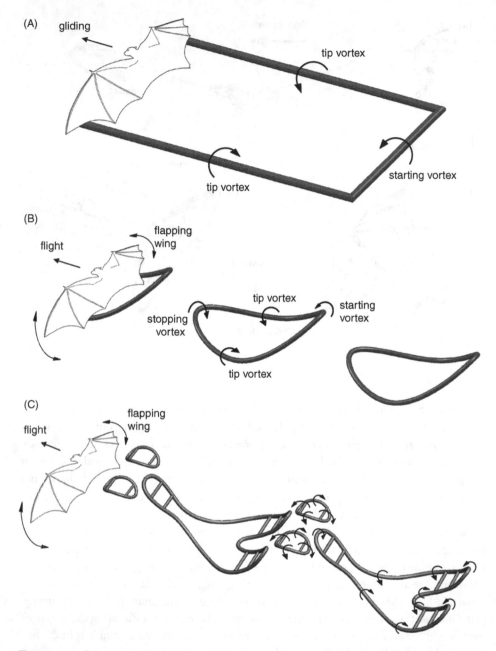

Figure 9.2 Schematic illustration of structure of vortex wakes for different kinds of flight. (A) Gliding or flight in fixed-wing aircraft produces relatively simple wakes that possess a starting vortex and a pair of nearly linear, parallel tip vortices where vorticity is shed from the wingtips as lift is produced by forward movement of the wing through the air. (B) In flapping flight, the path of the wingtip is much more complex spatially, such that even where the magnitude of vorticity is constant, the spatial

behind a flying animal. Most readers have some everyday experience that provides a useful heuristic for this concept; a fixed-wing airplane in flight leaves behind it two vapor trails, created by two *tip vortices*, one trailing the tip of each wing, that arise directly from the aerodynamic forces produced as the plane moves through the atmosphere. Newtonian mechanics assures us that for any force there is an equal and opposite reaction force, and the force generated on each wing is mirrored by its reaction force, experienced by the fluid surrounding the wing. The wake left behind the wing thus contains a complete "footprint" of its force production. Bats also leave an aerodynamic wake, albeit a wake that is much smaller and less intense than that of a jet aircraft, but, one which can persist for several meters, making it amenable to measurement using modern fluid mechanics diagnostic tools such as particle image velocimetry, or PIV.

An aerodynamic wake flow can be analyzed in terms of its *vortex structure* and its associated *circulation*. Vorticity is the local angular or rotational velocity of the fluid, and a vortex is somewhat subjectively defined as a concentration of vorticity. Tornados and the swirling motions of water draining from the bathtub are familiar everyday examples of vortices. These so-called "trailing vortices" are generated by every flying object, from large airplanes to birds and bats, although they have a very different character in small flyers, such as insects, due to the very low Reynolds numbers that characterize their flight. These vortices exist due to the fact that, to generate an upward force, lift, the animal uses its wings to direct air downwards, creating what is known as the "downwash." The downwash, in turn, interacts with the surrounding air to produce the trailing vortex wake (Figure 9.2). At high

Caption for Figure 9.2 (*cont.*) motion of the wingtip would lead to a more complex wake shape. However, lift changes continuously through the wingbeat cycle, hence the intensity of the wingtip vortex changes in parallel with its repositioning in space. One possible wake configuration for flapping flyers is a set of discrete vortex rings; this pattern would result if there is a period in each wingbeat cycle in which no lift is generated and vorticity falls to zero, producing a stopping vortex, leading to the closing of the trailing vortex into a ring. In this case, each wingbeat produces a ring with its own starting, wingtip and stopping vortices. This somewhat abstracted wake pattern has served as a starting point for discussions of the possibility of distinct gaits in animal flight, analogous to walking and running gaits in terrestrial locomotion. (C) Experimental techniques for wake visualization, such as particle image velocimetry, can be employed to describe natural wakes of flying bats and birds in detail to test hypotheses generated by theory, such as illustrated in (B). Here, a wake is generated by *Cynopterus brachyotis*, the lesser dog-faced fruit bat, flying at moderate speed, as documented by PIV (Hubel *et al.*, 2010). The realistic wake structure is far more complex than both the gliding and flapping models, showing many additional components in wake for each wingbeat than would have been predicted from theory alone.

Reynolds numbers, the dissipation of motion due to the viscosity of the air is weak, and these vortex structures can persist for a long time after the animal has flown by leaving a "footprint" in the air. The intensity and structure of these vortices directly reflects the way in which aerodynamic forces, including lift and thrust, were generated.

The total vortex strength, or circulation, Γ, of a vortex is directly related to the magnitude of the lift force of the vortex by the Kutta–Joukowski theorem as follows:

$$L = \rho U \Gamma w \qquad (9.4)$$

where L is lift, ρ is air density, U is the speed of the object relative to the surrounding fluid and w is the wingspan. Quantitative analysis of wake vortices can thus give very specific information about aerodynamic force production.

There is more to a vortex than lift magnitude, however. The geometry of vortices contains important information about aerodynamic conditions. At high Reynolds numbers, Kelvin's circulation theorem requires that a vortex must have constant strength, and can neither start nor end in the flow, and hence vortex lines must either extend forever or form closed rings (Kundu and Cohen, 2008). This fundamental constraint has far-reaching consequences for the geometry of the vortex wake. For steady gliding flight, it requires that the two trailing tip vortices must have a constant and fixed magnitude. Further-more, if the lift force increases and decreases as the wings flap down and up, the strength of the primary wake vortex must change accordingly. The technical constraints of Kelvin's theorem require that this waxing and waning of the vortex can only be accomplished by the introduction of "starting" and "stopping" vortices (Figure 9.2). In this way, the straight-line vortex pair that is characteristic of steady flight (e.g., gliding flight, or an airplane) can become a series of discrete vortex rings, characteristic of discrete wing flaps (Figure 9.2). More complex flapping kinematics, such as are common in bat flight, generate even more complex wake structures, and are the subject of active research at present (e.g., Hedenström *et al.*, 2007; Muijres *et al.*, 2008; Hubel *et al.*, 2009, 2010).

A comment on efficiency is in order at this point. Since only the vorticity that lies in the direction of flight, the streamwise vorticity, is associated with the lift force, any non-streamwise component of vorticity, such as the starting and stopping vortices, represents fluid motion generated by the animal that is not used for weight support and is, in some sense, wasted energy. These non-streamwise vortex components are, however, unavoidable consequences of flapping flight, and therefore, from the standpoint of energy efficiency, are inherent disadvantages to any flapping mechanism of lift generation, particularly for long-range flight, such as migration. However, energy is not the only relevant currency for an organism, and flapping clearly confers other advantages, most notably the abilities to maneuver with ease and to fly in complex environments, where rapid changes in aerodynamic forces are advantageous. Besides, until flying animals evolve propellers or jets, there is no way to produce thrust in the air without flapping.

9.2.2   Drag and thrust

It is almost impossible to measure drag empirically on flying animals. Estimates of drag from live animals are also notoriously inaccurate. This is because we can only directly measure the net horizontal acceleration of an animal, which is the sum of thrust, the force that accelerates the animal forward, and drag, the force that decelerates the animal, and not their independent contributions. Moreover, attempts to use wind-tunnel tests to assess drag using dead specimens or models that recreate geometries of flying animals cannot reproduce the subtleties of a living, flapping animal, and are so destined to overpredict drag forces.

Although it might seem convenient to think of drag as a single entity, drag arises from several distinct sources, and their relative importance varies, depending on the physical situation. The four primary types of drag that influence flight are: (1) *skin friction drag*, drag associated with the viscosity of fluid flowing over a body; (2) drag due to lift, the so-called *induced drag*; (3) *form drag*, drag due to large-scale separation of flow from the object experiencing aerodynamic forces; and (4) *parasitic drag*, a catch-all phrase associated with minor flow separation over non-streamlined appendages such as legs, ears etc. Skin friction is an unavoidable consequence of the viscosity of air, and even for a perfectly streamlined object, represents about 40% of total drag. Drag due to lift, "induced" drag, is also unavoidable, and is due to the fact that any three-dimensional object that generates lift must also generate drag along with the vortex wakes created with the production of lift, as discussed above (Anderson, 2005). The downwash generated by the wake vortices "tilts" the lift force, slightly reducing the lift and thereby adding a small contribution to drag. Form drag is due to large-scale separation of the flow and the generation of large vortices. For well-streamlined bodies, including most bats in flight, this is usually minimal during the downstroke, but may be important during the upstroke. For species that fly for extended periods of time, it is likely that selection has led to streamlined body and wing anatomy and efficient flapping motions for steady forward flight, and that energy losses associated with form drag are relatively small; this is much less true for maneuvering flight and flight in other extreme conditions such as hovering or very fast flight.

9.2.3   Unsteady flow effects

The trailing vortex wake is not the only aerodynamic effect that we need to consider for the quantitative analysis of bat flight. Other kinds of fluid motions, grouped under the designation of unsteady effects, can occur for a wide variety of reasons, complicating the study of animal flight (for an excellent discussion of this subject geared for biologists, see Dickinson, 1996). Examples of unsteady effects include *stall* or *separation* – flight conditions in which large vortices can be shed from the wings and body, resulting in unstable changes in aerodynamic forces. Even when the wings are stationary, complex fluid motions can cause unsteady effects in some situations. The Reynolds number range typical of bat flight coincides with a critical aerodynamic transition between smooth and predictable *laminar* flow and chaotic *turbulent* flow, and unsteady effects often occur at those transitions. Most important, however, are the unsteady fluid effects induced by the flapping of the wings, which are necessary for sustained powered flight. The flapping motion generates time-dependent variations in the aerodynamic forces, which typically increase in strength during the downstroke, which is responsible for the bulk of the lift and thrust force generation, and decrease in strength during the upstroke, which, for bats appears to be a relatively passive recovery stroke. These effects are complex in nature, and are an area of intense research at the present. Unsteady effects have been the subject of considerable attention in the insect flight community since the 1970s. This body of work has demonstrated that unsteady phenomena such as delayed or dynamic stall, the Wagner effect and wake capture play a crucial role in aerodynamics in insects (Ellington, 1975; Maxworthy, 1979; Dickinson, 1994; Van den Berg and Ellington, 1997; Sane, 2003). Any complete model of bat flight aerodynamics will require consideration of unsteady effects, in addition to wake analyses.

One way to assess unsteady effects in a fluid is by the *Strouhal* number, *St*, a non-dimensional number that describes the importance of unsteady effects in relation to steady, inertial forces. The Strouhal number is defined as:

$$St = \frac{fA}{U} \tag{9.5}$$

where *f* is flapping frequency, *A* is flapping amplitude and *U* is flight speed. *St* values close to zero suggest that the flow is quasi-steady, and that steady aerodynamic theories should be largely applicable. A high value signifies the dominance of unsteady effects, while a value in the range of 0.2–0.3 means that both steady and unsteady effects are important. Bat flight is typically in the range of St of 0.2 to 0.6 (Taylor *et al.*, 2003; Riskin *et al.*, 2010) implying that unsteady effects play an important role. However, both the importance and the specific nature of unsteady effects in bat flight are yet to be fully understood.

9.3 Morphology

The structure of the limbs of bats is their most obvious specialization, and generations of bat researchers have uncovered characteristics of wing structure that influence flight performance (e.g., Humphry, 1869; Macalister, 1872; Vaughan, 1959; Norberg, 1972; Hermanson and Altenbach, 1985; Meyers and Hermanson, 1994; Sears, 2006). We focus here on those aspects of wing morphology most directly relevant to flight mechanics and aerodynamics, with most attention to work carried out in the last ten years. An excellent review of older literature can be found in Norberg (1990).

9.3.1 Compliant wings

One critical difference between bats and human-engineered aircraft, and, indeed, to a lesser extent, between bats and the other flying animals, is the degree to which the wing surface is deformable. Virtually all human-made aircraft have possessed rigid wings, with the few exceptions of the slightly deforming wings of gliders and a small number of highly experimental micro air vehicles (Shyy *et al.*, 1999; Lian *et al.*, 2003a, 2003b; Ansari *et al.*, 2006). For birds, the combination of robust skeletal structure and relatively stiff feather shafts confers substantial rigidity on all but the tips of bird wings, such that there is little movement within the wing itself during flight, other than bending at synovial joints (Hedrick *et al.*, 2004; Usherwood *et al.*, 2005; Tobalske *et al.*, 2007). Although insect wings can change shape during flight to some degree, their deformation is limited, and insect wings lack any joints distal to the body hinge (Combes and Daniel, 2001, 2003; Daniel and Combes, 2002; Bergou *et al.*, 2007). Bat wings, in contrast, possess very little innate stiffness. The wing consists of a compliant membrane of skin stretched across jointed bones that are themselves poorly mineralized and thus flexible. Bat wings likely function at variable, but generally quite low levels of stiffness throughout the wingbeat cycle during typical forward flight (Figure 9.3). It is possible that the skin is rarely stretched tightly, even over a wide range of diverse flight behaviors; future studies that focus specifically on the mechanics of the skin during flight will be needed before we will be able to fully address the range of stiffness bat wing skin experiences during normal functions.

Figure 9.3 Left panel: *Choeronycteris mexicana*, the Mexican long-tongued bat, feeding at an agave flower, showing that even as the bat comes into the force-generating downstroke, the wing membrane is not taut, but experiences varying degrees of looseness depending on anatomical location. In this particular wingbeat, the plagiopatagium, the portion of the wing between the body and the hand skeleton, is so loose that a relatively large fold or flap is visible between the ankle and the tip of the fifth digit (white arrow). Photograph by Joseph Coelho, used with permission. See also color plate section.

Right panel, *Glossophaga soricina*, Pallas's long-tongued bat, flying up to a nectar feeder in the lab, showing relatively relaxed, wrinkled skin in the arm- and handwing even during the middle of the downstroke, the portion of the wingbeat cycle in which aerodynamic forces are greatest. Photograph by Caroline Harper, used with permission.

9.3.2   Skin contribution

The greatest part of the surface area of the bat wing comprises skin. The skin is supported, tensioned and moved through space in a highly controlled fashion by the bones of the body, forelimb and hindlimb, and by the muscles associated with these bones. In addition, the armwing skin, or plagiopatagium, contains intrinsic musculature that takes both origin and insertion within the connective tissue of the membrane itself (Gupta, 1967; Quay, 1970; Holbrook and Odland, 1978). Wing membrane skin is similar to that of most other mammals, but both the epidermis and dermis are exceptionally thin, and the dermis greatly enriched in highly organized elastic fibers (Quay, 1970).

There are numerous characteristics of wing membrane skin that appear to relate directly to the modification of the wing as a flight organ. The reduction of skin thickness is substantial enough that it is likely to contribute not only to determining mechanical characteristics of the skin, but to also provide some significant weight savings, particularly in the distalmost portion of the wing (Swartz, 1997). Nerve endings in the wing membrane skin are especially abundant and diverse (Quay, 1970), and the specialized sensory hairs project a fraction of a millimeter from the wing surfaces to provide the central nervous system with, it is hypothesized, a detailed map of the state of flow over the wing (Zook, 2007; Sterbing-D'Angelo et al., 2011). Wings carry out their aeromechanical roles at the same time as they play a central role in heat and water control (Basset and Studier, 1988; Thomson and Speakman, 1999). The reduction of skin thickness thus not only reduces the mass and thereby the energy required to accelerate and decelerate the wing during flapping, but also serves to greatly reduce surface-capillary diffusion distance, allowing for significant rates of skin gas exchange via the wing membrane. In this way, the wing may actually make a significant contribution to the oxygen budget of bats, with oxygen consumption and carbon dioxide production as much as 6–10% of whole-body values for resting, lightly anesthetized *Epomophorus wahlbergi* (Makanya and Mortola, 2007).

The mechanical properties of wing membrane skin are a major determinant of the behavior of bat wings as compliant airfoils (Swartz et al., 1996). In particular, wing membrane skin of all species tested to date show particularly low stiffness in the spanwise direction, the direction from the body to the wingtip, in both the plagiopatagium, and the dactylopatagium (Figure 9.4). In contrast, the skin is up to two orders of magnitude stiffer when stretched in the chordwise direction. The stress–strain relationship for wing membrane skin is highly non-linear, but in general, this trend holds true at all parts of the stress–strain curve – at low, intermediate and high strains.

One critical way in which the compliance of the bat wing membrane is functionally significant in comparison with rigid fixed wings is that compliant wings self-camber in the presence of a pressure difference between the upper and lower surface of the wing. This self-cambering produces a faster increase in lift with increasing angle of attack, along with increased resistance to stall and loss of lift at high angles (Song et al., 2008). These benefits likely offer bats and mammalian gliders an advantage in both lift generation and flight stability during rapid maneuvering, in comparison to the more rigid wings of birds or insects.

9.3.3   Bone contribution

The wing skeleton also contributes to the compliance of the bat wing, especially by the flexion, extension, abduction and adduction of the joints of the handwing. The primary mechanical function of bones in all vertebrates is to provide stiffness, however a few animals, bats among them, use bones of relatively low stiffness to perform locomotion via controlled deformation. In these cases, the low stiffness of the bone can arise by virtue of low mineralization, unusual geometry – such as extremely slender, elongated shapes – or both.

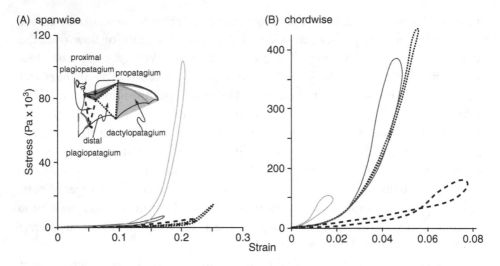

Figure 9.4 Mechanical characteristics of wing membrane skin of *Pteropus poliocephalus*, the gray-headed flying fox. The stiffness, given by the slope of the stress–strain trace, differs among regions of the wing membrane, and for each wing region differs greatly depending on whether the skin is tested from proximal to distal or spanwise (A) vs. from leading to trailing edge or chordwise (B). Adapted from Dumont and Swartz (2009). See also color plate section.

The avian furcula is one such example; the "wishbone" spreads laterally and then recoils with each wingbeat (Jenkins *et al.*, 1988). The hand skeleton of bats appears to be another example, in which relatively poorly mineralized bones that are also greatly elongated can undergo considerable deformations (Swartz *et al.*, 2005).

Typically, the mechanical properties of the compact bone tissue of mammalian long bones vary little among species (Currey, 1984, 2002). The bones of the bat wing, however, seem to represent a major exception to this pattern. Although the bat humerus is similar in mechanical properties to other mammals, the radius, metacarpals, proximal phalanges and distal phalanges each show progressively lower mineralization and hence stiffness (Papadimitriou *et al.*, 1996; Swartz *et al.*, 1998; Swartz and Middleton, 2008) (Figure 9.5). In combination with the structural geometry of the bones of the handwing (see below, wing bone cross-sectional geometry), the distal wing bones are therefore highly deformable, and preliminary evidence suggests that metacarpals and phalanges undergo significant bending during flight, even when animals do not attain high speeds or exhibit extreme maneuvers (Swartz *et al.*, 2005; Swartz and Middleton, 2008).

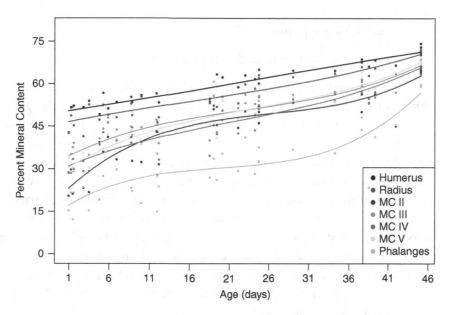

Figure 9.5 Ontogenetic pathway to adult variation in wing bone mineralization in *Tadarida brasiliensis*, the Brazilian free-tailed bat. Unlike typical mammalian limb skeletons, bat wing skeletons show great variation among bones in mineralization levels, with much greater mineralization in the proximal skeletal elements than the distal elements at all ages, from birth to skeletal maturity at 46 days. Within each bone, there is a steady increase in mineralization during lactation and eventual weaning. Adapted from Papadimitriou *et al.* (1996), Swartz and Middleton (2008). See also color plate section.

Although we do not yet have full understanding of the role of flexible bones on the mechanics and energetics of bat flight, evidence obtained to date suggests several intriguing possibilities. The flexibility of the distal wing bones arises through reduction in bone mineral, and as a consequence, the density and mass of these bones is reduced relative to their primitive condition. Because the metabolic cost of accelerating and decelerating limbs can be a significant portion of the total metabolic cost of locomotion, particularly for animals with large limbs, such as bats, the reduced mass of the distal wing skeleton that results from decreased mineralization significantly reduces wing mass and thus the energetic cost of locomotion, especially at high wingbeat frequencies. Flexible bones also may deform under aerodynamic loading, so it is possible that these bones can passively align with dynamically changing patterns of airflow, and the most distal portions of the wing, the region of the wing that moves most rapidly, could, by passive wing rotation, decambering and/or deformation, reconfigure in a manner that might decrease drag and local turbulence.

9.4   Wing geometry

9.4.1   Aspect ratio and wing loading

In aircraft aerodynamics, the wing loading and aspect ratio of a plane convey important information concerning an aircraft's energetics and ability to maneuver. To the extent that bats operate like fixed-wing aircraft, bats with higher aspect ratios, the mean ratio of wingspan to chord, should have decreased induced drag and therefore are predicted to enjoy a decreased energetic cost of flight (Norberg and Rayner, 1987; Norberg, 1990) – conversely, as aspect ratio increases, theory suggests that maneuverability of bats should decrease. As a result, the shapes of bat wings are often used to infer the relative importance of fast flight in open habitats (high aspect ratios) to maneuverability in cluttered habitats (low aspect ratios). Although some support for this relationship has been shown through field studies (Aldridge and Rautenbach, 1987), other studies have failed to demonstrate that relationship (Saunders and Barclay, 1992; Stockwell, 2001).

Wing loading, computed as body mass per unit of fully extended wing area, is also directly related to flight performance in aircraft in a manner that has invited comparison for winged animals. Animals with increased wing loading are expected to fly at higher speeds than animals with low wing loading, to generate enough lift to fly. Also, increased wing loading should increase the cost of flight and decrease maneuverability, so animals should have wing loadings as low as other biomechanical requirements of their lifestyles will allow. In general, wing loading scales positively with body size, so large animals have higher wing loading than small animals do (because weight increases faster than area as body size rises). Recent experiments with pteropodids demonstrate that the largest bats overcome their relatively higher wing loading by extending their wings more fully and using higher angles of attack during the downstroke than small bats do (Riskin *et al.*, 2010).

It is important to note, however, that many of the assumptions involved in the clear relationship of aspect ratio and wing loading on the one hand and aircraft flight performance on the other do not apply to flapping flight in bats. Not only do the large-scale changes in wing form produced by flapping dynamics fundamentally change the expectations of performance based on wing shape alone, bats fly at Reynolds numbers much lower than those of aircraft and therefore unsteady aerodynamic effects can be very important in their flight. As a consequence, simple extrapolation of aircraft performance expectations to bat wing shapes may not apply in a straightforward manner. We suggest that this is a subject that would benefit greatly from more attention as new studies seek to better understand how wing form and the details of flapping motion work synergistically to determine natural flight performance in bats.

9.4.2   Wing bone cross-sectional geometry

The cross-sectional geometry of a bone, like that of any other beam, plays a major role in determining the nature of its response to mechanical forces, along with its material stiffness or elastic modulus (Wainwright *et al.*, 1976; Currey, 1984). The shafts of typical limb long bones of mammals are elliptical in cross-sectional shape, varying from nearly circular to possessing a major axis roughly twice the minor axis, and the bone cortex is most often 25 to 75% of the bone diameter (Currey and Alexander, 1985). The wing bones of bats, however, differ from the customary mammalian pattern (Swartz *et al.*, 1992). The bones of the armwing are extremely thin-walled, with cortices less than 25% the magnitude of bone diameter, and with the outer diameter significantly expanded relative to those of non-volant mammals of comparable body size (Swartz *et al.*, 1992; Swartz and Middleton, 2008). In contrast, the metacarpals and phalanges are thick-walled or even completely solid (cortical thickness is 68–100% of bone diameter for phalanges). Although the metacarpals may be expanded in outer diameter relative to those of non-volant mammals, the phalanges, unlike the remainder of the wing bones, do not show this pattern (Swartz and Middleton, 2008).

These distinctive aspects of bone geometry suggest substantial functional differentiation in mechanics of the armwing and handwing. The geometry of the humerus and radius is most consistent with resisting loading in torsion, or bending loads applied from multiple different directions. Although there is little direct information concerning the loading of the bat wing during flight, the few hints available suggest that torsion and bending are indeed the predominant loading regimes in this part of the skeleton (Swartz *et al.*, 1992). In contrast, the low second moments of area, coupled with low stiffness, suggest that the bones of the handwing, unlike the long bones of terrestrial mammals, are specialized to maximize rather than minimize their deformation with respect to bending loads. As these elongated, slender bones interact with their fluid surroundings, their geometry will tend to promote deflection rather than resisting bending, perhaps contributing to an adaptive wing reconfiguration (see above).

9.4.3   Sensory hairs

One way in which the surface of the bat wing differs from the skin surface of other parts of the bat body and from the skin surface of the limbs of all other mammals is the presence of distinctive hairs that perform a somato-sensory function. These hairs, quite different in size and morphology from pelage hairs, emerge from small dome-shaped structures on both the dorsal and ventral wing surfaces, singly, but also in pairs or in small clusters (Crowley and Hall, 1994; Zook, 2007; Sterbing-D'Angelo *et al.*, 2011) (Figure 9.6). The domes comprise a cluster of supporting cells around the hair follicle, including Merkel cells, cells often known as "touch cells" that are believed to act as intermediates between an initial stimulus and afferent neuron impulses. The sensory hairs are distributed in a highly patterned fashion, with high densities along the wing bones, the intrinsic wing muscles (mm. plagio-patagiales), and in the regions of the wing's leading and trailing edges. Electrophysiological recordings from the primary afferent nerves of the hair-dome apparatus in *Antrozous pallidus* and *Eptesicus fuscus* demonstrate high sensitivity to air-puff and direct touch stimuli, but little or no response to direct touching of the wing membrane between the domes or stretching of the mem-brane (Zook, 2005; Sterbing-D'Angelo *et al.*, 2011). These responses are com-pletely surface specific; that is, ventral hairs show no response to stimuli on the dorsal wing surface, and vice versa, although wing membranes are extremely thin, usually between 0.03 and 0.08 mm (Studier, 1972; Swartz *et al.*, 1996).

This morphological and physiological information suggests that the function of the hair cell network is to provide bats with a detailed, real-time map of flow conditions on the wing during flight. Each hair is well suited to be able to monitor airflow in its immediate vicinity, albeit in a simple manner. A large number of simple measurements, however, obtained from relevant locations distributed throughout the wing's surface, may provide the central nervous system with the requisite raw data to produce an integrated map of airflow patterns over the wing as a whole, suggesting fine-scale adjustments to kin-ematics, wing membrane tension etc. must be made to deal with flow turbu-lence at particular anatomical locations on the wing (Dickinson, 2010).

9.4.4   Flight muscle

The distinctive anatomical specializations of the musculature of the wing for flight have been the subject of intense scientific interest since at least the middle of the nineteenth century (Humphry, 1869; Macalister, 1872; Miller, 1907; Vaughan, 1959, 1966; Norberg, 1970, 1972; Strickler, 1978; Altenbach, 1979). In addition to the unusual and, in some cases, unique gross morphology of bat musculature, the flight muscles of bats have notable physiological and/or biochemical characteristics (see Hermanson, 1998 for an excellent review).

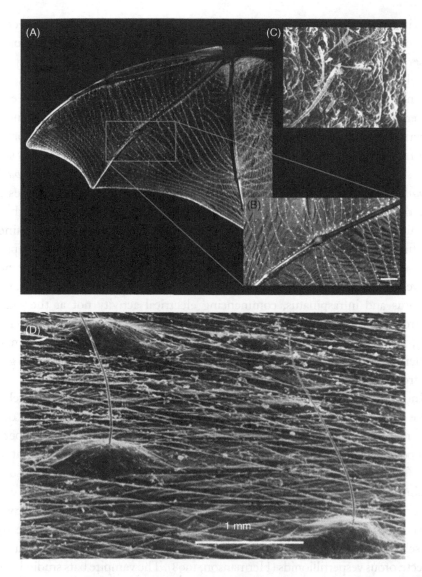

Figure 9.6 Sensory hair cells on the wing of *Antrozous pallidus*, the pallid bat (A)–(C) and *Pteropus poliocephalus* (D). (A)–(C) adapted from Zook (2007); (D) adapted from Crowley and Hall (1994). See also colour plates section.

Direct measurement of the patterns of muscle activation during normal flight in several bat species has shown complexities to these patterns that could not have been predicted from anatomical analysis alone. For example, Vaughan first hypothesized that the pectoralis, serratus ventralis and subscapularis muscles together drive the powered downstroke in bats, and although all subsequent studies have concurred, experimental measurement of muscle activity patterns have shown that electrical activity begins in the pectoralis not only before activity in the other muscles, but significantly before the onset of downstroke, leading the upstroke–downstroke transition by approximately 20 ms in *Antrozous pallidus* and *Artibeus jamaicensis* (Hermanson and Altenbach, 1981, 1983, 1985). It is likely that this pattern arises because the pectoralis undergoes eccentric contraction, electrical activity during lengthening and thus experiences enhanced force generation relative to an isometric or shortening contraction. In contrast to predictions that downstroke muscle activity should persist throughout the majority of the downstroke, measurements from *A. pallidus*, *A. jamaicensis* and *Eptesicus fuscus* show that the trio of downstroke muscles cease electrical activity approximately halfway through the downstroke, and that the remainder of the downstroke must therefore occur via the momentum gained in the first half of the cycle (Altenbach and Hermanson, 1987). Studies of upstroke musculature have demonstrated, similarly, that this portion of the wingbeat cycle is actively controlled by a combination of muscles, the trapezius, deltoideus, supraspinatus and infraspinatus, commencing electrical activity not at the onset of upstroke, but about halfway through downstroke (Hermanson and Altenbach, 1983, 1985). This allows bats to control upstroke in an active and precise manner, rather than passively, trading energetic costs of muscular activation for greater control of wing motions in this portion of the wingbeat cycle.

In comparison to comparable locomotor systems in other vertebrates, there is little variability in the composition of the flight muscle tissue of bats (Hermanson, 1998). There appear to be only two variants of pectoralis fiber type composition, a unitypic and a bitypic form. It has been proposed that these two forms could represent single and dual gear flight motors (Hermanson and Foehring, 1988; Hermanson *et al.*, 1993), with frugivores that occasionally employ hovering flight benefiting differentially from multiple fiber types in comparison to insectivores. This hypothesis is consistent with the presence of bitypic pectoralis muscles in *Artibeus jamaicensis* and *A. literatus*, and the unitypic character of insectivorous vespertillionids (Hermanson, 1998). The vampire bats studied to date, *Desmodus rotundus* and *Diaemus youngi*, and two mormoopids, *Mormoops megaphylla* and *Pteronotus parnellii*, however, are also bitypic. The complexity of the muscle fiber type distribution suggests that multiple factors may operate to exert selective pressure on fiber architecture, and that we have not yet fully explored the function of even the largest and most important of the flight muscles.

9.5 Flight performance

The kinematics of the bat wing are complex, among the most complex motions of those of any vertebrate limbs. Because there are so many joints in a bat wing, some of which, such as the shoulder, can be moved in more than one direction, bats have more than 20 degrees of freedom per wing, so even describing a wingbeat cycle is a considerable challenge (Riskin *et al.*, 2008). On the downstroke, bats extend their wings, and move them downward and forward, or ventrally and anteriorly in anatomical terminology (Norberg, 1976, 1990; Swartz *et al.*, 2005; Riskin *et al.*, 2008). On the upstroke, the wings are folded to varying degrees, and move dorsally and posteriorly. As the wings move through a locomotor cycle, the postures of the different finger bones change, causing complex changes in the three-dimensional curvature of the wing over the course of the cycle; these include changes in the degree to which the wing is cambered at different points along the length of the wing and at different times in the cycle (Swartz *et al.*, 2005). Furthermore, the wing bends and twists as it is flapped, so that by the time the wingtip has finished the downstroke, the wrist is already well into the upstroke (Norberg, 1976).

9.5.1   Changes in kinematics with speed and acceleration

Terrestrial animals employ distinct gaits at different speeds, like walking and running in humans, for example (Alexander, 2003; Biewener, 2003). In contrast, there is no evidence for an abrupt, discontinuous pattern of kinematic change with speed for flying bats. The kinematic transitions among speeds are gradual. For this reason, we prefer not to employ the term gait in the description of wing kinematics in bats.

One way to understand the link between kinematics and aerodynamic force production is to examine which kinematic parameters change during steady flight over a range of speeds. When an animal flies at a constant velocity, without climbing, falling, speeding up or slowing down, this is often referred to as steady flight. We note, however, that this does not imply that the operative aerodynamics are steady; steady flight, in this sense, can certainly involve unsteady aerodynamics. When a bat performs steady flight, over a complete wingbeat cycle, the amount of lift required is equal to its weight, and it must produce just enough thrust to overcome drag. Consider steady flight at a low speed compared with steady flight at a high speed. At both speeds, the amount of lift required is the same: it must equal the animal's body weight. At high speeds, however, the requirement for thrust is increased because the drag forces are substantially higher than during slow flights. As mentioned before, the absolute values of drag and thrust cannot be known, but they both increase exponentially with flight speed. Changes in kinematics with speed, therefore, can indirectly tell us how bats modulate the coefficient of lift, thrust and drag.

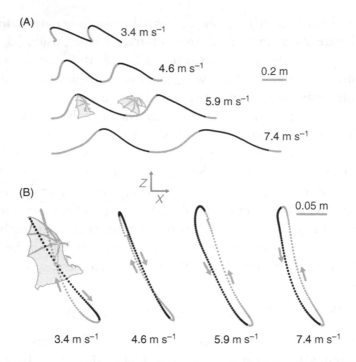

Figure 9.7 Lateral view of wingtip motions during flight at different speeds in *Cynopterus brachyotis*, the lesser dog-faced fruit bat. (A) In a world-based frame of reference, the distance traversed by both body and wingtip over each wingbeat increases with increasing forward flight speed; downstroke portions of wingtip motions given in black, upstroke in gray. (B) In the frame of reference of the bat's body, the change in the stroke plane angle with increasing flight speed is clear. At lower flight speed, to the left, stroke plane angle is relatively low, and as flight speed increases the stroke plane becomes increasingly vertical. Adapted from Iriarte-Díaz *et al.* (2011).

One aspect of wing kinematics that changes with flight speed in most bat species is the stroke plane angle (Figure 9.7). In lateral view, the movement of the wingtip of a bat with respect to its body traces a diagonal line, going posteriorly and dorsally on the upstroke, and ventrally and anteriorly on the downstroke. At lower speeds, there is considerable wingtip movement in the fore–aft direction, so the path of the wingtip is quite slanted from vertical. At higher speeds, the fore–aft component of that movement is diminished, and the line traced by the wingtip is almost vertical. Why does the stroke plane become more vertical as speed increases? Recall that lift is generated when the air moves over the wing with some speed U, and that the amount of lift generated increases with U^2. When a bat flies fast, the relative velocity of air with respect to the wing is high and high U generates substantial lift. At low forward speeds, much less lift is generated, and reaching an adequate lift magnitude to balance body weight is a physical challenge. Thus, by modulating wing kinematics to include a significant forward component in the downstroke during slow flight, there will be an additional component of relative forward motion of the wing with respect to the air that will generate extra lift in proportion to the square of its magnitude. When bats fly slowly, they also increase the camber and angle of attack of the wing, probably also to improve lift (Riskin *et al.*, 2010).

Of course, bats do not always fly at steady speeds. They often accelerate vertically or horizontally, and the magnitudes of those accelerations reflect, respectively, the total lift force or the sum of thrust and drag forces. Based on measurements of kinematic changes with those accelerations, one study on pteropodid bats demonstrated that bats increase lift forces by extending their wings more fully, increasing wingbeat frequency, increasing angle of attack and increasing camber. To increase horizontal acceleration, bats extended their wings more fully on the downstroke, drew their wings in more fully on the upstroke, increased stroke amplitude and decreased stroke plane angle (Riskin *et al.*, 2010).

9.5.2 Hovering flight

Few animals, such as some insects and a few bird species such as hummingbirds, have evolved the ability to sustain hovering flight, where the body is maintained still in air by the aerodynamic forces produced by flapping the wings. Nectar-feeding bats are also able to sustain hovering and they are among the largest animals to do so. Interestingly, among animals of comparable body size, the hovering flight of nectar-feeding bats is 40 and 60% less costly metabolically than that of hawkmoths and hummingbirds, respectively (Winter, 1998; Winter and von Helversen, 1998; Voigt and Winter, 1999), suggesting that bats have more efficient mechanisms of lift generation than members of the other groups.

The kinematics of hovering in bats differ from those of insects and hummingbirds. Insects and hummingbirds hover with fully extended wings during both downstroke and upstroke, and move the wings in a primarily horizontal stroke plane. Insects generate roughly equal amounts of lift between the upstroke and the downstroke, in what is called a "symmetrical" or "normal" hovering. Hummingbirds, however, perform "asymmetrical" hovering, producing only 25% of the lift during the upstroke, despite the relatively symmetrical up- and downstroke wing kinematics (Warrick *et al.*, 2005). Unfortunately, we still lack experimental measurements of lift generation during hovering in bats, but based on wing kinematics and relatively simple modeling of wing aerodynamics (Norberg, 1976), bats seem to also perform asymmetric hovering with most of the lift generated during the downstroke.

In the 1970s, studies of the mechanics of hovering flight in bats based on the use of kinematics and aerodynamic theory showed that the aerodynamic force generated during hovering could not be explained by quasi-steady aerodynamics, and that unsteady mechanisms must be used to produce enough lift to sustain flight (Norberg, 1976). Recently, studies using PIV (particle image velocimetry) methods conclusively document that slow-flying bats can increase lift generation as much as 40% by using attached leading-edge vortices (Muijres *et al.*, 2008) similar to those used by insects (e.g., Fry *et al.*, 2005) and hummingbirds (Warrick *et al.*, 2005), although the use of other unsteady mechanisms, such as rotational circulation and delayed stall could also be involved (Dickinson *et al.*, 1999). Why hovering flight in bats is energetically cheaper than that of insects and birds of similar size, however, is still unclear.

9.5.3   Thrust on the upstroke?

At any speed, the tip of the wing moves upwards and backwards relative to the body during the upstroke. But if the speed of the body is low enough, the tip can sometimes even move backward relative to the still air during the upstroke (e.g., Figure 9.7A, top illustration). This has been called "tip-reversal upstroke" (Aldridge, 1987a), and tip-reversal upstrokes have been thought by some to provide additional thrust to slow-flying bats. There are at least two possible mechanisms by which tip reversal may provide aerodynamic force; the wingtip could push air backward like a canoe paddle, or the bat may oversupinate the wingtip to produce a positive angle of attack at the tip and thereby produce lift locally. Support for the hypothesis that tip reversal is aerodynamically useful comes from the observation that when a bat performs a tip-reversal upstroke, markers on its body accelerate forward. If markers placed on a bat's body track the position of the center of mass, then their forward acceleration implies that net thrust exceeds drag during upstroke.

Recent work demonstrates that for *Cynopterus brachyotis* (Pteropodidae), although the trunk skeleton accelerates forward on the upstroke during slow flight, its acceleration can be partially explained by inertial effects due to the flapping motion of the wings (Iriarte-Díaz *et al.*, 2011). When the wings swing backward, approximately 20% of the bat's mass moves backward relative to the center of mass. To balance this, other portions of the body must move forward relative to the center of mass; this is reflected in the forward acceleration of the body markers. The detailed changes of the distribution of mass in the body and wings of a flying bat show that although markers on the body

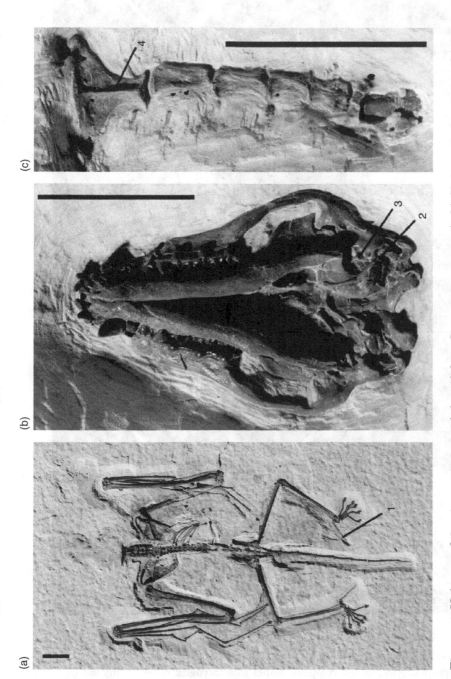

Figure 1.6 Holotype of *Onychonycteris finneyi* (adapted from Simmons *et al.*, 2008). (a) Skeleton in dorsal view. (b) Skull in ventral view. (c) Sternum in ventral view. Scale bars, 1 cm. Features labeled: 1, calcar; 2, cranial tip of stylohyal; 3, orbicular apophysis of malleus; 4, keel on manubrium of sternum.

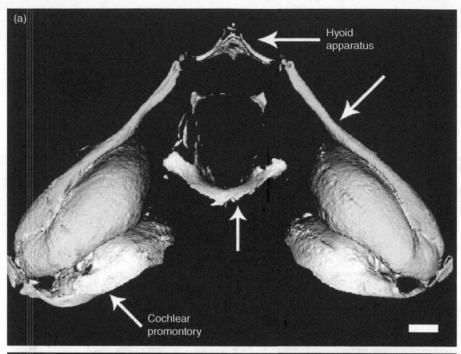

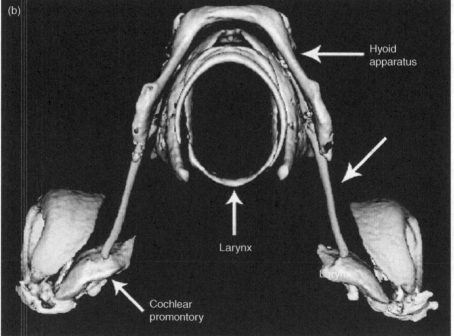

Figure 1.7 Shows articulation of stylohyal and tympanic bones in an echolocating versus non-laryngeal echolocating bat (adapted from Veselka *et al.*, 2010). a-b, Caudal view comparing stylohyal and tympanic bones in a laryngeal echolocating bat (a, *Desmodus rotundus*) and a pteropodid that lacks laryngeal echolocation (b, *Rousettus aegyptiacus*). The stylohyal bone is depicted in turquoise, the tympanic bone in yellow.

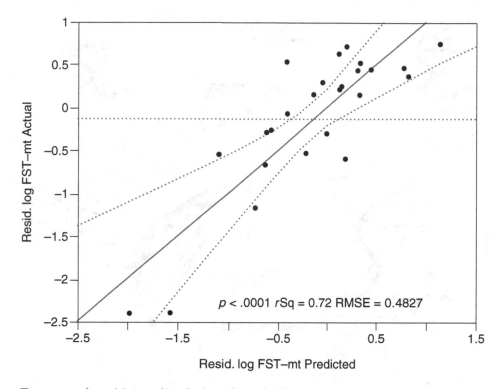

Figure 8.9 Actual by predicted plot of standard least squares fit for model with highest predictive ability for population genetic structure (explaining 72% of variance in F_{ST}_mt). Dashed lines represent 95% confidence limits. Model includes colony size, wing aspect ratio, diet and harem mating.

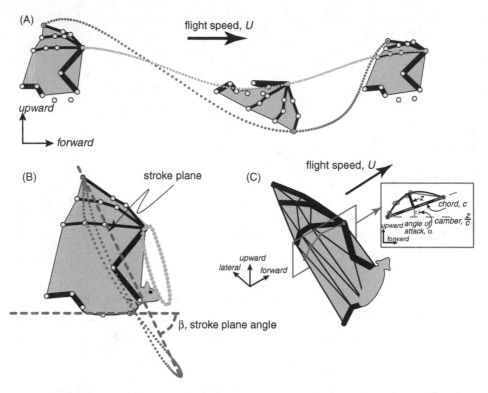

Figure 9.1 Schematic of bat in flight illustrating aerodynamic terms and concepts. (A) Lateral view of a bat wing moving through a wingbeat cycle, tracing out the motion of wingtip (red) and carpus (green) in frame of reference of external world. (B) In the frame of reference of the bat's body, motion of landmarks on the bat's wing can be seen as cyclical, tracing out a trajectory similar to a flattened, tilted ellipse. The movement of the wingtip defines, from its uppermost to lowermost positions, a stroke plane, which can be defined by β, the stroke plane angle, the angle between the line connecting these two points and the horizontal plane. (C) To define angle of attack, α, and camber, consider a parasagittal section though the wing, as outlined, and then shown on the right of schematic. The wing chord is the line connecting the frontmost, or leading, edge and rearmost, or trailing, edges of the wing in parasagittal section of interest. In flight, bat wings are typically curved in an upwardly convex fashion; the circles indicate the locations where an imaginary parasagittal cutting plane intersects the wing skin, and the lines connect those points, estimating the length of the wing in the cutting plane. Camber is then computed as the maximum height of the wing in the plane divided by the wing chord.

Figure 9.3 *Choeronycteris mexicana*, the Mexican long-tongued bat, feeding at an agave flower, showing that even as the bat comes into the force-generating downstroke, the wing membrane is not taut, but experiences varying degrees of looseness depending on anatomical location. In this particular wingbeat, the plagiopatagium, the portion of the wing between the body and the hand skeleton, is so loose that a relatively large fold or flap is visible between the ankle and the tip of the fifth digit (white arrow). Photograph by Joseph Coelho, used with permission.

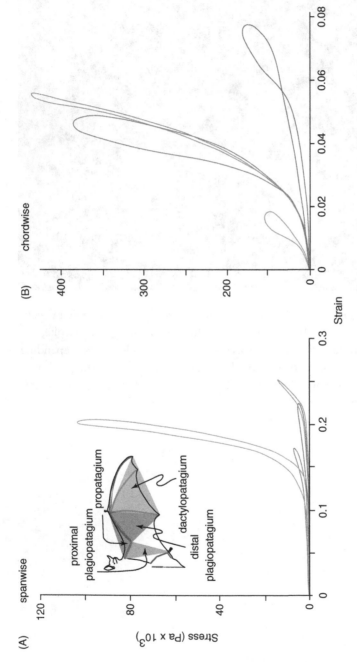

Figure 9.4 Mechanical characteristics of wing membrane skin of *Pteropus poliocephalus*, the gray-headed flying fox. The stiffness, given by the slope of the stress–strain trace, differs among regions of the wing membrane, and for each wing region differs greatly depending on whether the skin is tested from proximal to distal or spanwise (A) vs. from leading to trailing edge or chordwise (B). Adapted from Dumont and Swartz (2009).

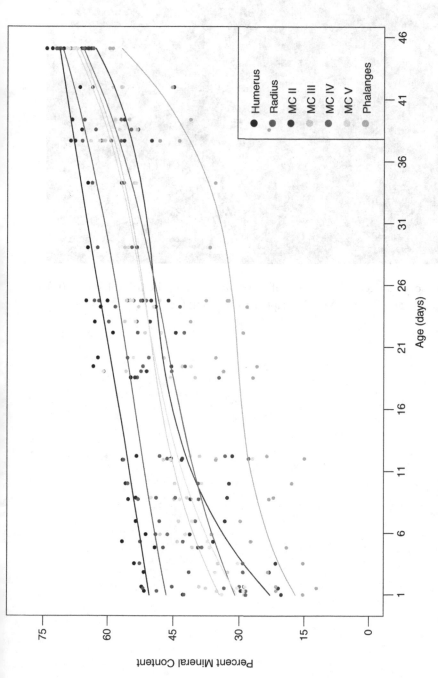

Figure 9.5 Ontogenetic pathway to adult variation in wing bone mineralization in *Tadarida brasiliensis*, the Brazilian free-tailed bat. Unlike typical mammalian limb skeletons, bat wing skeletons show great variation among bones in mineralization levels, with much greater mineralization in the proximal skeletal elements than the distal elements at all ages, from birth to skeletal maturity at 46 days. Within each bone, there is a steady increase in mineralization during lactation and eventual weaning. Adapted from Papadimitriou *et al.* (1996), Swartz and Middleton (2008).

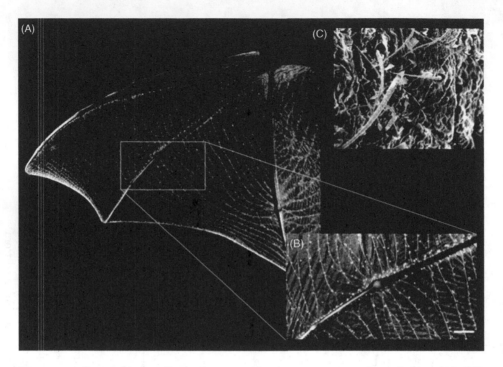

Figure 9.6 Sensory hair cells on the wing of *Antrozous pallidus*, the pallid bat (A)–(C) and *Pteropus poliocephalus* (D). (A)–(C) adapted from Zook (2007).

accelerate forward during tip-reversal upstroke, the center of mass does not (Iriarte-Díaz et al., 2011). For slow-flying C. brachyotis, there is no net thrust on the upstroke. Such mass models have not yet been computed for other species, but considering that the forward acceleration of markers on the sternum, taken as evidence of net thrust on the upstroke for other bats, is also present in C. brachyotis, it would seem that all of the net thrust and lift generated by bat wings during slow flight is on the downstroke.

9.5.4   Wake structure

In recent years, researchers have focused attention on the wakes left behind flying organisms (Rayner et al., 1986; Spedding, 1987; Spedding et al., 2003; Birch et al., 2004; Warrick et al., 2005; Johansson et al., 2008). As mentioned previously, a wake preserves, in its three-dimensional structure, the history of the forces generated as a solid object, such as an animal, moves through a fluid, such as air or water. Animal wakes have been studied to discern various aspects of force production, first qualitatively, using techniques such as smoke visualization (e.g., Vandenberg and Ellington, 1997; Srygley and Thomas, 2002) and photographic analysis of wakes created in volumes of neutrally buoyant helium bubbles (e.g., Rayner et al., 1986), and more recently using particle image velocimetry or PIV (e.g., Hedenström et al., 2007, 2009; Hubel et al., 2009, 2010). To date, only three bat species have been studied using PIV: two nectar-feeding bats, G. soricina and Leptonycteris curosae, and a fruit bat, C. brachyotis. The wake structure in bats seems to be consistent among species and shows considerable differences to wakes in birds. Similar to the wake observed behind birds during slow flight, bats shed trailing vortices from the wingtip, but in addition, there is a secondary streamwise vortex shed by the body, forming a vortex ring for each wing (Hedenström et al., 2007; Hubel et al., 2010) (Figure 9.2). This secondary vortex structure is much stronger in G. soricina (c. 50% of the tip vortex strength) than C. brachyotis (8% of the tip vortex strength). Wake structure analyses show a continuous change in circulation with speed, indicating a gradual change in aerodynamic force generation instead of the discontinuous change expected with distinct gaits (Hedenström et al., 2007).

9.5.5   Kinematic differences among species

It is well known that the wing shapes of bats vary among species (Norberg and Rayner, 1987), but very little is known about the diversity of wing kinematics among species. Most of what we know about bat wing kinematics comes from just a few species, but these span a range of body sizes and ecological roles. One important result from kinematic analyses carried out to date is that the flight performance of a bat species often differs from what is expected based on its morphology alone. This is an important point, but one that is rarely acknowledged, since it is often convenient to infer performance based on morphology alone.

For example, from arguments based on the aerodynamics of large fixed-wing aircraft, turning performance should be superior in bats with low wing loading compared to those with high wing loading. In keeping with this prediction, Aldridge (1987) found that maximum turn curvature was inversely proportional to wing loading among five insectivorous species that made banking turns. However, one of his species, *Rhinolophus ferrumequinum*, was able to perform a tighter turn than expected based on its wing loading, by dropping its flight speed to near zero while turning. Instead of making a banked turn, this species flew "like a helicopter." This, and a few other interspecific studies (e.g., Aldridge, 1987a; Stockwell, 2001; Riskin *et al.*, 2009, 2010) have demonstrated that variation in morphology can be modulated by kinematics.

Evidence that bat species vary substantially in the details of wing motions comes from data sources outside of kinematics per se (Hermanson and Altenbach, 1983, 1985). The timing of activity of the flight musculature in two species of similar body size, *Artibeus jamaicensis*, a frugivorous phyllostomid, and *Antrozous pallidus*, an insectivorous vespertilionid, directly assessed using electromyography, shows numerous differences. Many of the flight muscles fire once per wingbeat cycle in *Antrozous*, at the end of the downstroke only, but are biphasic in *Artibeus*, firing at the end of both downstroke and upstroke. This may reflect the use of a derived shoulder-locking mechanism, present in vespertilionids, but less well developed in phyllostomids, that is believed to arrest the motion of the wing at the end of upstroke passively. If the skeleton can block the abduction of the humerus, there is less need for muscle activity to actively resist the upward inertia of the wing. This kind of interspecific variation in muscle activity pattern clearly demonstrates that not all bats control their wing kinematics the same way and the investigators inferred that this kind of variation in activation pattern may also underlie patterns of variation in maneuverability in frugivorous compared with insectivorous bat species. It is clear that we have only begun to understand the motor control of the bat wing, and that further explorations of this subject may have a great deal to teach us about bat flight. In the meantime, even our limited understanding makes clear that it is worthwhile to remember that a bat's flight performance cannot be predicted from its wing shape alone.

9.5.6 Maneuvering: changing direction during flight

The ability to quickly alter flight direction and speed is fundamental for bats to successfully navigate three-dimensionally complex environments, to capture prey and to avoid predators. Despite this, maneuverability and the mechanisms underlying maneuvering abilities have yet to be thoroughly investigated. A flying organism has six degrees of freedom of movement: translation in three dimensions in space and rotation around three orthogonal axes centered on the center of mass. Flying maneuvers require the translation of the center of mass in space plus rotations of the body around three axes, termed yaw, pitch and roll.

In its most basic form, a turning maneuver requires that the forces experienced by the two sides of the body be asymmetrical; that is, the organism is subjected to a sideways or centripetal force that will drive it through the turn. Force asymmetry for turns can be achieved in multiple ways, but the most commonly described method of turning is the banked turn. In this kind of turn, an animal rolls into a bank, which reorients the lift vector by tilting it towards the direction of the turn (Figure 9.8). The tilting of the lift vector produces a centripetal force component that deflects the organism through the flight path. When the turn is complete, the body can reverse the process and return to an unbanked position such that centripetal force is no longer produced. Human-engineered aircraft employ this mechanism, it has been observed in insects and birds, and it has been proposed for bats. If a flying organism performs a banked turn, then for any given lift coefficient and bank angle, the turning radius depends directly on the wing loading or body weight per unit wing area; there is some evidence consistent with this relationship from bats in both field and obstacle-course settings (Aldridge, 1986; Aldridge and Rautenbach, 1987; Stockwell, 2001).

However, growing evidence suggests that differences in turning techniques (e.g., gliding vs. flapping turns; Aldridge, 1987b) and changes in wing posture throughout the turn (Lentink et al., 2007) can substantially alter the turning performance in ways that cannot be predicted by simple morphological parameters. The only study that has evaluated the kinematics of turning in bats suggests that turning performance is highly dependent on flight kinematics (Iriarte-Díaz and Swartz, 2008). Detailed analysis of the wings and body motion of fruit bats performing 90° turns showed that during the upstroke portion of the wingbeat cycle the body rotates in the direction of the turn, without significant changes in flight direction. This body rotation allows the bat to use part of the thrust produced during the downstroke to enhance centripetal force, allowing the bat to perform tighter turns than predicted by wing morphology alone (Iriarte-Díaz and Swartz, 2008). At least as importantly, turning is almost certainly one area of flight behavior in which the

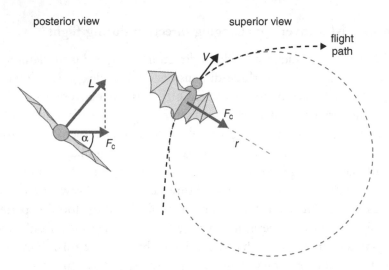

Figure 9.8 Forces required to turn during flight. If a bat banks during a turn, illustrated on left, the lift force, *L*, is rotated in such a way that is has a force component directed in the direction of the turn, *Fc*, which will propel the bat in the appropriate direction. The effect of this kind of turn, the turning radius, for a given velocity, *V*, and centripetal force, *Fc*, depend on the wing loading of the bat.

sensory input to the motor system is a crucial determinant of motor performance. The rate at which the bat receives information about changes in its three-dimensional position and accelerations, whether input to the central nervous system from the visual system, the auditory system, the vestibular apparatus or directionally sensitive wing hair sensors, has the potential to play a major role in controlling maneuvering abilities. One of the major challenges for future research is to separate sensory from mechanical and aerodynamic effects in limiting the capacity of bats to carry out extreme movements.

9.5.7   Landing: maneuvering to stop

In contrast to birds, which approach landing by simply reducing flight speed, bats face a biomechanical challenge as they must rotate their bodies in mid-air to roost head-under-heels, by performing elaborate acrobatic maneuvers. In a recent study, Riskin *et al.* (2009) found differences in the kinematics and kinetics of landing among three bat species: *Cynopterus brachyotis*, *Carollia perspicillata* and *Glossophaga soricina*. *C. brachyotis* employs a "four-point landing," using all four limbs to make contact with the substrate at impact by pitching up the body until the ventral side of the body faces the ceiling. The other two species use "two-point landings" to contact the ceiling using only the hindlimbs by simultaneously rotating their bodies in yaw, pitch and roll. Impact forces against the ceiling were four times larger during four-point landings than those recorded in two-point landings, and the authors postulated that the differences between these two landing styles could be explained by differences in the roosting habitats used by each species (Riskin *et al.*, 2009). *C. brachyotis* is a foliage-roosting species, whereas the other two species examined roost in caves. A surface such as leaves can absorb the impact energy of landing more effectively than can a rocky cave surface, and consistently, those species that roost in caves use two-point landings with low impact forces. However, differences in landing style might alternatively be explained by phylogenetic considerations: the four-landing species was a pteropodid while the two-point landing species were phyllostomids. Further analyses using more taxa distributed across the bat phylogeny are necessary to resolve this point.

9.6 Conclusion

We are fortunate to be at a turning point in the study of bat flight. Technological advances have made imaging the complex motion of bats far easier than at any time in the past, and are enabling researchers to gain insights into the aerodynamics of flight in ways that were unimaginable only a decade ago. Simultaneously, better collaborations among biologists and engineers are facilitating integrative research programs that are beginning to fruitfully apply the rich, powerful analysis methods of aeronautical engineering to the far more complex and subtle machinery of the bat flight apparatus. Once the early and most challenging steps are taken, we can look forward to an increased pace of advancement, and we predict that the next ten years will see a great increase in studies of bat flight. We can look forward to much better understanding of the ways in which bat flight is truly unique and unlike that of planes, birds or insects, and of the ways in which this special mode of flight has diversified over bat evolution.

9.7 Glossary

Angle of Attack. The angle at which the wing is inclined relative to the local air velocity. Lift increases as the angle of attack increases, up to a critical angle where the aerodynamic forces "stall" and lift declines and drag rises precipitously (Figure 9.1).

Camber. The front to back curvature of the wing, defined as the maximum height of the wing arc, divided by the chord length (Figure 9.1).

Circulation. The average vorticity contained in a defined area. In classical aerodynamics, all of the vorticity is confined to well-defined vortices, and hence the circulation of the flow is defined completely by the strength of vortices.

Lift, Drag. Aerodynamic forces acting perpendicular to, and parallel to the direction of the flow, respectively.

Reynolds Number. A dimensionless number, defined as $\rho U c/\mu$, where ρ is the fluid density, U is the flight speed, c the wing chord and μ the fluid viscosity. The Reynolds number indicates the relative importance of inertial forces compared with viscous forces. Bat flight is typically characterized by a Reynolds number of approximately 10 000–50 000, which is considered a low number for aerodynamic flows.

Starting and stopping vortices. Vortex structures, oriented in the spanwise direction, associated with the beginning of the downstroke and the end of the upstroke, respectively. These vortices connect to the trailing streamwise wake vortices.

Stroke plane angle. Angle between the lateral projection of the displacement of the wingtip with respect to the body and the horizontal, indicating the main direction of the flap.

Strouhal Number. A dimensionless number, defined as fA/U where f is the flapping frequency, U is the flight speed and A the flapping amplitude. The Strouhal number indicates the relative importance of unsteady fluid dyanamic effects. Small numbers ($St < 0.1$) usually indicate that the flow is relatively steady, while large numbers ($St > 3.0$) suggest that unsteady effects dominate. Bat flight, and indeed most biological fluid locomotion, is characterized by a Strouhal number between 0.2–0.4, indicating that steady aerodynamics are important, but that unsteady forces cannot be ignored.

Vortex. A concentration of aligned vorticity. Common examples of vortices are tornadoes and the "bathtub vortex," caused as water drains from a tank.

Vorticity. The local rotational or angular speed of the fluid.

9.9   REFERENCES

Aldridge, H. D. J. N. (1986). Kinematics and aerodynamics of the greater horseshoe bat, *Rhinolophus ferrumequinum*, in horizontal flight at various flight speeds. *Journal of Experimental Biology*, **126**, 479–497.

Aldridge, H. D. J. N. (1987a). Body accelerations during the wingbeat in six bat species: the function of the upstroke in thrust generation. *Journal of Experimental Biology*, **130**, 275–293.

Aldridge, H. D. J. N. (1987b). Turning flight of bats. *Journal of Experimental Biology*, **128**, 419–425.

Aldridge, H. D. J. N. and Rautenbach, I. L. (1987). Morphology, echolocation and resource partitioning in insectivorous bats. *Journal of Animal Ecology*, **56**, 763–778.

Alexander, R. M. (2003). *Principles of Animal Locomotion*. Princeton, NJ: Princeton University Press.

Altenbach, J. S. (1979). Locomotor morphology of the vampire bat, *Desmodus rotundus*. *Special Publication, American Society of Mammalogists*, **6**, 1–137.

Altenbach, J. S. and Hermanson, J. W. (1987). Bat flight muscle function and the scapulo-humeral lock. In *Recent Advances in the Study of Bats*, ed. M. B. Fenton, P. Racey and J. V. M. Rayner. Cambridge: Cambridge University Press, pp. 100–118.

Anderson, J. D. (2005). *Fundamentals of Aerodynamics*. New York: McGraw-Hill.

Ansari, S. A., Zbikowski, R. and Knowles, K. (2006). Aerodynamic modelling of insect-like flapping flight for micro air vehicles. *Progress in Aerospace Sciences*, **42**, 129–172.

Azuma, A. (2006). *The Biokinetics of Flying and Swimming*. Reston, VA: American Institute of Aeronautics and Astronautics.

Basset, J. E. and Studier, E. H. (1988). Methods for determining water balance in bats. In *Ecological and Behavioral Methods for the Study of Bats*, ed. T. H. Kunz. Washington, DC: Smithsonian Institution Press, pp. 373–386.

Bergou, A. J., Xu, S. and Wang, Z. J. (2007). Passive wing pitch reversal in insect flight. *Journal of Fluid Mechanics*, **591**, 321–337.

Biewener, A. A. (2003). *Animal Locomotion*. Oxford: Oxford University Press.

Birch, J. M., Dickson, W. B. and Dickinson, M. H. (2004). Force production and flow structure of the leading edge vortex on flapping wings at high and low Reynolds numbers. *Journal of Experimental Biology*, **207**, 1063–1072.

Combes, S. A. and Daniel, T. L. (2001). Shape, flapping and flexion: wing and fin design for forward flight. *Journal of Experimental Biology*, **204**, 2073–2085.

Combes, S. A. and Daniel, T. L. (2003). Into thin air: contributions of aerodynamic and inertial-elastic forces to wing bending in the hawkmoth *Manduca sexta*. *Journal of Experimental Biology*, **206**, 2999–3006.

Crowley, G. V. and Hall, L. S. (1994). Histological observations on the wing of the grey-headed flying fox (*Pteropus poliocephalus*) (Chiroptera: Pteropodidae). *Australian Journal of Zoology*, **42**, 215–231.

Currey, J. D. (1984). *The Mechanical Adaptations of Bones*. Princeton, NJ: Princeton University Press.

Currey, J. D. (2002). *Bones: Structure and Mechanics*. Princeton, NJ: Princeton University Press.

Currey, J. D. and Alexander, R. McN. (1985). The thickness of the walls of tubular bones. *Journal of Zoology, London*, **206**, 453–468.

Daniel, T. L. and Combes, S. A. (2002). Flexible wings and fins: bending by inertial or fluid-dynamic forces? *Integrative and Comparative Biology*, **42**, 1044–1049.

Dickinson, M. H. (1994). The effects of wing rotation on unsteady aerodynamic performance at low Reynolds numbers. *Journal of Experimental Biology*, **192**, 179–206.

Dickinson, M. H. (1996). Unsteady mechanisms of force generation in aquatic and aerial locomotion. *American Zoologist*, **36**, 537–554.

Dickinson, B. T. (2010). Hair receptor sensitivity to changes in laminar boundary layer shape. *Bioinspiration and Biomimetics*, **5**, 1–11.

Dickinson, M. H., Lehman, F. O. and Sane, S. P. (1999). Wing rotation and the aerodynamic basis of insect flight. *Science*, **284**, 1954–1960.

Ellington, C. P. (1975). Non-steady-state aerodynamics of the flight of *Encarsia formosa*. In *Swimming and Flying in Nature*, vol. 2, ed. T. Y. T. Wu, C. J. Brokaw and C. Brennen. New York: Plenum Press, pp. 783–796.

Fry, S. N., Sayaman, R. and Dickinson, M. H. (2005). The aerodynamics of hovering flight in *Drosophila*. *Journal of Experimental Biology*, **208**, 2303–2318.

Gupta, B. B. (1967). The histology and musculature of the plagiopatagium in bats. *Mammalia*, **31**, 313–321.

Hedenström, A., Johansson, L. C., Wolf, M. *et al.* (2007). Bat flight generates complex aerodynamic tracks. *Science*, **316**, 894–897.

Hedenström, A., Muijres, F. T., von Busse, R. *et al.* (2009). High-speed stereo DPIV measurement of wakes of two bat species flying freely in a wind tunnel. *Experiments in Fluids*, **46**, 923–932.

Hedrick, T. L., Usherwood, J. R. and Biewener, A. A. (2004). Wing inertia and whole-body acceleration: an analysis of instantaneous aerodynamic force production in cockatiels (*Nymphicus hollandicus*) flying across a range of speeds. *Journal of Experimental Biology*, **207**, 1689–1702.

Hermanson, J. W. (1998). Chiropteran muscle biology: a perspective from molecules to function. In *Bat Biology and Conservation*, ed. T. H. Kunz and P. A. Racey. Washington, DC: Smithsonian Institution Press, pp. 127–139.

Hermanson, J. W. and Altenbach, J. S. (1981). Functional anatomy of the primary downstroke muscles in the Pallid bat, *Antrozous pallidus*. *Journal of Mammalogy*, **64**, 795–800.

Hermanson, J. W. and Altenbach, J. S. (1983). The functional anatomy of the shoulder of the Pallid bat, *Antrozous pallidus*. *Journal of Mammalogy*, **64**, 62–75.

Hermanson, J. W. and Altenbach, J. S. (1985). Functional anatomy of the shoulder and arm of the fruit-eating bat, *Artibeus jamaicensis*. *Journal of Zoology*, **205**, 157–177.

Hermanson, J. W. and Foehring, R. C. (1988). Histochemistry of flight muscles in the Jamaican fruit bat, *Artibeus jamaicensis* – implications for motor control. *Journal of Morphology*, **196**, 353–362.

Hermanson, J. W., Cobb, M. A., Schutt, W. A., Muradali, F. and Ryan, J. M. (1993). Histochemical and myosin composition of vampire bat (*Desmodus rotundus*) pectoralis muscle targets a unique locomotory niche. *Journal of Morphology*, **217**, 347–356.

Holbrook, K. A. and Odland, G. F. (1978). A collagen and elastic network in the wing of a bat. *Journal of Anatomy*, **126**, 21–36.

Hubel, T. Y. and Tropea, C. (2009). Experimental investigation of a flapping wing model. *Experiments in Fluids*, **46**, 945–961.

Hubel, T. Y. and Tropea, C. (2010). The importance of leading edge vortices under simplified flapping flight conditions at the size scale of birds. *Journal of Experimental Biology*, **213**, 1930–1939.

Hubel, T. Y., Hristov, N. I., Swartz, S. M. and Breuer, K. S. (2009). Time-resolved wake structure and kinematics of bat flight. *Experiments in Fluids*, **46**, 933–943.

Hubel, T. Y., Riskin, D. K., Swartz, S. M. and Breuer, K. S. (2010). Wake structure and wing kinematics: the flight of the lesser dog-faced fruit bat, *Cynopterus brachyotis*. *Journal of Experimental Biology*, **213**, 3427–3440.

Humphry, G. M. (1869). The myology of the limbs of *Pteropus*. *Journal of Anatomical PhysiologyI*, **3**, 294–319.

Iriarte-Díaz, J. and Swartz, S. M. (2008). Kinematics of slow turn maneuvering in the fruit bat, *Cynopterus brachyotis*. *Journal of Experimental Biology*, **211**, 3478–3489.

Iriarte-Díaz, J., Riskin, D. K., Willis, D. J., Breuer, K. S. and Swartz, S. M. (2011). Whole-body kinematics of a fruit bat reveal the influence of wing inertia on body accelerations. *Journal of Experimental Biology*, **214**, 1546–1553.

Jenkins, F. A., Dial, K. P. and Goslow, G. E. (1988). A cineradiographic analysis of bird flight: the wishbone in starlings is a spring. *Science*, **241**, 1495–1498.

Johansson, L. C., Wolf, M., von Busse, R. *et al.* (2008). The near and far wake of Pallas' long tongued bat (*Glossophaga soricina*). *Journal of Experimental Biology*, **211**, 2909–2918.

Kundu, P. K. and Cohen, I. M. (2008). *Fluid Mechanics*. New York: Academic Press.

Lentink, D., Muller, U. K., Stamhuis, E. J. *et al.* (2007). How swifts control their glide performance with morphing wings. *Nature*, **446**, 1082–1085.

Lian, Y. S., Shyy, W., Ifju, P. G. and Vernon, E. (2003a). Membrane wing model for micro air vehicles. *American Institute of Aeronautics and Astronautics Journal*, **41**, 2492–2494.

Lian, Y. S., Shyy, W., Viieru, D. and Zhang, B. N. (2003b). Membrane wing aerodynamics for micro air vehicles. *Progress in Aerospace Sciences*, **39**, 425–465.

Macalister, A. (1872). The myology of the Cheiroptera. *Philosophical Transactions of the Royal Society of London*, **162**, 125–173.

Makanya, A. N. and Mortola, J. P. (2007). The structural design of the bat wing web and its possible role in gas exchange. *Journal of Anatomy*, **211**, 687–697.

Maxworthy, T. (1979). Experiments on the Weis-Fogh mechanism of lift generation by insects in hovering flight. Part 1. Dynamics of the "fling". *Journal of Fluid Mechanics*, **93**, 47–63.

Meyers, R. A. and Hermanson, J. W. (1994). Pectoralis muscle morphology in the little brown bat, *Myotis lucifugus*: a non-convergence with birds. *Journal of Morphology*, **219**, 269–274.

Miller, G. S. J. (1907). The families and genera of bats. *Bulletin of the United States National Museum*, **57**, 1–282.

Muijres, F. T., Johansson, L. C., Barfield, R. *et al.* (2008). Leading-edge vortex improves lift in slow-flying bats. *Science*, **319**, 1250–1253.

Norberg, U. M. (1970). Functional osteology and myology of the wing of *Plecotus auritus* Linnaeus (Chiroptera). *Arkiv for Zoologi*, **33**, 483–543.

Norberg, U. M. (1972). Functional osteology and myology of the wing of the dog-faced bat *Rousettus aegyptiacus* (É. Geoffroy) (Mammalia, Chiroptera). *Zoomorphology*, **73**, 1–44.

Norberg, U. M. (1976). Aerodynamics, kinematics and energetics of horizontal flapping flight in the long-eared bat *Plecotus auritus*. *Journal of Experimental Biology*, **65**, 179–212.

Norberg, U. M. (1990). *Vertebrate Flight: Flight Mechanics, Physiology, Morphology, Ecology, and Evolution*. Berlin: Springer-Verlag.

Norberg, U. M. and Rayner, J. M. V. (1987). Ecological morphology and flight in bats (Mammalia, Chiroptera) – wing adaptations, flight performance, foraging strategy and echolocation. *Philosophical Transactions of the Royal Society of London Series B*, **316**, 337–419.

Norberg, U. M., Kunz, T. H., Steffensen, J. F., Winter, Y. and von Helversen, O. (1993). The cost of hovering and forward flight in a nectar-feeding bat, *Glossophaga soricina*, estimated from aerodynamic theory. *Journal of Experimental Biology*, **182**, 207–227.

Papadimitriou, H. M., Swartz, S. M. and Kunz, T. H. (1996). Ontogenetic and anatomic variation in mineralization of the wing skeleton of the Mexican free-tailed bat, *Tadarida brasiliensis*. *Journal of Zoology*, **240**, 411–426.

Purcell, E. M. (1977). Life at low Reynolds number. *American Journal of Physics*, **45**, 3–11.

Quay, W. B. (1970). Integument and derivatives. In *Biology of Bats*, vol. II, ed. W. A. Wimsatt. New York: Academic Press, pp. 1–56.

Rayner, J. M. V. and Aldridge, H. D. J. N. (1985). Three-dimensional reconstruction of animal flight paths and the turning flight of microchiropteran bats. *Journal of Experimental Biology*, **118**, 247–265.

Rayner, J. M. V., Jones, G. and Thomas, A. L. R. (1986). Vortex flow visualizations reveal change in upstroke function with flight speed in bats. *Nature*, **321**, 162–164.

Riskin, D. K., Willis, D. J., Iriarte-Díaz, J. *et al.* (2008). Quantifying the complexity of bat wing kinematics. *Journal of Theoretical Biology*, **254**, 604–615.

Riskin, D. K., Bahlman, J. W., Hubel, T. Y., Ratcliffe, J. M., Kunz, T. H. and Swartz, S. M. (2009). Bats go head-under-heels: the biomechanics of landing on a ceiling. *Journal of Experimental Biology*, **212**, 944–953.

Riskin, D. K., Iriarte-Diaz, J., Middleton, K. M., Breuer, K. S. and Swartz, S. M. (2010). The effect of body size on the wing movements of pteropodid bats, with insights into thrust and lift production. *Journal of Experimental Biology*, **213**, 4110–4122.

Sane, S. P. (2003). The aerodynamics of insect flight. *Journal of Experimental Biology*, **206**, 4191–4208.

Saunders, M. B. and Barclay, R. M. R. (1992). Ecomorphology of insectivorous bats: a test of predictions using two morphologically similar species. *Ecology*, **73**, 1335–1345.

Sears, K. E., Behringer, R. R., Rasweiler, J. J. and Niswander, L. A. (2006). Development of bat flight: morphologic and molecular evolution of bat wing digits. *Proceedings of the National Academy of Sciences, USA*, **103**, 6581–6586.

Shyy, W., Berg, M. and Ljungqvist, D. (1999). Flapping and flexible wings for biological and micro air vehicles. *Progress in Aerospace Sciences*, **35**, 455–505.

Song, A., Tian, X., Israeli, E. *et al.* (2008). Aeromechanics of membrane wings, with implications for animal flight. *American Institute of Aeronautics and Astronautics Journal*, **46**, 2096–2196.

Spedding, G. R. (1987). The wake of a kestrel (*Falco tinnunculus*) in flapping flight. *Journal of Experimental Biology*, **127**, 59–78.

Spedding, G. R., Rosén, M. and Hedenström, A. (2003). A family of vortex wakes generated by a thrush nightingale in free flight in a wind tunnel over its entire natural range of flight speeds. *Journal of Experimental Biology*, **206**, 2313–2344.

Srygley, R. B. and Thomas, A. L. R. (2002). Unconventional lift-generating mechanisms in free-flying butterflies. *Nature*, **420**, 660–664.

Sterbing-D'Angelo, S., Chadha, M., Chiu, C., *et al.* (2011). Bat wing sensors support flight control. *Proceedings of the National Academy of Sciences, USA*, **108**, 11291–11296.

Stockwell, E. F. (2001). Morphology and flight manoeuvrability in New World leaf-nosed bats (Chiroptera: Phyllostomidae). *Journal of Zoology*, **254**, 505–514.

Strickler, T. L. (1978). Functional osteology and myology of the shoulder in the Chiroptera. In *Contributions to Vertebrate Evolution*, vol. 4, ed. M. K. Hecht and F. S. Szalay. New York: S. Karger, pp. 1–198.

Studier, E. H. (1972). Some physical properties of wing membranes of bats. *Journal of Mammalogy*, **53**, 623–625.

Swartz, S. M. (1997). Allometric patterning in the limb skeleton of bats: implications for the mechanics and energetics of powered flight. *Journal of Morphology*, **234**, 277–294.

Swartz, S. M. and Middleton, K. M. (2008). Biomechanics of the bat limb skeleton: scaling, material properties and mechanics. *Cells Tissues Organs*, **187**, 59–84.

Swartz, S. M., Bennett, M. B. and Carrier, D. R. (1992). Wing bone stresses in free flying bats and the evolution of skeletal design for flight. *Nature*, **359**, 726–729.

Swartz, S. M., Groves, M. S., Kim, H. D. and Walsh, W. R. (1996). Mechanical properties of bat wing membrane skin. *Journal of Zoology*, **239**, 357–378.

Swartz, S. M., Parker, A. and Huo, C. (1998). Theoretical and empirical scaling patterns and topological homology in bone trabeculae. *Journal of Experimental Biology*, **201**, 573–590.

Swartz, S. M., Bishop, K. L. and Ismael-Aguirre, M.-F. (2005). Dynamic complexity of wing form in bats: implications for flight performance. In *Functional and Evolutionary Ecology of Bats*, ed. Z. Akbar, G. McCracken and T. H. Kunz. Oxford: Oxford University Press, pp. 110–130.

Taylor, G. K., Nudds, R. L. and Thomas, A. L. R. (2003). Flying and swimming animals cruise at a Strouhal number tuned for high power efficiency. *Nature*, **425**, 707–711.

Thomson, S. C. and Speakman, J. R. (1999). Absorption of visible spectrum radiation by the wing membranes of living pteropodid bats. *Journal of Comparative Physiology B*, **169**, 187–194.

Tobalske, B. W. and Dial, K. P. (2007). Aerodynamics of wing-assisted incline running in birds. *Journal of Experimental Biology*, **210**, 1742–1751.

Torres, G. E. and Müller, T. J. (2004). Low-aspect-ratio wing aerodynamics at low Reynolds numbers. *American Institute of Aeronautics and Astronautics Journal*, **42**, 865–873.

Usherwood, J. R., Hedrick, T. L., McGowan, C. P. and Biewener, A. A. (2005). Dynamic pressure maps for wings and tails of pigeons in slow, flapping flight, and their energetic implications. *Journal of Experimental Biology*, **208**, 355–369.

van denBerg, C. and Ellington, C. P. (1997). The three-dimensional leading-edge vortex of a "hovering" model hawkmoth. *Philosophical Transactions of the Royal Society of London Series B*, **352**, 329–340.

Vaughan, T. A. (1959). Functional morphology of three bats: *Eumops, Myotis, Macrotus*. *Publications of the Museum of Natural History, University of Kansas*, **12**, 1–153.

Vaughan, T. A. (1966). Morphology and flight characteristics of molossid bats. *Journal of Mammalogy*, **47**, 249–260.

Vogel, S. (1981). *Life in Moving Fluids*. Princeton, NJ: Princeton University Press.

Voigt, C. C. and Winter, Y. (1999). Energetic cost of hovering flight in nectar-feeding bats (Phyllostomidae: Glossophaginae) and its scaling in moths, birds and bats. *Journal of Comparative Physiology B*, **169**, 38–48.

Warrick, D. R., Tobalske, B. W. and Powers, D. R. (2005). Aerodynamics of the hovering hummingbird. *Nature*, **435**, 1094–1097.

Warrick, D. R., Tobalske, B. W. and Powers, D. R. (2009). Lift production in the hovering hummingbird. *Proceedings of the Royal Society of London Series B*, **276**, 3747–3752.

Weis-Fogh, T. (1972). Energetics of hovering flight in hummingbirds and in *Drosophila*. *Journal of Experimental Biology*, **56**, 79–104.

Weis-Fogh, T. (1973). Quick estimates of flight fitness in hovering animals, including novel mechanisms for lift production. *Journal of Experimental Biology*, **59**, 169–230.

Winter, Y. (1998). Energetic cost of hovering flight in a nectar-feeding bat measured with fast-response respirometry. *Journal of Comparative Physiology B*, **168**, 434–444.

Winter, Y. and von Helversen, O. (1998). The energy cost of flight: do small bats fly more cheaply than birds? *Journal of Comparative Physiology B*, **168**, 105–111.

Zook, J. M. (2005). The neuroethology of touch in bats: cutaneous receptors of the bat wing. *Neuroscience Abstracts*, **78**, 21.

Zook, J. M. (2007). Somatosensory adaptations of flying mammals. In *Evolution of Nervous Systems: A Comprehensive Reference*, vol. 3: *Mammals*, ed. J. H. Kaas and L. Krubitzer. Boston, MA: Elsevier Academic Press, pp. 215–226.

10

Toward an integrative theory on the origin of bat flight

NORBERTO P. GIANNINI

In bats ... we perhaps see traces of an apparatus originally constructed for gliding through the air rather than for flight. Darwin (1859, p. 181)

10.1 Introduction

It is easy to grasp why bats are so successful: a small nocturnal mammal in possession of powered flight can explore resources in a relatively low-risk environment at spatial scales orders of magnitude larger than that of non-volant mammals of comparable size. As an example, the median home range of the 8–11 g vespertilionid *Chalinolobus tuberculatus* can be as large as 1500 ha (O'Donnell, 2001); this is the average area used, for instance, by a 300 kg herbivore, the Wapiti (*Cervus elaphus canadensis*; Calder, 1996). Acquisition of powered flight represented an immediate advantage to the bat lineage. As attested by the fossil record, bats reached nearly worldwide distribution early in their evolution. By the Early Eocene, bats suddenly appear in all the major landmasses they inhabit today (Gunnell and Simmons, 2005; Tejedor *et al.*, 2005; Eiting and Gunnell, 2009). This suggests that powered flight may have played a key role in the fast expansion of bats, thereby contributing to their spectacular diversification.

Beyond the presence of wings, adaptations to powered flight encompass most organ systems, including: full flexion and extension of whole wing (including hyperabduction of digits) automated via tendon rearrangements (Norberg, 1972); energy-saving locking mechanisms such as vertebral column rigidity (Vaughan, 1959); locking mechanisms in each forelimb joint to prevent hyperextension or rotation of the wing (Vaughan, 1959); concentration of forelimb muscle mass towards the center of gravity to reduce inertial power (Vaughan, 1970); leading-edge camber adjustment by pronation of hand, assisted by tension of propatagium via m. occipitopollicalis and stiffened dactilopatagium minus (Norberg, 1969); trailing-edge camber adjustment by

Evolutionary History of Bats: Fossils, Molecules and Morphology, ed. G. F. Gunnell and N. B. Simmons. Published by Cambridge University Press. © Cambridge University Press 2012.

flexion of digit V (Neuweiller, 2000); streamlining of the head-body to reduce parasitic drag, placing the head in between scapular blades by dorsiflexion of neck (Jepsen, 1970; Fenton and Crerar, 1984) and rotation of basicranial axis with respect to rostrum; rotation of hindlimb segments by modification of hip and ankle joints to spread laterally and caudally (Simmons, 1994); tendon-locking mechanism as energy-saving device used in hindlimb suspension (e.g., during roosting; see Simmons and Quinn, 1994); synchronization of wingbeat and respiratory cycles (Thomas, 1987; Speakman and Racey, 1991; Lancaster et al., 1995); mass-specific aerobic capabilities and cardiac outputs at least twice as high as those of running mammals (Thomas, 1987), supported by adaptive changes in genes involved in energy metabolism (mitochondrial- and nuclear-encoded oxidative phosphorylation (OXPHOS) genes) that trace to the bat common ancestor (Shen et al., 2010); specialized capillary circulation in the wing membrane (Neuweiller, 2000); and a number of other features.

Many of these characters represent un-reversed synapomorphies of Chiroptera (Simmons and Geisler, 1998). Nonetheless, exactly how bats achieved the morpho-functional conditions of flight, and how the many refinements of wing morphology and function were attained, remains poorly understood (see Swartz et al., 2005; Bishop, 2008). Indeed, the origin of bat flight challenged evolutionary biologists ever since *The Origin of Species*: bats figured prominently among the "Problems with the theory" because the transformation of a quadrupedal insect-ivore into a flying mammal by means of natural selection posed seemingly insurmountable difficulties (Darwin, 1859). The discovery of echolocation (Griffin, 1958) added a level of complexity to this problem. Once the importance of echolocation was fully realized, researchers focused on the potential intercon-nection of powered flight and echolocation in bat evolution. The debate fluctu-ated among three views: either echolocation evolved after powered flight was achieved ("flight-first" hypothesis); or flight evolved in a lineage of echolocating quadrupeds ("echolocation-first"); or flight and echolocation evolved together through a feedback interaction, with improvements in one system conditioning changes in the other ("tandem evolution" (reviewed by Simmons and Geisler, 1998; Speakman, 2001; Schnitzler et al., 2003; Jones and Teeling, 2006)).

The several proponents of hypotheses on the origin of bat flight (Table 10.1) have usually construed some sort of model organism to reflect their views on how the evolutionary transformations may have taken place in deriving a flying mammal from a non-volant precursor. Typically, these ideas have emphasized different aspects of bat biology (e.g., echolocation, anatomy, diet, habits or a combination thereof). Recent paleontological findings have contributed fresh evidence in a phylogenetic context. In particular, the morphology of *Onycho-nycteris finneyi*, an Early Eocene fossil recovered as sister to all other bats, supported the flight-first hypothesis of bat evolution (Simmons *et al.*, 2008, 2010; cf. Veselka *et al.*, 2010). The inquiry on the origin of echolocation then is placed on an evolved flying bat, redirecting the attention to powered flight as the major evolutionary change that distinguished the bat lineage from all other mammals. In the view of this writer, resolving the transition to flight amounts to understanding the origin of bats.

In this chapter I briefly review topics considered in previous hypotheses and attempt to provide further support to one particular bat precursor model – a version of Darwin's (1859) small, nocturnal insectivorous glider – incorporating in the discussion of bat flight origins recent developments in many fronts of bat research, including phylogenetic paleontology, gross anatomy and embryology, molecular control of organogenesis, as well as some basic aerodynamic and paleoecological aspects. Although it is certainly premature to materialize a comprehensive synthesis of all these aspects, here I advance ideas that may converge in an integrative theory on the origin of bat flight in the foreseeable future.

10.2   Brief overview of previous hypotheses

Modern hypotheses on bat flight origins were recently revised (e.g., Simmons and Geisler, 1998; Speakman, 2001), and most bat researchers tend to support a gliding precursor of the bat lineage (see Long *et al.*, 2003; Simmons *et al.*, 2008). Table 10.1 presents a synopsis of topics considered by selected proponents of ideas on the origin of bat flight, briefly commented on below. Throughout this chapter, bat precursors are identified with hypothetical lineages branching off from the line of descent that connects the ancestor of chiropterans with other laurasiatherian lineages. "Microbat" and "megabat" are names used informally in the traditional sense (i.e., megabat refers to any member of Pteropodidae, and microbat to any member of the remaining extant families).

Animals that glide or fly do so evolving in response to various needs. The proposed selective forces that have been linked to bat flight include trophic function (e.g., aerial hawking), transport advantage (of gliding or flying versus running or walking + climbing) and escape from (predominantly avian) predators and competitors. Transitional mechanisms included leaping, hovering and gliding. Comparatively, gliding is a low-cost transportation modality (e.g., Rayner, 1986; Scholey, 1986b); by contrast, hovering represents the most expensive and specialized mode of flight (Norberg, 1994) and thus has been discarded as a transitional mechanism (see Norberg, 1986; Speakman, 2001). Leaping is attractive because many mammalian gliders use a quadrupedal launch to impart a horizontal velocity to initiate a glide (Jackson, 1999; Bishop, 2008; Jackson and Schouten, in press). So leaping → gliding → flying has

Table 10.1 Summary of selected contributions to a theory on the bat flight origins and aspects treated by the different authors, arranged in comparative fashion.

Proponents	Known as	Emphasis on	Echolocation	Selective force	Transitional mechanism	Stratum used	Roosting sites	Land posture/ locomotion	Habit	Diet
Vaughan (1959, 1970)	–	body shape, wing, foot	–	transportation	–	rocky walls	crevices	crawling	nocturnal	insectivory
Jepsen (1966, 1970), Pirlot (1977)	Reach hunting[1]	wing, body shape, neck, dentition	present	trophic	leaping → hovering	ground → arboreal	crevices → cave walls	quadrupedal → hindlimb suspension	nocturnal	omnivory → insectivory
Smith (1977), Norberg (1986), Rayner (1986), Scholey (1986)	Flight first	wing, aerofoil peformance	absent	transportation	climbing + gliding	arboreal	trees	–	nocturnal	insectivory/ omnivory
Caple et al. (1983)	Reach hunting*	"birds"	–	trophic	leaping	rocky walls	–	cautious climbing	–	insectivory
Pettigrew (1986)	Diphyletic origin	sensory pathways	microbats yes/ megabats no	–	–	–	–	–	nocturnal/ diurnal	insectivory/ frugivory
Speakman (1993)	Tandem evolution	ventilation + echolocation	present	trophic	leaping → gliding	arboreal	caves	–	nocturnal	insectivory
Fenton et al. (1995)	Echolocation first	echolocation	present	trophic	climbing + gliding	arboreal	trees	–	nocturnal	insectivory

Speakman (2001)	–	sensory organs, diet	absent	escape	leaping → gliding	arboreal	–	–	diurnal	frugivory
Sears (2006), Cretekos et al. (2008), Weatherbee et al. (2008)	–	Dactilopatagium, digits, ulna	–	–	–	–	–	–	–	–
Bishop (2008)	–	wing	–	transportation	leaping → gliding	arboreal	–	cautious climbing, hindlimb suspension	–	–
Simmons and Geisler (1998)	Flight first	cochlea, limb skeleton, dentition	absent	transportation	gliding	arboreal	–	brachiation, four-leg underbranch suspension	nocturnal	insectivory
Simmons et al. (2008)	Flight first									
This study	–	wing, foot, patagia, paleoecology	absent	transportation	climbing + gliding	arboreal	tree cavities	hindlimb suspension	nocturnal	insectivory

* coined by Speakman (2001).

been proposed as a transitional sequence, varying on the substrate (either the ground, cliffs or branches).

Early authors (Vaughan, 1959, 1970; Jepsen, 1966, 1970) emphasized the compressed body shape of bats as an important adaptation in early bat evolution for using crevice-like retreats. In their view, this conditioned the evolution of flapping to one specific solution for powering the wings (many muscles involved, with a movable shoulder girdle) radically different from that evolved by birds (just one muscle involved, the pectoralis and its distinct supracoracoid part, with a rigid shoulder girdle). These authors also paid attention to neck, thorax and foot structure, and their implications for posture in land, as well as dentition as a conditioner of diet. Others have focused on auditory function, sensory pathways and, recently, molecular control of organogenesis.

Most authors have supported the use of an arboreal stratum by bat precursors. Jepsen (1970) suggested the ground, in addition to branches, for a transition from quadrupedal locomotion to hindlimb suspension. The latter may represent a mechanism for partially releasing forelimbs from the body-support function (Bishop, 2008). Caple *et al.* (1983) favored cliffs to initiate flight. Vaughan (1959) and Caple *et al.* (1983) noted that rocky walls provide suitable launching sites, as well as crevices for roosting, pointing out a retained ability for crawling in most extant bats. Simmons *et al.* (2008) noted that *Onychonycteris* had limb proportions intermediate between bats as a group and other forelimb-dominated, arboreal mammals. In an arboreal milieu, four-leg underbranch suspension for resting and moving along branches (Simmons, 1995; Bishop, 2008; Simmons *et al.*, 2008) may have preceded hindlimb suspension, the typical posture adopted by most bats for resting.

Insectivory has been generally accepted as the ancestral diet of bats, usually associated with nocturnality. However, Jepsen (1970) proposed omnivory, and Ferrarezi and Giménez (1996) and Speakman (2001) have suggested frugivory. To Speakman (2001) ancestral bats were diurnal frugivores taking advantage of the explosion of fruiting angiosperms during the Paleogene; nocturnality arose as an escape from avian predators. Teeling *et al.* (2005) used the same event – the Paleogene increase in angiosperm diversity (Wilf and Labandeira, 1999) – but in connection with the peak of Tertiary insect diversity, to propose a driver for the diversification of all four major, primarily insectivorous microbat lineages. Simmons *et al.* (2008) showed that the dentition of *Onychonycteris* was typical of an insectivore, as in all known Eocene bats. In fact, both insectivory and nocturnality are simply best explained by phylogeny, as these character states likely are plesiomorphic in mammalian behavior.

Finally, the echolocation capability and its connection with flight (see Suthers *et al.*, 1972; Speakman and Racey, 1991; Speakman, 1993; Lancaster *et al.*, 1995; Schnitzler and Kalko, 2001) has been the center of debate for the last three decades. "Echolocation-first" in particular is not a single hypothesis; instead, it has been stated either in terms of primitive echolocation originating as low-intensity broad-band clicks used for communication (e.g., Fenton *et al.*, 1995); or in terms of tonal signals used for spatial orientation (Schnitzler *et al.*, 2003). Origin of echolocation is of fundamental importance for bat biology, but it is not treated further here because the "flight-first" hypothesis was supported by the discovery of *Onychonycteris* (Simmons *et al.*, 2008, 2010; cf. Veselka *et al.*, 2010). The reader is directed to Simmons and Geisler (1998), Speakman (2001), Schnitzler *et al.* (2003), Jones and Teeling (2006), Simmons *et al.* (2008, 2010), Teeling (2009) and Veselka *et al.* (2010) for further discussion on the evolution of echolocation.

10.3 Anatomy of fossils

The anatomy of fossil bats has received considerable attention (for recent references see Simmons and Geisler, 1998; Gunnell and Simmons, 2005; Simmons *et al.*, 2008) and is not reviewed here *in extenso*. Discussed below are selected aspects directly linked with hypotheses of bat flight.

10.3.1 Axial skeleton

Skull – Highly valuable information from fossils is lost due to poor preservation of skulls. Still, two key regions are preserved in great detail: the auditory region and the part of the rostrum bearing the dentition (see details in Jepsen, 1966; Simmons *et al.*, 2008, 2010). The former is key in rejecting the "echolocation-first" hypothesis, given that *Onychonycteris* lacks character states associated with laryngeal echolocation (i.e., for emission of ultrasound and reception of echoes: enlarged cochlea, enlarged orbicular apophysis of malleus and paddle-like connection of the hyoid apparatus to the skull; Novacek, 1985; Habersetzer and Storch, 1987; Simmons *et al.*, 2008, 2010; cf. Veselka *et al.*, 2010). Note, however, that the morphological signature of primitive echolocation may not be as conspicuous as that of laryngeal echolocation and may be lost in fossils.

Second, mandible, maxilla and dentition are usually exquisitely preserved in Eocene fossils. The mandible of primitive bats is similar to the therian morphotype in overall shape. It is similar to microbats in having a hooked angular process, but differs in presenting a plesiomorphic coronoid process (relatively round and tall, directed posterodorsally). The dentition is complete in all key fossils, showing the primitive chiropteran dental formula with

38 elements. The unmodified dilambdodont molars are indicative of insectivory, which in Messel bats (e.g., *Archaeonycteris, Palaeochiropteryx, Hassianycteris*) is corroborated by preserved stomach contents (Habersetzer and Storch, 1987; Habersetzer *et al.*, 1994); thus, insectivory is parsimoniously predicted as the diet of bat precursors.

Vertebral column – Fenton and Crerar (1984) identified two distinct morpho-types of cervical vertebrae in bats, corresponding to the conditions of megabats and microbats. *Onychonycteris* and *Icaronycteris* exhibit a morphology consistent with the latter type, specifically: slender (versus robust) neural arches, short (versus long) vertebral bodies and saddle-shaped articular surface of vertebral body (versus flat (platycoelus vertebrae)). Microbats and primitive bats have a greater angle of dorsiflexion of the neck, which enables them to look up and scan the surroundings while roosting on a horizontal surface or hanging from a vertical surface (Fenton and Crerar, 1984), and allows placing the head almost between the scapulae during flight (Norberg, 1998).

Thoracic and lumbar vertebrae exhibit a trend towards rigidity in extant bats (Vaughan, 1959), either by fusion of vertebrae (see Simmons and Geisler, 1998) or by unusual articulations (Vaughan, 1959; T. Gaudin, personal communi-cation), neither of which have been so far reported in primitive bats. Finally, the long tail present in all Eocene fossils clearly suggests that reductions seen in some extant bats (e.g., most pteropodids and some phyllostomids) is a derived condition with specific function (e.g., to increase wing loading and thus flight speed, thereby reducing commuting time; see Norberg, 1994).

Thoracic skeleton – Fossil and extant bats exhibit a mesosternum with a shallow-to-moderate ventral crest, and a manubrium with a ventral process that is a shallow crest in the most primitive bats and can be prominent and bifurcated in some extant taxa (Simmons and Geisler, 1998; Simmons *et al.*, 2008). Extant bats also exhibit posterior, and sometimes anterior rib laminae, which apparently provide an increased area for muscle attachment on the rib cage that contribute to the ventilation, echolocation and flight functions (Vaughan, 1959; Lancaster *et al.*, 1995; Simmons and Geisler, 1998). Rib laminae are absent in *Onychonycteris* (Simmons *et al.*, 2008).

10.3.2   Appendicular skeleton

Scapula – Most mammals exhibit a flat, undivided infraspinous fossa. Eocene bats already possess a condition seen in many extant bats, the infra-spinous fossa with two subfossae, which likely increases area of origin of m. infraspinatus, m. subscapularis and other muscles (other extant bats present three subfossae; Simmons and Geisler, 1998; Simmons *et al.*, 2008). The relatively narrow scapula of primitive and many extant bats (e.g., pteropodids) resembles that of brachiating primates (e.g., *Hylobates*; Roberts, 1974, fig. 8), which move their forelimbs rapidly, but about specific ranges of circumduction.

Forelimb – Onychonycteris was capable of powered flight; however, its wing structure indicates a mixture of primitive and derived traits important in interpreting the evolution of flight (Simmons *et al.*, 2008). Bats are extreme forelimb-dominated mammals (see Simmons *et al.*, 2008, fig. 3). The humerus and radius are elongated; midshaft diameters of these bones, while scaling with mass at about the same rate as in other mammals (similar slope parameter), are larger than in non-volant mammals of comparable size (higher *y*-intercept; Swartz and Middleton, 2008). This effectively counteracts strong torsional forces during flight, also with the aid of thin walls that reduce bone mass and associated inertial forces (Swartz *et al.*, 1992; Papadimitriou *et al.*, 1996; Swartz, 1997, 1998). Swartz and Middleton (2008) have shown that bats possess thicker forelimb bones and thinner hindlimb bones than non-volant mammals, in correspondence with an extreme condition of forelimb domination. *Onycho-nycteris* was clearly intermediate in limb proportions between bats and arboreal mammals, also having the lowest aspect ratio of wings among all known bats (Simmons *et al.*, 2008).

Onychonycteris bore claws on all five digits (Simmons *et al.*, 2008). Remark-ably, the ungual phalanx of the third digit was about as large as expected for the estimated size of the bat (*c.* 40 g; Giannini *et al.*, Chapter 16, this volume) as compared with arboreal mammals (Figure 10.1). The regression line for bat fossils with ossified ungual phalanges (*Onychonycteris* and *Icaronycteris*) is much steeper (Figure 10.1). This suggests a sudden reduction in ungual phalanx size, possibly in connection with phyletic nanism during early bat evolution (see Giannini *et al.*, Chapter 16, this volume). Thus, a simple allometric process (e.g., deceleration *sensu* Reilly *et al.*, 1997) may be involved in the truncation of phalanx development. This has functional implications because intermediate phalanges of digits III to V are flexible in extant bats (Adams, 1992), bending with, rather than resisting against, the air flow around the wing. This may represent an adaptation to reduce bone strain from aerodynamic loads, and results from a specific mineralization pattern (Swartz and Middleton, 2008). Phalanges exhibiting this condition likely act as an elastic-energy storage (Norberg, 1990). By contrast, *Onychonycteris* and *Icaronycteris* exhibit straight, rigid intermediate phalanges (Figure 10.1); these phalanges probably were less efficient aerodynamically and not as safe against breakage. Swartz (1997, 1998) calculated that specializations that reduce the wing moment mass of inertia (e.g., reduced mineralization and bone mass distally in the wing) have a significant effect on the cost of flight; primitive

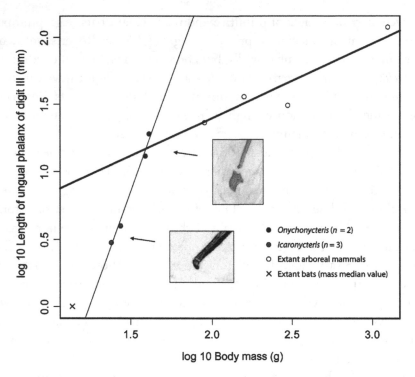

Figure 10.1 Regression of manual ungual phalanx length (third digit, measured along the dorsal side) on body mass. Bold line: in selected arboreal mammals (*Caluromys, Micoureus, Tupaia, Cynocephalus*) and two specimens of the fossil bat *Onychonycteris finneyi*; the estimated slope parameter (0.55) is greater than the value expected by geometric similarity (0.33), indicating that claws scale in a disproportionate fashion among these climbing mammals. Thin line: in five specimens of fossil bats, *Icaronycteris index* ($n = 3$) and *Onychonycteris finneyi* ($n = 2$); note that the median mass of extant Chiroptera (*c.* 14 g, not included in the regression) lies close to the line at $y = 0$. Mass of fossils from Giannini et al. (Chapter 16, this volume). Insets: images of ungual phalanx of manual digit III in *Icaronycteris* and *Onychonycteris*.

bats lack at least some of those specializations and are thus predicted to have flown at costs higher than those of modern bats.

The wings of both specimens of *Onychonycteris* are folded at the carpometa-carpal (hereafter C-MC) joints, just as in modern bats (see figs. 1C and S4 in Simmons et al., 2008). Full wing folding in these fossils indicates an early evolution of derived C-MC joints; in bats, these complex articulations function in guiding an orderly collapsing of the wing to the resting position and vary consistently across bat groups (personal observation). Based on some key elements that emerge from the rock matrix, the carpal morphology of *Onychonycteris* and *Icaronycteris* resembles more closely that of modern yangochiropteran bats (personal observation). C-MC joints belong to a complex set of extension-stopping, collapse-guiding or energy-saving locking devices that also include humeroscapular, radiocarpal, intracarpal and interphalangeal joints, which are essential for the automated functioning of the bat wing during flight and for folding and protection. These delicate structures have not been examined in detail in primitive fossils and deserve much attention. For instance, all fossil and extant bats, as well as the colugo, exhibit a strong radial carpal bone resulting from the fusion of specific proximal and central elements (the scaphocentralolunate; Simmons, 1994); this bone locks against the radius preventing overextension of the hand and therefore collapsing of the wing under aerodynamic forces. Thus, the scaphocentralolunate may have played a key role in the evolution of the wing.

Hindlimb – Onychonycteris differs from extant bats only in a few features of the hindlimb; e.g., relatively long stylopod and zeugopod, and short digit I relative to digits II–V (Simmons *et al.*, 2008). This fossil exhibits a femur reoriented 90° with respect to the standard quadrupedal position, a thin but complete fibula, a robust calcaneus with a long calcar articulating to the calcaneal tuber, and a partially hidden astragalus in which a dorsal trochlea is visible (indicating a plantigrade stance with the foot pointing laterally). The hindlimb in all fossil bats is strongly suggestive of hindlimb suspension, and it seems clear that the pes is advanced with respect to wings in terms of degree of similarity to modern bats (Simmons *et al.*, 2008). Thus, hindlimb suspension is predicted to be present in bat precursors. Finally, presence of a calcar in *Onychonycteris* refutes two hypotheses, that this structure is derived within Chiroptera, and that it originated as an aid in aerial hawking (Simmons *et al.*, 2008), and supports another one, that the calcar originated as a support for the uropatagium – just as analogous structures (styliform cartilages) support other patagial tracts in many mammalian gliders (see Jackson and Schouten, in press). Adams and Thibaud (2000) reported developmental differences in the calcar of diverse modern bats that are suggestive of functional differences across bat lineages.

10.3.3   Patagia

Specimens of *Onychonycteris* lack preserved soft tissues, although wing membranes likely were present in this taxon. The type specimen of *Icaronycteris index* (PU 18150) exhibits poorly preserved remains of dactilopatagium (Jepsen, 1966). By contrast, Eocene fossils from Messel preserve soft tissue, including patagia, in exquisite detail (Habersetzer and Storch, 1987). The patagia of

Messel bats are indistinguishable from those of extant bats: pro-, dactilo-, plagio- and uropatagium are all present and formed by thin, finely reticulated, hairless membranes. Therefore, these membranes probably possessed anisotropic variation of mechanical properties (e.g., stiffness) as seen in modern bats (Swartz *et al.*, 1996; Swartz, 1998). In addition, the m. occipitopollicalis is preserved in some Messel bats (e.g., *Archaeonycteris trigonodon*; Thewissen and Babcock, 1992), indicating that Eocene bats possessed a stiff leading edge array and probably anterior camber control much like that in extant bats (see Norberg, 1969).

10.4   Developmental clues

Recent advances in developmental science, from the level of molecules to gross embryology, offer rich new evidence on how ontogenetic changes shape morphology and the corresponding linked functions, in many organisms including bats. The sequence of appearance during development of important structures related to flight is synthesized in Figure 10.2. As expected, most of the qualitative changes in limb development occur during the embryonic stages (Figure 10.2). Using the *Carollia* staging (CS) system (Cretekos *et al.*, 2005), the hand- and footplate appear in stages CS 12 and 13, respectively. At CS 16, finger rays are visible in both the hand- and footplate, with the former showing anteroposterior (hereafter A-P) patterning (the A-P axis of the embryonic limb plates translates into the mediolateral axis of the adult quadruped limb). The webbing of the handplate is subsequently retained as one part of the flight membrane – the dactilopatagium (see below). The primordia of other patagial tracts appear during late CS 16 and 17, and become translucent (probably acquiring certain properties of compliant aerofoils, like elasticity) by CS 21. Patagial muscles become externally visible at CS 18–19. Claws of digits I, II and toes develop as folds in CS 18; claw primordia proper appear in CS 19 and become keratinized during CS 21–22. Digits flex, evincing functional interphalangeal joints, by CS 20; C-MC joints allow full wing folding at the beginning of the fetal period. Therefore, the wing folds when all its constituent elements are formed. Finally, epiphyseal discs of wing phalanges ossify during the transition to adulthood, thus completing wing development. The latter process varies widely across bats, but in many cases intermediate and distal phalanges remain cartilaginous for life (Adams, 1992; Simmons and Geisler, 1998).

10.4.1   Molecular control of organogenesis

The elongated, webbed handwing of bats results from changes in the spatiotemporal expression of specific signaling pathways shared with other

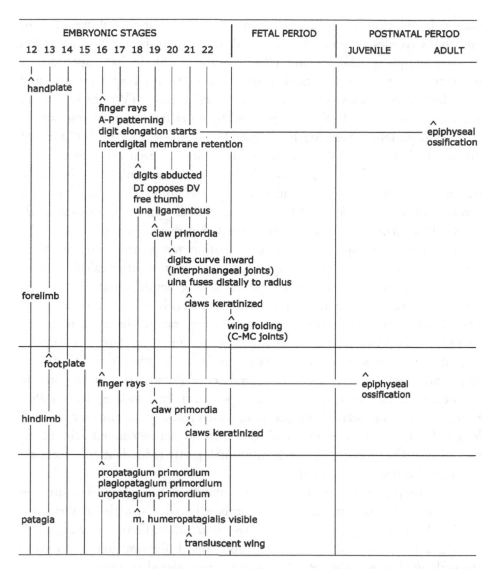

Figure 10.2 Sequence of major developmental events in the ontogeny of the forelimb, hindlimb and patagia in bats, based on *Rousettus* (Giannini *et al.*, 2006; Richardson *et al.*, 2008), with additional data from *Myotis* (Adams, 1998, 2000, 2008; Adams and Thibaud, 2000), *Carollia* (Cretekos *et al.*, 2005), *Miniopterus* (Hockman *et al.*, 2009) and *Pipistrellus* (Tokita, 2006). The developmental sequence is divided into the embryonic period (subdivided into *Carollia* stages CS 12 to 22 based on Cretekos *et al.*, 2005), the fetal period and postnatal period (subdivided into juvenile and adult stages). Each event represents either the appearance of a structure or the onset of a developmental process and is dated on this relative scale using an upward arrowhead. Abbreviations: A–P, anteroposterior; C–MC, carpometacarpal joints; D, digit; m., muscle.

mammals (Hockman *et al.*, 2008). In mammals, interdigital webbing of limb buds recedes by programmed cell death (apoptosis) via the expression of bone morphogenetic proteins (*Bmps*; Sears, 2008). In bats, apoptosis is prevented via expression in tandem of *Gremlin* (*Gre*, a *Bmp* repressor) and fibroblast growth factors (*Fgfs*) in the interdigital tissue (Weatherbee *et al.*, 2006). This scheme is further complicated by other factors, including expression of the protein Sonic hedgehog (*Shh*; Hockman *et al.*, 2008). *Fgf8* and *Shh* interact with *Bmp2* and *Gre*, activating or inhibiting each other in complementary (A-P vs. P-A) expression fields in the bat handplate. Handwing structure also depends on elongation of digits II to V. The specific process involves a massive alteration of the standard chondrocyte cycle (Farnum *et al.*, 2008a; Hockman *et al.*, 2008; Sears, 2008). Up-regulated *Bmp2* causes the chondrocytes to undergo accelerated proliferation, temporarily inhibiting their terminal differentiation. Chondrocyte death by apoptosis after calcification of their interterritorial matrices is postponed during development, resulting in very large hypertrophic zones in the bat digits (Sears *et al.*, 2006; Sears, 2008). Digital lengthening thus originates as an increased rate of chondrocyte formation at the epiphyses of long bones of bats (Farnum *et al.*, 2008a), continuing well after birth at accelerated pace (Farnum *et al.*, 2008b). In turn, forelimb lengthening, which is typical of bats (and some gliders; see Thorington and Heaney, 1981; McKay, 1989), has been experimentally associated with a *Prx1* transcriptional enhancer: *Prx1* expression, which partially accounts for forelimb length differences between mice and bats, maps to just two amino-acid substitutions in PRX1 proteins, although additional, partly redundant enhancers are required (Cretekos *et al.*, 2008).

These profound departures from the mammalian development are not observed in the bat hindlimb: the footplate is: (1) distinctly smaller than the handplate as early as CS 13 (Cretekos *et al.*, 2005; Giannini *et al.*, 2006; Tokita, 2006; Hockman *et al.*, 2008) and this difference greatly increases throughout development (*Prx1* enhancer up-regulated in forelimbs); (2) is symmetrically patterned (*Shh* expressed uniformly and at low levels along the A-P axis; Hockman *et al.*, 2008); (3) apoptosis is not prevented, so interdigital tissue disappears (*Bmp2* not inhibited by *Gre*, *Fgf8* not up-regulated in the interdigital tissue; Hockman *et al.*, 2008; Sears, 2008); and (4) digits do not experience dramatic elongation during development (*Bmp2* not up-regulated in specific metacarpals and phalanges; Sears *et al.*, 2006). However, by CS 16, opposite edges of the footplate expand lengthening the digital rays I and V, which results in all digits of similar length (Hockman *et al.*, 2008) rather than I and V shorter as in other (typical) mesaxonic mammals. Note that in *Onychonycteris* only digit I is shorter (Simmons *et al.*, 2008), which shows that two processes, one for digit I and another for digit V, may be acting in the footplate expansion, with lengthening of digit V predating that of digit I during evolution of the bat foot.

These differences are evidence of critical decoupling of regulatory mechanisms in fore- versus hindlimbs (as is the case of birds by means of selector genes that control limb identity; Logan, 2003). However, bats also exhibit patterns affecting both limbs in similar ways. Timing of major serially homologous ontogenetic events is not forelimb-advanced, as may be expected from the initial size of limb buds, but synchronous instead (based on *Rousettus*; Richardson *et al.*, 2009). Embryologically posterior zeugopod elements of both limbs (i.e., ulna and fibula in fore- and hindlimbs, respectively) exhibit initial mesenchymal condensations similar to those of the embryologically anterior elements (radius and tibia), but both ulna and fibula subsequently recede. Adams (1992) and Sears *et al.* (2007) described the reduction of the ulna in three stages: decelerating growth, becoming ligamentous distally and fusing to the distal diaphysis of radius. *Shh*, as a factor determining A-P patterning, and *Hox* genes, which are essential for the formation of limbs, have been involved in the reduction of "posterior" zeugopod elements (Sears, 2008).

10.4.2   Gross embryology

Macroscopic changes are especially illuminating with regard to development of structures that are key for the understanding of flight evolution, such as patagia. Development of dactilopatagium is radically different from that of the pro-, plagio- and uropatagium, whose underlying molecular mechanisms are yet unknown. I suggest that each of these patagial primordia originate as a dermal appendage, with its complement of muscles, nerves and vessels. The plagiopatagium develops from a digitiform primordium in the caudal edge of the forelimb bud (Figure 10.3; Giannini *et al.*, 2006; Hockman *et al.*, 2009). During development, this primordium reaches the tip of digit V; the m. humeropatagialis ends distally in this primordium (Giannini *et al.*, 2006). Pro- and uro-patagium also develop from primordia that contain patagial muscles (m. occipitopollicalis and m. uropatagialis, respectively) plus nerves and vessels. For instance, the propatagium carries on its leading edge the cephalic vein, a branch of the facial nerve, and nerves from the spinal cord (Thewissen and Babcock, 1992). Mesodermal elements may act as developmental inducers of these patagial tracts. It remains to be investigated whether in gliders the patagia originate as in bats, and if the vascular, neural and muscular associations of patagia are the same. It is significant that in bats (with few

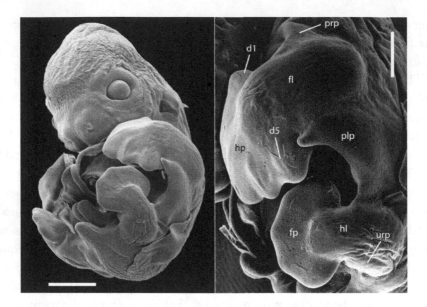

Figure 10.3 Left: scanning electron microscope image of a *Rousettus
amplexicaudatus* embryo at stage CS 16 Late (crown–rump length = 8.5 mm);
scale = 1 mm. Right: detail of the flank of the same specimen showing the
hand- and footplates (note size difference) and patagial primordia; scale =
0.5 mm. Abbreviations: d1, first digit; d5, fifth digit; fl, forelimb; fp, footplate;
hl, hindlimb; hp, handplate; plp, plagiopatagium primordium; prp,
propatagium primordium; urp, uropatagium primordium. Modified from
Giannini *et al.* (2006).

exceptions, e.g., lasiurine vespertilionids) patagial tracts develop as extensive
hairless membranes that never follow the head and trunk in acquiring a dense
hair cover (Giannini *et al.*, 2006), which probably represents a substantial
difference in developmental pathways between bats and gliding mammals. This
primary hairless condition of the primordial patagia is retained in adult bats in
striking contrast with gliders, which exhibit furred membranes.

Gross embryology also reveals the nature of bat limb position. In the adult,
the bat hand is rotated (supinated) 90° with respect to the position of the hand
in quadrupedal mammals (Vaughan, 1959; Simmons, 1994). However, as seen in
ontogenetic series (e.g., Cretekos *et al.*, 2005; Giannini *et al.*, 2006), the bat
hand retains the condition of the handplate (see Figure 10.3); i.e., the hand is
primarily supinated, never experiencing pronation as in quadrupeds. The foot
also skips the pronation process. By contrast, the proximal hindlimb and ankle
joint do depart from the fetal condition by effective 90° rotation of the hip joint
(femur directed laterally or caudally) and reorientation of the ankle joint (foot
pointing laterally or caudally; Vaughan, 1959; Simmons, 1994).

10.5   Evolving the handwing

In bats, high-AR wings obtain primarily from evolving the handwing, with less important contributions from elongated stylo- and zeugopod. As pointed out by Norberg (1986), the evolutionary extension of the handwing simultaneously decreased wing loading (WL), thereby decreasing air speed and enhancing maneuverability, and increased aspect ratio and hence lift-to-drag ratio (see below). An extended handwing is thus essential for powered flight in bats. Lack of fossil gliders related to bats keeps the evolution of the handwing in hypothetical grounds. At any rate, data from fossils, development and comparative anatomy can be articulated in such a way that a cumulative sequence of events is construed to explain the evolution of the handwing in a typical glider. This gliding bat precursor would be in possession of all the other patagial tracts, as is frequently observed in mammals, so the high-AR wings of bats evolves from the body center outwards (i.e., in *distal* direction; Norberg, 1986; *contra* Caple et al., 1983).

10.5.1   Potential evolutionary sequence

Interdigital membrane retention – Starting point of the gliding model (e.g., Smith, 1977; Rayner, 1986; Scholey, 1986a) involves blocking apoptosis in the interdigital tissue of the handplate. A stage represented by the colugo (in which the feet are also webbed).

Digital elongation – A stage unrepresented in known taxa, but a necessary intermediate step between low- and high-AR bat wings (e.g., Smith, 1977; cf. Caple *et al.*, 1983; Scholey, 1986a). Initially, slightly elongated digits may have allowed the hand to function as a winglet or endplate, diminishing wingtip vortices and hence improving performance of the aerofoil, just as in some extant gliders (Stafford, 1999; Bishop, 2008; Jackson and Schouten, in press). Digital elongation, the result of extremely high rates of chondrocyte formation chiefly due to up-regulated *BMPs*, is responsible for much of the increased AR of bat wings, and is a process continued through bat evolution.

Wing folding – A stage achieved by *Onychonycteris*. This novelty, dependent on the evolution of complex collapse-guiding C-MC joints, is key in reducing drag during the upstroke, protecting a long wing from damage and minimizing the physiological impact on metabolic rate of such extended, hairless surface. At the level of prenatal adaptations, wing folding may have been important in facilitating passage of a disproportionate forelimb through the birth canal. The developmental basis of the bat wing-folding mechanism is unknown, but recent research on joint development points to the expression of the fibronectin receptor (α5β1-integrin) and the acting of the Wnt/β-catenin signaling pathway. These play a role in determining the fate of chondrocytes at the control point where these cells decide between the cartilage versus joint differentiation programs, inducing the formation of sinovia and presumably determining the specific patterning of the joint (Garciadiego-Cázares *et al.*, 2004; Guo *et al.*, 2004).

Truncation of claw development – One of the finest instances of gradual, orderly reduction of an A-P-serial structure during evolution. *Onychonycteris* bears fully developed claws in all five hand digits. *Icaronycteris* provides a spectacular example of an intermediate condition (Figure 10.1), with claws in digits I and II and bony knobs in digits III–V. *Archaeonycteris*, *Archaeopteropus* and (primitively) most extant megabats exhibit claws in digits I and II and cartilaginous tips in digits III–V. Extant microbats achieved the maximal reduction of ungual phalanges in all digits but the pollex. As suggested above, heterochrony may suffice to explain the phalanx reduction process. The finding of an atavistic index claw in an individual of a living microbat species (*Lonchophylla robusta*) indicates that expression of claw structures in this digit is normally suppressed by a regulatory mechanism (Tschapka, 2008).

10.6   In the transition to flight

In strong contrast with birds (e.g., Dial *et al.*, 2008), gliding, as opposed to running and jumping, is the transitional mechanism favored by researchers investigating bat flight evolution (Long *et al.*, 2003; Bishop, 2008; Simmons *et al.*, 2008). In the gliding model, or "trees-down" theory, of bat flight evolution, climbing is followed by stages of gliding that are superseded by flapping flight (Smith, 1977; Rayner, 1986; Scholey, 1986a; Norberg, 1990; *contra* Pirlot, 1977; Caple *et al.*, 1983). The transition from climbing to gliding has evolved at least eight times in Mammalia – in the Mesozoic Volaticotheriidae (Meng *et al.*, 2006), three times in Marsupialia and three times in Placentalia (Jackson and Schouten, in press), in addition to the ancestor of bats. The extant diversity of mammalian gliders (66 extant species in 22 genera and 6 independent clades; Jackson and Schouten, in press) suggests that this life form is successful once it evolves. Among these taxa, there are cases of extremely similar non-gliding species that are sister to gliders (e.g., *Gimnobelideus* and *Petaurus*, respectively), and of initial stages of gliding in living forms that are closely related to gliders (*Hemibelideus* and *Petauroides*, respectively; Johnson-Murray, 1987). This suggests that equipping an agile climber with patagia is evolutionarily easy and happened frequently, likely representing most of what it takes to start air transport at low cost. Significantly, many problems of aerial locomotion are already resolved at the gliding stage (*contra* Caple *et al.*, 1983): take-off and landing, turning and braking, are operations these mammals execute in every glide.

The next step, the transition from gliding to flapping flight, has been considered highly problematic (see Bishop, 2008). Bat flight is currently understood in terms of unsteady, vortex-based aerodynamics (e.g., Hedenström *et al.*, 2007; Muijres *et al.*, 2008) and complex wingbeat kinematics (Watts *et al.*, 2001; Riskin *et al.*, 2008, 2010). Flapping flight seems exceedingly demanding without functional adaptations encompassing most organ systems, as attested by the highly derived anatomy and physiology of bats. Nevertheless, preconditions for flapping flight are likely met by mammalian gliders (Bishop, 2008) so stepwise increases in performance between the gliding regime and flapping flight are possible (e.g., Norberg, 1986, 1990; Rayner, 1986). However, "gliding and flapping flight have typically been considered separate adaptive zones, each with its own set of optimization parameters" (Bishop, 2008, p. 154). Thus, two objections have challenged the gliding model: gliders have never been observed to flap (Jackson and Schouten, in press), and bat wings do not seem to make good gliding airfoils. In fact, a catastrophic loss of lift is predicted when flapping low-AR wings (Caple *et al.*, 1993; note that Bishop (2008) pointed out that data in support of this prediction are lacking). At the speeds tested in wind tunnels with simple models, compliant wings with AR of about 2 (and supposedly, also higher AR) would stall at angles of attack around 17°, whereas gliders with lower AR wings typically operate at much higher angles (30–50°; Scholey, 1986b; Jackson, 1999; Bishop, 2008). In other words, intermediate-AR wings apparently exhibit poor aerodynamic performance. This creates a discontinuity problem in the gliding scenario.

The first objection to the gliding model – low-AR wings are not effective for flapping flight – is probably true. In principle, this is problematic because inception of the flight stroke – a precondition of flight in Bishop's (2008) criteria – would be seriously compromised. However, flapping a low-AR (< 2) wing is not a requirement in a gliding scenario. In Rayner's (1986) model, flapping is feasible (in terms of power) only after AR is higher than a certain mass-specific threshold, when it becomes compatible with other parameters, particularly stroke amplitude; e.g., AR > 3 for a mid-sized microbat, *Nyctalus noctula* (mean 27 g, around the size estimated for *Icaronycteris*; Giannini *et al.*, Chapter 16, this volume). If bat precursors were able to sustain a reasonable gliding performance with wings evolving towards high AR (see below), then the inception of the flight stroke is not required at the beginning of the glide-to-flight transition, when it would inevitably fail due to impaired aerodynamic force generation as these wings tend to lose lift capacity (Bishop, 2008). Rather, inception of flight may be placed in the middle of the evolutionary gradient of increasing AR, when conditions for flapping are first met and will continue to improve as the handwing elongates (Rayner, 1986). This requires that bat precursors continued gliding as usual until entering into the "flapping zone" of higher AR. This represents the second objection – that bat wings are not effective gliding aerofoils – but it can be decidedly rejected. Bats can be effectively trained to glide in a wind tunnel (Pennycuick, 1973, 1986); intermittent flight, with regular lapses of gliding, has been reported in *Rhinopoma* (e.g., Rayner, 1986) and studied in detail in *Pipistrellus* (Thomas *et al.*, 1990); and individuals of *Tadarida brasiliensis* have been observed gliding for long stretches at high altitude with strong tail wind (T. H. Kunz, personal communication). In addition, some species of *Pteropus* soar during daytime, with gliding being a major component of soaring (Lindhe-Norberg *et al.*, 2000).

In fact, birds that glide do so with mid-to-high-AR wings ($5 < AR < 15$; Norberg, 1990). Avian wings certainly operate at relatively low angles of attack; by contrast, bat wings have been shown to flap at angles as large as $50°$ or greater when flying at low speeds (e.g., 1.5 m s^{-1}; see fig. S4 in Hedenström *et al.*, 2007). Air speed in a large glider, the colugo, can be as low, with a reported minimum of 0.8 m s^{-1} landing velocity (mean 4.0 m s^{-1}; Byrnes *et al.*, 2008). Interestingly, best glide speeds calculated for flying foxes (*Pteropus*) of comparable size (> 1 kg) were 9.6–11.0 m s^{-1} (mean 10.1 m s^{-1}; Lindhe-Norberg *et al.*, 2000), which is strikingly similar to the reported average asymptotic horizontal velocity of 9.9–10.2 m s^{-1} (mean 10.1 m s^{-1}) for the colugo (Byrnes *et al.*, 2008). In addition, a 120 g megabat (*Rousettus aegyptiacus*) trained to glide in a wind tunnel exhibited local angles of attack varying between 20 and $38°$ along various wing sections at 5.5 m s^{-1} (Pennycuick, 1973), which is around minimum power speed for large bats (Rayner, 1986; Norberg, 1994).

Taken together, these data suggest that mammalian gliders and fliers can operate at comparable aerodynamic regimes with regard to speed and angle of attack. However, the problem of inception of flapping remains. Minimum power speed (easily achieved and surpassed by gliders at any given mass) has been suggested as the easiest "entry point" to introduce wing oscillation (Pennycuick, 1986). Low-amplitude wing oscillations would produce sufficient amounts of useful trust conducive to gradually achieving level flight (Norberg, 1986). In turn, Bishop (2008) proposed incipient flapping as a possible mechanism to effect the transition from launching (possibly impaired in bat precursors) to dropping from a hindlimb-suspended position. Yet another situation is conceivable in which flapping may be initiated just before landing as an aid in breaking. Here bats may generate some lift from incipient wing oscillation in a low-speed situation where AR, oscillation and angle of attack are all compatible, and the aerodynamic force produced would be useful for the immediate purpose: breaking. Subsequently bats may start flapping earlier in the glide path, progressively lowering the angle of attack and air speed as the handwing elongates and aerodynamic efficiency increases, eventually attaining level flight and independence of a landing site fixed upon departure.

10.7 Paleoecology

Some derived bat species make tents or modify bird or termite nests, but bats in general do not build nests themselves, so seek shelter in cavities of various sizes, physical conditions and permanency, as well as in foliage (Kunz and Lumsden, 2003). Roosting ecology is a key aspect of the biology of extant bats; so it was, we can safely assume, during the early history of bats. Bat flight evolved in the greenhouse conditions of the Paleocene-Eocene during which a substantial portion of the globe was covered with forests (Wilf and Labandeira, 1999; Lowenstein and Demicco, 2006). Eocene fossil sites were not the exception (Jepsen, 1970; Habersetzer and Störch, 1987), so trees must have been the dominant elements in the environment populated by bat ancestors. Product of accidental damage, fire, action of organisms, or decay, tree holes and crevices in standing or fallen trees are abundant in forests and may represent the single most important source of shelter for forest bats (Kunz and Lumsden, 2003). Tree cavities provide protection from predators and severe weather, exhibiting remarkable microclimatic properties. In warm regions, temperature inside tree cavities tends to be less variable along the circadian cycle, so tree holes are cooler during the day and warmer during the night than is outside (Dechman *et al.*, 2004). With their extensive, hairless membranes, bat precursors may have faced metabolic stress during resting periods. Just as for modern forest bats, roosting in tree cavities probably was advantageous for bat precursors both during the day (preventing overheating) and night between flights (diminishing heat loss).

10.8 Summary and conclusions

In this chapter I hope to have shown, if anything, that a theory on the origin of bat flight has perhaps greater potential than any other theme in bat biology for integrating results from diverse fields of research, including phylogenetic paleontology, molecular control of organogenesis, gross embryology and development, functional anatomy, mechanics of locomotion, and roosting and trophic ecology. This limited review left aside several aspects that deserve a great deal of attention, including: physiological impact of evolving a large, hairless flight membrane; impact of genomic approaches in understanding the evolution of gene systems that allow bat metabolic performance (e.g., Shen *et al.*, 2010); implications of specific muscle physiology of bats and their ontogeny (e.g., Hermanson, 1998, 2000) in the transition to flight; correspondence between ontogeny of flight (e.g., Elangovan *et al.*, 2004, 2007), proposed adaptive landscapes (Adams, 2000, 2008) and the sequence of morphological changes implied in current phylogenies; the possible connection between rapid growth of long bones (e.g., Farnum *et al.*, 2008a, 2008b) and precociality (e.g., Jepsen, 1970) with morphological changes in evolutionary time; a thorough analysis of aerodynamic models in light of previously unknown morphologies now available from new fossils; and a full-scale allometric analysis of structural change (e.g., of wing development; Kunz and Robson, 1995; Wyant and Adams, 2007) applied to interspecific variation in the phylogeny of bats.

However, as summarized below, a substantial body of knowledge from numerous fields of bat biology exists, and the current challenge consists of elaborating a comprehensive synthesis of all such knowledge, perhaps starting from the oldest fossils, the Early Eocene bats. These bats exhibited a mosaic of derived (e.g., greatly elongated, foldable handwing) vs. primitive wing features (e.g., clawed digits with rigid bony phalanges), as well as embryonic retentions (e.g., hand webbing, supinated hand) vs. divergences from the mammalian developmental standard (e.g., elongated digits). These character states changed rapidly during the course of early bat evolution, but left morphological signatures in fossils that allow an increasingly finer level of detail in reconstructing the transition experienced by bats as the only volant mammals. In this way, fossils reveal not only the morphology of extinct organisms but also the ontogenetic stages attained in diverse developing systems. For instance, the claws of *Onychonycteris* show that this taxon underwent the standard mammalian process of ungual phalanx development, whereas this process is arrested to varying degrees in more derived bats. In general, bats seem particularly useful in showing how paleontological and developmental evidence can be combined in ways that illuminate each other and the evolution of a lineage.

Evolution of the handwing is key in the transition to flapping flight. In this chapter, a cumulative sequence of four major events is proposed: interdigital membrane retention, digital elongation, wing folding and truncation of distal phalanx development. In terms of developmental control, the first two processes, which are linked in bats, are understood at the molecular level (e.g., Hockman *et al.*, 2008) and indicate that specific changes in spatiotemporal expression of regulatory genes played a major role in shaping the bat handwing (as in derived limbs of other mammals; Honeycutt, 2008). The extraordinary rate of ontogenetic digital elongation reported for bats (Farnum *et al.*, 2008a, 2008b) may give clues about the question of Cooper and Tabin (2008) on how fast, and how intergraded, was the evolution of a novelty such as the bat handwing. Pattern of expression of specific receptor proteins and signaling pathways may help explain the structure of C-MC joints, hence the origin of the complex wing-folding mechanism of bats. In turn, a simple allometric process (e.g., deceleration) may account for the truncation of phalanx development.

In spite of the evolution of the handwing and flapping flight, the center of lift of the bat wing is in the plagiopatagium (Swartz *et al.*, 2005) as in all gliders (Jackson and Schouten, in press). The plagiopatagium and the other patagial tracts likely evolved before the handwing and independently of each other. Each tract develops as a separate entity representing a distinct dermal append-age. These appendages appear to be already fully evolved in known fossils, so character states of the patagia, as well as hindlimb and some wing traits (e.g., C-MC-folding, faceting of scapular fossae), indicate that primitive bats were functionally as advanced as some of their modern descendants. However, primitive bats also exhibited functionally poor character states in wing propor-tions (e.g., relatively low AR), muscle attachments (e.g., ribs without laminae) and traits probably affecting aerodynamic performance (e.g., clawed, rigid finger tips). Fossils also show that, as compared to the wing, the hindlimb was more advanced towards the character states exhibited by modern bats, so bat precursors likely used hindlimb suspension before being fully capable of flapping flight (e.g., reconstructions in Smith, 1977).

In a gliding, or Darwinian, model of bat flight origin, the hypothetical transition sequence is from climbing and leaping to gliding to flapping. The first step is unproblematic as there are living representatives documenting the relative ease with which gliding has been achieved in independent clades of arboreal mammals. The transition to flapping flight has been questioned mainly because bats only seldom glide and gliders do not flap their aerofoils, also differing in other aspects like their use of high angles of attack as compared to flying bats. All of these point to a wide performance gap between fliers and gliders (see Bishop, 2008). However, bats are capable of gliding and, at low speed, they flap their wings at angles of attack typical of gliders. Therefore the transition envisioned by Norberg (1986), Rayner (1986), Scholey (1986a) and others, driven by improvements in cost of transport, is much less difficult than currently thought as long as: (1) a performance gradient of increasing AR and decreasing WL is sustained in the evolving wing (see above) and (2) low-amplitude flapping is introduced later in the process, when the wing reaches an intermediate AR and a low WL. Interestingly, the onset of flapping in developing bats occurs when maximum AR and minimum WL combine favorably (Elangovan *et al.*, 2004, 2007; Adams, 2008), so there could be a strong ontogenetic component correlated with evolving powered flight. AR and WL are oversimplified characters which may not describe bat flight with the necessary functional accuracy (Swartz *et al.*, 2005; Riskin *et al.*, 2010); still, these parameters allow depiction of a coarse-grained gradient of possible changes during the evolution of bat flight.

The morphology of bats likely posed problems at the ecophysiological level, particularly the combination of being small-sized homeotherms with extensive, highly vascularized, hairless membranes. While these membranes may be greatly advantageous to dispose of waste heat during the nighttime (Pennycuick, 1986; Neuweiler, 2000), by the same principle of heat exchange bat precursors may have experienced problems of excessive heat loss during resting. Hindlimb suspension using a tendon-locking mechanism may have allowed bats to exploit tree cavities and take advantage of this energy-saving roost, as they do today in forests. Subsequent evolution of wing folding greatly improved control of heat loss at the roost, with additional advantages (e.g., wing protection). Other, somewhat contentious ecological aspects of bat evolution, particularly activity patterns and diet, are resolved by recourse to phylogenetic inertia given that nocturnality and insectivory are plesiomorphic in mammals so no change is required at the root of the chiropteran subtree.

In view of the above exposed, hypothetical bat precursors are here envisioned as relatively small mammals with a forest-glider body plan, but with naked membranes and some behavioral character states of manifest evolutionary inertia (insectivory, nocturnality), initially evolving hindlimb suspension and a tree-cavity dwelling habit, next rapidly evolving size reduction (phyletic nanism), an elongated handwing and a wing-folding mechanism, initiating flapping at some low-speed situation such as breaking before landing, and continuing in a path of increasing AR and wing refinement (e.g., reduction of claws) that allowed an explosive adaptive radiation and invasion of all continents by the Early Eocene.

Finally, if the gliding model is correct, the performance of just any gliding mammal would be enhanced by, for instance, wings evolving towards high AR. So why only bats among mammals developed powered flight when there are many independent gliding lineages, each of which represents a potential "source" of bat-like descendants? The answer probably is historical and lays on the nature of changes required to elaborate the bat handwing. Cretekos *et al.* (2008) suggested that evolutionary changes like forelimb elongation may require many changes in regulatory genes with minor or transient, but cumulative effect, which would produce sudden, large-scale modifications once combined in a bat ancestor. In addition, some of the regulatory effects that generate an elongated and webbed hand are intrinsically antagonistic if not for the rather elaborate patterning of expression in the developing handplate. For instance, an up-regulated *Bmp2* is required to elongate the digits, but its expression causes cell death in the interdigital tissue; so specific patterning and additional regulators are required to produce elongated digits *and* dactilopatagium (see Hockman *et al.*, 2009); notice that a possible regulatory mechanism becomes evident for the colugo, with webbed autopods *but* short digits (e.g., similar to webbed aquatic birds; see Weatherbee *et al.*, 2006). These effects, to be compatible overall, are orchestrated in ways that just do not favor chances of replication in other lineages, even those lineages that could be selected by these changes (e.g., other gliders). Moreover, those extraordinary transformations were not heavily penalized by selection probably because they did not happen in isolation; it is possible that pre-existing habits (e.g., hindlimb suspension and roosting in energy-saving cavities) and additional and timely adaptations (e.g., C-MC wing folding) enabled bat precursors to endure a seemingly difficult transition, ultimately crossing over from gliding to powered flight.

10.9   Acknowledgments

Earlier versions of this chapter, both wider and shallower, were presented as talks in the celebrations of the 200th anniversary of Charles Darwin and 150th anniversary of *The Origin of Species* held at the Universidad de Tucumán, Argentina, and at the National Museum of Natural History, Smithsonian Institution, Washington, DC. I thank the organizers of those events (H. R. Grau and K. M. Helgen, respectively). I am indebted to the editors of this book, Gregg F. Gunnell and Nancy B. Simmons, who invited me to contribute this chapter. Nancy B. Simmons granted access to key specimens at the American Museum of Natural History (AMNH). This chapter greatly benefited from the insightful suggestions from K. L. Bishop, W. C. Lancaster and one anonymous reviewer. This research originated during my Coleman and Vernay postdoctoral fellowships at the AMNH, and subsequently continued with support from CONICET, Argentina, and National Science Foundation AToL Grant 0629959.

10.10   REFERENCES

Adams, R. A. (1992). Comparative skeletogenesis of the forearm of the little brown bat *(Myotis lucifugus)* and the Norway rat *(Rattus norvegicus)*. *Journal of Morphology*, **214**, 251–260.

Adams, R. A. (1998). Evolutionary implications of developmental and functional integration in bat wings. *Journal of Zoology (London)*, **246**, 165–174.

Adams, R. A. (2000). Wing ontogeny, shifting niche dimensions, and adaptive landscapes. In *Ontogeny, Functional Ecology and Evolution of Bats*, ed. R. A. Adams and S. C. Pedersen. Cambridge: Cambridge University Press, pp. 275–316.

Adams, R. A. (2008). Morphogenesis in bat wings: linking development, evolution and ecology. *Cells Tissues Organs*, **187**, 13–23.

Adams, R. A. and Thibaud, K. M. (2000). Ontogeny and evolution of the hindlimb and calcar: assessing phylogenetic trends. In *Ontogeny, Functional Ecology and Evolution of Bats*, ed. R. A. Adams and S. C. Pedersen. Cambridge: Cambridge University Press, pp. 317–332.

Bishop, K. L. (2008). The evolution of flight in bats: narrowing the field of plausible hypotheses. *Quarterly Review of Biology*, **83**, 153–169.

Byrnes, G., Lim, N. T.-L. and Spence, A. J. (2008). Take-off and landing kinetics of a free-ranging gliding mammal, the Malayan colugo (*Galeopterus variegatus*). *Proceedings of the Royal Society of London Series, B*, **275**, 1007–1013.

Calder, W. A. (1996). *Size, Function, and Life History*. New York: Dover Publications.

Caple, G., Balda, R. P. and Willis, W. R. (1983). The physics of leaping animals and the evolution of preflight. *American Naturalist*, **121**, 455–476.

Cooper, K. L. and Tabin, C. J. (2008). Understanding of bat wing evolution takes flight. *Genes and Development*, **22**, 121–124.

Cretekos, C. J., Weatherbee, S. D., Chen, C.-H. *et al.* (2005). Embryonic staging system for the short-tailed fruit bat, *Carollia perspicillata*, a model organism for the mammalian Order Chiroptera, based upon timed pregnancies in captive-bred animals. *Developmental Dynamics*, **233**, 721–738.

Cretekos, C. J., Wang, Y. Green, E. D. *et al.* (2008). Regulatory divergence modifies limb length between mammals. *Genes and Development*, **22**, 141–151.

Darwin, C. R. (1859). *On the Origin of Species by Means of Natural Selection, or the Preservation of Favoured Races in the Struggle for Life*. London: John Murray.

Dechman, D. K. N., Kalko, E. K. V. and Kerth, G. (2004). Ecology of an exceptional roost: energetic benefits could explain why the bat *Lophostoma silvicolum* roosts in active termite nests. *Evolutionary Ecology Research*, **2004**, 1037–1050.

Dial, K. P., Jackson, B. E. and Segre, P. (2008). A fundamental avian wing-stroke provides a new perspective on the evolution of flight. *Nature*, **451**, 985–990.

Eiting, T. P. and Gunnell, G. F. (2009). Global completeness of the bat fossil record. *Journal of Mammalian Evolution*, **16**, 151–173.

Elangovan, V., Raghuram, H., Yuvana Satya Priya, E. and Marimuthu, G. (2004). Wing morphology and flight performance in *Rousettus leschenaulti*. *Journal of Mammalogy*, **85**, 806–812.

Elangovan, V., Yuvana Satya Priya, E., Raghuram, H. and Marimuthu, G. (2007). Wing morphology and flight development in the short-nosed fruit bat *Cynopterus sphinx*. *Zoology*, **110**, 189–196.

Epstein, J. H., Olival, K. J., Pulliam, J. R. C. *et al.* (2009). *Pteropus vampyrus*, a hunted migratory species with a multinational home-range and a need for regional management. *Journal of Applied Ecology*, **46**, 991–1002.

Farnum, C. E., Tinsley, M. and Hermanson, J. W. (2008a). Forelimb versus hindlimb skeletal development in the big brown bat, *Eptesicus fuscus*: functional divergence is reflected in chondrocytic performance in autopodial growth plates. *Cells Tissues Organs*, **187**, 35–47.

Farnum, C. E., Tinsley, M. and Hermanson, J. W. (2008b). Postnatal bone elongation of the Manus versus Pes: analysis of the chondrocytic differentiation cascade in *Mus musculus* and *Eptesicus fuscus*. *Cells Tissues Organs*, **187**, 48–58.

Fenton, M. and Crerar, L. (1984). Cervical vertebrae in relation to roosting posture in bats. *Journal of Mammalogy*, **65**, 395–403.

Fenton, M. B., Audet, D., Obrist, M. K. and Rydell, J. (1995). Signal strength, timing, and self-deafening: the evolution of echolocation in bats. *Paleobiology*, **21**, 229–242.

Ferrarezi, H. and Giménez, E. A. (1996). Systematic patterns and the evolution of feeding habits in Chiroptera (Mammalia: Archonta). *Journal of Comparative Biology*, **1**, 75–95.

Garciadiego-Cázares, D., Rosales, C., Katoh, M. and Chimal-Monroy, J. (2004). Coordination of chondrocyte differentiation and joint formation by α5β1 integrin in the developing appendicular skeleton. *Development*, **131**, 4735–4742.

Giannini, N. P., Goswami, A. and Sánchez-Villagra, M. (2006). Development of integumentary structures in *Rousettus amplexicaudatus* (Mammalia: Chiroptera: Pteropodidae) during late-embryonic and fetal stages. *Journal of Mammalogy*, **87**, 993–1001.

Griffin, D. R. (1958). *Listening in the Dark*. New Haven, CT: Yale University Press.

Gunnell, G. F. and Simmons, N. B. (2005). Fossil evidence and the origin of bats. *Journal of Mammalian Evolution*, **12**, 209–246.

Guo, X., Day, T. F., Jiang, X. *et al.* (2004). Wnt/β-catenin signaling is sufficient and necessary for synovial joint formation. *Genes and Development*, **18**, 2404–2417.

Habersetzer, J. and Storch, G. (1987). Klassifikation und funktionelle Flügelmorphologie paläogener Fledermäuse (Mammalia, Chiroptera). *Courier Forschungsinstitut Senckenberg*, **91**, 11–150.

Habersetzer, J., Richter, G. and Storch, G. (1994). Paleoecology of early middle Eocene bats from Messel, FRG. Aspects of flight, feeding and echolocation. *Historical Biology*, **8**, 235–260.

Hedenström, A., Johansson, L. C., Wolf, M. *et al.* (2007). Bat flight generates complex aerodynamic tracks. *Science*, **316**, 894–897.

Hermanson, J. W. (1998). Chiropteran muscle biology. A perspective from molecules to function. In *Bat Biology and Conservation*, ed. T. H. Kunz and P. A. Racey. Washington, DC: Smithsonian Institution Press, pp. 127–139.

Hermanson, J. W. (2000). Ontogeny of flight muscles: an evolutionary model. In *Ontogeny, Functional Ecology and Evolution of Bats*, ed. R. A. Adams and S. Pedersen. Cambridge: Cambridge University Press, pp. 333–361.

Hockman, D., Cretekos, C. J., Mason, M. K. *et al.* (2008). A second wave of *Sonic Hedgehog* expression during the development of the bat limb. *Proceedings of the National Academy of Sciences, USA*, **105**, 16982–16987.

Hockman, D., Mason, M. K., Jacobs, D. S. and Illing, N. (2009). The role of early development in mammalian limb diversification: a descriptive comparison of early limb development between the Natal long-fingered bat (*Miniopterus natalensis*) and the mouse (*Mus musculus*). *Developmental Dynamics*, **238**, 965–979.

Honeycutt, R. (2008). Small changes, big results: evolution of morphological discontinuity in mammals. *Journal of Biology*, **7**, 9, doi:10.1186/jbiol71.

Jackson, S.M. (1999). Glide angle in the genus *Petaurus* and a review of gliding in mammals. *Mammal Review*, **30**, 9–30.

Jackson, S.M. and Schouten, P. (in press). *Gliding Mammals*. Melbourne: CSIRO Publishing.

Jepsen, G.L. (1966). Early Eocene bat from Wyoming. *Science*, **154**, 1333–1339.

Jepsen, G.L. (1970). Bat origins and evolution. In *Biology of Bats*, vol. 1, ed. W.A. Wimsatt. New York: Academic Press, pp. 1–64.

Johnson-Murray, J.L. (1987). The comparative myology of the gliding membranes of *Acrobates*, *Petauroides* and *Petaurus* contrasted with the cutaneous myology of *Hemibelideus* and *Pseudocheirus* (Marsupialia: Phalangeridae) and with selected gliding Rodentia (Sciuridae and Anamoluridae). *Australian Journal of Zoology*, **35**, 101–113.

Jones, G. and Teeling, E.C. (2006). The evolution of echolocation in bats. *Trends in Ecology and Evolution*, **21**, 149–156.

Kunz, T.H. and Lumsden, L.F. (2003). Ecology of cavity and foliage roosting bats. In *Bat Ecology*, ed. T.H. Kunz and M.B. Fenton. Chicago, IL: University of Chicago Press, pp. 3–89.

Kunz, T.H. and Robson, S.K. (1995). Postnatal growth and development in the Mexican free-tailed bat, *Tadarida brasiliensis mexicana*: birth size, growth rates and age estimation. *Journal of Mammalogy*, **76**, 769–783.

Lancaster, W.C., Henson, Jr., O.W. and Keating, A.W. (1995). Respiratory muscle activity in relation to vocalization in flying bats. *Journal of Experimental Biology*, **198**, 175–191.

Lindhe-Norberg, U.M., Brooke, A.P. and Trewhella, W.J. (2000). Soaring and non-soaring bats of the family pteropodidae (flying foxes, *Pteropus* spp.): wing morphology and flight performance. *Journal of Experimental Biology*, **203**, 651–664.

Long, C.A., Zhang, G.P., George, T.F. and Long, C.F. (2003). Physical theory, origin of flight, and a synthesis proposed for birds. *Journal of Theoretical Biology*, **224**, 9–26.

Logan, M. (2003). Finger or toe: the molecular basis of limb identity. *Development*, **130**, 6401–6410.

Lowenstein, T.K. and Demicco, R.V. (2006). Elevated Eocene atmospheric CO_2 and its subsequent decline. *Science*, **313**, 1928.

McKay, G.M. (1989). Family Petauridae. In *Fauna of Australia, Vol. 1B: Mammalia*, ed. D.W. Walton and B.J. Richardson. Canberra: Australian Government Publishing Service, pp. 665–678.

Meng, J., Hu, Y., Wang, Y., Wang, X. and Li, C. (2006). A Mesozoic gliding mammal from northeastern China. *Nature*, **444**, 889–893.

Muijres, F.T., Johansson, L.C., Barfield, R. *et al.* (2008). Leading-edge vortex improves lift in slow-flying bats. *Science*, **319**, 1250–1253.

Neuweiler, G. (2000). *The Biology of Bats*. Oxford: Oxford University Press.

Norberg, U. M. (1969). An arrangement giving a stiff leading edge to the hand wing of bats. *Journal of Mammalogy*, **50**, 766–770.

Norberg, U. M. (1972). Functional osteology and myology of the wing of the dog-faced bat *Rousettus aegyptiacus* (É. Geoffroy) (Mammalia, Chiroptera). *Zeitschrift für Morphologie und Okologie der Tiere*, **73**, 1–44.

Norberg, U. M. (1986). On the evolution of flight and wing forms in bats. In *Bat Flight/Fledermausflug, BIONA Report 5*, ed. W. Nachtigall. Stuttgart: Gustav Fischer, pp. 13–26.

Norberg, U. M. (1990). *Vertebrate Flight: Mechanics, Physiology, Morphology, Ecology and Evolution*. Berlin: Springer-Verlag.

Norberg, U. M. (1994). Wing design, flight performance, and habitat use in bats. In *Ecological Morphology: Integrative Organismal Biology*, ed. P. C. Wainright and M. Reilly. Chicago, IL: University of Chicago Press, pp. 205–239.

Norberg, U. M. (1998). Morphological adaptations for flight in bats. In *Bat Biology and Conservation*, ed. T. H. Kunz and P. A. Racey. Washington, DC: Smithsonian Institution Press, pp. 93–108.

Novacek, M. J. (1985). Evidence for echolocation in the oldest known bats. *Nature*, **315**, 140–141.

O'Donnell, C. F. J. (2001). Home range and use of space by *Chalinolobus tuberculatus*, a temperate rainforest bat from New Zealand. *Journal of Zoology (London)*, **253**, 253–264.

Papadimitriou, H. M., Swartz, S. M. and Kunz, T. H. (1996). Ontogenetic and anatomic variation in mineralization of the wing skeleton of the Mexican freetailed bat, *Tadarida brasiliensis*. *Journal of Zoology*, **240**, 411–426.

Pennycuick, C. J. (1973). Wing profile shape in a fruit-bat gliding in a wind tunnel, determined by photogrammetry. *Periodicum Biologorum*, **75**, 77–82.

Pennycuick, C. J. (1986). Mechanical constraints on the evolution of flight. In *The Origin of Birds and the Evolution of Flight*, ed. K. Padian. *Memoirs of the California Academy of Sciences*, **8**, 83–98.

Pirlot, P. (1977). Wing design and the origin of bats. In *Major Patterns in Vertebrate Evolution*, ed. M. K. Hecht, P. C. Goody and B. M. Hecht. New York: Plenum Press, pp. 375–410.

Rayner, J. (1986). Vertebrate flapping mechanics and aerodynamics, and the evolution of flight in bats. In *Bat Flight/Fledermausflug, BIONA Report 5*, ed. W. Nachtigall. Stuttgart: Gustav Fischer, pp. 27–74.

Reilly, S. M., Wiley, E. O. and Meinhardt, D. J. (1997). An integrative approach to heterochrony: the distinction between interspecific and intraspecific phenomena. *Biological Journal of the Linnean Society*, **60**, 119–143.

Richardson, M. K., Gobes, S. M. H., Van Leeuwen, A. C. *et al.* (2009). Heterochrony in limb evolution: developmental mechanisms and natural selection. *Journal of Experimental Zoology (Molecular Development and Evolution)*, **312B**, 639–664.

Riskin, D. K., Willis, D. J., Iriarte-Díaz, J. *et al.* (2008). Quantifying the complexity of bat wing kinematics. *Journal of Theoretical Biology*, **254**, 604–615.

Riskin, D. K., Iriarte-Díaz, J. Middleton, K. M., Breuer, K. S and Swartz, S. M. (2010). The effect of body size on the wing movements of pteropodid bats, with insights into thrust and lift production. *Journal of Experimental Biology*, **213**, 4110–4122.

Roberts, D. (1974). Structure and function of the primate scapula. In *Primate Locomotion*, ed. F. A. Jenkins. New York: Academic Press, pp. 171–200.

Schliemann, H. and Schlosser-Sturm, E. (1999). The shoulder joint of the Chiroptera – morphological features and functional significance. *Zoologischer Anzeiger*, **238**, 75–86.

Schnitzler, H.-U. and Kalko, E. K. V. (2001). Echolocation by insect eating bats. *Bioscience*, **51**, 557–569.

Schnitzler, H.-U., Moss, C. F. and Denzinger, A. (2003). From spatial orientation to food acquisition in echolocating bats. *Trends in Ecology and Evolution*, **18**, 386–394.

Scholey, K. (1986a). The evolution of flight in bats. In *Bat Flight/Fledermausflug, BIONA Report 5*, ed. W. Nachtigall. Stuttgart: Gustav Fischer, pp. 1–12.

Scholey, K. (1986b). The gliding and climbing locomotion of the giant red flying squirrel *Petaurista petaurista*. In *Bat Flight/Fledermausflug, BIONA Report 5*, ed. W. Nachtigall. Stuttgart: Gustav Fischer, pp. 187–204.

Sears, K. E. (2008). Molecular determinants of bat wing development. *Cells Tissues Organs*, **187**, 6–12.

Sears, K. E., Behringer, R. R., Rasweiler, IV, J. J and Niswander, L. A. (2006). Development of bat flight: morphologic and molecular evolution of bat wing digits. *Proceedings of the National Academy of Sciences, USA*, **103**, 6581–6586.

Sears, K. E., Behringer, R. R., Rasweiler, IV, J. J and Niswander, L. A. (2007). The evolutionary and developmental basis of parallel reduction in mammalian zeugopod elements. *American Naturalist*, **169**, 105–117.

Shen, Y.-Y., Liang, L., Zhu, Z.-H. *et al.* (2010). Adaptive evolution of energy metabolism genes and the origin of flight in bats. *Proceedings of the National Academy of Sciences, USA*, **107**, 8666–8671.

Simmons, N. B. (1994). The case for chiropteran monophyly. *American Museum Novitates*, **3103**, 1–54.

Simmons, N. B. (1995). Bat relationships and the origin of flight. In *Ecology, Evolution, and Behaviour of Bats, Symposia of the Zoological Society of London*, vol. 67, ed. P. A. Racey and S. M. Swift. New York: Oxford University Press, pp. 27–43.

Simmons, N. B. and Geisler, J. H. (1998). Phylogenetic relationships of *Icaronycteris*, *Archaeonycteris*, *Hassianycteris*, and *Palaeochiropteryx* to extant bat lineages, with comments on the evolution of echolocation and foraging strategies in Microchiroptera. *Bulletin of the American Museum of Natural History*, **235**, 1–182.

Simmons, N. B. and Quinn, T. H. (1994). Evolution of the digital tendon locking mechanism in bats and dermopterans: a phylogenetic perspective. *Journal of Mammalian Evolution*, **2**, 231–254.

Simmons, N. B., Seymour, K. L., Habersetzer, J. and Gunnell, G. F. (2008). Primitive early Eocene bat from Wyoming and the evolution of flight and echolocation. *Nature*, **451**, 818–821.

Simmons, N. B, Seymour, K. L, Habersetzer, J. and Gunnell, G. F. (2010). Inferring echolocation in ancient bats. *Nature*, **466**, E8–E10.

Smith, J. D. (1977). Comments on flight and the evolution of bats. In *Major Patterns in Vertebrate Evolution*, ed. M. K. Hecht, P. C. Goody and B. M. Hecht. New York: Plenum Press, pp. 427–437.

Speakman J. R. (1993). The evolution of echolocation for predation. In *Mammals as Predators. Symposia of the Zoological Society of London 65*, ed. N. Dunstone and M. L. Gorman. New York: Oxford University Press, pp. 39–63.

Speakman, J. R. (2001). The evolution of flight and echolocation in bats: another leap in the dark. *Mammal Review*, **31**, 111–130.

Speakman, J. R. and Racey, P. A. (1991). No cost of echolocation for bats in flight. *Nature*, **350**, 421–423.

Stafford, B. J. (1999). Taxonomy and ecological morphology of the flying lemurs (Dermoptera, Cynocephalidae). Unpublished Ph.D. thesis, City University of New York.

Suthers, R. A., Thomas, S. P. and Suthers, B. J. (1972). Respiration, wing-beat and ultrasonic pulse emission in an echolocating bat. *Journal of Experimental Biology*, **56**, 37–48.

Swartz, S. M. (1997). Allometric patterning in the limb skeleton of bats: implications for the mechanics and energetics of powered flight. *Journal of Morphology*, **234**, 277–294.

Swartz, S. M. (1998). Skin and bones. Functional, architectural, and mechanical differentiation in the bat wing. In *Bat Biology and Conservation*, ed. T. H. Kunz and P. A. Racey. Washington, DC: Smithsonian Institution Press, pp. 109–126.

Swartz, S. M. and Middleton, K. M. (2008). Biomechanics of the bat limb skeleton: scaling, material properties and mechanics. *Cell Tissues Organs*, **187**, 59–84.

Swartz, S. M., Bennett, M. B. and Carrier, D. R. (1992). Wing bone stresses in free flying bats and the evolution of skeletal design for flight. *Nature*, **359**, 726–729.

Swartz, S. M., Groves, M. S., Kim, H. D. and Walsh, W. R. (1996). Mechanical properties of bat wing membrane skin. *Journal of Zoology*, **239**, 357–378.

Swartz, S. M., Bishop, K. and Aguirre, M.-F. I. (2005). Dynamic complexity of wing form in bats: implications for flight performance. In *Functional and Evolutionary Ecology of Bats*, ed. A. Zubaid, G. F. McCracken and T. H. Kunz. Oxford: Oxford University Press, pp. 110–130.

Teeling, E. C. (2009). Hear, hear: the convergent evolution of echolocation in bats? *Trends in Ecology and Evolution*, **24**, 351–354.

Teeling, E. C., Springer, M. S., Madsen, O. *et al.* (2005). A molecular phylogeny for bats illuminates biogeography and the fossil record. *Science*, **307**, 580–584.

Tejedor, M. F., Czaplewski, N. J., Goin, F. J. and Aragón, E. (2005). The oldest record of South American bats. *Journal of Vertebrate Paleontology*, **25**, 990–993.

Thewissen, J. G. M. and Babcock, S. K. (1992). Distinctive cranial and cervical innervation of wing muscles: new evidence for bat monophyly. *Science*, **251**, 934–936.

Thomas, A. L. R., Jones, G., Rayner, J. M. V. and Hughes, P. M. (1990). Intermittent gliding flight in the pipistrelle bat (*Pipistrellus pipistrellus*) (Chiroptera: Vespertilionidae). *Journal of Experimental Biology*, **149**, 407–416.

Thomas, S. P. (1987). The physiology of bat flight. In *Recent Advances in the Study of Bats*, ed. M. B. Fenton, P. Racey and R. M. V. Rayner. Cambridge: Cambridge University Press, pp. 75–99.

Thorington, Jr., R. W. and Heaney, L. R. (1981). Body proportions and gliding adaptations of flying squirrels (Petauristinae). *Journal of Mammalogy*, **62**, 101–114.

Tokita, M. (2006). Normal embryonic development of the Japanese pipistrelle, *Pipistrellus abramus*. *Zoology*, **109**, 137–147.

Tschapka, M. (2008). Rudimentary finger claws in a flower-visiting phyllostomid bat. *Acta Chiropterologica*, **10**, 177–178.

Vaughan, T. A. (1959). Functional morphology of three bats: *Eumops*, *Myotis*, *Macrotus*. *University of Kansas Publications, Museum of Natural History*, **12**, 1–153.

Vaughan, T. A. (1970). The skeletal system. In *Biology of Bats 1*, ed. W. A. Wimsatt. New York: Academic Press. pp. 98–139.

Veselka, N., McErlain, D. D., Holdsworth, D. W. *et al.* (2010). A bony connection signals laryngeal echolocation in bats. *Nature*, **463**, 939–942.

Watts, P., Mitchell, E. J. and Swartz, S. M. (2001). A computational model for estimating the mechanics of horizontal flapping flight in bats: model description and validation. *Journal of Experimental Biology*, **204**, 2873–2898.

Weatherbee, S. D., Behringer, R. R., Rasweiler, J. J. and Niswander, L. A. (2006). Interdigital webbing retention in bat wings illustrates genetic changes underlying amniote limb diversification. *Proceedings of the National Academy of Sciences, USA*, **103**, 15103–15107.

Wilf, P. and Labandeira, C. C. (1999). Response of plant-insect associations to Paleocene-Eocene warming. *Science*, **284**, 2153–2156.

Wyant, K. A. and Adams, R. A. (2007). Prenatal growth and development in the Angolan free tailed bat, *Mops condylurus* (Chiroptera: Molossidae). *Journal of Mammalogy*, **88**, 1248–1251.

11

Molecular time scale of diversification of feeding strategy and morphology in New World Leaf-Nosed Bats (Phyllostomidae): a phylogenetic perspective

ROBERT J. BAKER, OLAF R. P. BININDA-
EMONDS, HUGO MANTILLA-MELUK, CALVIN A.
PORTER AND RONALD A. VAN DEN BUSSCHE

11.1 Introduction

Diversification of feeding strategies within each of the 19 chiropteran families (Hoofer and Van Den Bussche, 2003; Van Den Bussche and Hoofer, 2004; Simmons, 2005) typically is limited to one (13 families) or two (five families) food sources. The family Phyllostomidae, however, represents an exception to this pattern with six distinct feeding strategies: sanguivory, insectivory, frugivory, nectivory, carnivory (feeding on vertebrates) and omnivory.

Among families of bats, phyllostomids comprise the largest number of genera (56) and the third largest number of species (160+) (Simmons, 2005). They are distributed throughout tropical and subtropical regions of North and South America and have been highly successful in exploiting a diverse array of life-history strategies. Included among its members are three species of obligate sanguivores, a feeding strategy unknown in vertebrates other than fish (Figure 11.1). Among phyllostomids additional examples of feeding specialization exist, including subsisting exclusively on insects, as well as primarily on fruit, nectar, frogs, rodents and other vertebrates. Such specializations are remarkable when viewed in the context of the concomitant suite of adaptations associated with the sensory apparatus, locomotion, digestion, dentition, kidney function and reproduction, among others (Griffiths, 1982; Greenhall and Schmidt, 1988: Fleming *et al.*, 2005) that must be favored by directional natural selection for successful exploitation of new ecological opportunities. No other clade of mammals with roots in the Eocene displays such radical evolutionary modifications.

Evolutionary History of Bats: Fossils, Molecules and Morphology, ed. G. F. Gunnell and N. B. Simmons. Published by Cambridge University Press. © Cambridge University Press 2012.

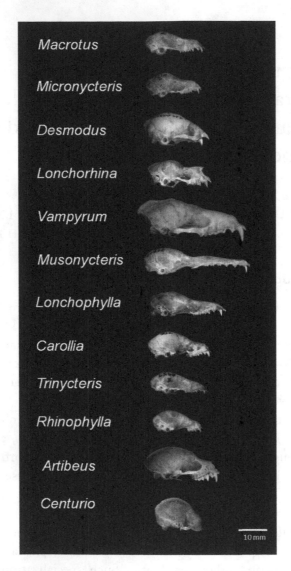

Figure 11.1 Crania of 12 genera representing the 11 subfamilies of Phyllostomidae (see Baker *et al.*, 2003). Shape and morphology of crania provide a perspective of the magnitude of variation present in this family. *Macrotus, Micronycteris, Lonchorhina* and *Trinycteris* are primarily insectivores, but take some fruit. *Desmodus* feeds on blood. *Phyllostomus* and *Carollia* are omnivores and primarily eat fruit, but also insects and nectar. *Musonycteris* and *Lonchophylla* are primarily nectivores; *Rhinophylla* and *Artibeus* are primarily frugivores, but take some insects; *Centurio* is an obligate frugivore.

Understanding the inter-relationships of genomics, ecosystem dynamics, and morphological and physiological adaptation associated with different feeding strategies in phyllostomid bats has implications to furthering our understanding of the mode and tempo of the evolution of adaptive radiation. For example, there has been a series of proposals on the genetic mechanisms that drive morphological and physiological change. Proposed mechanisms (modified from Sutter *et al.*, 2007) include demographic features producing inbreeding resulting in fixation of chromosomal rearrangements (Wilson *et al.*, 1975); mobile DNA altering gene expression patterns (Pascale *et al.*, 1993; Furano *et al.*, 1994); increased recombination or mutation rates (Thompson, 1917; Eldredge and Gould, 1972; Wilson *et al.*, 1974); a unique role of short repeat loci near genes (Eldredge and Gould, 1972); expansion of specific interspersed nuclear elements (Carroll, 2000); timing variation in regulatory genes (Kirschner and Gerhart, 1998; Wren *et al.*, 2000); readily altered developmental programs (Darwin, 1859; Kirschner and Gerhart, 1998); gene duplication (Kawasaki *et al.*, 2007; Ohno *et al.*, 1968); gene recruitment to new tissues/cell expression sites (Phillips *et al.*, 1993); gene sharing (Phillips *et al.*, 1993; Phillips, 1996). This research focus has seen increasing attention of late, especially with regard to the relative importance of mutational vs. regulatory changes, as the main driving force behind evolutionary change (Hoekstra and Coyne, 2007).

Most of the experimental studies of rapid morphological evolution are based on cultivars, lab animals or human pets that have been analyzed in the context of artificial selection, where genetic mutations have a disproportionate probability of survival, reproduction (Fondon and Garner, 2004, 2007; Sutter *et al.*, 2007) and relative fitness. Artificial selection has been valuable in understanding the potential for genetic change and has been a common theme since Darwin's *The Origin of Species* (Darwin, 1859). However, to better understand the evolution of biodiversity and adaptive radiation, more model systems are needed to study the significance of the above proposed mechanisms for facilitating rapid morphological and physiological change under the constraints and rigors of natural selection. With the advent of genome sequencing, phylogenetic comparative methods and measures of magnitude of morphological and physiological change, the significance of these proposed mechanisms can be tested by mapping the presence-absence of each of their expected genetic footprints onto a phylogenetic tree, together with multiple examples of character stasis and change. We propose that phyllostomid bats represent such a model system.

The ever-increasing application of whole-genome sequencing, bioinformatics, candidate gene identification etc. will make it increasingly likely to determine what genes or combination of genes dictate specific adaptations and how genome organization can facilitate the extreme examples of adaptational change. Furthermore, a perspective of geological time will provide insight into the temporal requirements for transformation (e.g., of a generalized insectivorous bat into respective lineages of bats specialized for frugivory, carnivory or sanguivory), as well as providing insights into past ecological and environmental conditions and events that existed at the time the feeding strategies evolved.

The study of when and where evolutionary change occurred can only be accomplished within a robust phylogenetic framework, providing a priori knowledge of the evolutionary relationships of the group in question and thus the order of diversification of the focal trait(s). Numerous studies (reviewed in Wetterer *et al.*, 2000) inferred phyllostomid relationships for more than a century. However, prior to the molecular studies, trees derived from external and skeletal morphology showed either little agreement among studies and/or a lack of resolution. Two studies examined the origin and diversification of feeding strategies in phyllostomid bats within a phylogenetic context (Ferrarezzi and Gimenez, 1996; Wetterer *et al.*, 2000). Although these studies utilized the most robust hypotheses of phyllostomid phylogeny known at that time, both were limited by a lack of statistically supported resolution for the deep branching order within the family.

Since then, two studies have examined variation in DNA sequences of the nuclear protein-coding *RAG2* gene (1.3 kilobases; Baker *et al.*, 2000) and mitochondrial ribosomal genes (2.6 kilobases; Baker *et al.*, 2003). Results from separate analysis of these two unlinked data sets were highly congruent, yet radically different from all previous morphologically based estimates of phyllostomid relationships. Furthermore, these genetic data provide well-supported resolution to the primary divergences within the family, and thus to the order of diversification among phyllostomid feeding strategies. A study by Datzmann *et al.* (2010) using sequences involving ten 396 base pairs from an additional set of genes (von Willebrand factor (*vwf*), recombination activating gene 2 (*rag2*), axonιι of the breast cancer susceptibility gene (*brca1*), non-coating nuclear loci of the phospholipase C beta 4 gene (*plbc4*) and short intron of the phospheonopyruvate carbonxykinase gene (*pepck*), a mitochondrial fragment of NADH (*ndι*) and tRNA Valin, plus published sequences of five mitochondrial loci (COι, Cytb, 12S rRNA, 16S rRNA and tRNA Valin), produced a phylogenetic tree for 37 species of 29 genera. The focus of their paper was the origin of nectar feeding, but their analysis supported the general branching order of clades in phyllostomid bats shown in Figure 11.2.

In this chapter, we use the molecular phylogenetic hypothesis of Baker *et al.* (2003): (i) to estimate geologic time of shared ancestry within phyllostomid clades and (ii) to re-examine the origin and diversification from strict insectivory to omnivory, carnivory, nectivory, frugivory and sanguivory. Our

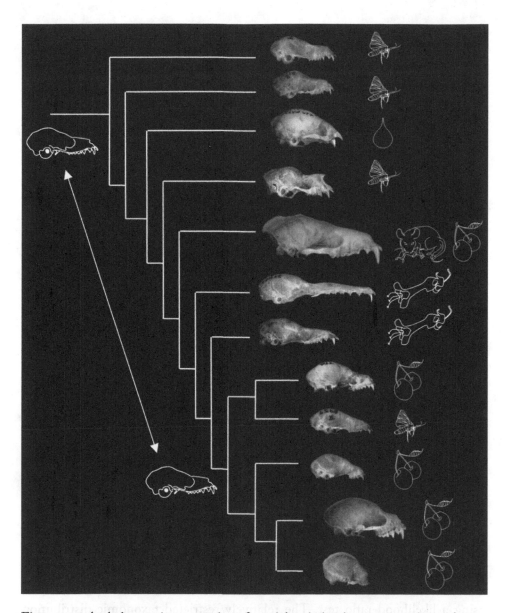

Figure 11.2 A phylogenetic perspective of cranial variation in extant species at the terminal branches of the tree and a stylized image of a *Macrotus-Micronycteris-Glyphonycteris*-like common ancestor for nodes 1–8. That a *Macrotus-Micronycteris-Glyphonycteris*-like bat that was primarily insectivorous, but took some plant material gave origin to the highly derived respective feeding strategies in the Phyllostomidae is hypothesized based on parsimony. Icons to the right of skulls represent the primary food of each extant genus depicted in the pictures: insectivory (moth); sanguivory (droplet); vertebrates (opossum); nectivory (flower); frugivory (fruit).

format is (1) to estimate the primitive condition for ancestors at basal nodes, in a time order of establishment of clades that gave rise to stasis and change; (2) to establish the nature of clades that result in different specific feeding strategies; and (3) to describe the ecological conditions that existed during the origin of clades that exploited the different feeding strategies.

11.2   Methods and materials

In the following discussion, we follow the feeding habits based on Ferrarezzi and Gimenez (1996) as modified by Wetterer *et al.* (2000, pp. 164–165). Clades in Figure 11.3 are numbered using Baker *et al.* (2003: 9–10, table 1, and figure 5). Outgroup clades in Figure 11.3 are numbered with letters A–E. Specimens examined and Genbank sequences utilized are listed in the appendix of Baker *et al.* (2003). Divergence times were determined using the relDate procedure (Bininda-Emonds *et al.*, 2007), whereby sequence data for genes (nDNA: *ADRA2B*, *RAG1*, *RAG2* and *VWF*; mtDNA: *MT-CYB*, *MT-ND1*, *MT-TP* (tRNA-proline), and the ribosomal gene sequence 12S rRNA, MT-TV (tRNA-valine) and 16S rRNA) were fitted to the tree topology of Baker *et al.* (2003) under the assumption of a local molecular clock (see Purvis, 1995) and calibrated using the fossils employed by Jones *et al.* (2005) to estimate the geological times of divergence for the families of bats. Using the same priors as Jones *et al.* (2005), there is a link between the ages in their tree of bat families and dates in our tree of phyllostomid bats. Genetic data were derived from the data sets of Baker *et al.* (2003) and Bininda-Emonds *et al.* (2007).

The optimal model of evolution was inferred for all genes under the AIC as implemented in ModelTEST v3.6 (Posada and Crandall, 1998) in combination with PAUP* 4.0b10 (Swofford, 2002), with the reference tree being the Baker *et al.* (2003) topology pruned to the specific taxon set present for each gene (in place of the default NJ tree). At the same time, the applicability of a global molecular clock for each gene was investigated in PAUP* for the optimal model using a likelihood-ratio test. The relatively restricted taxonomic distribution meant that a global clock could not be rejected for most genes, the exceptions being *RAG2*, *VWF* and the ribosomal gene sequence.

Thereafter, the sequence data for each gene were fitted to the Baker *et al.* (2003) topology (again, pruned to only the relevant species) according to the optimal model of evolution under a maximum-likelihood criterion using PAUP*. Following Purvis (1995) and Bininda-Emonds *et al.* (2007), the relative branch lengths for each relevant branch on the Baker *et al.* (2003) topology were determined individually for each gene tree using the Perl script relDate v2.3.

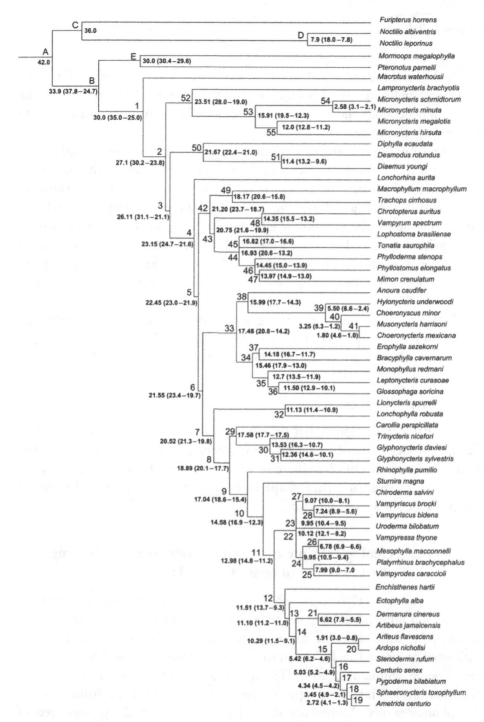

Figure 11.3 Estimates of geological age mapped on phylogenetic tree of mitochondrial and nuclear genes after Baker *et al.* (2003). Geological time-frame estimates are firsts with confidence lower and upper limits in the parenthesis. Outgroups (non-phyllostomids) are labeled A through E, and nodes (1–55) for phyllostomids follow the numbering system in figure 5a of Baker *et al.* (2003, p. 10). Branch lengths reflect genetic distance scores in Baker *et al.* (2003) rather than scores of the geological age. Confidence limits for nodes A and C are unavailable.

Using this method, the age of a node is taken to be some percentage of the age of an ancestral node based on the height of the node relative to that of the ancestral node. Only the gene trees for the clock-like genes were considered to be rooted at this stage.

The relative branch lengths were then calibrated against a set of eight fossil dates with the initial divergence date for any given node being the maximum of either the fossil date or median of all fossil plus molecular date estimates (i.e., the fossil date acted as a minimum age constraint). Upper and lower bounds on these estimates were obtained from the 95% confidence intervals of all individual gene and/or fossil estimates for that node. Finally, any negative branch lengths arising through this procedure (e.g., due to conflict between date estimates from different genes) were corrected for, using the Perl script chronoGrapher v1.3.3.

More details regarding this dating procedure, including its strengths and weaknesses with respect to other relaxed molecular clock methods (recently reviewed in Renner, 2005) can be found in Bininda-Emonds *et al.* (2007).

11.3   Results

Estimates of geological time for each node are shown in Figure 11.3. Feeding strategies associated with various members of respective clades and other characteristics of diversification are reviewed in Table 11.1. Examples of differences in level in cranial shape in extant forms and proposed common ancestors are shown in Figure 11.2.

11.4   Geological time frame and development of feeding characteristics in respective clades

Teeling *et al.* (2005) used a phylogenetic tree generated from both nuclear and mitochondrial genes to establish the deep branching nodes with a molecular time scale for the evolution of bats. The origin of bats was estimated to be at the KT boundary, and diversification of families was in the Eocene (Simmons, 2005; Teeling *et al.*, 2005). The molecular gene tree was divided into four superfamilies, Rhinolophoidea, Emballonuroidea, Noctilionoidea and Vespertilionoidea. By contrast, the dated supertree of Bininda-Emonds *et al.* (2007), which is based on the bat supertree of Jones *et al.* (2002), pushes both the origin of bats (84.8 mya) and the first, basal split of the crown group (71.2 mya) well into the Cretaceous.

For this chapter, it is valuable to analyze the evolution of feeding habits in an explicit phylogenetic framework to better understand what feeding habits the

Table 11.1 Characteristics of nine clades of phyllostomid bats that are hypothesized to have evolved from a common ancestor with *Micronycteris*-like characteristics as present in *Macrotus*, *Micronycteris* and *Glyphonycteris*. Clade numbers are from node numbers in Figure 11.2, where respective clades are separated from the remainder of the family. Intraclade age is defined as the time that the clade separated from all other phyllostomid bats until the initial divergence within that clade of extant genera present in that clade. Change from common ancestor is based on an overview of general knowledge of the bats of the clade.

Clade	Food Source	Age (mya)	Intraclade age (mya)	Time in common ancestor	No. of genera	No. of species	Change from ancestor
1	Primarily insects, some plant	29.5	n/a	n/a	1	2	Minimal
2	Primarily insects, some plant	26.7	23.2	3.5	2	8?	Minimal
3	Blood	25.9	21.7	4.2	3	3	Maximum
4	Primarily insects, some fruit	23.3	n/a	n/a	1	6	Moderate
5	Omnivores, vertebrates, insects	22.3	21.1	1.2	9	20	Moderate to High
6	Primarily nectar, pollen, some fruit, insects	21.4	17.4	4.0	13	31	High
7	Primarily nectar, pollen, some fruit, insects	18.5	11.0	7.5	4	10	High
8	Primarily insects, some fruit or primarily fruit, some insects	18.5	17.2	1.3	3	13	Minimal (*Trinycteris*) Moderate (*Carollia*)
9	Mostly fruit or obligate fruit	18.5	16.2	2.3	19	68	High
					55	161	

common ancestor of the sister clade to the Phyllostomidae and the last common ancestor of all phyllostomid bats possessed. Establishing the latter in particular represents the first step in reconstructing the order and relationships of change of feeding habits in phyllostomids. Within the 19 families of bats, 13 are strict insectivores, four include strict insectivores and carnivores (Megadermatidae, Nycteridae, Noctilionidae, and Vespertilionidae), one includes fruit as well as nectar feeders (Pteropodidae) and one, the focus of this paper, includes species that are strict insectivores, omnivores, nectivores, carnivores, sanguivores, and frugivores.

The widespread occurrence of strict insectivory as a feeding strategy, together with its distribution, results in it being reconstructed as the primitive strategy for bats as a whole plus most major clades within bats, including all nodes connecting families within Noctilionoidea. Carnivory instead represents a derived state, usually within individual genera, with other members of the genus retaining the primitive state of strict insectivory. For instance, nearly all species of Vespertilionoidea are strict insectivores, and those that take fish (carnivores) are congeneric with strict insectivore species. Likewise, nearly all species of Emballonuroidea are strict insectivores with the exception of the genus *Nycteris*, where some species take some vertebrates but where others are strict insectivores. In the Rhinolophoidea, most species are strict insectivores except for the two carnivorous genera *Megaderma* and *Macroderma*. For Noctilionoidea, five families in this superfamily are strict insectivores, with the sixth family Noctilionidae comprising only two species; one is a strict insectivore and the other takes fish and insects. However, molecular data indicate that the fish-eater evolved within the last ten million years (Lewis-Oritt *et al.*, 2001; Bininda-Emonds *et al.*, 2007). Given this pattern and also that all extant members of the Mormoopidae, the sister family to Phyllostomidae, are strict insectivores, the most parsimonious explanation is that the ancestral condition for the last common ancestor of phyllostomid and mormoopid bats was a strict insectivore (Figure 11.3).

Molecular data calibrated against geologic estimates place the divergence of the last common ancestor for mormoopids and phyllostomids between 36 (42–32) mya (Teeling *et al.*, 2005) and 42.5 (46.7–38.3) mya (Bininda-Emonds *et al.*, 2007). Our estimates for this node are younger than both at 33.9 (37.8–24.7) mya. Both our estimate and that of Teeling *et al.* (2005) fall within the Oligocene, although the confidence limits for the latter overlap the boundaries of the Eocene. This brings into question whether the families Phyllostomidae and Mormoopidae existed in the Eocene as proposed by Simmons (2005) and supported by the dates of Bininda-Emonds *et al.* (2007).

Using the information available in the molecular phylogenetic tree and the geological dates proposed for each node, we hypothesize that the successful diversification within Phyllostomidae is a result of the ancestral stock of the family shifting from strict insectivory to also including plant material in the diet. Such a shift in diet would have reduced competition with all other bat species known to be present in the New World from the Early Cenozoic forward. The transition from strict insectivory to a diet with plant material in extant bat families has been accomplished only twice (Phyllostomidae and Pteropodidae). However, given that plant material is abundantly available in all ecosystems exploited by bats, the logical conclusion is that the transition from insectivory to consuming plant material (or even other feeding habits) must be difficult, possibly requiring the availability of a series of ecological determinants and ancestral genetic variation that facilitate directional selection of appropriate characters. The benefits of making this transition are apparently exceptional if the diversity within phyllostomid and pteropodid bats are a reliable indicator. These two families account for over 350 species or 30% of the total number of bats in all 19 families.

Teeling *et al.*'s (2005) phylogeny is designed primarily to address evolution among the bat families and does not provide much detail within the Phyllostomidae where only four of the 56 recognized genera were examined. Hoffman *et al.* (2008) focused on the subfamily Phyllostominae (Baker *et al.*, 2003) to provide a geological time scale of 18 genera and 24 species. Despite the different taxonomic foci of the two studies and different underlying data sets (both in terms of the genes and outgroups used), their tree shows generally good agreement with the dates presented in this chapter. Topological differences tend to be present for the more poorly supported nodes in the respective trees. Importantly, there is also generally substantial overlap in the confidence limits for the date estimates for nodes shared between Hoffman *et al.* (2008) and our studies but see the variation in date estimates in table 2 of Datzmann *et al.* (2010, p. 34).

In the estimates of the geological time frame in Datzmann *et al.* (2010) their dates were substantially older in their figure 4. However, in their table 2 of model comparisons of alternative molecular clocks and priors their third molecular clock model (Datzmann *et al.*, 2010, p. 34) estimated the origin of phyllostomid bats to be 29.07 mya (23.4–37.0) as compared to the estimate in this chapter of 30.0 mya (35.0–25.0). Clearly the use of different priors and models give substantially different results. We note that our priors and use of fossils are the same as those reported in Jones *et al.* (2005). Therefore our dates as discussed below are most relevant to those shown in Jones *et al.* (2005) for other bat families.

11.5   From insectivory to herbivory: prevailing trends in feeding diversification

An overview of the extent to which clades (Figure 11.3) within Phyllostomidae used plant material as a food source is outlined in Table 11.1. The first five extant lineages include no species that are obligate plant feeders and few that are predominantly plant feeders. Three lineages (1, 2 and 4, recognized as subfamilies Macrotinae, Micronycterinae and Lonchorhininae, respectively) do take some plant material, but are primarily insectivores. The fourth lineage (clade 3, subfamily, Desmodontinae) feeds entirely on animal tissue (blood), and it seems likely that the species in this lineage never took much plant material after its divergence from other phyllostomids. The final basal lineage contains species that are true omnivores, but also another that is a strict insectivore (*Macrophyllum*) as well as others that are carnivores (*Vampyrum* and *Trachops*). Therefore, we hypothesize that, up to this point, the ancestral condition for the first five clades was primarily an insectivore that took some plant material.

The origins for each of the first five clades is prior to the end of the Oligocene, and much, if not all, of the diversification in the fifth lineage (node 42 forward) occurred during the Miocene. The remainder of the diversification within these clades that are mostly dependent on plant material as a food source (nodes 6–15) also occurred during the Miocene (Table 11.1, Figures 11.2–11.3).

Each of the 31 and 14 extant species belonging to clades 6 and 7, respectively, are primarily consumers of plant material (Figure 11.2). Although each of these species are typically referred to as nectar feeders (Gardner, 1977; Ferrarezzi and Gimenez, 1996), the available data suggest that they probably all take pollen, fruit and insects as well. Thus, both of these clades have made the transition to feeding primarily on plant material, with several species showing extreme adaptations (Figure 11.1) for retrieving nectar from flowers such as *Musonycteris* (Phillips, 1971), *Anoura fistulata* (Muchhala, 2006) and *Lonchophylla* (Griffiths, 1982). In addition, both clades have coadapted as pollinators such that certain flowering plants and bats are symbiotic in a complex community structure (Fleming *et al.*, 2005).

Clade 8 comprises two lineages with different feeding strategies. One includes the genera *Glyphonycteris* and *Trinycteris*, which are so similar morphologically to members of *Micronycteris* (clade 2) and *Macrotus* (clade 1) that they have been held to be congeners (Jones *et al.*, 2002). We propose instead that the observed morphological similarities (Figure 11.2) represent shared primitive features present in *Macrotus* (clade 1), *Micronycteris* (clade 2) and *Glyphonycteris* and *Trinycteris* (clade 8) rather than evidence of a monophyletic group as implied in the supertree of Jones *et al.* (2002). All extant species resulting from nodes 1 and 2, and node 30, which is a subset of the diversity in node 8, share a feeding strategy (primarily insectivorous with some plant material) that was retained in the respective common ancestors of nodes 1–8 (Figure 11.2). If this hypothesis is true, then the primitive state for the respective common ancestors for nodes 1–8 was *Macrotus-*, *Micronycteris-* and *Glyphonycteris*-like, with this phenotype giving rise to all the specialized and unique feeding strategies and derived phenotypes found in the family Phyllostomidae (Figure 11.2) except for the obligate fruit feeding of node 15, the subtribe Stenodermatina which represents additional specialization of a primarily frugivore ancestor.

The other member of clade 8 is the genus *Carollia* whose species specialize on fruits (Fleming *et al.*, 2005), especially those of *Piper*, a plant that has coadapted its fruit presentation to provide easy access for bats. However, *Carollia* also employs a mixed feeding strategy, with the primary food source being insects for times when fruit availability is limited (Fleming, 1988).

The extant species that have evolved from clade 9 have a feeding strategy ranging from primary to obligate frugivores. This successful transition to fruit as the primary or sole food source has resulted in a clade with the greatest relative number of genera (20) and species (68) among phyllostomids (Baker *et al.*, 2003). Our results indicate that the final stage of the transition (node 15), obligate frugivory, occurred less than 10 mya with the lineage diversifying into eight monotypic genera in the last 5 mya (Dávalos, 2007; 5.9 (7.6–4.5) mya according to Bininda-Emonds *et al.*, 2007). The magnitude of morphological change and evolutionary plasticity that is present among the extant members of clade 8 suggests both that extensive vacant niches were present and that the respective genomes were exceptional at producing the morphological variation necessary to facilitate directional selection to exploit these new niches. This radiation to exploit obligate frugivory relative to the generic divergence within that observed in clades 1–7 is the most recent burst of diversification present in the family. The hypothesis that obligate frugivory evolved in the Antilles Islands and successfully invaded the competitive landscape of bats on the continental mainland of the New World Tropics (Dávalos, 2007) is also contra to most proposed scenarios of evolution on islands.

11.6   Evolution of unique feeding strategies

Many papers have presented hypotheses concerning the intermediate stages involved in the origin of the multiple feeding strategies present in Phyllostomidae (Gillete, 1975; Fenton, 1992; Ferrarezi and Gimenez, 1996).

As proposed above, character evolution along the phylogeny shown in Baker *et al.* (2003) is that all derived feeding strategies evolved independently from a *Macrotus, Micronycteris, Glyphonycteris*-like morphology characterized by adaptations to an insectivorous diet that also allowed the consumption of some plant material. Gillete (1975) emphasized the role of the feeding habit duality in the process of evolution from primitive insectivory to some other kinds of feeding habit specialization, a hypothesis accepted by Ferrarezi and Gimenez (1996) for phyllostomids. Our scenarios of diversification of feeding strategies among phyllostomids are compatible with the feeding habit duality model of Gillette (1975). Our conclusion is that the basal ancestor for all derived feeding strategies and resulting evolutionary consequences was primarily insectivorous, but taking some fruit. No highly derived feeding strategy evolved from another highly derived feeding strategy, as proposed for the origin of vampires from nectar feeders (Baker, 1979; Straney *et al.*, 1979), from fruit eaters (Slaughter, 1970) or from omnivorous phyllostomines (Schutt, 1998). Feeding habit duality is present in all phyllostomid bats except vampires, *Macrophyllum* which is a strict insectivore and the obligate frugivores of the Stenodermatinae (node 15, Figure 11.3).

11.6.1   Sanguivory

Obligate blood feeding has evolved only once in tetrapods. The lineage that gave rise to the vampires diverged (node 3) from the remainder of the family about 26 (31.1–21.1) mya, and the three extant species in the clade are obligate vampires. As currently recognized, each is a distinct genus, with two being specialists on bird blood, and the third, the common vampire bat (*Desmodus rotundus*), feeding primarily on mammals. The most basal genus of vampire, *Diphylla* (node 50), diverged from the two remaining forms 21.67 (22.4–21.0) mya. Because bird blood contains nucleated red blood cells and a higher level of sugar content, it is probable that the first successful vampire activities involved feeding on birds (Schutt, 1998), which the current distribution of vampire prey would also support. During the transition period to sanguivory, it was probably important to provide food in quantities that were small enough to not kill the prey, suggesting possibly a preference for larger prey objects (e.g., large ratites or flying species weighing a kilo or more). Flight and the ability to search large areas for sources of blood meals would also be critically important. Similarly, nocturnal feeding would be advantageous because the high diurnal activity of birds would likely have resulted in avoidance behaviors of the activities that are typical of vampire bats during feeding.

To address what types of birds were available for initial exploitation by the vampires requires knowledge of the avian fauna of the Neotropics from the geological time frame between 31.1 and 21.0 mya, when our study suggests that sanguivory presumably evolved (including outer 95% confidence intervals). A varied bird population has been documented for the middle to upper part of the Eocene, including a number of currently existing bird genera such as pelicans, ibises, marabous, ducks, cormorants and flamingos (Osborn, 1910). Many species in these genera are of a relatively large size and also form large migrating flocks that hypothetically could have populated the typical marshy environments that surrounded common inland water masses during the Aquitanian (23.03–20.43 mya), the narrower time frame during which our results suggest an evolutionary origin of sanguivory (26.11–21.67 mya).

We hypothesize that the common ancestor of the vampires was *Micronycteris*-like and in a window of a little more than four million years made the myriad of necessary changes to be a successful obligate blood feeder (e.g., the evolution of anticoagulants, changes in dentition necessary to obtain blood, adaptation to a highly specialized diet of blood cells and proteins, kidneys that can facilitate flight weight restriction by quickly reducing the amount of water in a blood meal, the anatomical changes necessary to be sufficiently agile in non-flight locomotion to obtain a blood meal from roosting birds and the sensor adaptation to find a blood meal). The evolution of this suite of traits undoubtedly involved many areas of the genome and changes in many single copy loci.

Indeed, the magnitude of evolution required to be a successful sanguivore (Greenhall and Schmidt, 1988) may be the greatest outlier from the shared features of all other bats. It is also a candidate to be the mammalian lineage that has undergone the largest magnitude of directional selection within a restricted time frame (4 mya) resulting in an extremely modified morphology, physiology, behavior and all the other unique features required to be a sanguivore.

11.6.2   Carnivory

Carnivory is defined herein as feeding on other vertebrates. The transition to vertebrates as food has occurred in bats independently at least six times. In addition to phyllostomids, examples of carnivory are present in megadermatids, noctilionids, nycterids and vespertilionids (at least twice in *Myotis*), *M. vivesi* and *M. macrodactylus*. Each of these four families is primarily insectivorous with isolated examples of carnivorous species (Simmons and Conway, 2003). Within phyllostomids, *Trachops* has evolved as a specialist for small frogs, and the sister clade to *Trachops/Macrophyllum* includes two other genera (*Vampyrum* and *Chrotopterus*) that take a variety of foods but are primarily specialists for feeding on vertebrates. The *Vampyrum/Chrotopterus* clade (node 43) diverged from the remainder of the phyllostomid bats 20.75 (21.6–19.9) mya in our tree and 19.5 (±2) mya according to Hoffmann *et al.* (2008), with the two genera diverging from each other (node 48) 14.35 (15.5–13.2) mya in our tree and 14.4 (+/–1.9) mya in Hoffmann *et al.* (2008). *Vampyrum* is the largest bat in the New World and feeds on other bats, rodents, birds and possibly also some fruit and insects (Gardner, 1977). It is known to take bats in flight, but probably also catches rodents by passive acoustic detection from sounds made by them on the forest floor (Siemers *et al.*, 2001). *Chrotopterus*, like *Vampyrum*, takes small vertebrates (e.g., geckos, opossums, birds and bats), but also takes insects and fruit.

11.6.3   Nectivory

Within our tree, there are 12 genera in the subfamily Glossophaginae (*sensu* Baker *et al.*, 2003; node 6) that are primarily nectar-feeding bats, with the morphological variation ranging from species with relatively short rostra (*Brachyphylla*) to those with rostra and tongues of extreme length, Figure 11.1 (*Musonycteris* and *Anoura fistulata*, Muchhala *et al.*, 2005). Glossophagines diverged from the other phyllostomid bats 21.55 (23.4–19.7) mya, with the last common ancestor of all 12 genera being present 17.48 (20.8–14.2) mya. Because all extant genera are primarily nectar feeders, it is parsimonious to assume that the last common ancestor for all members of node 33 was a nectar feeder as well. This results in a maximal estimate of nine million years to evolve from a primarily insectivorous omnivore (node 6) into a nectar feeder (node 33).

A second group of nectar-feeding bats (node 7) diverged from the remainder of Phyllostomidae about 1 mya after the divergence from those present in node 6, 20.52 (21.3–19.8) mya (although the confidence intervals for both clades overlap). Within our data set, the last common ancestor for *Lionycteris* and *Lonchophylla* diverged 11.13 (11.4–10.9) mya, yielding a similar maximal estimate (ten million years) of time as proposed for the Glossophaginae during which this clade would have evolved all the necessary adaptations to be a nectivore. However, not all genera within this clade have been sequenced and dated such that the date estimate of the last common ancestor for the extant members of the group may be nearer to the time of divergence than all other phyllostomids. In the Datzmann *et al.* (2010) estimates of the origin of the two nectar-feeding groups, more than two million years separate their independent origins.

There has been extensive debate concerning how many times nectar feeders have evolved within the phyllostomid assemblage, as well as a wide array of proposed monophyletic assemblages (Baker, 1967; Baker and Bass, 1979;

Griffiths, 1982; Haiduk and Baker, 1982; Warner, 1983; Smith and Hood, 1984; Koopman, 1993; Wetterer *et al.*, 2000; Carstens *et al.*, 2002; Datzmann *et al.*, 2010). If there had been morphological canalization for nectar feeding, it is likely that all nectar feeders would have shared a substantial number of synapomorphies in the hyoid and lingual regions. However, Griffiths' (1982) substantial data set, concerning the morphology associated with lingual aspects of nectar feeding, documents two alternative modifications of these regions to facilitate nectar feeding. This result is compatible with the hypothesis derived from our molecular reconstructions and those of Datzmann *et al.* (2010) that there was independent evolution of the two lineages from a primarily insectivorous bat taking some fruit into highly derived nectar feeders.

   The only nectar-feeding bat fossil known from the northern part of South America is *Palynephyllum antimaster*, a form intermediate between *Lonchophylla handleyi* and *Anoura caudifer* (Czaplewski *et al.*, 2003), recovered from the Miocene deposits of La Venta, Colombia and dated between 13 and 12 mya. This fossil, together with our proposed dates for the last common ancestor of nectivorous forms, ties the origin of nectivory with significant global climate changes that resulted in abrupt episodes of global cooling causally linked to major expansions of ice cover over Antarctica and accentuated due to a preceding interval of extreme global warmth in the latest Oligocene (Zachos *et al.*, 2001). Furthermore, the arid conditions associated with the cooler environments of the MI-1 Glaciation coincide with the appearance of the Neotropical endemic Cactaceae (Hershkovitz and Zimmer, 1997), a group typically pollinated by nectar-feeding bats. Morphological, physiological and behavioral changes in both bats and plants involved in pollination syndromes are not trivial and reflect a parallel evolutionary history (Heithaus, 1982).

11.6.4 Frugivory

   Species that are primarily frugivorous rather than nectivorous or insectivorous first appeared in the tree at nodes 29 (*Carollia*), 9 (*Rhinophylla*) and 10 (*Sturnira*), and all other taxa derived from node 11. Obligate frugivory characterizes the species descended from node 15. Node 9 represents the most successful radiation in the family in number of genera (20) and number of species (67). Our date estimates suggest that predominant frugivory evolved towards the end of the Oligocene (17.58 (17.7–17.5) mya for *Carollia* and 18.89 (20.1–17.7) mya for the last common ancestor of all members of node 9) and, therefore, slightly later than the other bat family that also evolved frugivory and nectivory (Pteropodidae) (24 (29–20) mya, Teeling *et al.*, 2005; 25.1 (30.3–22.7) mya, Bininda-Emonds *et al.*, 2007). The relative congruence of the estimates for the origins of frugivory in phyllostomids and pteropodids suggests that a global transtropical floral shift facilitated the evolution of frugivory in bats. This hypothesis is supported by data from paleobotany, which suggest that the Neotropical rainforest appeared in North America in the Early Paleocene and in South America during the Early Eocene (Burnham and Johnson, 2004 and citations therein).

11.6.5   Strict insectivory

Strict insectivory is the *status quo* for essentially all other non-phyllostomid bats that were present in the Neotropics during the Oligocene/Miocene time frame. Therefore, it is not surprising that phyllostomid bats did not commonly revert to the highly competitive feeding strategy of strict insectivory, and then apparently only if they could occupy a unique niche that would avoid competition with other strictly insectivorous species.

The only species of phyllostomid bat that is a strict insectivore, *Macrophyllum macrophyllum* (clade 49), arose in the Early Miocene, 18.1 (19.3–16.8) mya, as part of a clade (42) comprising a broad array of omnivores. Unlike any other strict insectivorous bats present in the Neotropics, *Macrophyllum* feeds from the surface of the water by gleaning insects. Otherwise, *Macrophyllum* is morphologically a typical insectivorous as well as carnivorous bat, except that its hindlimbs are uniquely modified for gleaning from the surface of the water. We hypothesize that *Macrophyllum* was successful at becoming a strict insectivore because this method of collecting insects avoided competition with other strict insectivorous as well as other phyllostomid bats. There are two noteworthy observations relative to *Macrophyllum* competing as a strict insectivore. First, the sister taxon to *Macrophyllum* is *Trachops*, which feeds on frogs and frequently takes them from the water's surface while they are making mating calls (Gardner, 1977), and second, that gaffing insects from the surface of the water has not commonly evolved in other insectivorous bat assemblages. *Macrophyllum* is another example of a member of the phyllostomid complex exploiting a relatively unique unfilled niche.

11.7   Conclusion and implications for future work

We propose that phyllostomid bats present the most radical adaptive radiation of feeding strategies from a common ancestor for any monophyletic group of mammals. Further, the time frame for this radiation, and the evolution of the individual feeding strategies, has been relatively short when compared to other phylogenetically defined examples of diversification under the rigors of natural selection (Teeling *et al.*, 2005). An appropriate question is why did this level of diversification occur only in this group of bats? This is a difficult question. Rapid radiation is typically associated with short generation times, rapid sexual maturity and large numbers of offspring per breeding cycle. Relative to most mammals, however, bats do not embody these characteristics. Female bats typically reach sexual maturity after one year, the number of offspring is usually one per breeding cycle, and bats are notoriously long-lived relative to other groups in the mouse-to-elephant curve of mammalian life-history characteristics. Nonetheless, since the beginning of the Oligocene, phyllostomids have not only changed radically in a broad array of phenotypic character states, but have also successfully evolved adaptation to more feeding strategies, sanguivores, frugivores, nectivores, carnivores and omnivores, than are represented collectively in all other 19 families of bats. No other bat family has more than two feeding strategies. We propose that this successful radiation is a result of successfully including plant material in addition to insects in the diet in concert with the environmental opportunities present in the Oligocene/Miocene in the Neotropics. It is not obvious how these lineages have overcome their relatively long generation time and other life-history characteristics usually adjacent with slower rates of evolution to facilitate the extreme directional selection needed, but there must have been involvement of genetic and/or genomic mechanisms to allow expression of the broad array of phenotypes to be acted upon by natural selection.

In the Introduction, we proposed that mapping genetic changes onto the clades of a phylogenetic tree can be used to test which genetic mechanisms were active in accomplishing rapid morphological and physiological change. Assuming that the common ancestor for nodes 1–8 was morphologically and physiologically a relatively typical insectivorous bat that fed on some plant material, and that this ancestral phenotype, through multiple cladogenic events, gave rise to clades that ultimately evolved into the diversity of highly specialized feeding types (blood feeders, nectar feeders, fruit and foliage eaters, and carnivores), such a test should be possible.

Within our phylogenetic tree (Figures 11.2–11. 3) there are three independent examples of species (clades leading to *Macrotus* (node 1), *Micronycteris* (node 2) and *Glyphonycterinae*, node 30) that today are quite similar to the proposed ancestral conditions (Jones *et al.*, 2002). These lineages are examples of morphological stasis and the genomic comparisons of extant members of these three respective groups should reflect stabilizing selection for comparison to lineages that have undergone extensive directional selection. Seven other examples (lineages derived from clades 50, 42, 33, 32, 9 and those leading to *Carollia* and *Lonchorhina* (Figures 11.2–11.3) have undergone exceptional morphological and physiological evolution, which permits multiple tests to better understand the extent to which the proposed mechanisms (see Introduction, Fondon and Garner, 2004, 2007, and Sutter *et al.*, 2007 for reviews) of genetic change have been involved in these examples of substantial directional evolution as compared to minimal genetic change observed is the extant species comprising the subfamilies *Macrotinae, Micronycterinae* and *Glyphonycterinae*. Interestingly, the results from Bininda-Emonds (2007), who examined substitution rates across all mammals based on 18 nuclear and 26 mitochondrial genes would seem to rule out a global change in substitution rate across the genome for phyllostomids in that no branch or clade within the group showed a significant deviation from the mammalian average. This result, however, does not exclude the possibility of genetic mechanisms such as gene duplications, mutations in tandem repeats, bursts of transposable element activity, variation in mutation rates of single copy genes and timing of gene expression, which were not examined by Bininda-Edmonds (2007).

Finally, it is important to experimental design that there are multiple examples of stasis, as well as of independent clades whose members have different phenotypes. In initial studies of genetic mechanisms that potentially underlie rapid and extensive evolutionary change (Fondon and Garner, 2004, 2007; Sutter *et al.*, 2007), candidate genes or mechanisms of genomic rearrangement etc. have mostly been compared for single examples of change. In the phyllostomid model, by contrast, it will be possible to compare the variation present in each of the proposed mechanisms across multiple examples of independent evolution to different phenotypes. Thus, both this model system and the database it provides should be invaluable, providing a better understanding of how mechanisms that promote rapid evolution under the constraints of natural selection over a geological time frame successfully accomplish the directional evolution required to exploit a new feeding niche or evolutionary grade. Once patterns of change in different morphological evolution are revealed, such patterns can be explored for possible convergence in the context of the recent discovery of convergent sequence evolution between echolocating bats and dolphins (Li *et al.*, 2010; Lui *et al.*, 2010). Examples would include nectar-feeding mutations shared by members of the Glossophaginae and the Lonchophyllinae for presence in independently evolved Old World nectar feeders such as the genera *Eonycteris, Macroglossus, Megaloglossus* and *Syconycteris*. Rostral elongation and tongue feeding have evolved numerous times in the class Mammalia. It will be fascinating to determine commonality in the genetic basis of each of these events in the diversification of bats and other mammals for successful nectar feeding, as was proposed for echolocation across mammalian orders (Jones, 2010). Further, using these sequence alterations in genes to be transplanted for expression in the *Mus* model system will be quite an experiment to synthesize an understanding of the alternatives and diversity of how the genome can successfully function to facilitate evolutionary change.

11.8   Acknowledgments

A special thanks to Kate Jones for assistance in writing Methods and Materials. We thank Federico Hoffmann, Peter Larsen, Caleb Phillips, Holly Wichman, Jim Bull, Steve Hoofer, Robert Bull and Bobby Baker for reviews of the manuscripts and for discussions on evolution of bats. We thank Lisa Torres for editorial assistance. Supported by the Biological Database program of Texas Tech University.

11.9   REFERENCES

Baker, R. J. (1967). Karyotypes of bats of the family Phyllostomidae and their taxonomic implications. *Southwestern Naturalist*, **12**, 407–428.

Baker, R. J. (1979). Karyology. In *Biology of Bats of the New World Family of Phyllostomatidae, Part III*, ed. R. J. Baker, J. K. Jones and D. C. Carter. *Special Publication of the Museum of Texas Tech University*, **16**, 107–155.

Baker, R. J. and Bass. R. A. (1979). Evolutionary relationship of the Brachyphyllinae to glossophagine genera *Glossophaga* and *Monophyllus*. *Journal of Mammalogy*, **60**, 364–372.

Baker, R. J., Porter, C. A., Patton, J. C. and Van Den Bussche, R. A. (2000). Systematics of bats of the family Phyllostomidae based on *RAG2* DNA sequences. *Occasional Papers, Museum of Texas Tech University*, **202**, 1–16.

Baker, R. J., Hoofer, S. R., Porter, C. A. and Van Den Bussche, R. A. (2003). Diversification among New World Leaf-Nosed Bats: an evolutionary hypothesis and classification inferred from digenomic congruence of DNA sequence. *Occasional Papers, Museum of Texas Tech University*, **230**, 1–32.

Bininda-Emonds, O. R. P. (2007). Fast genes and slow clades: comparative rates of molecular evolution in mammals. *Evolutionary Bioinformatics*, **2007**, 59–85.

Bininda-Emonds, O. R. P., Cardillo, M., Jones, K. E. *et al.* (2007). The delayed rise of present-day mammals. *Nature*, **446**, 507–512.

Burnham, R. J. and Johnson, K. R. (2004). South American palaeobotany and the origins of neotropical rainforests. *Philosophical Transactions of the Royal Society of London B*, **359**, 1595–1610.

Carroll, S. B. (2000). Endless forms: the evolution of gene regulation and morphological diversity. *Cell*, **101**, 577–580.

Carstens, B. C., Lundrigan, B. L. and Myers, P. (2002). A phylogeny of the Neotropical nectar-feeding bats (Chiroptera: Phyllostomidae) based on morphological and molecular data. *Journal of Mammalian Evolution*, **9**, 23–53.

Czaplewski, N. J., Takai, M., Naeher, T. M., Shigehara, N. and Setoguchi, T. (2003). Additional bats from the middle Miocene La Venta Fauna of Colombia. *Revista de la Academia Colombiana de Ciencias Exactas, Físicas, y Naturales*, **27**, 263–282.

Darwin, C. (1859). *On the Origin of Species by Means of Natural Selection, or the Preservation of Favoured Races in the Struggle for Life*, 4th edn. London: John Murray.

Datzmann, T., von Helversen, O. and Mayer, F. (2010). Evolution on nectivory in phyllostomid bats (Phyllostomidae Gray, 1825, Chiroptera: Mammalia). *BMC Evolutionary Biology*, **10**, 165.

Dávalos, L. M. (2007). Short-faced bats (Phyllostomidae: Stenodermatina): a Caribbean radiation of strict frugivores. *Journal of Biogeography*, **34**, 364–375.

Eldredge, N. and Gould, S. J. (1972). Punctuated equilibria: an alternative to phyletic gradualism. In *Models in Paleobiology*, ed. T. J. M. Schopf. San Francisco, CA: Freeman & Cooper, pp. 305–332.

Fenton, M. B. (1992). Wounds and the origin of blood-feeding in bats. *Biological Journal of the Linnean Society*, **47**, 161–171.

Ferrarezi, H. and Gimenez, E. A. (1996). Systematic patterns and evolution of feeding habits in Chiroptera (Archonta: Mammalia). *Journal of Comparative Biology*, **1**, 75–94.

Fleming, T. H. (1988). *The Short-Tailed Fruit Bat: A Study in Plant-Animal Interactions*. Chicago, IL: University of Chicago Press.

Fleming, T. H., Muchhala, N. and Ornelas, J. F. (2005). New World nectar-feeding vertebrates: community pattern and processes. In *Contribuciones Mastozoologicas en Homenaje a Bernardo Villa*, ed. V. Sanchez-Cordero and R. A. Medellin. Mexico: Instituto de Biologia e Instituto de Ecologia, UNAM, pp. 163–185.

Fondon, J. W. and Garner, H. R. (2004). Molecular origins of rapid and continuous morphological evolution. *Proceedings of the National Academy of Sciences, USA*, **101**, 18058–18063.

Fondon, J. W. and Garner, H. R. (2007). Detection of length-dependent effects of tandem repeat alleles by 3-D geometric decomposition of craniofacial variation. *Development Genes and Evolution*, **217**, 79–85.

Furano, A. V., Hayward, B. E., Chevret, P., Catzeflis, F. and Usdin, K. (1994). Amplification of the ancient murine Lx family of long interspersed repeated DNA occurred during the murine radiation. *Journal of Molecular Ecology*, **38**, 18–27.

Gardner, A. L. (1977). Feeding habits. In *Biology of Bats of the New World Family Phyllostomidae, Part II*, ed. R. J. Baker, J. K. Jones and D. C. Carter. *Special Publications, Museum of Texas Tech University*, **13**, 293–350.

Gillete, D. D. (1975). Evolution of feeding strategies in bats. *Tebiwa*, **18**, 39–48.

Greenhall, A. M. and Schmidt, U. (1988). *Natural History of Vampire Bats*. Boca Raton, FL: CRC Press.

Griffiths, T. A. (1982). Systematics of the New World nectar-feeding bats (Mammalia, Phyllostomidae), based on the morphology of the hyoid and lingual regions. *American Museum Novitates*, **2742**, 1–45.

Haiduk, M. W. and Baker, R. J. (1982). Cladistical analysis of the G-banded chromosomes of nectar-feeding bats (Glossophaginae: Phyllostomidae). *Systematic Zoology*, **31**, 252–265.

Hershkovitz, M. A. and Zimmer, E. A. (1997). On evolutionary origins of the cacti. *Taxon*, **46**, 217–232.

Hiethaus, E. R. (1982). Coevolution between bats and plants. In *Ecology of Bats*, ed. T. H. Kunz. New York: Plenum Press, pp. 327–367.

Hoekstra, H. E. and Coyne, J. (2007). The locus of evolution: evo devo and the genetics of adaptation. *Evolution*, **61**, 995–1016.

Hoffmann, F. G., Hoofer, S. R. and Baker, R. J. (2008). Molecular dating of the diversification of Phyllostominae bats based on nuclear and mitochondrial DNA sequences. *Molecular Phylogenetics and Evolution*, **49**, 653–658.

Hoofer, S. R. and Van Den Bussche, R. A. (2003). Molecular phylogenetics of the chiropteran family Vespertilionidae. *Acta Chiropterologica*, **5** (Suppl.), 1–63.

Jones, G. (2010). Molecular evolution: gene convergence in echolocating mammals. *Current Biology*, **20**, 62–64.

Jones, K. E., Purvis, A., MacLarnon, A., Bininda-Emonds, O. R. P. and Simmons, N. B. (2002). A phylogenetic supertree of the bats (Mammalia: Chiroptera). *Biological Reviews*, **77**, 223–259.

Jones, K. E., Bininda-Edmonds, O. R. P. and Gittleman, J. L. (2005). Bats, clocks, and rocks: diversification patterns in Chiroptera. *Evolution*, **59**, 2243–2255.

Kawasaki, K., Buchanan, A. V. and Weiss, K. M. (2007). Gene duplication and the evolution of vertebrate skeletal mineralization. *Cells Tissues Organs*, **186**, 7–24.

Kirschner, M. and Gerhart, J. (1998). Evolvability. *Proceeding of the National Academy of Sciences, USA*, **95**, 8420–8427.

Koopman, K. F. (1993). Order Chiroptera. In *Mammals Species of the World: A Taxonomic and Geographic Reference*, 2nd edn., ed. D. E. Wilson and D. M. Reeder. Washington, DC: Smithsonian Institution Press, pp. 137–241.

Lewis-Oritt, N., Van Den Bussche, R. A. and Baker, R. J. (2001). Molecular evidence for evolution of piscivory in *Noctilio* (Chiroptera: Noctilionidae). *Journal of Mammalogy*, **82**, 748–759.

Li, Y., Lui, S., Shi, P. and Zhang, J. (2010). The hearing gene *Prestin* unites echolocating bats and whales. *Current Biology*, **20**, 55–56.

Lui, Y., Cotton, J. A., Shen, B. *et al.* (2010). Convergent sequence evolution between echolocating bats and dolphins. *Current Biology*, **20**, 53–54.

Muchhala, N. (2006). Nectar bat stows huge tongue in its rib cage. *Nature*, **444**, 701–702.

Muchhala, N., Mena, P. and Albuja, L. (2005). A new species of *Anoura* (Chiroptera: Phyllostomidae) from the Ecuadorian Andes. *Journal of Mammalogy*, **86**, 457–461.

Ohno, S., Wolf, U. and Atkin, N. B. (1968). Evolution from fish to mammals by gene duplication. *Hereditas*, **59**, 169–187.

Osborn, H. F. (1910). *The Age of Mammals in Europe, Asia, and North America*. New York: The Macmillan Company.

Pascale, E., Liu, C., Valle, E., Usdin, K. and Furano, A. V. (1993). The evolution of long interspersed repeated DNA (L1, LINE 1) as revealed by the analysis of an ancient rodent L1 DNA family. *Journal of Molecular Evolution*, **36**, 9–20.

Phillips, C. J. (1971). The dentition of glossophagine bats: development, morphological characteristics, variation, pathology, and evolution. *Miscellaneous Publications of the Museum of Natural History, University of Kansas*, **54**, 1–138.

Phillips, C. J. (1996). Cells, molecules, and adaptive radiation in mammals. In *Contributions in Mammalogy: A Memorial Volume Honoring Dr. J. K. Jones, Jr.*, ed. R. J. Baker and H. H. Genoways. Lubbock, TX: Museum of Texas Tech University, pp. 1–24.

Phillips, C. J., Tandler, B. and Nagato, T. (1993). Evolutionary divergence of salivary gland acinar cells: a format for understanding molecular evolution. In *Biology of Salivary Glands*, ed. K. Doborosieski-Vergona. Boca Raton. FL: CRC Press, pp. 39–80.

Posada, D. and Crandall, K. (1998). MODELTEST: testing the model of DNA substitution. *Bioinformatics*, **14**, 817–818.

Purvis, A. (1995). A composite estimate of primate phylogeny. *Philosophical Transactions of the Royal Society of London B*, **348**, 405–421.

Renner, S. S. (2005). Relaxed molecular clocks for dating historical plant dispersal events. *Trends in Plant Science*, **10**, 550–558.

Schutt, Jr., W. A. (1998). Chiropteran hindlimb morphology and the origin of blood feeding in bats. In *Bat Biology and Conservation*, ed. T. H. Kunz and P. A. Racey. Washington, DC: Smithsonian Institution Press, pp. 157–168.

Siemers, B. M., Stilz, P. and Schnitzler, H. (2001). The acoustic advantage of hunting at low heights above water: behavioural experiments on the European "trawling" bats *Myotis capaccinnii*, *M. dasycneme* and *M. daubentonii*. *Journal of Experimental Biology*, **204**, 3843–3854.

Simmons, N. B. (2005). An Eocene big bang for bats. *Science*, **307**, 527–528.

Simmons, N. B. and Conway, T. M. (2003). Evolution of ecological diversity in bats. In *Bat Ecology*, ed. T. H. Kunz and M. B. Fenton. Chicago, IL: University of Chicago Press, pp. 493–535.

Slaughter, B. H. (1970). Evolutionary trends of chiropteran dentitions. In *About Bats*, ed. B. H. Slaughter and D. W. Walton. Dallas, TX: Southern Methodist University Press, pp. 51–83.

Smith, J. D. and Hood, C. S. (1984). Genealogy of the New World nectar-feeding bats reexamined: a reply to Griffiths. *Systematic Zoology*, **33**, 435–460.

Straney, D. O., Smith, M. H., Greenbaum, I. F. and Baker, R. J. (1979). Biochemical genetics. In *Biology of Bats of the New World Family of Phyllostomatidae, Part III*, ed. R. J. Baker, J. K. Jones and D. C. Carter. *Special Publication of the Museum of Texas Tech University*, **16**, 157–176.

Sutter, N. B., Bustamante, C. D., Chase, K. *et al.* (2007). A single *IGF1* allele is a major determinant of small size in dogs. *Science*, **316**, 112–115.

Swofford, D. L. (2002). PAUP*. Phylogenetic analysis using parsimony (*and other methods). Version 4. Sunderland, MA: Sinauer Associates, Inc.

Teeling, E. C., Springer, M. S., Madsen, O. *et al.* (2005). A molecular phylogeny for bats illuminates biogeography and the fossil record. *Science*, **307**, 580–584.

Thompson, D 'A. W. (1917). *On Growth and Form*. New York: Dover.

Van Den Bussche, R. A. and Hoofer, S. R. (2004). Phylogenetic relationships among recent chiropteran families and the importance of choosing appropriate out-group taxa. *Journal of Mammalogy*, **85**, 321–330.

Warner, R. M. (1983). Karyotypic megaevolution and phylogenetic analysis: New World nectar-feeding bats revisited. *Systematic Zoology,* **32,** 279–282.

Wetterer, A. L., Rockman, M. V. and Simmons, N. B. (2000). Phylogeny of phyllostomid bats: data from diverse morphological systems, sex chromosomes, and restriction sites. *Bulletin of the American Museum of Natural History,* **248,** 1–200.

Wilson, A. C., Sarich, V. M. and Maxon, L. R. (1974). The importance of gene rearrangement in evolution: evidence from studies on rates of chromosomal, protein, and anatomical evolution. *Proceedings of the National Academy of Sciences, USA,* **71,** 3028–3030.

Wilson, A. C., Bush, G. L., Case, S. M. and King, M. C. (1975). Social structuring of mammalian populations and rate of chromosomal evolution. *Proceedings of the National Academy of Sciences, USA,* **72,** 5061–5065.

Wren, J. D., Forgacs, E., Fondon, III, J. W. *et al.* (2000). Repeat polymorphisms within gene regions: phenotypic and evolutionary implications. *American Journal of Human Genetics,* **67,** 345–356.

Zachos, J., Pagani, M., Sloan, L., Thomas, E. and Billups, K. (2001). Trends, rhythms, and aberrations in global climate 65 Ma to present. *Science,* **292,** 686–693.

12

Why tribosphenic? On variation and constraint in developmental dynamics of chiropteran molars*

IVAN HORÁČEK AND FRANTIŠEK ŠPOUTIL

* Dedicated to Professor Dr. Vladimír Hanák on the occasion of his 80th birthday

12.1 Introduction

Teeth and dentitions are key evolutionary novelties of vertebrates – much of the success of that clade can be traced to just these structures. The extreme ecological efficiency, rapid rate of adaptive rearrangement and growth dynamics, and large body size, as well as the finely tuned developmental mechanisms characterizing the vertebrates have one commonality: all are closely linked to a very high rate of energetic turnover. The core of the circuit lies in the rate of energetic flux from outside to inside vertebrate bodies. Teeth and dentitions act as its powerful amplifiers, the appearance of which may have played a decisive role in triggering a great deal of the current scope of vertebrate adaptations.

The dentition is not only a physical interface between the exterior and interior of an organism, but also a very complex interface between the energetic demands of the body, characteristics of diet, food availability and foraging. The form of dentition is influenced by selection related to these factors, as well as the phylogenetic history of a taxon and pathways of its past adaptive efforts. Theoretically, the state of dental characters may provide condensed and relevant information on any of these variables.

The core agent of the interface is the highly mineralized enamel cap, which covers the dentine body of the tooth crown with a rigid layer of hydroxyapatite crystallites. It is the most pertinent attribute of any tooth and the ultimate source of the extraordinary resistance of teeth to external influences, either chemical or mechanical. The exclusive role of teeth in vertebrate evolution is directly linked to this enamel crown, the resistence of which is, alongside

Evolutionary History of Bats: Fossils, Molecules and Morphology, ed. G. F. Gunnell and N. B. Simmons. Published by Cambridge University Press. © Cambridge University Press 2012.

others, also responsible for excellent preservation of teeth in sediments and, hence, for the particular richness of the vertebrate fossil record.

Dental characters present the essential (and often the only) source of information available from the vertebrate fossil record. Supplemented with comparative information on contextual variables accompanying the respective character states in extant forms (for which detailed ecological or phylogenetic information often is available), these characters can provide information about the phylogenetic and ecological attributes of fossil forms. Yet, to extract such information necessitates not only a careful descriptive screening of the states of particular dental characters, but also an understanding of their variation and the functional, structural and developmental constraints that modify the pattern of their variation (Gould, 1989) and which may essentially influence the phenotypic effects of phylogenetic rearrangements (Maynard Smith et al., 1985).

The variation dynamics of dental characters and the developmental mechanisms producing teeth are constrained by spatial limitations (including spatial demands of neighboring structures) and the structural integrity requirements of the whole orofacial complex. In addition, the life-history traits (essentially unrelated to them) may impose temporal constraints on development. Thus, the implicit demand to maximize the size and functional versatility of the dentition has been moderated by the capacity of available developmental mechanisms, by spatial competition, the developmental requirements of the morphological structures surrounding it and by competition of the factors contributing to biomechanical and structural integrity of all its elements (such as shape and size of particular teeth, modes of their articulation, construction morphology of jaws and jaw musculature etc.). The stronger the constraint upon variation of dental characters, the more acute the pressure to evolve innovations that would release the constraining effects and enlarge the variation. The inherent feedback control of the respective innovations by developmental constraints and functional demands is perhaps the major path of dental phylogeny and source of the conspicuous continuity of dental evolution.

All these factors somehow influence the dental phenotype and should be taken into account when analyzing the phylogenetic and ecological information that dental morphology provides. This chapter is intended to explore some of these influences and to discuss the basic setting of developmental and construction constraint upon chiropteran molars, the patterns of innovation and the methodological consequences for study of phylogenetic transformations of dental characters and bat evolution.

The objectives of this work are to: (1) analyze structural and functional specificities of tribosphenic molars, including their developmental background and a possible mechanism leading to precise occlusion (2) document patterns of variation in chiropteran molars and relate this to taxon-specific developmental differences; and (3) argue that the current state of knowledge of dental characters in bats is insufficient for their direct use in phylogenetic studies, unless detailed information on their developmental dynamics is taken in account.

12.2   The essentials

Probably the most influential constraint upon dental variation arises from the essential functional quality of a tooth, i.e., the extreme rigidity of the enamel due to its complete mineralization, which provides the tooth with enormous resistance to mechanical stress. Of course, for obvious reasons: (1) the enamel must resist mechanical stress of any act of occlusion and hence be completely hardened even prior to the first use; (2) the completely hardened enamel does not allow any plastic changes, reshaping etc.; (3) any rearrangement, however minute, in tooth shape and size can be thus achieved exclusively via rearrangements of the developmental processes forming the tooth prior to its first use; and (4) once the tooth is mineralized and appears above the gingiva no changes in its size and shape appear (except by wear caused by usage).

The state of dental characters is strictly controlled by capacity of dentition to respond to functional demands appearing during the course of an individual's life (such as actual efficiency of food processing, correspondences between dietary preferences and capacity of respective food resources, and ad hoc variations in their availability). Such a *prospective regulation* would undoubtedly promote enlargement of capacity to modify the state of particular characters in response to ad hoc variation of the respective contextual variables, the enlargement of variation and plasticity of dental characters. Yet, everything in that respect must be produced prior to the first occlusion event, which would be beyond the direct effect of the prospective regulation. In other words, the ultimate effects responsible for all fine dental variations arise from a domain of *retrospective regulation* via the rigorous developmental machinery that produces the teeth. More strictly than in any other systems of the vertebrate body, in the dentition any influences of prospective regulation can be imposed only in the form of a phylogenetic event, i.e., via rearrangements of the inherited programs of retrospective regulation. For the practice of odontologic analysis this means the following: the state of a particular dental character, while undoubtedly having something to do with foraging specificities of a given taxon, does not reflect this directly, nor does it illustrate the best response to prospective regulation – rather it expresses the capacity of taxon-specific retrospective regulation to respond to the respective cues. A dental character is produced by a superposition of developmental rearrangements implemented into the recurrent pathways of dental ontogeny. It is a palimpsest rather than a message easily readable with the help of an odontologic dictionary.

Nevertheless the information provided by dental characters is undoubtedly quite important, and in some cases, such as often in fossil taxa, the only one which is available. To exploit it means to adapt our reading in regards to what its direct phylogenetic or functional meaning might be – the fact that the proper triggers of the phylogenetic divergences and/or the factors responsible for taxon-specific variation or constraints on particular dental characters are situated somewhere in the developmental machinery producing teeth and any rearrangements in dental phenotype (regardless of whether it is significant for a phylogenetic or functional point of view) are the epiphenomena of developmental processes. Yet, to support these claims is, for more than one reason, still a bit complicated.

Teeth are produced by specifically regulated interactions of the ectodermal epithelium and the neural crest mesenchyme, the most progressive cell populations of the vertebrate body. Recurrent steps of the interaction are triggered by cascades of signaling proteins specifically expressed either in the epithelium or in mesenchyme, which mostly effect growth rate and metabolic profiles in neighboring cell populations (Thesleff and Sharpe, 1997; Thesleff, 2003; Tucker and Sharpe, 2004). A signaling function was established for about 130 of c. 280 proteins for which genes were identified to express in tooth primordia (comp. http://bite-it.helsinki.fi for a detailed account). Despite extensive redundancy in signaling effects, even a minute variation in expression of a single factor may cause extensive phenotypic effects (Kassai et al., 2005; Peterková et al., 2005). In that respect, the genocentric optics of contemporary biology would orient our search for the triggering factors of phylogenetic divergences into the sphere of comparative molecular biology of dental development – a field both conceptually and methodically very distant from the scope of the primary interest of this book. Nevertheless, as demonstrated below, relevant information on the developmental background of dental variation in bats can be obtained beyond molecular optics.

The process of odontogenesis starts with the initiation stage by local thickening of the odontogenetic epithelium and local condensation of mesenchyme, and continues by extension of an epithelial bud into the mesenchyme. A dramatically increased growth rate of the epithelium in the next stage results in lateral expansion of the epithelial bud which subsequently encapsulates the mesenchymatic condensation in the form of the bilaminar epithelial cap. The outer epithelial layer forms (in interaction with the neighboring mesenchyme) the dental sac separating the spatial domain of the tooth primordium from outside effects, while the inner layer shows a deep evagination of the three-dimensional growth of the encapsulated mesenchymatic papilla. The accelerated growth of that stage results in the spatial interplay between two-dimensional growth of the inner epithelium and three-dimensional growth of mesenchyme, which characterizes the next step of dental development – the stage of the dental bell.

In the bell stage, enlargement of the tooth primordium and shaping of the future tooth by locally specifying folding of the inner enamel epithelium (tooth morphogenesis) is later accompanied by histogenetic changes in the cells interacting along the basal membrane of the inner enamel epithelium: epithelial cells change into cylindrical ameloblasts, mesenchymal cells into odontoblasts equipped with protruding, densely branched odontoblastic processes. Both ameloblasts and odontoblasts present a mineral compound into the bordering zone. With their secretory processes the ameloblasts increase concentration of the mineral matrix and preform hydroxyapatite crystallites, which subsequently aggregate into a highly mineralized enamel, while the odontoblasts organize a complex redistribution of primary crystallites along the collagene scaffold by which they produce a tubular, highly organized, but less mineralized dentin. The mineralization begins from the inflexion points of the deepest epithelial evagination – tip of the adult tooth cusp – and subsequently spreads down to the cervical loop separating the inner and outer enamel epithelium. Following complete enamel mineralization, the morpho-differentiation of the tooth terminates and the completely mineralized tooth erupts (nothwithstanding derived conditions in some mammals as discussed below). The above described pattern presents an invariant phenomenon common to all gnathostomes (Reif, 1982; Smith and Hall, 1990; Stock, 2001; Huysseune et al., 2009), and this holds also for the proximate developmental machinery and the signaling cascades specific for particular stages (Stock, 2001; Thesleff, 2003; Tucker and Sharpe, 2004).

The essential qualities of adult teeth, such as size or thickness and rigidity of enamel depend upon the time available for tooth development (and the particular steps through which it proceeds e.g., morphogenesis, histogenesis and mineralization). Prospective control (e.g., deficiencies in these qualities in adult teeth) can quite strongly promote any innovation prolonging the period of dental development or establishing specific heterochronies accelerating the performance of the respective processes or heterotopies acting in a similar way (e.g., simultaneous development of different dental generations or synchronous appearance of different odontogenetic processes at divergent loci of a single tooth primordium).

The odontogenesis of mammals, both in diversity of teeth types and their structural or functional complexity, is characterized by rigorous exploitation of all these possibilities. It is particularly well illustrated by the major dental apomorphies of mammals: large monophyodont multicuspidate posterior teeth (molars) and prismatic enamel. The prolonged early tooth development and accelerated growth of the oral region in mammals is the outcome of accelerated epithelial growth that results in the morphogenesis of mammalian teeth starting with very large cell populations, both epithelial and mesenchymal. Then, thanks to a long gestation period, mammalian teeth can grow large, are disposed to intricate shapes and their enamel coat can be quite thick, at least at tooth tips where amelogenesis starts first. For teeth of the permanent generation and monophyodont molars, the developmental time is even further prolonged, with a postnatal lactation period that allows these teeth to grow very large and respond to requirements of large adult body sizes and/or specialized foraging and longevity.

12.3   Tribosphenic molar: an enigmatic phenomenon

The vast majority of the information on dental development has been obtained from just a single mammalian model, unfortunately one that is extremely derived in almost all dental characters, the mouse. Which of the odontogenetic mechanisms described in detail for mice (e.g., Caton and Tucker, 2009) operate in dental ontogeny of other mammals cannot be answered completely due to scarcity of relevant comparative data. Unfortunately, surprisingly little embryological information (except perhaps for Marshall and Butler, 1966) is available on development of the tribosphenic molar – the molar type uniformly distributed in bats (except for Pteropodidae and some Phyllostomidae). This fact is noteworthy in particular because the tribosphenic molar is generally considered to represent the phylotypic stage of mammalian dental evolution, the molar type common to ancestors of all extant clades of mammals. Such a view was proposed first by Cope (1874, 1883) and Osborn (1888) within the framework of the trituberculate theory of mammalian dental evolution. In its final form (Osborn, 1897, reprinted in Osborn, 1907) the theory provided an integrated concept of the evolutionary history of mammalian molar teeth and a paradigm for comparative study of mammalian dentitions. The theory declared the pathways of phylogenetic transformation from a presumed ancestral condition – a single-cone tooth of reptiles to derived multicuspidate molars of extant groups – predicted a set of homologous relations among teeth of mammals and among the structural elements of mammalian teeth in phylogenetic, structural, developmental and functional respects. These assumed homologies were further expressed the form of a detailed nomenclature of structural elements of mammalian molars – cusps, crests, cingula, styles etc. The identical terms applied for particular elements of the upper and lower molars (with the suffix -id used for the latter) stressed a complete homology between them. The theory also explained the obvious difference in design of the upper and lower molars, which it stressed even by different terms denoting the respective designs, i.e., trituberculate for the former and tuberculosectorial for the latter.

The term *tribosphenic* replacing the Cope–Osborn term trituberculate-tuberculosectorial was introduced for the respective molar type by Simpson (1936), perhaps in response to criticism of verbal confusions of the Osbornian nomenclature by Friant (1933, see Simpson, 1933). The term (*tribein* = to rub, *sphenic* = wedge) stressed integrity of upper and lower molar design in their complex function, combining effects of shearing and compression at every single occlusion event, suggestive of a mortar and pestle (with talonids of the lower molars as mortar and protocone of the upper ones as pestle).

It should be remembered that particular issues of the Cope–Osborn concept were almost immediately criticized, first using embryogenetic data (Woodward, 1896), and later by fossils as well (Gidley, 1906), both demonstrating convincingly that the primary cusp supposedly homologous to a single-cone tooth of reptiles is not the protocone (in the upper molars) but the antero-external cusp, i.e., paracone or parastyle. The condition in lower molars (i.e., protoconid as the primary cusp), in contrast, fits well with the prediction, which seems to disprove the predicted model of upper-lower molar homology. In his later survey, Osborn (1907) partly accepted the criticism, but did not respond to it with any changes in dental nomenclature, apparently in order to prevent possible confusion.

Subsequently, of course, nearly all particular conceptual issues and predictions of the Cope–Osborn concept (reviewed in detail by Hershkovitz, 1971) were refuted and the multiple homologies among all elements of mammalian teeth disproved (Butler, 1956; Vandebroek, 1961; Butler, 1978; Van Valen, 1982, 1994). Several attempts were made to innovate the nomenclature of dental elements in response to changes in the concept of dental homology, the most conclusive by Vandebroek (1961, 1967), who proposed a radical departure from the Osbornian terminology. Perhaps due to deeply ingrained usage of the Osbornian terms in practice of odontologic descriptions and doubts about alternative homology predictions, the nomenclatural innovations were not widely accepted, except perhaps for the terms eoconus (= paraconus)/eoconid (= protoconid) that are alternatively used for the Osbornian paraconus/protoconid.

Despite all doubts and refutations, the basic idea and key assumption of Cope–Osborn theory, i.e., that the trituberculate (tribosphenic) molar represents the phylotypic stage of mammalian dental organization – a platform from which the diversification of molar teeth began, retained its role in the paradigmatic framework of mammalian comparative odontology and in recent decades it has received even more robust empirical support. The earliest mammals (Flynn *et al.*, 1999; Luo *et al.*, 2002; Rauhut *et al.*, 2002) all show the same structurally basic pattern of molar shape: V-shaped crest centered with major cusps positionally identical with those characterizing tribosphenic molars in extant mammals. In other words, despite the possible dual origin of tribosphenic design (Australosphenida vs. Boreosphenida – Luo *et al.*, 2001) or variations in pretribosphenic stages exceeding states in extant mammals (Luo *et al.*, 2007), tribosphenic molars and the structural design characterizing them presents a robust common feature of mammalian organization and perhaps the most influential triggering factor in the radiation of mammalians.

Both with respect to taxonomic distribution and in terms of number of species, the tribosphenic molar is also the most common molar type among the extant mammals. An important question that (particularly concerning the enormous divergence of derived tooth types in mammals) necessarily arises is why just tribosphenic? What makes that design so successful, and at the same time so progressive and so constrained that it survived nearly unchanged for almost 180 million years of mammalian evolution?

A default answer to these questions would perhaps first address the functional qualities of that tooth type. A combination of shearing, grinding and transporting of food particles, simultaneously during a single occlusal action, provides nearly ideal condition for rapid food processing and enhancement of mastication effort. The geometry of shearing blades that contribute to these effects was recently analyzed in detail by Evans and Sanson (1998, 2003, 2006) and Evans (2005, 2006) in terms of bio-engineering models of optimal tools. The chiropteran tribosphenic molar was found to be a "tooth of perfection" because its shearing blades and the arrangement of its leading and rake angles correspond exactly to the functionally most effective design. The study was criticized for underestimation of the role of crushing in occlusal effort (Weil, 2003), but this objection, in the case of tribosphenic molars, is not well supported. The tribosphenic molar is obviously designed for insectivory or microfaunivory, i.e., feeding on tissues of animals, the diet that is essentially easily digestible without any mechanical treatment prior to the digestion (e.g., reduction of molars and secodont rearrangements in the specialized flesh-eaters or complete disappearance of teeth in specialized myrmecophagous mammals). Feeding on insects is constrained by the fact that the easily digestible food is covered by a chitinous cuticula that must be mechanically broken apart. Not compression, but shearing of the cuticula is the most efficient way to proceed, and correspondingly, the shearing tools (the crests) are by far the most essential attributes of the tribosphenic design. Insectivory further differs from macrofaunivory in that insect food resources are quite common, widely available, but also in that insects are rather small, and variegated in spatial distribution, which promotes tactics of opportunistic feeding with a need for rapid processing of large amounts of a small-sized vagile prey.

The basic design of tribosphenic molars is quite uniform and remains generally unchanged except in clades that abandoned insectivory. Nevertheless, taxon-specific arrangements of particular dental elements (e.g., proportions of molars and their crests) show considerable variation, although some common trends can be easily identified. For instance, teeth are higher and more robust when larger and more heavily armored prey is consumed the arrangement often correlated with enlargement of temporalis musculature and its insertion areas on the mandibular ramus (Storch, 1968; Freeman, 1981a, 1981b, 1984, 1988). Strait (1993) argued that dietary choice inversely scales with total length of molar crests. Accordingly, soft-object feeders exhibit more shearing development than do closely related taxa that have more generalized feeding patterns and/or those feeding on hard-bodied insects (but see Evans and Sanson, 1998).

The phylogenetic rearrangements *within* the frame of generalized micro-faunivory and tribosphenic molar design are characterized by little selection for crushing and food compression, but very strong selection for efficiency of shearing and rapid food processing. Correspondingly, relatively large variation in compression surfaces (protoconal fossa, talonid basin) can be expected with rather constrained variation in the components contributing to shearing (i.e., crest sharpness, resistance of the crest edges to mechanical stress of attrition and the total length of the crest system of the dentition). Indeed, the crests present the most obvious component of the tribosphenic molar and are true agents of its function. At least from the functional point of view, tribosphenic molars can be viewed as mere *bases for promoting shearing crests*. Then, of course, the morphology of the tooth would present a balance of two structural and functional units quite different in their structural and developmental demands: (1) the crest system requiring extensive enlargement and potential flexibility (at least from a phylogenetic perspective) and (2) a tooth base requiring reliable physical integration of all components of the tooth into a rigid framework of spatial organization (ensuring an even redistribution of the residual energy of shearing and counterbalancing local strain). Yet, these partly opposing require-ments and the structural elements disposed to respond to them share the same limited space of the tooth primordium and the same limited time available for tooth development.

In addition there is still another factor that may play a role in ultimate constraint on tribosphenic design, the factor that scales efficiency of shearing and strictly controls functionality of the whole dentition: the pattern of inter-locking among occluding blades and, hence, among all occluding elements of the dentition – crests, cusps, walls, fossae etc. To achieve a proper shearing effect, the interlocking must be quite precise, and the smaller the prey the more precise interlocking is required. The interlocking precision also acts as a key factor in the domain of life-history traits. It can considerably reduce the effects of attrition of occluding structures (i.e., a predominant component of tooth wear in insectivore's insect food is essentially non-abrasive), prolong functionality of dentition and promote longevity.

The functional complexity of the crest system and its precision interlocking requirements integrate the whole dentition into a single synergistic functional unit. This arrangement is highly advantageous and increases the functional effect of every single occlusal action. Nevertheless, at the same time it is extremely risky because any small misalignment in cusp position, crest angle or cusp height may dramatically reduce the functionality of the whole dentition and can potentially be lethal. For that reason, answering how interlocking is achieved and which developmental mechanisms prevent malocclusion is the paramount step in answering, Why tribosphenic?

Marshall and Butler (1966), who specifically searched for answers to that question, demonstrated that at every stage of embryonic development the lower and upper molars develop in perfect correlation and are potentially occludable. They concluded that "the mechanism by which this unity is brought about remains mysterious." The contemporary paradigm of odontogenetic mechanics suggests that there is a fine-tuning of respective signaling cascades and synchronous timing in both upper and lower molars.

Yet, the developmental context of upper and lower molars is very different and the differences in shape and size of potentially occluding crests are also different. Indeed, it is very hard to imagine a molecular mechanism capable of tuning the interlocking pattern over the whole dentition with precision at a micrometer level. In any case, despite certain variation (Polly et al., 2005), respective interlocking precision is the essential quality of tribosphenic dentition. This precision necessarily must be produced by developmental mechanisms and we are obliged to look for its explanation there.

12.4 Mammalian molars and developmental traits

Salazar-Ciudad and Jernvall (2004) proposed two models of regulation of tooth development: morphodynamic (with the appearance of apomorphic regulatory cues, the establishment of an autonomous regulation domain etc.) and morphostatic (where the morphogenetic variation is exclusively due to heterotopic or heterochronic modulation of growth). Morphodynamic regulation characterizes the early stages of molar development, while later development falls entirely in the domain of morphostatic regulation. The former is particularly linked to the specific apomorphy appearing during development of multicuspidate mammalian teeth: the temporary condensations of cells at the outer surface of inner enamel epithelium, first beginning at cap stage, called enamel knots. Since Ahrens (1913), the enamel knot is considered as the primary source of morphogenetic information or, more strictly, an organizing agent of the morphogenetic field of cusp formation (Butler, 1956, 1995).

Histogenetic techniques demonstrate that the enamel knot plays the role of an autonomous signaling organ controlling initiation of cusp formation via expression of signaling proteins. In turn this influences mitotic activity of cell populations that surround the knot resulting in specific folding of the basal membrane of the enamel epithelium and shaping the future enamel–dentine junction (Jernvall et al., 1994; Thesleff et al., 2001; Tucker and Sharpe, 2004). In molar primordia, the primary enamel knot activates infoldings of the epithelium into mesenchyme papilla, then disappears. Later, at the beginning of the bell stage, the secondary knots appear de novo (Matalova et al., 2005), at the tips of the cusps. Butler (1956) proposed and Jernvall et al. (1994), Jernvall (1995) and Salazar-Ciudad and Jernvall (2002, 2004) demonstrated in detail that additional cusps within tooth primordium are produced by heterotopic repetition of one and the same signaling module characterizing the enamel knot (e.g., Shh, Bmp-4 signals from mesenchyme, mass production of Fgf-4 by cells of the enamel knot, besides a large number of other signaling molecules). Correspondingly, the phylogenetic transformations in tooth shape or in the proportions of its structural components are established by specific spatial and temporal changes in triggering and switching of particular modules of dental development (and, last but not least, a different response of epithelial cells to these signals at different stages of their histogenesis). A high resolution three-dimensional map of these heterotopies and heterochronies overlain on a presumptive design of molar tooth (say a phylotypic blueprint of the mammalian molar) would be the best way to express taxon-specific dental characteristics. Unfortunately, this is beyond the scope of available methods to achieve directly via embryogenetic studies.

Nevertheless, a source of indirect but finely scaled information on that subject is available and it seems to be relatively easy (at least in comparison to methical requirements of embryogenetic studies) to gather it from a fairly broad spectrum of taxa. The source of this information can be found in spatial variation of enamel microarchitecture (e.g., enamel thickness or prism orientation). In this context it is important to remember that: (1) enamel production is enormously time consuming and has to cover a considerable part of the developmental time; (2) when an enamel crystallite and/or enamel rod is established it becomes an invariant component of a morphospace and preserves, by its physical qualities, the information on conditions under which it originated; and (iii) local differences in crystallite and/or enamel rod size reflect local differences in amelogenetic activity (similarly, local variations in orientation of enamel rods with respect to enamel–dentine junctions or another plane of comparison inform on the variations in orientation of the secretory front of ameloblasts and on local variations in morphogenetic dynamics within the tooth primordium). In short, a detailed three-dimensional map of the inner architecture of the enamel cap of a molar tooth can provide an informative reference on heterotopic variations in tooth morphogenesis during the mineralization stage and a reliable source of comparative information.

12.5   Enamel architecture of some chiropteran tribosphenic molars

Perhaps Loher (1929) was the first to study enamel microstructure in bats (*Myotis myotis*) in detail – then as a model taxon of primitive mammalian enamel. He provided detailed data on prism and crystallite size and relevant histological evidence suggesting the essential role of "Tomes' fibers" (odontoblast processes) in the organization of prismatic enamel. Other than descriptive notes by Friant (1964) and Boyde (1965), further data appeared much later. Lester and Boyde (1987) and Lester and Hand (1987) examined the basic enamel patterns in *Hipposideros diadema* and compared its occlusal surface enamel patterns to *Rhinolophus euryotis*, *Macroderma gigas*, *Taphozous georgianus*, *Nycteris* sp., *Artibeus* sp., *Mystacina tuberculata*, *Chalinolobus morio*, *Macroglossus minimus*, *Pteropus scapulatus*, *Dobsonia* sp. and two fossil species, *Palaeochiropteryx tupaiodon* and *Hipposideros nooraleebus*. Lester *et al.* (1988) supplemented that information with several species of phyllostomid bats and discussed certain specificities of that group.

Since then the topic has been addressed only rarely. Koenigswald (1997a), who provided a comprehensive survey of diversity in enamel patterns and "schmelzmuster" (Koenigswald and Clemens, 1992) in mammals, summarized the situation in bats (both with respect to previous studies and his own examinations) by noting that mainly radial enamel (prism orientation perpendicular to the enamel–dentine junction) is present with a thin prismless outer enamel, and that the crystallites of interprismatic matrix are mostly parallel to the prisms. In contrast to most other mammalian clades, neither any derived state of these characters (such as prism decussation, Hunter–Schreger bands, tangential or zipper enamel) nor any pronounced variability of enamel structure at the dentition level (Koenigswald, 1997b) was found in any bat. Considering the enormous evolutionary radiation in enamel structure among mammals typically accompanied by modification from simple to more complex structures within each particular clade and the uni-directionality of the trends in enamel differentiation (Koenigswald, 1997b), the situation in bats is quite exceptional. Bats are one of the few high-ranking clades that retain a number of dental plesiomorphies, including characters like an incomplete sheath of enamel prisms, otherwise only found in early mammals (Wood and Rougier, 2005).

The concluding textbook summary on bat molars (and tribosphenic molars in general) is that, in contrast to derived molar types with thick and hetero-topically divergent enamel, chiropteran molars are characterized by a primitive mammalian condition that includes thin and evenly distributed radial enamel with a cingular rim. This rim presumably compensates for the functional scantiness of the enamel cover by redistribution of mechanical stress to the crown base (Rensberger, 2000; Lucas *et al.*, 2008).

To re-examine these conclusions we developed a simple technique of serial sectioning of molar teeth and post hoc quantitative analyses of SEM images and applied it in several model species of vespertilionid, rhinolophid and molossid bats. Particularly detailed records enabling us to reconstruct three-dimensional patterns of enamel architecture for all teeth were obtained from *Myotis myotis*, a large-sized vespertilionid bearing a complete generalized dentition. In that species we also studied details of enamel maturation and eruptional dynamics of molars using a postnatal series of 89 individuals of known age from 0 to 65 days (previously reported, mostly in regard to external and postcranial characters by Sklenar, 1962, Sigmund, 1964, and Krátký, 1970).

12.5.1   Adult molars

In all species under study the results show that: (1) there is a greatly pronounced spatial heterogeneity in enamel thickness with a markedly thickened layer at shearing (i.e., the convex) walls of major crests, both in the upper and lower molars (i.e., palatal side of para-, meta- and protocristas in upper molars, labial side of protocristid, cristid obliqua and hypocristid in lower molars, Figures 12.1, 12.3); (2) in contrast, the reverse (concave) sides of shearing walls (i.e., paraconal and metaconal foveae, and the bottom of protocone fossa in the upper, trigonid fovea and talonid fossid in the lower molars) are covered by a very thin enamel layer; (3) while in the shearing walls the enamel thickness increases towards the cusp tip, the reverse is true for the reverse sides, so that the most contrasting differences in enamel thickness are just along the crest edges; (4) the enamel coat is built of three types of enamel, prismatic, interprismatic (IPM) and aprismatic (AP), which differ not only in mode of spatial integration of structural crystallites, but also in their mean length and volume (Figure 12.2); (5) aprismatic enamel forms a surface of the

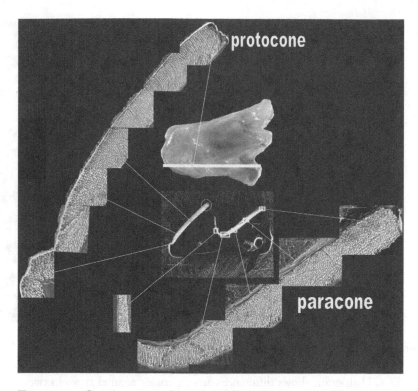

Figure 12.1 Cross section of an adult M1 *(Myotis myotis)*. Note spatial heterogeneity of enamel thickness and differences between protocone and paracone in enamel microarchitecture at their convex walls.

enamel coat and it appears as a denser layer in the labial cingular ridge of the upper molars and at the lingual base of the lower molars; (6) prismatic enamel is the dominant component of the enamel layer throughout the tooth – its relative volume (to that of IPM and AP) is consistently the highest around the middle of the column thickness; (7) despite the prismatic enamel being of the radial type only, it is far from being structurally homogeneous – namely (8) the patterns of deviations of prismatic axes from the plane perpendicular to the enamel–dentine junction (declination) and/or the vertical section plane (inclination) show conspicuous differences between spatial domains of particular major cusps (Figure 12.3); (9) at the same time, declination angles of the enamel prisms exhibit clear altitudinal changes along cusp walls from bottom to cusp tips, and the differences in these characteristics between particular cusps are even more conspicuous (Figure 12.1, the detailed quantitative data are published elsewhere – Špoutil *et al.*, 2010); (10) in contrast to differences among the major cusps of the upper molars in these characteristics, the shearing walls of lower molars, both trigonid and talonid, show a uniform

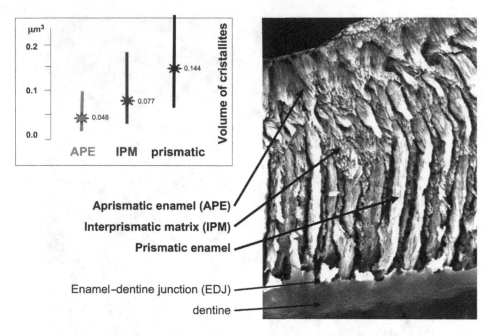

Figure 12.2 Cross section of the enamel layer (*Myotis myotis*: slope of adult M1 fossa) illustrating differences among three enamel types in size and spatial arrangements of crystallites. Histogram shows differences among major enamel types in two-dimensional volume of crystallites (estimated as s = length × width), stars = mean values, bars = span of recorded values (*n* = 1367). For further characteristics see Špoutil *et al.* (2010).

pattern, and consequently, it seems that there is no massive heterotopy in lateral expansion of particular structures of the lower molars during the late morphogenesis and mineralization of their enamel coat.

12.5.2 Ontogenetic aspect

The above-mentioned spatial heterogeneity in declination or inclination dynamics of enamel rods undoubtedly has something to do with the responses of the enamel epithelium to the morphogenetic movements of the tooth primordium. An essential question therefore is which part of tooth morphogenesis is recorded by these characteristics. To answer this question requires first to explain at which stage of mineralization the enamel prisms are formed. This is no easy task, of course, because the histological techniques applied in the study of earlier stages of tooth morphogenesis include decalcification, by which the enamel crystallites are removed. Nevertheless, a micro-dissection of embryonic tooth primordia (Figure 12.4) reveals that from the

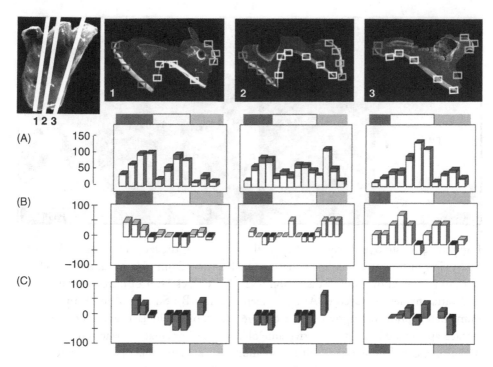

Figure 12.3 Enamel thickness (A), declination (B) and inclination (C) of enamel prisms at three vertical sections of adult M1 of *Myotis myotis*. Note considerable differences between cusps in combination of these characteristics.

beginning of mineralization the hydroxyapatite is primarily organized into prismatic form and also shows that the volume of embryonic enamel crystallites corresponds to that of mature prismatic enamel (see Figure 12.2). Consequently, the spatial rearrangements mapped by inclination/declination dynamics refer to the whole period of mineralization that, in main cusp tips, begins synchronously with the onset of major morphogenetic activity during the first half of the bell stage. During most of the tooth development the enamel rods remain unconsolidated and take the form of a loose bundle of hydroxyapatite nodules, the surface of which may resemble shrinkage cracks when dehydrated (Figure 12.5). The infilling of interprismatic matrix which consolidates them into a compact layer takes place much later – just shortly prior to eruption, and the surface aprismatic enamel (with the smallest and least spatially coordinated crystallites) is produced just below the gingiva immediately prior to eruption perhaps by the end of life of individual ameloblasts and under a reduced effect of organic intervention (Figure 12.4B).

Until consolidation by interprismatic matrix and the surface aprismatic layer, the prismatic enamel cover of the crown is flexible, the rods are disposed to

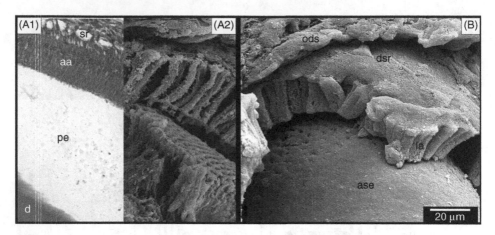

Figure 12.4 Histological section of enamel formation at the late bell stage (A1) and an SEM view of the corresponding situation at M1 primordium (*Myotis nattereri*, embryo 14 mm) sectioned by a micro-dissection (A2). Note thick layer of unconsolidated prismatic enamel (removed in A1 by decalcification). (B): Surface of tooth tip (P4) immediately prior to eruption – note changed form of ameloblasts and secretion of surface aprismatic enamel. aa, active ameloblasts, ase, aprismatic surface enamel, d, dentine, dsr, disintegrated stellate reticulum, la, late ameloblasts ods, outer dental sac, pe, prismatic enamel (removed in A1), sr, stellate reticulum.

plastic deformations and the enamel cover still does not present an ultimate mechanical constraint upon growth and shaping movements of the underlying mesenchyme–dentine core of the tooth. The completely matured enamel first appears at crest edges of the paracone, metacone and parastyle of upper molars, followed by the edge of the protocrista, and mesial and buccal cingulum. The enamel of the protocone complex (including its cingulum) matures much later, while the distal margin of the protocone complex, together with the bottom of the protocone fossa, are the zones where the enamel matures last. In the lower molars, the enamel maturation begins from the edge of the protocristid (including para- and metaconids) and trigonid cingulum, followed by the entoconid and lingual base of the tooth, while enamel maturation in the talonid is apparently delayed (it starts with the entocristid, followed by the tip of hypocristid, while the bottom of the talonid basin and cristid obliqua mature later).

Even during the time when the crest edges and mesial cingula are completely matured and performing initial occlusion actions, the deeper zones of the crown covered by unconsolidated enamel, as well as the foveae of major crests and bottom of the fossa, remain flexible and capable of responding to three-dimensional growth of the dentine core of the tooth. At that stage, the mineralized segments of the tooth crown can be looked upon as autonomous

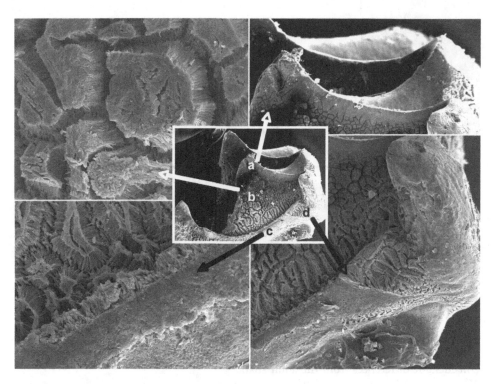

Figure 12.5 Mesial view of the enamel surface of M1 (*Myotis myotis*) at the onset of peri-eruption (postnatal stage 3, *c.* 14 days): note (a) the completely hardened enamel at crest edges and at (c) cingula, (b) thick layer of still unconsolidated prismatic enamel at walls of crown, still lacking interprismatic matrix and surface aprismatic enamel, and (d) suture zones between cingular elements (here mesial cingulum and parastyle).

structural units at least in the context that their mutual position and size are further modified during tooth eruption and the zones interconnecting them are until then almost without enamel cover.

The period from the first appearance of mineralized cusp tips above the gingiva to the complete mineralization of whole crown is rather long. In *Myotis myotis*, the first tips of molars appear by the end of the first week of life, while the bottom of the fossa is not completely mineralized until after weaning. The duration of molar eruption is about 65 days. During that time the size of molars more than doubles (Figure 12.6).

The detailed analysis of the eruption dynamics (46 metrical and non-metrical variables in a series of 89 juveniles covering that period) further demonstrated that: (1) the rate of (peri)eruptional growth of a particular dental element is neither isometric nor isochronous; (2) M1/m1 are the first permanent teeth which appear in occlusion (first with protoconids of the lower and paracones of

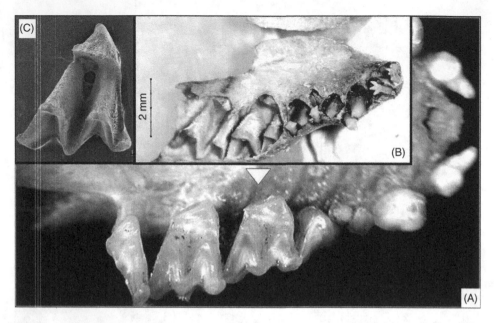

Figure 12.6 An upper jaw of adult *Myotis myotis* (A) compared to (B) a juvenile (postnatal stage 2, *c.* 12 days, same scale as in A). (C) M2 at roughly same stage showing considerable differences from the adult shape.

the upper molars); (3) the tips of the teeth of the unicuspid row appear after the eruption of the major cusps of M1/m1 and M2/m2 are complete. This is synchronous with the disappearance of deciduous teeth accompanying an accelerated growth of the I–P4 length; (4) the lingual axis of lower molars (paraconid–entoconid length, established in m1) appears in nearly an adult dimension at eruption and shows a minimum growth rate after that; (5) the trigonid width growth rate is nearly constant (particularly in m1) and relatively low, while the growth of the width of talonid shows a clear acceleration, both at the beginning and at the terminal stage of eruption (Figure 12.7); (6) the longitudinal dimensions of upper molars show an accelerated growth, particularly during the first half of the eruption period, with a certain delay to acceleration of lower molar growth (see Figures 12.7–12.8); (7) the asymmetric growth of paracone–protocone and metacone–protocone distances and the fluctuations in their growth rate (Figure 12.8) suggest a dynamic growth of the protocone complex and changes in its relative position during the second half of the eruption period; (8) despite the asymmetries, the angle between the postparacrista and premetacrista steadily increases, as does the areas of the paraconal and metaconal foveae; (9) nearly all variables show a constant decrease in variation during the eruption period, with minimum variation at

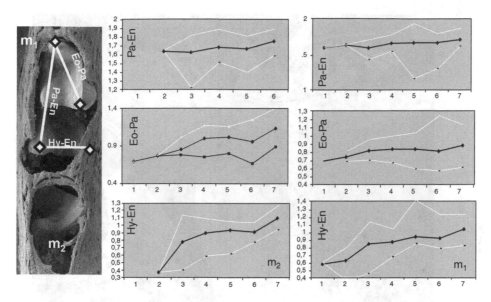

Figure 12.7 Biometry of the postnatal development of m1 and m2 in *Myotis myotis* (*n* = 89, 1–65 days): mean and min-max values of three metric variables (ordinates) for postnatal stages 1–7 of dental development (abscissa). In terms of postnatal days the stages can be roughly delimited as follows: stage 0 = day 0–5, stage 1 = 6–9, stage 2 = 9–12, stage 3 = 12–16, stage 4 = 16–21, stage 5 = 21–31, stage 6 = 32–45, stage 7 = 45–65.

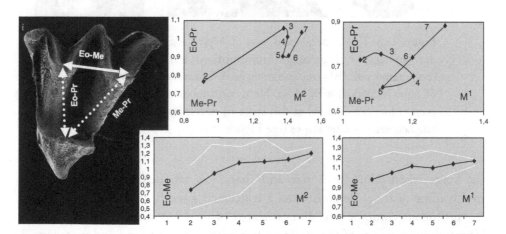

Figure 12.8 Biometry of the postnatal development of M1 and M2 in *Myotis myotis* (*n* = 89, 1–65 days). Upper: eo (= para) cone to protocone distance vs. metacone to protocone distances in individual stages of dental development (1–7) in M2 and M1. Lower: mean and min–max values of eo (= para) cone to metacone distance at postnatal stages 1–7. In terms of postnatal days the stages can be roughly delimited as follows: stage 0 = day 0–5, stage 1 = 6–9, stage 2 = 9–12, stage 3 = 12–16, stage 4 = 16–21, stage 5 = 21–31, stage 6 = 32–45, stage 7 = 45–65.

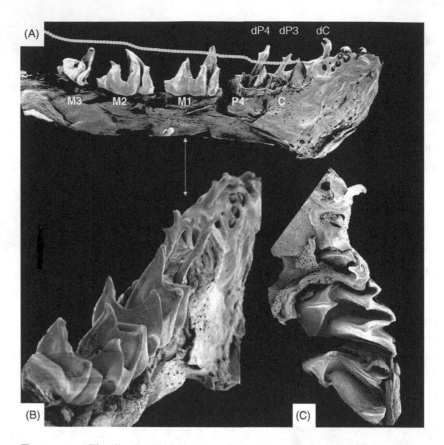

Figure 12.9 Totally decalcified jaws of *Myotis myotis* (stage 1 of postnatal development, i.e., beginning of m1 eruption): (A) lower jaw, lingual view (bone removed), (B) same specimen as A in distal occlusal view, (C) upper jaw of same individual with M1–2. Note continuous collagen surface of the teeth even at loci not calcified at this stage of development and not covered by enamel, even in later stages (see Figure 12.6).

the stage when adult tooth size and shape are attained; (10) the results of multivariate analyses demonstrated a concerted growth of the components of the protocone complex (including the area of fossa), but only faint relation to longitudinal variables of the upper molars, either the paracone and metacone – also the lateral enlargement of talonid exhibiting the fastest growth rate during the eruption period (particularly in m2) is only faintly related to growth of the trigonid (which is, conversely, almost isometric to growth of the lingual base of lower molars) – the lingual axis of the lower molars (m1 in particular) takes a central position of the factor space (in CCA representation), suggesting its indexing role in the geometry of the growth field and indicating that it may be an ordinational factor of peri-eruptional reshaping of the molariform row; (11) the application of the newly developed technique of total decalcification (which removes all enamel cover on the teeth) to embryonic dentitions (Figure 12.9) demonstrated that all teeth (even those in which the enamel cover is still not completely established and that split into separate cusps in a dry native preparation – such as m3, M3) are completely designed with all fine details of their mature crowns in dentine collagen. The collagen surface of dentin visible in SEM images of completely decalcified teeth shows in detail even those structures still not established in the enamel (deep roots of crests, cingula etc.); (12) against expectation, the collagen blueprints of teeth do not bear any vestigial or ancient structures (such as unresolved position of postcristid of lower molars, form of mesostyle or presence of para- and metalophs in the upper molars). Rather than the expected role of enamel formation, this suggests that the mesenchymatic papilla preforms design of the adult crown as a whole, without leaving the unresolved alternatives to be accomplished by enamel formation.

12.6   Conclusions

In fact, we are unable (because of scarcity of comparative data) to distinguish exactly which of the above-mentioned observations are specific just for the model species, which refer to specificities of bats and which refer to constitutional qualities of the tribosphenic molar organization in terms of essential apomorphy of mammals. Nevertheless, we believe that at least some of them may address topics of general significance. Here we will focus just on four of them.

12.6.1   Crest sharpness

An essential precondition of shearing efficiency results not only from occlusal sharpening and the geometric design of blade edges, as proposed by Popowics and Fortelius (1997) and Evans (2005), but also from constitutional asymmetry of enamel thickness along the crest edges, with very thick enamel at the convex and a very thin enamel at the concave sides. Inversely, in terms of developmental causality of the adult phenotype, the crests as the constitutional elements of tribosphenic design are preformed by extensive epithelial-mesenchymal evagination associated with heterotopic histo-differentiation of inner enamel epithelium along presumptive crest edges. The process starts at the bell stage and apparently represents an essential part of the developmental program that produces tribosphenic molars (Figure 12.10A). Hypothetically, this can appear as a result of strong lateral strain during early growth of a cusp, which then will produce convex ridges rather than straight vertical cones. Then

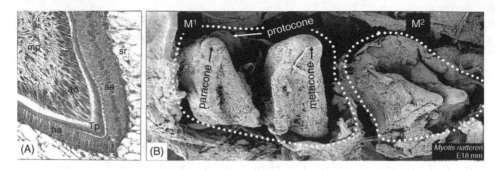

Figure 12.10 (A) Heterotopic histo-differentiation along presumptive crest edges at early bell stage is likely common to all tribosphenic teeth (here M1 *Monodelphis domestica*, postnatal day 4). (B) Topology of major cusps in embryonal primordia of M1 and M2 (*Myotis nattererii*, embryo length 18 mm): note position of protocone different from adult arrangement. ha, unmatured ameloblast, Tp, odoblast processes, other abbreviations as in Figure 12.4.

the above-mentioned differences in size and histo-differentiation of ameloblasts can simply respond to heterotopy in the growth of their spatial domains, with the expansion of the surface of the convex side and compression of that on the concave side. We can further hypothesize that the geometry of contextual rearrangements may produce a steady effect upon the shaping of the tooth primordia (cf. dorsoventral growth of the lateral wall of dentale vs. medial growth of palatal extension of maxilles towards palatal fissure, and palatal enlargement of the upper molars).

12.6.2   Modular structure of molar teeth

Although the common pattern of enamel thickness along major crests suggests that all are produced by the same developmental machinery, the greatly pronounced differences among the cusps in the angularity pattern of the prisms (at least in the upper molars) suggest considerable differences in their positional history during the formation of enamel prisms, i.e., movements of the enamel organs within the tooth primordium relative to vertical growth of the tooth. Uneven distribution of enamel thickness with the zones of thick enamel at cingula and shearing walls (convex sides of crests of major cusps) mutually separated by the zones uncovered by enamel (or covered by a very thin enamel layer with incomplete or delayed mineralization) provides to the tooth primordium considerable flexibility. Until the enamel of the latter zones is completely mineralized, the former can develop more or less independently and (as also confirmed by dissections of embryonic teeth) in positions different from the adult arrangement (Figure 12.10B). In the form of densely compressed structural modules, the major structures of a tooth can grow large in the relatively small spaces available in the embryonic jaw. The positional competition among primordial centers of the tribosphenic crown during early morphogenesis is perhaps responsible both for the excessive vertical growth of cusps and the lateral strain producing the V-shaped crest pattern. The lateral elements of the crown base may take the key role in that respect, i.e., parastyle and mesostyle in the upper molars and lingual base of the tooth (paraconid–metaconid–entoconid) in the lower molars.

In that respect the tribosphenic molar seems to be composed of several more or less *autonomous developmental units* whose mutual integration is postponed to the later stages of tooth morphogenesis. The large potential variation provided by that arrangement is, of course, strictly constrained by the demands of the functional integrity of the adult tooth and perfect interlocking of occluding crests over the entire dentition. The integrity is achieved by refined interaction of growth dynamics of these structural modules. The particular modules can be subdivided by their functional and developmental characteristics into three distinct groups: (1) modules of the crown base (providing a socket for the crest system); (2) modules of shearing crests; and (3) modules of the spaces "in between," including suppressed enamel formation. The modules composing these groups are, in the *upper* molars: (1a) parastyle, (1b) mesostyle, (1c) palatal cingulum, (2a) paracone, (2b) metacone, (2c) protocone, (3a) paracone fovea, (3b) protocone fovea and fossa, and, in the *lower* molars: (1a) metaconid, (1b) entoconid, (1c) hypoconulid, (1d) labial cingulum, (2a) protoconid, (2b) paraconid, (2c) hypoconid, (2d) cristid obliqua, (3a) trigonid fovea, (3b) talonid fossid.

12.6.3   Delayed enamel maturation, peri-eruption growth and reshaping of teeth

We demonstrated that the enamel prisms (at shearing walls) remain unconsolidated (with soft enamel) until eruption when the fine crystallites of interprismatic matrix and surface aprismatic enamel are inserted among them. Until then the respective zones of the tooth remain flexible and can be reshaped by heterotopic growth of the dentine core of the crown. In full it holds true for the enamel-less zones "in between" which can expand and redistribute the neighboring structures to distant adult positions. Thanks to that and further aspects of modular structure (see above), the tribosphenic molar can expand its size during eruption and the prolonged eruption period in bats allows an increase in rate of expansion to as much as 200% of initial size. The cusp tips and the crest edges, which first appear in eruption, are completely hardened (as are the major socket elements which fix the tooth position in the jaw – the mesiolabial base of the tooth and palatal cingulum in the upper molars and lingual axis in the lower molars) and can act in real occlusion events. The soft coated surface of the crown allows, at the same time, the tooth primordium to respond sensitively to the contextual signals, such as improper crest interlocking or ad hoc cues of actual morphogenetic and ecological contexts. Stress of imperfect peri-eruptional occlusion responded to by mesenchymatic papilla with heterotopic growth and apposition of dentine provides then a simple, yet apparently quite reliable and precise, mechanism of interactive fine-tuning of the crest interlocking pattern. The reshaped tooth resulting from this process is subsequently fixed into the adult tooth form by peri-eruptional enamel hardening.

In agreement with previous observations (e.g., Phillips, 2000), the eruption dynamics of the permanent dentition follows a hard-wired scheme of sequence and timing of particular events. M1/m1 are the first elements that start the process and establish the functional center of the dentition. The lingual axis of lower molars delimits the space of interaction – further steps include the interaction of the postprotocristid/preparacrista, while the talonid/protocone interaction and reshaping terminate the peri-eruptional reshaping of the molar.

12.6.4   Cope–Osborn concept of dental homology exhumed?

In correspondence with earlier discussions on the primary cusp of tribosphenic molars (Woodward, 1896; Osborn, 1907 etc.) and consensual opinion on that subject, Marshall and Butler (1966) centered their projections of growing stages of upper molars on the paracone as the singular point remaining apart from the lateral morphogenetic movements within the tooth primordium. Above we demonstrated dramatic changes in prism declination (Figures 12.1 and 12.3), which suggest massive lateral movements of the paracone during early morphogenesis, indicating that the paracone is not the true singular point of the upper molar primordium. In contrast, the expected characteristics seem to appear in the parastyle, the cusp that occupies exactly the most anterolateral position predicted for the primary cusp by Woodward (1896). In the lower molars, the invariant axial element of the lingual tooth socket, the metaconid, seems to represent the probable candidate for the most ancient structure. These conclusions address the conceptual issues of dental homology and should be, perhaps, briefly discussed in that context.

The traditional Cope–Osborn concept of dental homology can be summarized as follows (see Hershkovitz, 1971; Butler, 1978; Van Valen, 1982, 1994 for details): (i) all mammalian teeth are homologous to the single cone tooth of reptiles supplemented by (2) the additional structures which appeared first as secondary cusps at mesial and distal sides of the major cusp; (3) the additional structures (cusps, cingula, crests etc.) arise within a single tooth primordium by budding, not by concrescence; (4) the enlarged secondary cusps (mesial and distal) push the primary cusp out of the axis of occlusion, which results in a triangular arrangement of the three basic cusps; (5) the position of the primary cusp is then palatal in the upper molars, and labial in the lower molars, the basic cusp triangles of maxillary and mandibular molars (trigon and trigonid) are then inversely homologous; (6) the primary cusps retain their essential functional significance and structural dominance; (7) the further derived elements arise by budding on the buccal side of the tooth in the upper molars (styles) and distal to the principal cusp triangle, i.e., (7) talon and hypocone in the upper molars is homologous to talonid of the lower molars. The paradigmatic statements (1)–(4) refer to rather hypothetical situations, however, they present a default core framework of the dental homology concept and as such they can be accepted by definition. The others already have been refuted by empirical evidence to the contrary (Van Valen, 1982). Here we reconsider them with respect to the above conclusions. Statement (5) could be in a good accord with our findings if statements (6), (7) and (8) are replaced with the following: (6) the primary cusps are subsequently reduced, though they may play an essential polarizing role in early stages of tooth morphogenesis (e.g., as the agents contributing to development of convexity of crests of the secondary cusp), while (7) the dominant functional role is taken by the apomorphic elements (i.e., those situated palatally in the upper and labially in the lower molars). In that respect, (8) the paracone is homologous to the protoconid, and the talonid (or hypoconid) to the protocone of upper molars, while the talon and hypocone represent an apomorphic structure that appeared independently in various clades (Hunter and Jernvall, 1995) and cannot be directly homologized with the structures of lower molars. In any case, the developmental processing of these relations falls in the domain of morphodynamic regulation and for the above-mentioned reasons hypotheses on that matter can hardly be tested by direct morphological comparisons.

12.7   Implications

With the appearance of the tribosphenic molar and the apomorphies constituting its developmental background, mammals were released from the essential constraints forming the barriers of dental evolution in other vertebrates.

The structural modularity of the tribosphenic molar, including the possibility of peri-eruptional growth and late reshaping, provides a sensitive substrate for considerable variation – but within the framework of the tribosphenic design and insectivory, the feeding specialization constrains the life-trait pattern in many respects. Except for specialized foraging for social insects, feeding on insects requires agile locomotion and the ability to search for spatially dispersed food that is seldom abundant. Consequently, small body size, excessive spatial activity and high metabolic rate are directly related to insectivory, and apparently promote the r-selection effects upon the life-history traits, including shortening of the gestation and lactation period (the trend just the reverse of the demands of complex tooth development). Nevertheless, thanks to their capability for powered flight, bats, at least partly, succeeded in escaping from these constraints and evolved a pronounced K-strategy life-history pattern. The associated rearrangements of life-history traits, such as extensivelly enlarged parental investment, including prolonged gestation and lactation, also provided a developmental release for bat tooth morphogenesis.

Thus, the basic setting of dental morphogeny of bats (redirections of certain constraints, and specific heterochronies and heterotopies modifying the original developmental program) was most probably established within the frame of the complex constitutional rearrangements composing the chiropteran organization. The release of some developmental constraints (provided by prolongation of developmental time) could at the same time become an essential driving factor for increased variation in dental characters and adpative radiation of the early stages of chiropteran evolution. Yet, as any rearrangement in domain of the retrospective regulation of dental morphogeny can be fixed only when passed through a rigorous control of the prospective regulation (by the efficiency of respective dental rearrangement under actual ecological context) we are obliged to supplement the above hypothesis on the early dental evolution of bats with a brief comment on its contextual factors.

The environmental conditions of the Early Paleogene were undoubtedly much different from modern analogies and the contextual factors that promoted a transition from a non-volant insectivore to bats and the abrupt diversfication of the night aerial insectivory niche were probably quite unique. In any case it seems obvious that the night sky at the time of chiropteran early evolution must have been enormously rich in flying insects (say, the effect of pressure by diverse groups of daytime insectivorous birds, including aerial foragers and passeriforms – for the corresponding fossil record and/or time setting see, e.g., Mlíkovský, 2002; Beresford et al., 2005; Mayr, 2009). A large capacity of such a food resource – dense clouds of insects swarming at crepuscular and night hours – could eventually play the role of a sufficient driving factor promoting capability of flight and the rearrangements in sensory systems characterizing the earliest stages of bat evolution. Yet specialization for just such a kind of food resource also necessarily had an influence on the setting of the dental rearrangements. The clouds of swarming insects are charaterized by (1) enormous abundance and high density and (2) small body size and thin cuticula of their members. These qualities would strongly select (1) large capacity and rapid performance of each occlusing effort and (2) an extremely precise shearing effect. No wonder that the basic dental design of bats is characterized by the enlarged molariform part of the dentition, prolonging crests and refining their complex interlocking pattern.

12.7.1   Tribosphenic molars: developmental sources of variation and phylogenetic signal of dental characters

As demonstrated above, the perfection of interlocking occlusion, the key prerequisite for proper functioning of the tribosphenic dentition, is achieved neither by genetic regulation or by any other "mysterious" mechanism of intricate retrospective regulation, but is achieved exclusively within the domain of prospective and morphostatic (in the sense of Salazar–Ciudad and Jernvall, 2004) regulation. The regulation proceeds via the mechanical forces exerted from physical contact between the upper and lower teeth after initial eruption. Since the enamel on the non-erupted parts of the tooth is not hardened, the shape of the tooth can be modified – most probably under active contribution by the mesenchymatic papilla responding to uneven mechanical stress by heterotopic production of dentin (for the respective capabilities of dental papilla and included mechanisms see Yoshiba *et al.*, 2006, 2007; Lee *et al.*, 2010). The essential factors that make it possible are these: (1) "simple" radial prismatic enamel combined with (2) delayed enamel maturation via last minute apposition of IPM/APE; (3) constituent heterotopy of enamel thickness closely linked to appearance of crests; (4) modular structure of the tooth primordium and (5) positional flexibility of structural modules; and (6) a prolonged period of tooth eruption accompanied by (7) peri-eruptional growth and (8) late reshaping of the tooth crown by heterotopic activity of mesenchymatic papilla and apposition of dentine, subsequently fixed by peri-eruptional enamel hardening. We believe that the combination of these factors (or at least some of them) characterizes the developmental background of tribosphenic molars in general, and together with the morphodynamic mechanisms accelerating developmental heterotopy of the tooth primordium (enamel knots), it can be looked upon as the most essential apomorphy of the mammalian dental organization.

Yet, it seems well substantiated to expect considerable variation in each of these factors and to hypothesize that a combination of specific states presents the most pertinent odontologic characteristics of particular clades. Accordingly, particular grades of dental traits can be expressed in terms of differences in spatial and temporal settings of these factors and/or heterochronies in growth dynamics of particular structural units, and it could be hypothesized that the bifurcation points or transitional states between the respective grades of adult organization can be found in earlier stages of dental development. For instance, the essential grades of organization of bat lower molar, nyctalodonty and myotodonty (Menu and Sigé, 1971; Menu, 1985) can be, in terms of the developmental background, looked upon as a difference in growth rate of the modules composing the distal margin of the tooth, i.e., posthypocristid (= postcristid), entoconid, hypoconulid and talonid fossid. The early enlargement of the talonid fossid, delayed growth of the entoconid with early catchment of the distal posthypocristid at the hypoconulid margin of the tooth will produce nyctalodonty, while a delayed lateral to vertical growth of the hypoconid, delayed lateral to longitudial enlargement of the fossid and rapid vertical growth of the entoconid with early catchment of the hypocristid will produce myotodonty. Yet, according to the standard view of dental development (e.g., Bhaskar, 1991), the morpho- and histogenesis of inner enamel epithelium and enamel mineralization are to be looked on as the true agents of the differences in tooth morphology and ultimate processing of heterochronies should appear just in that domain. Against that expectation, the analysis of completely decalcified teeth in various stages of development demonstrates that the adult design of crown surface (as later visualized in its enamel coat) is entirely preformed in a collagenous blueprint produced by the mesenchymatic papilla. The expected transitional states may possibly appear as virtual steps of morpho- and histogenesis of mesenchymatic papilla, which finally produces the tooth as a single, completely integrated unit. In accordance with the conclusions of Loher (1929), this suggests that the dental papilla is the major agent of molar morphogenesis and tooth design, while the actual role of enamel epithelium in tooth morphogenesis can be much less than commonly proposed, a fact also stressed by further recent studies (see, e.g., Soukup et al., 2008; Koentges, 2008).

12.7.2   Common trends of phylogenetic morphoclines

The tribosphenic molar and heterodont dentition are designed for a generalized faunivory and generalized aerial insectivory is a plesiomorphic and widespread state in foraging biology of bats (Jones and Rydell, 2003).

Nevertheless, as the local diversity of bats is promoted by the diversity of their diet and foraging preferences (Findley, 1993; Patterson et al., 2003), it might be expected that this factor could have played a key role in the driving mechanism of early phylogenetic divergence in bats (Habersetzer and Storch, 1989). Specialization within faunivory is typically associated with an elementary dilemma: to forage larger or smaller (poorer but widespread) prey? Larger prey provides high energy intake per foraging action, but necessitates appropriate morphological and behavioral adaptations for immobilization and mechanical processing of the large prey. The dentition typically responds to this with increased size of canines, increased height of dentition and robustness of individual teeth combined with shortening of the postcanine elements and relative jaw length (which increases transposition of the power of mandibular adduction to canines, see Figure 12.13). Smaller prey, in contrast, provides less energy intake per foraging action, but compensates for this by requiring less energy expenditure, which can be further reduced with specialized foraging for mass prey aggregations (as is the case in fast aerial hawkers). The dentition is then required to increase the mean rate of food processing, which is typically achieved by enlargement of total occlusion space and maximizing the length of crests via enlargement of the molariform part of the dentition, including molarization of P4/p4. In both the cases, each specialization is accompanied by a reduction in premolar number and, particularly in those taxa that specialize on larger prey, a reduction in distal dental elements (M3/talonid of m3).

The fine molar traits controlled by a sophisticated set of constraining factors probably respond more slowly to specialization tendencies than do unicuspid teeth. Nevertheless, under selection for larger prey, which promotes high and robust teeth, the selection for the crest length can be suppressed (also the rate of "cuticula to be sheared relative to energy intake" is here much lower than in microfaunivory). In contrast, the specialization for smaller prey will undoubt-edly further enlarge crest length, either via relative enlargement of molar size or by specific enlargement of major crests by reshaping molar teeth or by enlarge-ment of crest-like structures within the connecting spaces (as para- and metaloph).

The reduction of premolars and M3/m3 obviously has contributed to phylo-genetic morphoclines in almost all clades of bats. The responses of the molar traits are, of course, quite diverse, far less uniform and not directly linked to degree of premolar reduction. A vast majority of chiropteran families show radical reduction of unicuspids (often with complete disappearance of P2/p2 and/or P3/p3) and a surprisingly high degree of premolar size reduction is well pronounced, even in the earliest and most primitive known bat, Onychonycteris finneyi (Simmons et al., 2008). Nevertheless, several clades of extant bats

(Natalidae, Thyropteridae, Myzopodidae) show opposite trends: enlargement of premolars, prolongation of the unicuspid row, evenly sized molars with unreduced M3/m3, and the state of these characters exceeds even those of the earliest bats. One of them (Kerivoulini – *Phoniscus* in particular) appears in the largest family of bats, Vespertilionidae, for which the generalized state of dental characters is particularly characteristic (Simmons and Geisler, 1998). All three premolars in both the upper and lower jaw are retained in *Cistugo*, Kerivoulinae and its sister clade, Myotinae, with the most diverse genus of the order, *Myotis*, showing broad variations in degree of premolar size reduction. The other genera of Vespertilionidae are characterized by further reduction terminating with the disappearance of P3/p3 and/or P2/p2 or both, often in specific combinations for particular genera (Miller, 1907). A random distribution of loss of premolars over all branches of the family tree (Hoofer and Van Den Bussche, 2003; Hoofer *et al.*, 2006; Stadelmann *et al.*, 2006) suggests that the process has developed convergently in different lineages of the clade, in a manner similar to that of other characters such as myotodonty/nyctalodonty (Horáček and Hanák, 1985). It seems quite probable that the respective rearrangements of dental system in each particular clade passed through a phylotypic stage of a complete but moderately reduced *Myotis*-like dentition. Correspondingly, the Oligocene and Early Miocene record consists almost exclusively of the forms bearing *Myotis*-like dentitions (Horáček, 2001; Ziegler, 2003) though robust molecular evidence (Stadelmann *et al.*, 2006) suggests a much younger date of about 15 Ma for basal divergence of *Myotis sensu stricto*.

Accordingly, despite common trends in premolar and M3/m3 reduction, the fine molar characters show considerable divergence. None of the molar characters is directly linked to degree of premolar reduction and the variation patterns of particular molar traits are mostly mutually independent (at least in a sample of 65 species of extant vespertilionid bats – see Figure 12.11). Despite absence of any significant correlation among particular dental characters, the clusters of species produced by the same analysis are quite consistent with standard generic and suprageneric affiliations. This suggests that the particular combinations of these dental characters are obviously clade specific, and may be relevant for phylogenetic analyses.

12.7.3   Grades, clades and dental phylogeny of bats

It is beyond the scope of this chapter to discuss dental phylogeny of particular chiropteran clades in detail. Nevertheless, it should be remembered that this was the primary motivation for our study and such a step should be taken in order to properly integrate the fossil record into chiropteran

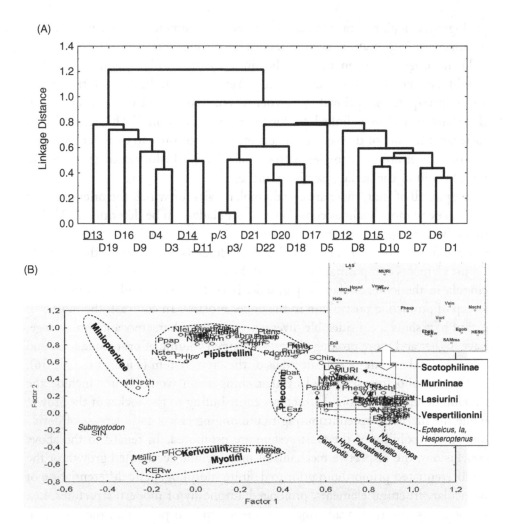

Figure 12.11 (A) UPGMA clustering of 24 non-metric dental characters in 65 species of vespertilionid bats based on the $1 - r$ distance (dynamic coding of character states scaled to categories 1 to 5). Note: absence of strong correlation among state of particular dental characters (except for P3/p3 reduction) and conspicuous independence among molar characters (underlined). (B) Factors I and II scores resulting from PCA based on state of respective dental characters. Note good correspondence with current taxonomic grouping (Hoofer *et al.*, 2006). For list of species and sample specimen see appendix in Hoofer *et al.* (2006). Variables, postcristid lingual termination (D1); lower incisor spatial integration (D2); I/3 relative size (D3); I/3 cusp number (D4); C/1 relative height (D5); P/2 mesio-distal compression (D6); P/2 height (D7); M/3 talonid size (D8); coronoid height (D9); Msup/paralophus reduction (D10); Msup/metalophus reduction (D11); Msup/transcrista reduction (D12); Msup/hypoconal extension of protocone complex (D13); M sup/distocrista reduction (D14); M3/reduction (D15); I1/distal cusp size (D16); I2/relative height (D17); I2/position, serial/displaced (D18); C1/shape (D19); P2/reduction (D20); P4/mesiolateral cingulum shape (D21); dentition elongate or compressed (D22); P3 reduction (p3/); p3 reduction (p/3).

phylogeny, complementary to analyses based on completely different sources of phylogenetic information, such as molecular phylogenetics.

Delineating phylogenetic morphoclines for particular dental characters should be here perhaps the first step. Yet, the straightforward trends are apparently quite exceptional in bat dental characters, and this concerns even the characters closely linked to foraging specialization. Rather typical is a variegated mosaic of derived characters often conforming to divergent special-izations, illustrating a complex pathway of adaptive history oscillating around the centroid of the aerial insectivory niche.

Despite all of that, there are certain molar traits for which the orientation of phylogenetic morphoclines common to most clades can be determined. Molars of early bats appear to be more compact, proportional in respect to the structural roles of particular functional modules (see above), with markedly robust elements composing the socket of the crown (such as the lingual axis and cingula in the lower molars or parastyle, buccal styles, cingula and talon-like sweep of protocone postvallum in the upper molars). In contrast, the molars of extant bats show a considerable divergence in these characters and, on average, have higher and more prolonged crests, enlarged volume of protocone fossa and talonid fossid and more delicate walls of the crest system (Figures 12.14–12.16).

The morphocline suggested by that comparison would hence include the developmental supression of structures contributing to the socket of the crown, while the mechanisms contributing to prolonging crests and increase of devel-opmental flexibility of the crest system are promoted. In regard to the above conclusions, the respective mechanisms may include accelerated growth of the undifferentiated primordium prior to definite morphogenetic differentiation of particular structural elements, prolonged autonomy of modern structures, i.e., major crests and their walls (modules 2a–2c) and, in particular, the zones "in between them" not covered by enamel (modules 3) for which late enlargement allows an extension of the overall size of the tooth and fine tuning of its shape during the peri-eruptional period. In terms of developmental dynamics it means a redirection of developmental energy to derived components of the tooth crown and enlargement of the domain of the prospective regulation in the latest stage of tooth development. As a result, the effects of modularity of molar structure (and its consequences for dynamics of dental adaptations) increases the developmental freedom for increased variation in pathways of mutual integration of particular structural modules.

Thus, on a phylogenetic scale, variation in structures interconnecting the basic modules and/or extending the essential tribosphenic design (e.g., meso-style, protofossa, para- and metalophs, hypocone and talon, hypoconid fossid, mesial end of cristid obliqua, distal postcristid, entoconid crest) may extensively

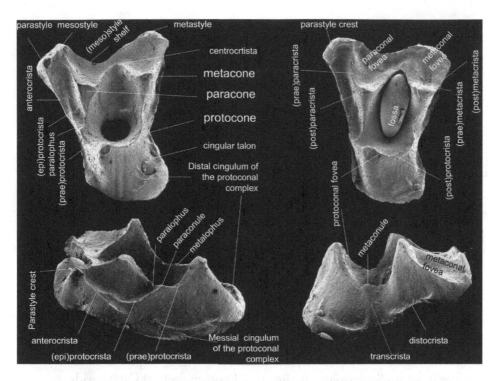

Figure 12.12 Upper molars of archaic bat *Ageina tobieni* (MP7 Mutigny, France) and topology of cusp and crests: M1 occlusal and mesial view (left, UCMP Louis 409Mu), M2 occlusal and distal view (right, UCMP Louis 375Mu), not to scale. Note roughly proportional volume of major modules, including excessively large parastyle with a separate parastylar crest, small volume of protocone fossa and shallow loop of centrocrista.

increase. Clade-specific divergences in the states of these characters illustrate this phenomenon quite convincingly and confirm that these characters and their combinations may be the best candidates for dental traits that provide the most robust phylogenetic signals (see Figures 12.12–12.16 for examples).

12.8 Concluding remarks

As illustrated by the above discussions, an immediate application of dental characters in phylogenetic analyses is not yet feasible for many groups. This is particularly true for studies of the relationship among distant clades (say at the family level), where the dental phenotype is typically characterized by a large number of autapomorphies or pleisomorphies, while the synapomorphies are often limited to just a few characters, often those for which a detailed taxonomic mapping of their character states is not available. Establishment of

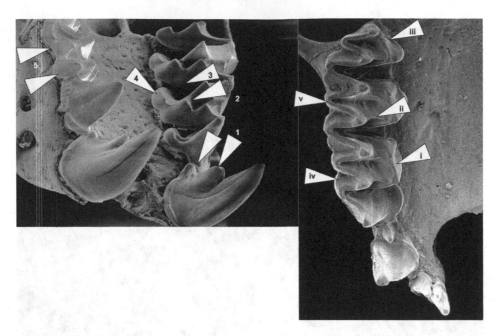

Figure 12.13 Innovations in upper molars related to specialization for larger (left: Megadermatidae: *Cardioderma cor*) and smaller prey (right: Vespertilionidae: *Pipistrellus pipistrellus bactrianus* feeding on Chironomidae). (1) Canine with supplementary distal cusps, (2) reduction of protocone and protocrista, (3) upper molar cingulum absent, (4) emancipation of a low distal talon extension, not related to crest system, (5) extensive increase in height of ectoflexus via enlargement of para- and mesostyle. (i) High and extended protocrista, mesio-distally prolonged, (ii) fossa broad and spacious, subdivided by para- and metalophs (often absent in derived dentitions), but (iii) closed with fusion of protocrista and metalophus at M3, (iv) crest high and delicate, fovea deeply incised, and (v) centrocrista tapered above mesostyle catchment.

higher taxa and/or inferences on interfamilial relations based on such weak evidence may start controversies. The current discussion accompanying the description of the fossil chiropteran family Mixopterygiidae Maitre, 2008 can illustrate such a situation quite well. The particular source of this controversy is the proposed position of the new family near the root of divergence of emballonurids and Rhinolophoidea, suggesting (corresponding to the traditional view – e.g., Van Valen, 1979) a sister group relationship between these clades (Maitre *et al.*, 2008). We do not wish to discuss here the actual content of the new clade nor will we comment on the relevance of its familial status. However, we feel it important to mention that this case is an example of a phylogenetic conclusion entirely based on odontologic evidence (certain similarities in dental characters among *Vespertiliavus*, Emballonuridae, some of the forms composing the new taxon and rhinolophoids) that necessarily comes in

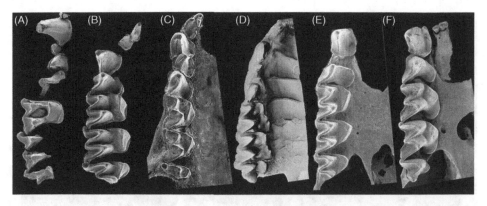

Figure 12.14 Primitive vs. derived upper dentition in selected examples (not to scale).
(A) *Palaeochiropteryx spiegeli* (Palaeochiropterygidae): primitive with molar row not
enlarged (in comparison to anterior teeth), protocone complex occupies most of tooth
volume, para- and metacone crests relatively small, centrocrista do not reach buccal
margin, complete anterior tooth row with unusual reduction of P_2 and small unicuspid
incisors. (B) *Hypsugo savii* (Vespertilionidae): modern with unicuspid row reduced,
moderately large C and P_4, relatively large incisors (I_1 bicuspid) and unreduced molar
row exhibiting derived characters (molar row occupies greater than half toothrow
length, crest system prolonged, para- and metacone fovea spacious and deep,
centrocrista tapered behind buccal margin of tooth, protocone complex enlarged
distally, transcrista integrated to postprotocrista, paraloph and metaloph absent).
(C) *Murina leucogaster* (Vespertilionidae): unique features include: very robust teeth,
extremely reduced M_3, heavy mesial premolar (?P_3), very large M_1 and M_2 with narrow
protocone, flat fovea without para- and metaloph, no transcrista, broad and fused
paracone-metacone fovea distinctly separated from buccal crown base, centrocrista
shallow, not reaching buccal margin, mesostyles absent. (D) *Emballonura alecto*
(Emballonuridae): combination of primitive and unique derived states including:
unicuspid section longer than molar row, but unicuspids relatively small, incisors conic
and very small, canine mesio-distally tapered, with incomplete cingulum, mesial
premolar reduced, P_4 molarization modest, molars prolonged but para- and metacones
remain relatively narrow while protocones are enlarged with extensive talons bearing
distinct cingular cusps, all crests present, conules and transcrista divide flat talon basin.
(E) *Rhinolophus* aff. lepidus ("kirgisorum" *sensu* Horáček *et al.*, 2000) (Rhinolophidae):
derived rhinolophoid arrangement with large canines, reduced premolars, distinct
mesio-distal compression of molarized P_4, enlarged molars, molar crests elongate,
unreduced M_3, para- and metalophs and transcrista absent. (F) *Nycteris thebaica*
(Nycteridae): derived as in (E), but exhibiting differences: M_3 reduced, P_4 less
compressed and with a different protocone crest pattern (extensive disto-palatal
extension of protocone base, small protocone fossa and incomplete postprotocrista).

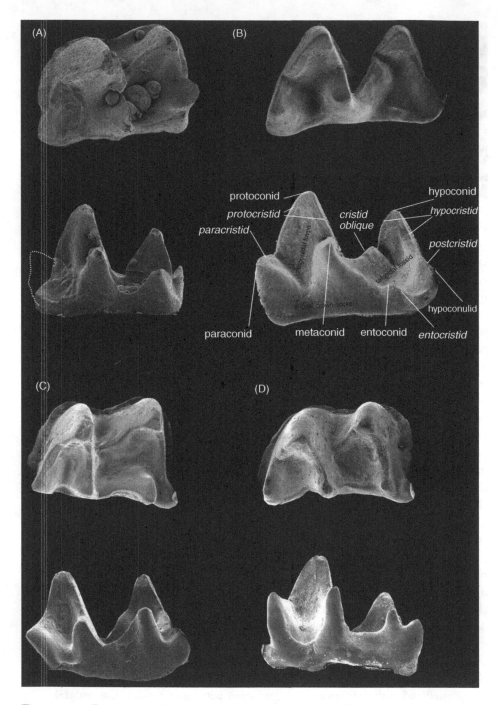

Figure 12.15 Primitive and derived states in m1: examples of topology of basic structures in occlusal and lingual views. (A) *Ageina tobieni* (MNHN-Mu5112) – note robustness, almost square in outline, massive labial cingulid, continuous lingual wall,

direct confrontation with the current phylogenetic hypotheses of the order based on robust support from molecular studies (e.g., Teeling *et al.*, 2000, 2002, 2005; Hulva and Horáček, 2002). Regardless of the factual validity of the newly proposed family, a formal comparison of the supporting arguments will clearly indicate the inferiority of odontological evidence compared to molecular information. Though we can believe in the relevance of the odontologic information and point out the empirical arguments of its importance for phylogenetic studies, we must admit that the immediate output of odontologic comparisons often fails to establish a sufficient platform for rigorous testing of alternative phylogenetic hypotheses corresponding to implicit usage of molecular data. Simply put, we need more robust odontologic information. To achieve this requires refining the definition of dental characters in regard to both their developmental setting and actual variation, identifying the orientation of their phylogenetic morphoclines and gathering a large amount of comparative data on dental character states in all clades of bats, both fossil and extant. We hope that the present chapter illustrates not only that such a task is unfortunately a bit complicated, but that – at the same time – there are ways to manage it.

Caption for Figure 12.15 (*cont.*) narrow trigonid basin, relatively low crowned, pre-entocristid not contacting base of metaconid, highly medially situated hypoconulid (necromantodonty see Sigé *et al.*, Chapter 13, this volume), rounded postcristid separated from postentocristid and posthypocristid, cristid obliqua labial and short, joins postvallid almost at right angle. (B) *Nycteris thebaica*: mostly the apomorphies, nyctalodonty (*sensu* Menu and Sigé, 1972) combined with reduction of entoconid, absence of pre-entocristid due to enlargement of metaconid, cristid obliqua joins high on distal wall of metaconid and not on postvallid (rare characters in bats but common in Didelphimorphia), disproportional enlargement of trigonid, particularly the paraconid and metaconid, talonid relatively reduced with low hypoconid, all suggesting adaptation to larger prey and carnivore-like food processing. (C) *Pipistrellus pipistrellus*: nyctalodonty with broad talonid, deep and wide trigonid fossa with narrow paracristid and metacristid, high entoconid, cristid obliqua joins postvallid labial of center, hypoconid high, distal position of hypoconulid along with integration of postcristid and (post)hypocristid results in enlargement of talonid basin (characteristic of molossids and vespertilionids that utilize small aerial prey). (D) *Otonycteris hemprichii*: myotodonty (*sensu* Menu and Sigé, 1972) and light construction (narrow crests) combined with adaptations for larger prey and carnivore-like food processing such as relatively enlarged and broadened trigonids, high para- and metaconids, and low hypoconid with sharp and high (post)hypocristid and indistinct (pre-) entocristid.

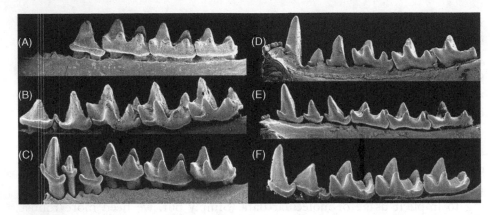

Figure 12.16 Lingual view of lower dentition illustrating some clade-specific rearrangements. (A) *Stehlinia* sp. (Eocene, Coll. IPHG Munich): generalized state dentition with large unreduced premolars, nyctalodont molars with heavily built lingual crown base and a massive high wall. (B) *Myzopoda aurita* (Recent, BM (NH)): a unique vertical tapering of molar crests, open trigonids, paraconids oriented mesially and not completely integrated into lingual crown base, which is formed by high and lingually protruding metaconid and entoconid separated by a deep incision, lingual wall continues as low cingular strip along lingual base of paraconid. (C) *Vespertilio murinus* (Recent, ISZ Prague): derived dentition with extreme premolar reduction and mesio-distal compression of the unicuspid row, large molars reducing in size from m1 to m3, major cusps of lingual crown edge equally developed and integrated into common wall. (D) *Rhinopoma cystops* (Recent, NM Prague): robust unicuspids with incomplete cingulids, large, unreduced m3, heavily built lingual base of trigonids and relatively small entoconids lacking well-marked entoconid crests. (E) *Emballonura alecto* (Recent, ISZ Prague): similar to *Rhinopoma* but differs in having molars with enlarged trigonids, relatively small para-, meta- and entoconids that are integrated into a moderately high lingual wall with a well developed entoconid crest (see Figure 12.14D). (F) *Nycteris thebaica* (Recent, NM Prague): unique mode of premolar reduction: with minute p4 displaced from tooth row, larger p3 not molarized but with sharp mesial shearing edge (for details of unique molar morphology see Figure 12.15B).

12.9   Acknowledgments

We are particularly obliged to Gregg F. Gunnell for inviting us to the project and for his careful reading of the manuscript which improved the chapter in essential ways. The detailed revisions of the manuscript by Alistar Evans and an anonymous reviewer were also enormously helpful. For important materials and repeated discussions on these topics we also thank Gerhard Storch, Bernard Sigé, Elodie Maitré, P. David Polly, Petr Benda, Vladimír Hanák and Robert Cerny. The study was supported by grants COST B23.3, GACR206/09/1245 and MSM 0021620828 (all to IH).

12.10   REFERENCES

Ahrens, K. (1913). Die Entwicklung der menschlichen Zähne. *Arbeiten aus dem anatomisches Institut, Wiesbaden*, **48**, 169–266.

Beresford, P., Barker, F. K., Ryan, P. G. and Crowe, T. M. (2005). African endemics span the tree of songbirds (Passeri): molecular systematics of several evolutionary "enigmas". *Proceedings of the Royal Society of London B*, **272**, 849–858.

Bhaskar, S. N. (ed.) (1991). *Orban's Oral Histology and Embryology*, 11th edn. St. Louis: Mosby Year Book.

Boyde, A. (1965). The structure of developing mammalian dental enamel. In *Tooth Enamel, its Composition, Properties, and Fundamental Structure*, ed. M. G. Starck and R. W. Fearnhead. Bristol: John Wright and Sons, pp. 163–196.

Butler, P. M. (1939). Studies of the mammalian dentition. Differentiation of the postcanine dentition. *Proceedings of the Zoological Society of London*, **109B**, 329–356.

Butler, P. M. (1956). The ontogeny of molar pattern. *Biological Reviews of the Cambridge Philosophical Society*, **31**, 30–70.

Butler, P. M. (1978). Molar cusp nomenclature and homology. In *Development, Function and Evolution of Teeth*, ed. P. M. Butler and K. A. Joysey. London: Academic Press, pp. 439–453.

Butler, P. M. (1995). Ontogenetic aspects of dental evolution. *International Journal of Developmental Biology*, **39**, 25–34.

Caton, J. and Tucker, A. S. (2009). Current knowledge of tooth development: patterning and mineralization of the murine dentition. *Journal of Anatomy*, **214**, 502–515.

Cope, E. D. (1874). On the homologie and origin of the types of molar teeth in Mammalia Educabila. *Journal of the Academy of Natural Sciences Philadelphia*, **8**, 71–89.

Cope, E. D. (1883). The tritubercular type of superior molar tooth. *Proceedings of the Academy of Natural Sciences Philadelphia*, **35**, 56.

Crompton, A. W. (1971). The origin of the tribosphenic molar. In *Early Mammals*, ed. D. M. Kermack and K. A. Kermack. London: The Linnean Society of London, pp. 65–87.

Crompton, A. W. and Sita-Lumsden, A. (1970). Functional significance of the therian molar pattern. *Nature*, **227**, 197–199.

Crompton, A. W., Wood, C. B. and Stern, D. N. (1994). Differential wear of enamel: a mechanism for maintaining sharp cutting edges. In *Advances in Comparative and Environmental Physiology*, vol. 18., ed. V. L. Bels, M. Chardon and P. Vandewalle. Berlin: Springer-Verlag, pp. 321–346.

Dumont, E. R. (1995). Enamel thickness and dietary adaptation among extant Primates and Chiropterans. *Journal of Mammalogy*, **76**, 1127–1136.

Evans, A. R. (2005). Connecting morphology, function and tooth wear in microchiropterans. *Biological Journal of the Linnean Society*, **85**, 81–96.

Evans, A. R. (2006). Quantifying relationship between form and function and the geometry of the wear process in bat molars. In *Functional and Evolutionary Ecology of Bats*, ed. A. Zubaid, G. F. McCracken and T. H. Kunz. New York: Oxford University Press, pp. 93–109.

Evans, A. R. and Sanson, G. D. (1998). The effect of tooth shape on the breakdown of insects. *Journal of Zoology*, **246**, 391–400.

Evans, A. R. and Sanson, G. D. (2003). The tooth of perfection: functional and spatial constraints on mammalian tooth shape. *Biological Journal of the Linnean Society*, **78**, 173–191.

Evans, A. R. and Sanson, G. D. (2006). Spatial and functional modeling of Carnivora and Insectivora molariform teeth. *Journal of Morphology*, **267**, 649–662.

Findley, J. S. (1993). *Bats. A Community Perspective*. Cambridge: Cambridge University Press.

Flynn, J. J., Parrish, J. M., Rokotosamimanana, B., Simpson, W. F. and Wyss, A. R. (1999). A Middle Jurassic mammal from Madagascar. *Nature*, **401**, 57–60.

Freeman, P. W. (1979). Specialized insectivory: beetle-eating and moth-eating molossid bats. *Journal of Mammalogy*, **60**, 467–479.

Freeman, P. W. (1981a). A multivariate study of the family Molossidae (Mammalia, Chiroptera): morphology, ecology, evolution. *Fieldiana, Zoology*, **7**, 1–173.

Freeman, P. W. (1981b). Correspondence of food habits and morphology in insectivorous bats. *Journal of Mammalogy*, **62**, 166–173.

Freeman, P. W. (1984). Functional cranial analysis of large animalivorous bats (Microchiroptera). *Biological Journal of the Linnean Society*, **21**, 387–408.

Freeman, P. W. (1988). Frugivorous and animalivorous bats (Microchiroptera): dental and cranial adaptations. *Biological Journal of the Linnean Society*, **33**, 249–272.

Friant, M. (1933). Contribution a l'etude de la differenciation des dents jugales chez les mammiferes. Essai d'une theorie de la dentition. *Publications diverses du Muséum national d'Histoire naturelle*, **1**, 1–132.

Friant, M. (1964). Sur le dévelopment de l'émail tubulé des Chiroptera. *Acta Zoologica*, **45**, 133–138.

Gidley, J. W. (1906). *Evidence Bearing on Tooth-Cusp Development*. Washington, DC: Washington Academy of Sciences.

Gould S. J. (1989). A developmental constraint in *Cerion*, with comments on the definition and interpretation of constraint in evolution. *Evolution*, **43**, 516–539.

Habersetzer, J. and Storch, G. (1989). Ecology and echolocation of the Eocene Messel bats. In *European Bat Research 1987*, ed. V. Hanák, I. Horáček and J. Gaisler. Praha: Charles University Press, pp. 213–233.

Hershkovitz, P. (1971). Basic crown patterns and cusp homologies of mammalian teeth. In *Dental Morphology and Evolution*, ed. A. A. Dahlberg. Chicago, IL: University of Chicago Press, pp. 95–150.

Hoofer, S. R. and Van Den Bussche, R. A. (2003). Molecular phylogenetics of the chiropteran family Vespertilionidae. *Acta Chiropterologica*, **5** (Suppl.), 1–63.

Hoofer, S. R., Van Den Bussche, R. A. and Horáček, I. (2006). Generic status of the American pipistrelles (Vespertilionidae) with description of a new genus. *Journal of Mammalogy*, **87**, 981–992.

Horáček, I. (2001). On the early history of vespertilionid bats in Europe: the Lower Miocene record from the Bohemian Massif. *Lynx*, **33**, 123–154.

Horáček, I. and Hanák, V. (1985). Generic status of *Pipistrellus savii* and comments on classification of the genus *Pipistrellus* (Chiroptera, Vespertilionidae). *Myotis*, **23–24**, 9–16.

Horáček, I., Gaisler, J. and Hanák, V. (2000). The bats of the Palearctic region: a taxonomic and biogeographic review. In *Proceedings of the 8th European Bat Research Symposium*, ed. B. W. Woloszyn. Krakow: Polish Academy of Science, pp. 11–157.

Hulva, P. and Horáček, I. (2002). *Craseonycteris thonglongai* (Chiroptera, Craseonycteridae) is a rhinolophoid: molecular evidence from cytochrome b. *Acta Chiropterologica*, **4**, 107–120.

Hunter, J. P. and Jernvall, J. (1995). The hypocone as a key innovation in mammalian evolution. *Proceedings of the National Academy of Sciences, USA*, **92**, 10718–10722.

Huysseune, A., Sire, J. Y. and Witten, P. E. (2009). Evolutionary and developmental origin of the vertebrate dentition. *Journal of Anatomy*, **214**, 465–476.

Jernvall, J. (1995). Mammalian molar cusp patterns: developmental mechanisms of diversity. *Acta Zoologica Fennica*, **198**, 1–61.

Jernvall, J. (2000). Linking development with generation of novelty in mammalian teeth. *Proceedings of the National Academy of Sciences, USA*, **97**, 2641–2645.

Jernvall, J. and Thesleff, I. (2000). Reiterative signaling and patterning in mammalian tooth morphogenesis. *Mechanisms of Aging and Development*, **92**, 19–29.

Jernvall, J., Ketene, P., Karavanova, I., Martin, L. B. and Thesleff, I. (1994). Evidence for the role of the enamel knot as a control center in mammalian tooth cusp formation: non-dividing cells express growth stimulating Fgf-4 gene. *International Journal of Developmental Biology*, **38**, 463–469.

Jones, G. and Rydell, J. (2003). Attack and defense: interactions between echolocating bats and their insect prey. In *Bat Ecology*, ed. T. H. Kunz and M. B. Fenton. Chicago, IL and London: University of Chicago Press, pp. 301–345.

Kangas, A. T., Evans, A. E., Thesleff, I. and Jernvall, J. (2004). Nonindependence of mammalian dental character. *Nature*, **432**, 211–214.

Kassai, Y., Munne, P., Hotta, Y. *et al.* (2005). Regulation of mammalian tooth cusp patterning by ectodin. *Science*, **309**, 2067–2070.

Koenigswald, W. von. (1997a). Brief survey of enamel diversity at the schmelzmuster level in Cenozoic placental mammals. In *Tooth Enamel Microstructure*, ed. W. von Koenigswald and P. M. Sander. Rotterdam: Balkema, pp. 137–162.

Koenigswald, W. von. (1997b). Evolutionary trends in the differentiation of mammalian enamel. In *Tooth Enamel Microstructure*, ed. W. von Koenigswald and P. M. Sander. Rotterdam: Balkema, pp. 203–236.

Koenigswald, W. von and Clemens, W. A. (1992). Levels of complexity in the microstructure of mammalian enamel and their application in studies of systematics. *Scanning Microscopy*, **6**, 195–218.

Koenigswald, W. von and Sander, P. M. (eds.). (1997a). *Tooth Enamel Microstructure*. Rotterdam: Balkema.

Koenigswald, W. von and Sander, P. M. (1997b). Glossary of terms used for enamel microstructures. In *Tooth Enamel Microstructure*, ed. W. von Koenigswald and P. M. Sander. Rotterdam: Balkema, pp. 267–280.

Koentges, G. (2008). Teeth in double trouble. *Nature*, **455**, 747–748.

Krátký, J. (1970). Postnatale Entwicklung des Grossmausohrs, *Myotis myotis* (Borkhausen, 1797). *Acta Societatis Zoologicae Bohemoslovacae*, **23**, 202–218.

Lee, S.-K., Lee, C.-Y., Kook, Y.-A., Lee, S.-K. and Kim, E.-C. (2010). Mechanical stress promotes odontoblastic differentiation via the heme oxygenase-1 pathway in human dental pulp cell line. *Life Sciences*, **86**, 107–114.

Lester, K. S. and Boyde, A. (1987). Relating developing surface to adult ultrastructure in chiropteran enamel by SEM. *Advances in Dental Research*, **1**, 181–190.

Lester, K. S. and Hand, S. J. (1987). Chiropteran enamel microstructure. *Scanning Microscopy*, **1**, 421–436.

Lester, K. S., Hand, S. J. and Vincent, F. (1988). Adult phyllostomid (bat) enamel by scanning electron microscopy – with note on dermopteran enamel. *Scanning Microscopy*, **2**, 371–383.

Lester, K. S. and von Koenigswald, W. (1989). Crystalite orientation discontinuities and the evolution of mammalian enamel – or, when is a prism? *Scanning Microscopy*, **3**, 645–663.

Loher, R. (1929). Beitrag zum groberen und feineren (submikroskopischen) Bau des Zahnschmelzes und der Detinfortsatze von *Myotis myotis*. *Zeitschrift für Zellforschung*, **10**, 1–37.

Lucas, P., Constantino, P., Wood, B. and Lawn, B. (2008). Dental enamel as dietary indicator in mammals. *Bioessays*, **30**, 374–385.

Luo, Z.-X., Cifelli, R. L. and Kielan-Jaworowska, Z. (2001). Dual origin of tribosphenic mammals. *Nature*, **409**, 53–57.

Luo, Z.-X., Kielan-Jaworowska, Z. and Cifelli, R. L. (2002). In quest for a phylogeny of Mesozoic mammals. *Acta Palaeontologica Polonica*, **7**, 1–78.

Luo, Z.-X., Ji, Q. and Yuan, C.-X. (2007). Convergent dental adaptations in pseudo-tribosphenic and tribosphenic mammals. *Nature*, **450**, 93–97.

Maitre, E., Sigé, B. and Escarguel, G. (2008). A new family of bats in the Paleogene of Europe: systematics and implications for the origin of emballonurids and rhinolophoids. *Neues Jahrbuch des Geologie u. Paläontologie – Abhandlungen*, **250**, 199–216.

Marshall, P. M. and Butler, P. M. (1966). Molar cusp development in the bat, *Hipposideros beatus*, with reference to the ontogenetic basis of occlusion. *Archives of Oral Biology*, **11**, 949–965.

Matalova, E., Antonarakis, G. S., Sharpe, P. T. and Tucker, A. S. (2005). Cell lineage of primary and secondary enamel knots. *Developmental Dynamics*, **233**, 745–759.

Maynard Smith, J., Burian, R., Kauffman, S. *et al.* (1985). Developmental constraints and evolution. *Quarterly Review Biology*, **60**, 265–287.

Mayr, G. (2009). *Paleogene Fossil Birds*. Berlin, Heidelberg: Springer-Verlag.

Menu, H. (1985). Morphotypes dentaires actuels et fossiles des Chiropteres vespertilioninés. Ie partie: Étude des morphologies dentaires. *Palaeovertebrata*, **15**, 71–128.

Menu, H. and Sigé, B. (1971). Nyctalodontie et myotodontie, importants caractères de grades évolutifs chez les chiroptères entomophages. *Comptes rendus de l'Académie des sciences Paris*, **272**, 1735–1738.

Miller, G. S. (1907). The families and genera of bats. *Bulletin of the United States National Museum*, **57**, 1–282.

Míšek, I. and Witter, K. (2002). Die ontogenetische Entwicklung des Gebisses beim Grossen Mausohr (*Myotis myotis*, Chiroptera). *Wien Tierärztliche Monatschriften*, **89**, 1–6.

Mlíkovský, J. (2002). *Cenozoic Birds of the World. Part 1: Europe.* Praha: Ninox Press.

Osborn, H. F. (1888). The evolution of mammalian molars and form of the tritubercular type. *American Naturalist*, **26**, 1067–1079.

Osborn, H. F. (1897). Trituberculy: a review dedicated to the late Professor Cope. *American Naturalist*, **31**, 993–1016.

Osborn, H. F. (1907). *Evolution of Mammalian Molar Teeth To and From the Triangular Type.* New York and London: Macmillan.

Patterson, B. D., Willig, M. R. and Stevens, R. D. (2003). Trophic strategies, niche partitioning, and patterns of ecological organization. In *Bat Ecology*, ed. T. H. Kunz and M. B. Fenton. Chicago, IL and London: University of Chicago Press, pp. 536–579.

Peterková, R., Lesot, H., Viriot, L. and Peterka, M. (2005). The supernumerary cheek tooth in tabby/EDA mice – a reminiscence of the premolar in mouse ancestors. *Archives of Oral Biology*, **50**, 219–225.

Phillips, C. J. (2000). A theoretical consideration of dental morphology. Ontogeny, and evolution of bats. In *Ontogeny, Functional Ecology, and Evolution of Bats*, ed. R. A. Adams and S. C. Pedersen. Cambridge: Cambridge University Press, pp. 275–316.

Polly, P. D., Le Comber, S. C. and Burland, T. M. (2005). On the occlusal fit of tribosphenic molars: are we understanding species diversity in the Mesozoic? *Journal of Mammalian Evolution*, **12**, 283–299.

Popowics, T. E. and Fortelius, M. (1997). On the cutting edge: tooth blade sharpness in herbivorous and faunivorous mammals. *Acta Zoologica Fennica*, **34**, 73–88.

Rauhut, O. W. M., Martin, T., Ortiz-Jaureguizar, E. and Puerta, P. (2002). A Jurassic mammal from South America. *Nature*, **416**, 165–168.

Reif, W. E. (1982). Evolution of dermal skeleton and dentition in vertebrates: the odontode-regulation theory. *Evolutionary Biology*, **15**, 287–368.

Rensberger, J. M. (2000). Pathways to functional differentiation in mammalian enamel. In *Development, Function and Evolution of Teeth*, ed. M. F. Teaford, M. M. Smith and M. W. J. Ferguson. Cambridge: Cambridge University Press, pp. 252–268.

Russell, D. E., Louis, P. and Savage, D. E. (1973). Chiroptera and Dermoptera of the French Early Eocene. *University of California Publications. Geological Sciences*, **95**, 1–55.

Salazar-Ciudad, I. and Jernvall, J. (2002). A gene network model accounting for development and evolution of mammalian teeth. *Proceedings of the National Academy of Sciences, USA*, **99**, 8116–8120.

Salazar-Ciudad, I. and Jernvall, J. (2004). How different types of pattern formation mechanisms affect the evolution of form and development. *Evolution and Development*, **6**, 6–16.

Sigé, B., Maitre, E. and Hand, S. J. (2007). The primitive condition of lower molars among bats. *Journal of Vertebrate Paleontology*, **27** (Suppl. to no. 3), 46A.

Sigmund, L. (1964). Relatives Wachstum und intraspezifische Allometrie der Grossmausohr (*Myotis myotis* Borkh.). *Acta Universitatis Carolinae – Biologica,* **1964,** 235–303.

Simmons, N. B. and Geisler, J. H. (1998). Phylogenetic relationships of *Icaronycteris, Archaeonycteris, Hassianycteris,* and *Palaeochiropteryx* to extant bat lineages, with comments on the evolution of echolocation and foraging strategies in Microchiroptera. *Bulleting of the American Museum of Natural History,* **235,** 1–182.

Simmons, N. B., Seymour, K. L., Habersetzer, J. and Gunnell, G. F. (2008). Primitive Early Eocene bat from Wyoming and the evolution of flight and echolocation. *Nature,* **451,** 818–821.

Simpson, G. G. (1933). Critique of a new theory of mammalian dental evolution. *Journal of Dental Research,* **13,** 261–272.

Simpson, G. G. (1936). Studies of the earliest mammalian dentitions. *Dental Cosmos,* **78,** 2–24, 791–800, 940–953.

Sklenar, J. (1962). Notes on biology and postnatal development in *Myotis myotis. Časopis Národního musea – prirodovedny oddil,* **131,** 147–154 (in Czech).

Smith, M. M. and Hall, B. K. (1990). Development and evolutionary origins of vertebrate skeletogenic and odontogenic tissues. *Biological Reviews of the Cambridge Philosophical Society,* **65,** 277–373.

Soukup, V., Epperlein, H.-H., Horáček, I. and Černý, R. (2008). Dual epithelial origin of vertebrate oral teeth. *Nature,* **455,** 795–798.

Špoutil, F., Vlček, V. and Horáček, I. (2010). Enamel microarchitecture of a tribosphenic molar. *Jornal of Morphology,* **271,** 1204–1218.

Stadelmann, B., Lin, L.-K., Kunz, T. H. and Ruedi, M. (2006). Molecular phylogeny of New World *Myotis* (Chiroptera, Vespertilionidae) interred from mitochondrial and nuclear DNA genes. *Molecular Phylogenetics and Evolution,* **43,** 32–48.

Stock, D. W. (2001). The genetic basis of modularity in the development and evolution of the vertebrate dentition. *Philosophical Transactions of the Royal Society of London B,* **356,** 1633–1653.

Storch, G. (1968). Funktionmorphologische Untersuchungen an der Kaumuskulatur und an Korrelierten Schadelstrukturen der Chiropteren. *Abhandlungen der Senckenbergischen Naturforschenden Gesellschaft,* **517,** 1–92.

Strait, S. G. (1993). Molar morphology and food texture among small-bodied insectivorous mammals. *Journal of Mammalogy,* **74,** 391–402.

Teeling, E. C., Scally, M., Kao, D. J. *et al.* (2000). Molecular evidence regarding the origin of echolocation and flight in bats. *Nature,* **403,** 188–192.

Teeling, E. C., Madsen, O., Stanhope, M. J. *et al.* (2002). Microbat paraphyly and the convergent evolution of a key innovation in Old World rhinolophoid microbats. *Proceedings of the National Academy of Sciences, USA,* **99,** 1432–1436.

Teeling, E. C., Springer, M. S., Madsen, O. *et al.* (2005). A molecular phylogeny for bats illuminates biogeography and the fossil record. *Science,* **307,** 580–584.

Thesleff, I. (2003). Epithelial-mesenchymal signaling regulating tooth morphogenesis. *Journal of Cell Science,* **116,** 1647–1648.

Thesleff, I. and Sharpe, P. (1997). Signalling networks regulating dental development. *Mechanisms of Aging and Development,* **67,** 111–123.

Thesleff, I., Keranen, S. and Jernvall, J. (2001). Enamel knots as signaling centers linking tooth morphogenesis and odontoblast differentiation. *Advances in Dental Research*, **15**, 14–18.

Tucker, A. and Sharpe, P. (2004). The cutting-edge of mammalian development; how the embryo makes teeth. *Nature Reviews Genetics*, **5**, 499–508.

Vandebroek, G. (1961). The comparative anatomy of the teeth of lower and non specialized mammals. In *International Colloquium on the Evolution of Lower and Non Specialized Mammals*, ed. G. Vanderbroek. Brussels: Koninklijke Vlaamse Akademie voor Wetenschappen, Letteren en schone Kunsten, pp. 215–320.

Vandebroek, G. (1967). Origin of the cusps and crests of the tribosphenic molar. *Journal of Dental Research, Suppl.5*, **46**, 796–804.

Van Valen, L. M. (1966). Deltatheridia, a new order of mammals. *Bulletin of the American Museum of Natural History*, **132**, 1–125.

Van Valen, L. M. (1979). The evolution of bats. *Evolutionary Theory*, **4**, 103–121.

Van Valen, L. M. (1982). Homology and causes. *Journal of Morphology*, **173**, 305–312.

Van Valen, L. M. (1994). Serial homology: the crests and cusps of mammalian teeth. *Acta Paleontologica Polonica*, **38**, 145–158.

Weil, A. (2003). Teeth as tools. *Nature*, **422**, 128.

Wood, C. B. and Rougier, G. W. (2005). Updating and recoding enamel microstructure in Mesozoic mammals: in search of discrete characters from Phylogenetic reconstruction. *Journal of Mammalian Evolution*, **12**, 433–460.

Wood, C. B., Dumont, E. R. and Crompton, A. W. (1999). New studies of enamel microstructure in Mesozoic mammals: a review of enamel prisms as a mammalian synapomorphy. *Journal of Mammalian Evolution*, **6**, 177–213.

Woodburne, M. O., Rich, T. A. and Springer, M. S. (2003). The evolution of tribosphery and the antiquity of mammalian clades. *Molecular Phylogenetics and Evolution*, **28**, 360–385.

Woodward, M. F. (1896). Contributions to the study of mammalian dentition, Pt. II., on the teeth of certain Insectivora. *Proceedings of the Zoological Society of London*, **1896**, 557–594.

Yoshiba, N., Yoshiba, K., Stoetzel *et al.* (2006). Differential regulation of TIMP-1, -2, and -3 mRNA and protein expressions during mouse incisor development. *Cell and Tissue Research*, **324**, 97–104.

Yoshiba, N., Yoshiba, K., Hosoya, A. *et al.* (2007). Association of TIMP-2 with extracellular matrix exposed to mechanical stress and its co-distribution with periostin during mouse mandible development. *Cell and Tissue Research*, **330**, 133–145.

Ziegler, R. (2003). Bats (Chiroptera, Mammalia) from Middle Miocene karstic fissure fillings of Petersbuch near Eichstätt, southern Franconian Alb (Bavaria). *Geobios*, **36**, 447–490.

13

Necromantodonty, the primitive condition of lower molars among bats

BERNARD SIGÉ, ELODIE MAITRE AND
SUZANNE HAND

13.1 Introduction

Two dominant structural types in the lower molars of insectivorous
bats have been described and their evolutionary implications interpreted: these
are known as the nyctalodont and myotodont conditions (Menu and Sigé,
1971). Although previously noted as differential characters by some authors
(e.g., Lavocat, 1961), these structures had not been the subject of extensive
study among bats. Since then, intermediate conditions, interpreted as transitional evolutionary steps between the two patterns, have been reported in
natural populations of both living and fossil bats.

Among the oldest known bats, a different but characteristic pattern in the
posterior structure of the lower molar is exhibited by various species, and is
interpreted here to represent the primitive condition of chiropteran lower
molars. It is the pattern displayed by the most archaic bats, notably within,
but not restricted to, archaeonycterids, although not all of them. These archaic
bats, known only as fossils, are reported from the Early Eocene of various and
presently disjunct regions of the world. The condition is less commonly
displayed by younger, more derived and taxonomically diverse fossil bats.
The classic fossil bat genus *Necromantis* Weithofer, 1887, now more accurately
dated as Middle to Late Eocene in age (Maitre *et al.*, 2007; Maitre, 2008;
Hand *et al.*, Chapter 6, this volume), well exemplifies this archaic lower molar
structure, and the name necromantodonty is used here to typify it.

In our synthetic attempt, we review the previous known molar patterns of
nyctalodonty, myotodonty and intermediate structures, before discussing
necromantodonty in detail, as well as the evolution of chiropteran lower molar
structure through time and space. Camera lucida drawings are used to illustrate
the major patterns in chiropteran lower molars. For detailed illustrations of the

Evolutionary History of Bats: Fossils, Molecules and Morphology, ed. G. F. Gunnell and
N. B. Simmons. Published by Cambridge University Press. © Cambridge University Press 2012.

teeth of the individual taxa discussed, readers are referred to the primary descriptions by the cited authors.

13.2 Nyctalodonty

The most common lower molar condition (best shown by m1 and m2 rather than the usually reduced posterior part of m3) is the nyctalodont pattern. This pattern is exhibited by most Recent bats that retain the original insectivorous diet of chiropterans, as well as their known fossil representatives and other fossil bats without known survivors. It consists of the distal border of the talonid being formed by the oblique or transverse postcristid issuing from the hypoconid and directly joining the generally small hypoconulid at the posterolingual corner or close to it. The nyctalodont type shows noticeable variation in the differentiation of its elements and also in relation to species size with the pattern less discernible among the smallest bats. This condition is typically illustrated by *Nyctalus* species (Figure 13.1 (1)) and is widely shared among living and fossil insectivorous bats of various families (e.g., *Natalus*, Figure 13.1 (2)) some among them, like hipposiderids or emballonurids, being nyctalodont with very few exceptions.

13.3 Myotodonty

A more advanced pattern, myotodonty, is where the postcristid directly joins the hypoconid to the entoconid. The postcristid is often tall and sharp, so forming a blade of more efficient cutting function (against the mesial face of the paracone in the following upper molar). In this pattern, the hypoconulid at the distolingual corner is often much reduced in height and volume in comparison to the entoconid, and is disjunct or linked to it by a tiny independent crest. This kind of structure is typically shown by several taxa within vespertilionoids, for example, among *Myotis* species and related taxa (leuconoformes *sensu* Menu, 1987) (Figure 13.1 (3)–(4)). Although more derived and less well represented than nyctalodonty among bats, the myotodont condition was independently acquired in some Tertiary bat lineages in different families (but mostly among vespertilionoids), as soon as late Early Eocene times for one lineage of still unknown relationships. Well-marked myotodonty occurs among species of *Philisis* from the Lower Oligocene of Fayum, Egypt (Sigé, 1985; Gunnell *et al.*, 2008) and Taqah, Oman (Sigé *et al.*, 1994).

13.4 Intermediate structures

Structures exhibiting intermediate conditions between nyctalodonty and myotodonty have been described in some bats. A submyotodont condition

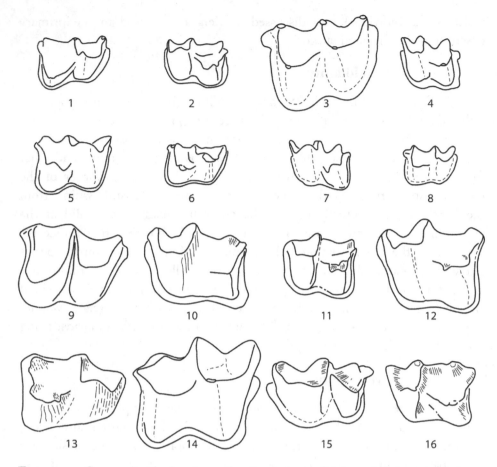

Figure 13.1 Camera lucida drawings of occlusal views of right or left lower molars m1 or m2 exposing different types of distal structure among various bat genera and families, mostly recent taxa and one fossil (*Necromantis*), and an extant tree shrew (Scandentia) *Tupaia*. Scale is ×25 for all except *Macroderma* and *Tupaia* which are ×12. 1. *Nyctalus leisleri*. 2. *Natalus tumidirostris*. 3. *Myotis myotis*. 4. *Kerivoula argentata*. 5. *Murina hilgendorfi intermedia*. 6. *Murina tubinaris*. 7. *Phoniscus javana*. 8. *Chalinolobus morio*. 9. *Noctilio albiventris*. 10. *Tonatia bidens*. 11. *Tonatia brasiliense*. 12. *Tonatia sylvicola*. 13. *Tonatia sylvicola*. 14. *Necromantis gezei*. 15. *Macroderma gigas*. 16. *Tupaia belangeri*.

described by Legendre (1984) among fossil molossids is where the postcristid divides into two parts, one joining the hypoconulid at the distolingual corner or slightly close to it, and the other linked to the entoconid. This represents an intermediate condition, which can sometimes be generalized in a specific natural population, but is absent in other spatial and/or chronological populations of the same lineage. A similar condition assessed to be intermediate between nyctalodonty and myotodonty has been described under the name murinodonty (Rossina, 2004). As found among different species of the extant Asian *Murina* genus, the lingual part of the postcristid is divided or not, between full nyctalodont to submyotodont conditions (Figure 13.1 (5)–(6)), like the situation described by Legendre (1984), with a branch meeting the tall, strong entoconid, and another joining the small and low posterolingual hypoconulid. Variously advanced steps towards myotodonty can be seen among several bats, such as *Phoniscus, Chalinolobus, Noctilio* and others (Figure 13.1 (7)–(9)).

13.5 Intraspecific transitional structures

The co-occurrence of nyctalodont, submyotodont and myotodont specimens in local populations of a single species has been observed and reported in living and fossil bats. Among the various extant noctilionid *Tonatia* species, specimens displaying all types, including transitional ones, occur (Figure 13.1 (10)–(13)). In the early Late Oligocene emballonurid *Vespertiliavus gerscheli*, nyctalodont, submyotodont and myotodont specimens occur together in the same population and locality in the Quercy area, southwestern France (Sigé, 1995). Nevertheless, the structure observed among the other numerous described *Vespertiliavus* fossil species and lineages (Maitre, 2008) is nyctalodont, and the highly variable condition shown by *V. gerscheli* appears to represent a genetically unstable condition, perhaps facilitating adaptive evolution under changing conditions.

13.6 Necromantodonty

When nyctalodonty and myotodonty were originally described (Menu and Sigé, 1971), the postulated primitive condition of the posterior structure of the lower molar was that of a postcristid comprised of two subequal parts meeting in an obtuse angle, one part linking the hypoconid to the hypoconulid, the other linking the hypoconulid to the entoconid. Rather than providing an example among fossil bats, this primitive condition was exemplified by a small eutherian predator with generalized molar morphology, described as species of the leptictid genus *Labes* from the Late Cretaceous of Spain and France (Sigé in Pol *et al.*, 1992). However, the fossil bat record now provides among its oldest representatives good examples of the early lower molar condition in this mammalian order. In these early bats, the hypoconulid is variably developed, acute or expanded, forming a median posterior projection, as a large inflated cusp or else a simple acute corner. Bats sharing this primitive condition are mostly archaeonycterids *sensu lato* and relatives, now known from the Early and Middle Eocene of most continents, e.g., *Australonycteris* (1), *Ageina* (2), *Honrovits* (3), *Eppsinycteris* (4), *Protonycteris* (5), as well as some unnamed and undescribed bats, and also some species tentatively attributed to *Icaronycteris* (6), *Archaeonycteris* (7) and *Hassianycteris* (8), such as *menui* (Russell *et al.*, 1973), *brailloni* (Russell *et al.*, 1973), *joeli* (Smith and Russell, 1992) and *praecursor* (Tabuce *et al.*, 2009). The numbers following these taxon names refer to their origin and authors as follows: (1) Murgon, southern Queensland, Australia, earliest Eocene, Hand *et al.*, 1994; (2) Mutigny and Avenay localities, eastern Paris Basin, Ypresian Early Eocene, Russell *et al.*, 1973; (3) central Wyoming, North America, Middle Ypresian Eocene, Beard *et al.*, 1992; (4) London Basin, Early Eocene, Hooker, 1996; (5) Vastan Mine, western India, middle Early Eocene, Smith *et al.*, 2007; (6) area and age as in (3), Jepsen, 1966, 1970; (7) Hesse area, western Germany, early Middle Eocene, Revilliod, 1917, emended in Russell *et al.*, 1973, Russell and Sigé, 1970; (8) same location, Smith and Storch, 1981.

Because of the difficulty in observing the occlusal surface of the lower molars in fossil specimens that remain partly imbedded in matrix or hidden by bone, the exact postcristid morphology in the type species of *Icaronycteris*, *Archaeonycteris*, *Hassianycteris*, *Tachypteron* (the latter regarded to be an emballonurid, Storch *et al.*, 2002) and *Onychonycteris* are not fully known. This makes the referral of new species to these genera uncertain or even inaccurate since they may be revealed ultimately to be significantly different in this character. This has occurred, and caused subsequent taxonomic confusion, for rare specimens of the type species *Archaeonycteris trigonodon* Revilliod, 1917 from Messel, and those of *A. revilliodi* Russell and Sigé, 1970, from the same locality. In the type specimen of *A. revilliodi* N° 4294, Me 16 (Russell and Sigé, 1970, fig. 19), the lower toothrow is well exposed in occlusal view: on m1 the hypoconulid appears to be large, extended posteriorly, well shifted lingually and posterior to the entoconid, thus clearly nyctalodont. Another specimen, an isolated m1 or m2 (HH 244), shows an expanded posterolingual lobe made of the fused hypoconulid and entoconid. From dentary specimens also exposed in lingual view, the dominant condition among Messel *Archaeonycteris* species seems to be nyctalodonty or subnyctalodonty, never necromantodonty. This situation is suggested by the lingual views of *Archaeonycteris* presented nearly a century ago (Revilliod o.c.). From the available evidence, several necromantodont species from various, geographically widely separated localities require reassignment to genera other than *Archaeonycteris*, species of which are characterized by nyctalodont or subnyctalodont lower molars.

Other fossil bats, geologically younger and morphologically more derived in various aspects, preserve some archaic features, including the plesiomorphic postcristid structure in their lower molars. Among these younger taxa is *Palaeophyllophora*, of Middle Eocene to Early Oligocene age, widely recognized as a peculiar hipposiderid lineage and abundantly represented in the paleokarstic Quercy deposits, southwestern France (Maitre, 2008). The necromantodont condition is variably marked among various *Palaeophyllophora* species (Maitre, 2008, pl. 19–21), most often with an expanded hypoconulid, more or less lingually shifted, sometimes up to a subnyctalodont pattern.

The genus which well illustrates the necromantodont type is *Necromantis* Weithofer, 1887 itself, so far only known from the French Quercy Phosphorites and presently recorded *in situ* from only four localities, of Middle to Late Eocene age (Maitre, 2008; Hand *et al.*, Chapter 6, this volume). Among the available *Necromantis* specimens, the hypoconulid is somewhat variable in position, most often occurring as an inflated median lobe (Figure 13.1 (14)), and in some specimens as large, but somewhat lingually shifted. Understanding about the phylogenetic relationships of *Necromantis* remains unclear, and its previously accepted or inferred attribution to megadermatids (e.g., Revilliod, 1920; Hand, 1985) now appears to be incorrect (Hand *et al.*, Chapter 6, this volume).

An unusual derived evolutionary step from necromantodonty, but nevertheless also distinct from nyctalodonty, is that of an inflated hypoconulid, posteriorly projecting, often taller and more voluminous than the entoconid, or even fused with it, then forming a rounded or lamellar posterolingual lobe on the molar (on m1 and 2, and even better developed on m3). This unusual type is well represented among megadermatids, including described fossil and recent *Megaderma* and *Macroderma* species (Figure 13.1 (15)), but with the latest Eocene *Saharaderma pseudovampyrus* from the Fayum Depression, Egypt, being nyctalodont, although otherwise clearly referable to Megadermatidae (Gunnell *et al.*, 2008). Among various other megadermatids, the reported *Megaderma* species from Beni Mellal, Middle Miocene of Morocco (Sigé, 1976) and some from Miocene localities in Australia (Hand, 1985) provide examples of this somewhat peculiar structure, presumably derived from necromantodonty.

Some older bats, such as *Archaeonycteris? storchi* and *Hassianycteris kumari* from the middle Early Eocene Vastan Mine, northwestern India (Smith *et al.*, 2007) display a lingually shifted hypoconulid that is clearly smaller and lower than the entoconid. This condition can be seen as nyctalodont or subnyctalodont and is shared by various bats, and may also occur as a dominant type in some taxa as intraspecific variation.

13.7   Bat lower molar evolution through space and time

Currently, the oldest described bats are earliest Eocene in age, very close to the Paleocene–Eocene boundary, and appear almost simultaneously in vastly distant areas, specifically southwestern Europe and southern Queensland, Australia. Given this wide geographic distribution, it is clear that continental areas continuously or temporarily linked to these regions were similarly involved in the Late Paleocene rapid worldwide dispersal of early bats, including large continental areas as yet without a significant fossil bat record. The known localities in southern France are Rians, with rare bat remains (Godinot, 1981), Fordones and Fournes, with more significant material including necromantodont lower molars (Marandat, 1991) and Silveirinha in Portugal, with the clearly necromantodont *Archaeonycteris? praecursor* (Tabuce et al., 2009; clearly not referable to *Archaeonycteris* because of the postcristid structure, as well as being very different in age). These southern European localities belong to the earliest Eocene MP 7 reference level of the European Paleogene mammal scale.

Somewhat younger, but still within the Early Eocene, are the Middle Ypresian bat records of North America and Europe, with the most informative skeletons being those from the Green River Formation, Wyoming – first found was *Icaronycteris index* (Jepsen, 1966, 1970, revisited by Simmons and Geisler, 1998), then more recently *Onychonycteris* (Simmons et al., 2008). Of similar age are bat teeth from Paris Basin localities, described as *Icaronycteris? menui*, *Archaeonycteris brailloni* and *Ageina tobieni* (Russell et al., 1973). Most of these bats, possibly including *Icaronycteris* and *Onychonycteris* (although direct observation is still unavailable), are generally necromantodont (with noticeable subnyctalodont variants included among *I.? menui* molars) and not of the *Archaeonycteris* type. The proposed emballonurid bat *Eppsinycteris* from Abbey Wood, London Basin (Hooker, 1996) is necromantodont. The Middle Ypresian *Honrovits tsuwape* from central Wyoming, which exhibits a mixture of archaic and derived features (Beard et al., 1992), is also necromantodont, as is *Hassianycteris joeli* from the Ypresian Evere locality, Belgium (Smith and Russell, 1992).

From available data, a significant change in chiropteran lower molar structure appears to be widespread geographically as early as the Middle-Late Ypresian. The best and most recently uncovered example of this is provided by the Vastan Mine bat fauna from western India (Smith et al., 2007, without information from the available upper teeth) where seven species are reported to date, among which four are nyctalodont (*Icaronycteris sigei*, *Cambaya complexus*, *Jaegeria cambayensis*, *Microchiropteryx folieae*), two subnyctalodont (*Archaeonycteris? storchi*, *Hassianycteris kumari*) and one necromantodont (*Protonycteris gunnelli*). Another significant global record is that of a clearly nyctalodont bat from a late Early Eocene locality in the Chubut province, Argentina (Tejedor et al., 2005), while the same condition occurs rather close in time but far in distance in *Dizzya exsultans* at Chambi, Tunisia (Sigé, 1991a). Reported Late Ypresian and early Middle Eocene bat faunas show an increasing proportion of subnyctalodont and nyctalodont taxa among them, including all Messel bat taxa (see below). The fauna from Prémontré (Late Ypresian, eastern Paris Basin), undescribed to date, contains several species of various sizes, mostly represented by isolated teeth: nearly all the lower molars are nyctalodont, but one species is myotodont, the earliest yet recorded (Dégremont *et al.*, 1985), well before the Early Oligocene myotodont bats from the Fayum, Egypt, and Taqah, Oman (Sigé, 1985; Sigé *et al.*, 1994).

The latest Early Eocene Vielase paleokarstic Quercy fauna (Legendre *et al.*, 1992), MP10 of the European mammal scale, has produced bat remains, among them an archaeonycterid known as the "Vielase bat." This bat has several markedly plesiomorphic features, including a lower molar structure varying between necromantodont and subnyctalodont, and molariform milk molars (Sigé, 1991b), features shared with another early bat, the much older *Icaronycteris? menui* from the eastern Paris Basin (Russell *et al.*, 1973, fig. 5). It is noteworthy that the remains of the Vielase bat were found in a paleokarstic context, which seems to provide the oldest evidence of subterranean roosting habits for bats.

Here we note two important paleokarstic mammalian faunas of Paleocene age: one from Walbeck in northern Germany (e.g., Russell, 1964) and the other from Itaborai in western Brazil (e.g., Bergqvist and Ribeiro, 1998, although the Paleocene age may be in doubt – this locality is now presumed to be closer to that of some of the earliest Eocene southern formations, such as the Patagonian Las Flores or the southeast Queensland Tingamarra (Murgon) Formation; see Gelfo and Sigé, in press). These fillings (Walbeck and Itaborai) have produced abundant vertebrate remains of various sizes, including numerous small mammals that have been intensively studied and published (Itaborai) or only partly described (Walbeck), both so far without reported bats. It would seem then that either bats were (improbably) not yet differentiated at this early time, or they had evolved, but were still unable to colonize karstic voids as roosting places (see below).

The most famous fossil bat fauna, among abundant other organisms represented, is that of the early Middle Eocene Messel oil shale deposits in western Germany. These shales have yielded an extremely rich collection of intensively studied bat skeletons (e.g., Revilliod, 1917; Russell and Sigé, 1970; Smith and Storch, 1981; Habersetzer and Storch, 1987; Storch *et al.*, 2002). Although only few specimens allow good occlusal views of lower toothrows, some at least are available and most of the reported genera (*Palaeochiropteryx, Hassianycteris, Tachypteron*) appear to be more or less markedly nyctalodont, with *Archaeonycteris* most commonly subnyctalodont (see above). Clearly, the necromantodont lower molar condition becomes significantly less common after the Late Ypresian, continuing the trend earlier shown by India's Vastan Mine bats. The Early Middle Eocene bats reported from the Geiseltal coal mines of northern Germany, representing several genera and species (e.g., Sigé and Russell, 1980) including *Palaeochiropteryx tupaiodon*, *Cecilionycteris prisca*, *Matthesia germanica* and *M.? insolita*, display nyctalodont lower molars in rarely available occlusal views. In eastern Asia, Middle Eocene bat remains described by Tong (1997) show the occurrence of nyctalodont bats, such as *Lapichiropteryx* sp. and other bat remains reported as "archaeonycterids?" Nevertheless, other figured isolated lower molars (Tong, 1997, plate 1; figs. 6, 7, 9), not recognized there as chiropterans, may well in fact represent necromantodont bats with some peculiar characters.

From the Middle and Late Eocene up to the Early Oligocene, only rare bat taxa such as *Necromantis* (Hand *et al.*, Chapter 6, this volume), and the peculiar hipposiderid *Palaeophyllophora*, retain the plesiomorphic necromantodont postcristid structure, or its accented condition described here among fossil and extant megadermatids. Otherwise, the remaining insect-feeding bats are mostly nyctalodont. Some lineages in some families, mostly vespertilionoids, more rarely emballonurids (e.g., the extant *Rhynchiscus naso*) and noctilionoids, evolved towards myotodonty. This derived condition appears in the record as soon as the late Early Eocene (Prémontré) and continues through to Late Tertiary and Recent times. Among noctilionoids, more varied dental structures evolved in concert with an extended array of feeding specializations.

The necromantodont structure of the earliest bats was evidently inherited from an ancestral non-volant arboreal insectivorous quadruped. This structural type is shared by various mammal groups, and is clearly shown by several small, archaic eutherian taxa (presumably predators of small terrestrial arthropods) as early as the latest Cretaceous and Paleocene. Among these are leptictids (e.g., the above-cited *Labes* in Europe), many lipotyphlans, such as nyctitheriids (e.g., the Paleocene and Early Eocene *Leptacodon*), various adapisoriculids, as well as various early primates. In strong contrast, extant *Tupaia* (Scandentia) exhibits a well-expressed nyctalodont condition (Figure 13.1 (16)). Most of the various small "insect-eating" eutherians of the Paleocene inhabited tropical environments and were probably arboreal quadrupeds, a habitat that may be corroborated by tarsal anatomy (such as argued by Hooker (2001) for nyctitheriids), and most were presumably nocturnal, strongly favoring echolocation in bats.

The relative morphological homogeneity of the earliest yet reported bats (archaeonycterids) leads us to presume that they represent a unique, monophyletic lineage, with their earliest evolution occurring relatively rapidly under the effect of strong selection pressure. Their earliest evolution probably occurred during Paleocene times within a restricted paleotropical forest environment, where conditions favoring the success of the group may have included tolerable predator pressure, hiding places for diurnal rest and abundant insect prey, caught perhaps first from the support of tree trunks and branches, but soon by jumping and then by the earliest short flights. These conditions would have favored rapid development of adaptations necessary for survival, then success, as a nocturnal aerial small mammal with advanced acoustic abilities. Further refinement of adaptations for flight would have occurred quickly, but also progressive differentiation of other traits, such as complex metabolic adaptations to temperate climates and hibernation.

The embryological stages of extant bats (e.g., Mohr, 1932) provide good insight into the earliest evolutionary physical changes in early bats, without the need for highly speculative reconstruction, for example, as regards upper limb (wing) proportions. In parallel, adaptation, refinement and diversification of the articular joints would have occurred, as well as in the muscular system and sensory functions. Along with these progressive evolutionary steps in functional morphology were realized, the available habitat had been colonized step by step, while adaptive diversification continued – in size, morphology, behavior and specialization in many traits. Lineages that were still relatively generalized rapidly spread over all accessible continental areas with more or less continuous supporting conditions for such mammals, i.e., dense, wet tropical forests. This earliest wide dispersal occurred during the Paleocene, well before its close, since at the very beginning of the Eocene, efficiently flying true bats had already colonized both southern Europe and Australia.

Although no Late Paleocene nor earliest Eocene bats have yet been recorded from there, the primary center for dispersal could have been from North America towards Europe via a North Atlantic route, and from there towards Arabia, Africa and Asia; towards oriental then southern Asia through the Bearing route (although as yet no bats of this age are reported from there); and towards Australia through Central then South America, then Antarctica, although again relevant records for such large areas are still lacking. The colonization by bats of the Indian plate, by the middle Early Eocene (*c.* 52–53 Ma), is possibly related to that of western Europe, as recently recognized for the biogeographic relationships of rodents (Rana *et al.*, 2008). From monotreme and marsupial fossil records (Pascual *et al.*, 1992; Sigé *et al.*, 2009), a trans-Antarctic route appears to have been used in the Late Cretaceous to Early Tertiary by terrestrial mammals moving between South America and Australia, *a fortiori* for flying ones.

This hypothetical, but evidently wide and rapid dispersal by early bats would have required almost continuous or at least temporary warm and wet tropical conditions supporting dense forest and abundant arthropod prey. The hunt for flying arthropods, first over tree branches and trunks, and later in full flight, would have honed the echolocation ability of early bats through intensive refinement in spatial location, hunting searches and recognizing sounds, including calls from other bats. The capacity for such precise hearing allows a mother *Tadarida braziliensis*, back from her nightly hunt for food, to relocate her baby from among the calls of thousands crowding the cave walls!

The first diurnal resting places for bats were presumably in trees and hidden places in vegetation (e.g., Charles-Dominique *et al.*, 2001 for present-day examples). As noted above, it seems that natural underground voids were not colonized by bats during their earliest evolution and paleogeographic dispersal. This happened later, but at least before the late Early Eocene. From that time, different kinds of bats began to exploit alternative diurnal roosting places (Sigé and Legendre, 1983), including various hiding places within open habitats, and subterranean sites such as natural caves, used for daily rest, or for long seasonal resting periods.

13.8 Acknowledgments

For a long time, many specimens of the recent world bat fauna, like most of those considered in the present study, have been made available thanks to the Museum National d'Histoire Naturelle in Paris (Mammals and Birds). As regards reference collections, various colleagues, including the present authors, involved themselves in field collecting and providing useful comparative material and/or systematic checking, such as Omar Linares, Nancy Simmons, Gérard Dubost, Joaquin Arroyo-Cabrales, Nick Czaplewski and the late Henri Menu. Bonnie Miljour redrew and improved Figure 13.1. Efficient assistance and cooperation has been offered over many years by Dr. Jean-Pierre Aguilar in Montpellier.

13.9 REFERENCES

Beard, K. C., Sigé, B. and Krishtalka, L. (1992). A primitive vespertilionoid bat from the early Eocene of central Wyoming. *Comptes Rendus de l'Académie des Sciences, Paris*, **314**, 735–741.

Bergqvist, L. P. and Ribeiro, A. M. (1998). A paleomastofauna das bacias eotertiarias brasileiras e sue importância na datação das Bacias de Itaborai e Itaubaté. *Asociatión Paleontológica Argentina, Publicación Especial*, **5**, 19–34.

Charles-Dominique, P., Brosset, A. and Jouard, S. (2001). Atlas des chauves-souris de Guyane. *Patrimoines Naturels*, **49**, 1–172.

Dégremont, E., Duchaussois, F., Hautefeuille Laurain, M., Louis, P. and Tétu, R. (1985). Paléontologie: découverte d'un gisement du Cuisien tardif à Prémontré (Aisne). *Bulletin d' Information des Géologues du Bassin de Paris*, **22**, 11–18.

Gelfo, J. N. and Sigé, B. (in press). A new didolodontid mammal from the late Paleocene – earliest Eocene of Laguna Umayo, Peru. *Acta Paleontologica Polonica*.

Godinot, M. (1981). Les mammifères de Rians (Eocène inférieur, Provence). *Palaeovertebrata*, **10**, 43–126.

Gunnell, G. F., Simons, E. L. and Seiffert, E. R. (2008). New bats (Mammalia, Chiroptera) from the late Eocene and early Oligocene, Fayum Depression, Egypt. *Journal of Vertebrate Paleontology*, **28**, 1–11.

Habersetzer, J. and Storch, G. (1987). Klassifikation und funktionelle Flügelmorphologie paläogener Fledermäuse (Mammalia, Chiroptera). *Courier Forschungsinstitut Senckenberg*, **91**, 117–150.

Hand, S. J. (1985). New Miocene megadermatids (Chiroptera: Megadermatidae) from Australia with comments on megadermatid phylogenetics. *Australian Mammalogy*, **8**, 5–43.

Hand, S. J., Novacek, M., Godthelp, H. and Archer, M. (1994). First Eocene bat from Australia. *Journal of Vertebrate Paleontology*, **14**, 375–381.

Hooker, J. J. (1996). A primitive emballonurid bat (Chiroptera, Mammalia) from the earliest Eocene of England. *Palaeovertebrata*, **25**, 287–300.

Hooker, J. J. (2001). Tarsals of the extinct insectivoran family Nyctitheriidae (Mammalia): evidence for archontan relationships. *Zoological Journal of the Linnean Society*, **132**, 501–529.

Jepsen, G. L. (1966). Early Eocene bat from Wyoming. *Science*, **154**, 1333–1339.

Jepsen, G. L. (1970). Bat origins and evolution. In *Biology of Bats*, vol. 1, ed. W. A. Wimsatt. New York and London: Academic Press, pp. 1–64.

Lavocat, R. (1961). Le gisement des vertébrés miocènes de Beni Mellal (Maroc). Etude systématique de la faune de mammifères. *Notes et Mémoires du Service des Mines du Maroc*, **155**, 29–94.

Legendre, S. (1984). Etude odontologique des représentants actuels du groupe *Tadarida* (Chiroptera, Molossidae). Implications phylogéniques, systématiques et zoogéographiques. *Revue Suisse de Zoologie*, **91**, 399–442.

Legendre, S., Marandat, B., Sigé, B. *et al.* (1992). La faune de mammifères de Vielase (phosphorites du Quercy, Sud de la France): preuve paléontologique d'une karstification du Quercy dès l'Eocène inférieur. *Neues Jahrbuch für Geologie und Paläontologie*, **7**, 414–428.

Maitre, E. (2008). Les Chiroptères paléokarstiques d'Europe occidentale, de l'Eocène moyen à l'Oligocène inférieur, d'après les nouveaux matériaux du Quercy (SW France): systématique, phylogénie, paléobiologie. Unpublished Ph.D. thesis, Université Claude Bernard – Lyon 1.

Maitre, E., Sigé, B. and Hand, S. (2007). *Necromantis*, new data and relationships. *Journal of Vertebrate Paleontology*, **27** (Suppl. to no. 3), 111a.

Marandat, B. (1991). Mammifères de l'Ilerdien moyen (Eocène inférieur) des Corbières et du Minervois (Bas-Languedoc, France). Systématique, Biostratigraphie, Corrélations. *Palaeovertebrata*, **20**, 55–144.

Menu, H. (1987). Morphotypes dentaires actuels et fossiles des chiroptères Vespertilioninés, 2ème Partie: implications systématiques et phylogéniques. *Palaeovertebrata*, **17**, 77–150.

Menu, H. and Sigé, B. (1971). Nyctalodontie et myotodontie, importants caractères de grades évolutifs chez les chiroptères entomophages. *Comptes Rendus de l'Académie des Sciences, Paris*, **272**, 1735–1738.

Mohr, E. (1932). In *Die Fledermäuse Europas*, ed. W. Schober and E. Grimmberger. Stuttgart: Kosmos Verlag, p. 51.

Pascual, R., Archer, M., Ortiz-Jaureguizar, E. O. *et al.* (1992). First discovery of monotremes in South America. *Nature*, **356**, 704–706.

Pol, C., Buscalinoni, A. D., Carballeira, J. *et al.* (1992). Reptiles and mammals from the Late Cretaceous new locality Quintanilla del Coco (Burgos Province, Spain). *Neues Jahrbuch für Geologie und Paläontologie, Abhandlungen*, **184**, 279–314.

Rana, R. S., Kumar, K., Escarguel, G. *et al.* (2008). An ailuravine rodent from the lower Eocene Cambay Formation at Vastan, western India, and its paleobiogeopgraphic implications. *Acta Palaeontolologica Polonica*, **53**, 1–14.

Revilliod, P. (1917). Fledermäuse aus der braunkohle von Messel bei Darmstadt. *Abhandlungen der Hessischen Geologischen Landesanstalt, Darmstadt*, **7**, 157–201.

Revilliod, P. (1920). Contribution à l'étude des Chiroptères des terrains tertiaires. 2° part. *Mémoires de la Société Paléontologique suisse*, **44**, 63–129.

Rossina, V. V. (2004). Murinodonty as the special type of lower molars of Chiroptera. *Bat Research News*, **45**, 146.

Russell, D. E. (1964). Les Mammifères Paléocènes d'Europe. *Mémoires Museum National d'Histoire Naturelle*, **13**, 1–324.

Russell, D. E. and Sigé, B. (1970). Révision des chiroptères lutétiens de Messel (Hesse, Allemagne). *Palaeovertebrata*, **3**, 83–182.

Russell, D. E., Louis, P. and Savage, D. E. (1973). Chiroptera and Dermoptera of the French early Eocene. *University of California Publications, Bulletin of the Department of Geological Sciences*, **95**, 1–57.

Sigé, B. (1976). Les Megadermatidae (Chiroptera, Mammalia) miocènes de Beni Mellal. *Annale de l'Université de Provence, Géologie méditerranéenne*, **3**, 71–86.

Sigé, B. (1985). Les Chiroptères oligocènes du Fayum, Egypte. *Geologica et Palaeontologica*, **19**, 161–189.

Sigé, B. (1991a). Rhinolophoidea et Vespertilionoidea (Chiroptera) du Chambi (Eocène inférieur de Tunisie). Aspects biostratigraphique, biogéographique et paléoécologique de l'origine des chiroptères modernes. *Neues Jahrbuch für Geologie und Paläontologie, Abhandlungen*, **182**, 355–376.

Sigé, B. (1991b). Morphologie dentaire lactéale d'un chiroptère de l'Eocène inférieur-moyen d'Europe. *Geobios*, **13**, 231–236.

Sigé, B. (1995). Le Garouillas et les sites contemporains (Oligocène, MP 25) des phosphorites du Quercy, Lot, Tarn & Garonne, France, et leurs faunes des vertébrés. 5: Chiroptères. *Palaeontographica A*, **236**, 77–124.

Sigé, B. and Legendre, S. (1983). L'histoire des peuplements de chiroptères du bassin méditerranéen: l'apport comparé des remplissages karstiques et des dépôts fluvio-lacustres. *Mémoires de Biospéologie*, **10**, 209–225.

Sigé, B. and Russell, D. E. (1980). Compléments sur les chiroptères de l'Eocène moyen d' Europe. Les genres *Palaeochiropteryx* et *Cecilionycteris*. *Palaeovertebrata, Mémoire Jubilaire en Hommage à René Lavocat*, 81–126.

Sigé, B., Thomas, H., Sen, S. *et al.* (1994). Les Chiroptères de Taqah (Oligocène inférieur, Sultanat d'Oman). Premier inventaire systématique. *München Geowissenschaftliche Abhandlungen (A)*, **26**, 35–48.

Sigé, B., Archer, M., Crochet, J.-Y. *et al.* (2009). *Chulpasia* and *Thylacotinga*, late Palaeocene-earliest Eocene trans-Antarctic Gondwanan bunodont marsupials: new data from Australia. *Geobios*, **42**, 813–823.

Simmons, N. B. and Geisler, J. H. (1998). Phylogenetic relationships of *Icaronycteris*, *Archaeonycteris*, *Hassianycteris* and *Palaeochiropteryx* to extant bat lineages, with comments on the evolution of echolocation and foraging strategies in Microchiroptera. *Bulletin of the American Museum of Natural History*, **235**, 1–182.

Simmons, N. B., Seymour, K. L., Habersetzer, J. and Gunnell, G. F. (2008). Primitive early Eocene bat from Wyoming and the evolution of flight and echolocation. *Nature*, **451**, 818–821.

Smith, J. D. and Storch, G. (1981). New Middle Eocene bats from "Grube Messel" near Darmstadt, W-Germany (Mammalia: Chiroptera). *Senckenbergiana biologica*, **61**, 153–167.

Smith, R. and Russell, D. E. (1992). Mammifères (Marsupialia, Chiroptera) de l'Yprésien de la Belgique. *Bulletin de l'Institut Royal des Sciences Naturelles de Belgique*, **62**, 223–227.

Smith, T., Rana, R. S., Missiaen, P. *et al.* (2007). High bat (Chiroptera) diversity in the Early Eocene of India. *Naturwissenschaften*, **94**, 1003–1009.

Storch, G., Sigé, B. and Habersetzer, J. (2002). *Tachypteron franzeni* n. gen., n. sp., earliest emballonurid bat from the Middle Eocene of Messel (Mammalia, Chiroptera). *Palaeontologische Zeitschrift*, **76**, 189–199.

Tabuce, R., Telles Antunes, M. and Sigé, B. (2009). A new primitive bat from the earliest Eocene of Europe. *Journal of Vertebrate Paleontology*, **29**, 627–630.

Tejedor, M. F., Czaplewski, N. J., Goin, F. and Aragon, E. (2005). The oldest record of South American bats. *Journal of Vertebrate Paleontology*, **25**, 990–993.

Tong, Y. (1997). Middle Eocene small mammals from Liguan Qiao Basin of Henan Province and Yuanqu Basin of Shanxi Province, Central China. *Palaeontologia Sinica*, **18**, 187–256.

Weithofer, A. (1887). Zur Kenntnis der fossilen Chiropteren der französischen Phosphorite. *Mathematisch-naturwissenschaftlich Classe*, **96**, 341–360.

14

Echolocation, evo-devo and the evolution of bat crania

SCOTT C. PEDERSEN AND DOUGLAS W. TIMM

The geneticists are trying to make evolution fit the genes rather than to make the genes fit evolution. (Osborn, 1932)

14.1 Introduction

Despite all other cranio-dental adaptations (Covey and Greaves, 1994; Dumont and Herrel, 2003), the microchiropteran head must function as an efficient acoustical horn during echolocation. This becomes infinitely more interesting when one considers that echolocation calls are either emitted directly from the open mouth (oral emitters), or forced through the confines of the nasal passages (nasal emitters). Given that oral emission is the primitive state (Starck, 1954; Wimberger, 1991; Schneiderman, 1992), the advent of nasal emission is viewed as a complex morphological innovation that required a substantial redesign of the microchiropteran rostrum: the nasal passages must be reoriented and aligned with the direction of flight, and they must have dimensions that provide for the efficient transfer of sound (resonance) through the adult skull. Once the acoustical axis of the head is established, bats emit a remarkable array of echolocation calls that reflect a great deal of behavioral plasticity. In the following treatment, we draw examples from developmental studies and functional morphology to illustrate how evolution has solved this intriguing design problem associated with nasal emission of the echolocation call.

14.2 Terminology: operational definitions

The term echolocation has been broadly applied to the Microchiroptera and to some members of the Megachiroptera. Despite evidence that shows that *Rousettus aegyptiacus* is able to navigate quite well by tongue clicking (Waters and Vollrath, 2003), there is no clear neuroanatomical, dental, developmental or physiological data whatsoever suggesting that pteropodids ever

Evolutionary History of Bats: Fossils, Molecules and Morphology, ed. G. F. Gunnell and N. B. Simmons. Published by Cambridge University Press. © Cambridge University Press 2012.

had the capacity for laryngeal echolocation or were derived from bats that did echolocate. Herein, the term "echolocation" will refer only to ultrasound produced by the larynx. It is our opinion that to do otherwise will confuse the understanding of the evolution of chiropteran communication, navigational skills and neural processing, i.e., ultrasound and tongue clicking should be considered separately during taxonomic analyses.

There is some confusion in the recent literature concerning mode of echolocation and call design (Pedersen, 2000; Eick *et al.*, 2005; Jones and Teeling, 2006; Jones and Holderied, 2007). Pedersen used the phrase "mode of echolocation" to refer to oral or nasal emission of the echolocation call (Pedersen, 1993, 1995, 1996, 2000), but more recently, others have used the term "mode" to refer to duty-cycle and band-width of a call (e.g., Fenton *et al.*, 1998). Herein, we will follow the latter use of the term and will specify oral or nasal emission separately.

There is an interesting dichotomy that is either overlooked or ignored in the taxonomic literature – that is, narrow-band, high-duty-cycle (CF) calls emitted by oral-emitting bats are quite different from those emitted by rhinolophids and hipposiderids; such oral emitted calls emphasize the fundamental frequency of the call, whereas nasal emitted calls emphasize the second harmonic and significantly reduce the fundamental (see discussion below).

This is a conceptual chapter. We are not testing phylogenetic hypotheses. We follow the taxonomy proposed by Simmons and Geisler (1998) with regard to nasal-emitting bats: Rhinolophoidea – nycterids, megadermatids, rhinolophids and hipposiderids; Noctilionoidea – phyllostomids.

14.3 Packaging of the head

The dynamic nature of the developing skeletal system is all too frequently overlooked in phylogenetic reconstructions wherein the skull is presented as an immutable structure into which the brain, ears and eyes are stuffed during development. Rather, the converse is more accurate; head growth and form are soft tissue phenomena affected only secondarily by osteological development (Hanken, 1983). As such, there is a great need to critically re-evaluate morphological data sets to see if epigenetic characters (e.g., location of various foramina, muscular processes, joint surfaces, linear dimensions of various squama) might be replaced by more conservative characters driven by the growth of cavities and spatial relationships amongst the various components of the head. That is, some characters or processes that were considered independent may not be as independent as they first seemed (Kangas *et al.*, 2004).

Indeed, early in development, it is differential growth of the brain and pharynx that governs the shape of the chondrocranium. Later, differential volumetric changes (brain, brainstem, eyes, tongue, teeth and pharynx) together with the ensuing mechanical competition for space within the confines of the growing head effect a cascade of modifications (often distant) to the shape, position and orientation of other structures throughout the growing skull via forces transmitted through the dura and periosteum to adjacent bones and sutures. These packaging concerns are accommodated/restricted by the developmental plasticity of each system in proportion to tissue composition, material availability, compliance in growth rates, the gross translation-distortion of elements *in situ* and *in utero* neonatal function (Haines, 1940; Moss, 1958; Burdi, 1968; Hanken, 1983, 1984; Smit-Vis and Griffioen, 1987; Müller, 1990; Hanken and Thorogood, 1993; Ross and Ravosa, 1993; Pedersen, 1995; reviewed by Pedersen, 2000).

14.4   Cephalometry

14.4.1   Rotation of the rostrum

Radiographic study of the angular relationships among the various skull components provides a size-free description of the basic internal arrangement of the head using internal landmarks and anatomical planes that are otherwise unavailable for morphometric analysis during a developmental study (Figure 14.1). As is typical in mammals, bat heads begin growth tucked firmly against the chest wall from where they rotate dorsad about the cervical axis (Figure 14.2). Simultaneously, the facial component of the bat skull rotates dorsad about the braincase. Certainly, rotation of the rostrum is limited in rate and direction by the ability of adjacent structures to get out of each other's way (Starck, 1952; Gaunt, 1967; Radinsky, 1968; Spatz, 1968; Sperry, 1972; Thilander and Ingervall, 1973; Moss, 1976; Tejada-Flores and Shaw, 1984; Smit-Vis and Griffioen, 1987; Schachner, 1989; Pedersen, 1993; Ross and Ravosa, 1993; Ostyn *et al.*, 1995; reviewed by Pedersen, 2000).

The motive forces behind these rotations are complex, but the brain has been identified as the primary driving force in primate skulls (Sperry, 1972; Thilander and Ingervall, 1973; Moss, 1976). However, it would appear that brain volume in bats has not played a strong role in the craniofacial form; rather, the relative size and differential development of the chiropteran brain is associated with the occupation of specific aerial/feeding niches (Eisenberg and Wilson, 1978; Stephan *et al.*, 1981; Jolicoeur *et al.*, 1984, Pedersen, 1993, 2000; Reep and Bhatnagar, 2000; Hutcheon *et al.*, 2002; Safi and Dechmann, 2005; Dechmann and Safi, 2009) leaving us to search for other forcing elements such as the olfactory bulbs, eyes and larynx. Pedersen (2000) argued that there was only enough room in the bat rostrum to deploy (anatomically speaking) two of the three sensory modalities available to bats at any one time (visual, olfactory, echolocation).

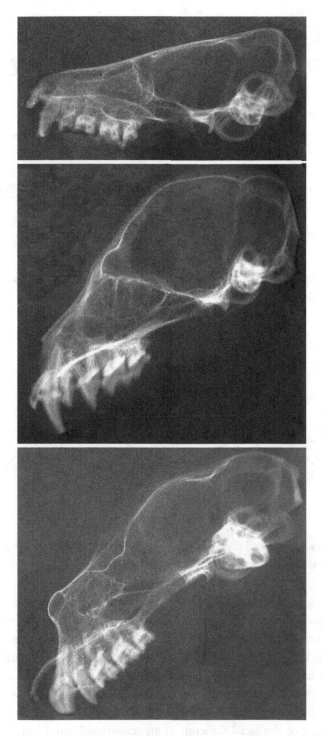

Figure 14.1 Representative skull radiographs (top to bottom): oral emitter (*Eptesicus*); phyllostomid nasal emitter (*Artibeus*); rhinolophid nasal emitter (*Rhinolophus*). Each skull is oriented such that the lateral semicircular canals share a similar orientation with the horizontal (approx. 15°).

EPTESICUS

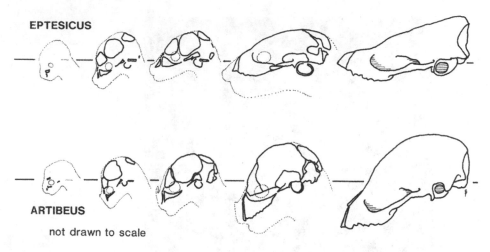

ARTIBEUS

not drawn to scale

Figure 14.2 The distinctive dorsad rotation of the orofacial complex in oral emitters is illustrated by *Eptesicus*. The orofacial complex in nasal emitters as exemplified by *Artibeus* remains "tucked" throughout development (from Pedersen, 1993).

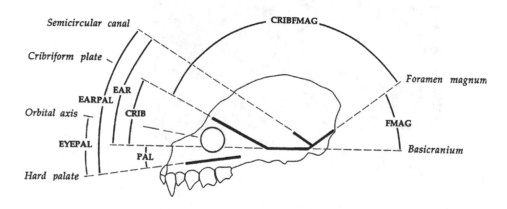

Figure 14.3 Anatomical planes and cephalometric angles are shown superimposed on a tracing of an *Artibeus jamaicensis* skull (from Pedersen, 1993).

During his analyses, Pedersen identified four anatomical planes that were readily located in radiographs of bat skulls: lateral semicircular canals, palate, foramen magnum and cribriform plate. Two angles relate these four anatomical planes in a functional context (EARPAL and CRIBFMAG; Figure 14.3; Pedersen, 1993, 1995, 2000), delineate the inertial and acoustic axes of the head and relate the general organization of the braincase to the rest of the body. These data clearly show that microchiropteran skulls follow a unique set of constructional rules based on the use of either the oral cavity or the facial skeleton as an acoustical horn. However, the fetal heads of oral-emitting and nasal-emitting taxa are grossly indistinguishable early in development. Species-specific skull morphology becomes increasingly

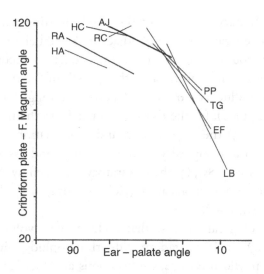

Figure 14.4 Bivariate plot of EARPAL vs. CRIBFMAG for fetuses and juveniles – ontogenetic trajectories of oral-emitting and nasal-emitting developmental series are presented in this scatterplot. Each trajectory moves from left to right across the plot. Nasal-emitting taxa (*Hipposideros armiger, Hipposideros galeritus, Rhinolophus affinus, Artibeus jamaicensis*) are clustered together in the upper left-hand corner of the plot, whereas the trajectories of oral-emitting taxa (*Eptesicus fuscus, Lasiurus borealis, Pteronotus parnellii, Taphozous georgianus*) "migrate" away from this nasal-emitting morphospace down towards the lower right-hand corner into an oral-emitting morphospace (from Pedersen, 1993).

more apparent, but always within the framework of either the nasal-emitting or oral-emitting cranial form (Figures 14.3–14.4). Remarkably, the internal dichotomous arrangement of the two "kinds" of head is well established before the skull has begun to ossify and well before the forces of mastication begin to affect skull morphogenesis. In all oral-emitting taxa, the hard palate rotates dorsally until it is aligned with or elevated above the basicranium and the echolocation call is forced directly out through the mouth. This skull form is the plesiomorphic condition for mammals (Starck, 1954; Wimberger, 1991; Schneiderman, 1992) and is clearly exemplified by mormoopids, emballonurids and vespertilionids.

Conversely, palates of nasal-emitting bats are retained ventral to the basicranial axis (Starck, 1952; Freeman, 1984; Pedersen, 1993). This dichotomy between oral- and nasal-emitting baupläne has imposed dramatic changes in general head posture and compensatory rotation of the otic capsules to align them with the inertial axis of the head. As a result, fetuses of oral-emitting species follow a very different developmental trajectory than do fetuses of nasal-emitting species. Skulls of nasal-emitting taxa remain within a well-defined morphospace through both ontogeny and phylogeny (Figure 14.4).

The innovation and evolutionary potential of the nasal-emitting baupläne have relied upon the morphogenetic plasticity of adjacent skeletal elements to accommodate changes throughout development and then function adequately in the adult. This balancing act is difficult because the growth of the mammalian rostrum and pharynx are influenced by many factors including: (1) tooth eruption (Lakars and Herring, 1980); (2) the tissue pressures from the muscles, lips and tongue (Proffit, 1978); (3) the organization and coordination of each muscle mass in proportion to the complexity of the dentition and associated dynamics of mastication (Herring, 1985); (5) the respiratory tidal airflow (Solow and Greve, 1979); and (6) phonation/echolocation (Roberts, 1972, 1973; Hartley and Suthers, 1988; Suthers et al., 1988).

Given this dynamic, it is of great interest that at least two evolutionary lineages (Rhinolophoidea – nycterids, megadermatids, rhinolophids, hipposiderids and Noctilionoidea – phyllostomids; sensu Simmons and Geisler, 1998) exhibit the anatomical requirements for the emission of calls through the nostrils (Simmons, 1980; Simmons and Stein, 1980; Hartley and Suthers, 1987, 1988, 1990; Pye, 1988). The developmental and the cephalometric data both suggest a classic example of convergent evolution on nasal-emitting baupläne driven by a developmental shift involving the rotation of the rostrum and inner ear. However, it is naïve to think that this can translate into a simple character state (see below).

Taken in isolation, nasal-emitting baupläne have evolved at least twice by retention of a developmental construct reminiscent of the fetal shapes of oral-emitting taxa (neoteny rather than hypo-morphosis). Because of the precise anatomical and physiological requirements needed for the efficient emission of ultrasound (Simmons and Stein, 1980; Pye, 1988; Suthers et al., 1988), intermediate states would be quickly weeded out, suggesting that the shift from oral emission to nasal emission must have occurred swiftly, both in developmental and evolutionary terms (Lewin, 1986; Price et al., 1993). Certainly, the divergence between oral- and nasal-emitting forms is an exaptation (Gould and Vrba, 1982) resulting from selective forces acting upon echolocation rather than the result of selection on cranial shape or head posture per se.

14.4.2 Anatomical landmark data

Multivariate analyses of landmark data allowed identification of developmental paths by which these taxonomically distinct clades arrived at their nasal-emitting baupläne. Cranial landmark data were taken from developmental series of bats (see Pedersen, 1995, 2000 for details; Figure 14.5) and these measurements were grouped into distinct suites of variables

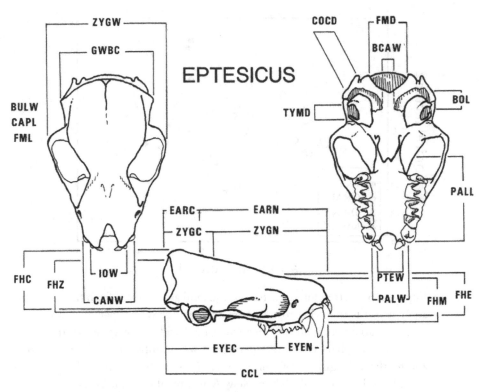

Figure 14.5 Anatomical measurements used in the canonical function analysis are shown on a tracing of an *Eptesicus fuscus* skull (from Pedersen, 1993).

according to function and/or anatomical relatedness. Each grouping was subjected to canonical analysis to identify covariance patterns among variables (Figure 14.6). In post-hoc identification of individuals, nasal-emitting and oral-emitting fetuses were rarely mistaken for each other and although early fetuses were frequently misclassified to the incorrect species, they were always assigned to the correct oral- or nasal-emitting types. For the most part, the clarity of these groupings can be attributed to the functional integration within each of the two major skull components (neuro- and viscerocrania).

Some unique features deserve discussion. Megachiroptera are clustered apart from the other developmental series because of their relatively large, albeit unspecialized, choanae and pterygoid complexes. Oral emitters are equally cohesive. The skull of rhinolophid bats, however, is characterized by a short, hard palate, large-bore choanae and a relatively long nasopharynx. Therein, the unique laryngo-nasal junction between the soft palate and the cartilages of the larynx (Matsumura, 1979; Hartley and Suthers, 1988) has forced a repositioning

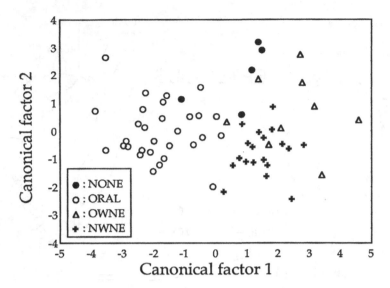

Figure 14.6 Scatterplot of the first and second axes derived from the canonical analysis of the following rostral characters: canine width, palate width, external narial width, choanal diameter, TMJ width, pterygoid width, hard palate length, pharynx length, interorbital width, infraorbital foramen width and toothrow length. Taxonomic abbreviations are as follows: OWNE = Old World nasal emitter, NWNE = New World nasal emitter, ORAL = oral emitter, NONE = Megachiroptera.

of all musculoskeletal elements associated with the soft palate and larynx (i.e., pterygoid width, choanal bore, pharyngeal length and otic capsule separation). Phyllostomids cluster near other nasal emitters, but exhibit some affinity for the megachiropteran morphospace most likely due to the fact that they both have well-developed visual and olfactory acuities. Given the diversity in phyllostomid dentition and diet, it would be very interesting to re-evaluate the packaging of the rostrum in this diverse group.

Therefore, the simplistic dichotomy between oral- and nasal-emitting skulls was revised to include the observation that there are at least two distinct developmental paths that generate a nasal-emitting skull and there are four fundamental assembly plans for the chiropteran head in general: (1) skulls relatively unmodified from the basic mammalian plan (oral emitters); (2) highly derived nasal-emitting forms built around expansive nasal cavities (rhinolophids, see discussion below); (3) nasal-emitting forms in which olfactory/visual modalities dominate the facial component of the skull (i.e., phyllostomids); and (4) skulls of non-echolocating megachiropterans that exhibit a wide range of skull shapes not restricted by the demands of ultrasonic echolocation.

14.4.3   Ossification rates and skull design

Have these gross reformulations of the bat skull altered the number of ossification centers or their sequence of appearance in any systematic pattern? We know that the shape and orientation of each element is strongly influenced by the enclosed volume, but ossification is usually independent of capsular growth. This permits epigenetic remodeling of the skull to accommodate developmental variation in the enclosed soft tissues and changing directive interactions in the mechanical environment, e.g., functional loading patterns during weaning (Haines, 1940; Washburn, 1947; Spyropoulos, 1977; Buckland-Wright, 1978; Alberch and Alberch, 1981; Herring and Lakars, 1981; Hanken, 1983, 1984; Griffioen and Smit-Vis, 1985; Herring, 1985; Von Schumacher *et al.*, 1986, 1988; Hoyte, 1987; Smit-Vis and Griffioen, 1987; Starck, 1989; Byrd, 1988; Kylamarkula, 1988; Ross and Ravosa, 1993; Ross and Henneberg, 1995).

Previous work suggested that bat ossification patterns follow the common mammalian pattern (de Beer, 1937; Pedersen, 1996, 2000): basicranial elements ossify in the correct posterior–anterior sequence, the auditory bullae and the ossicular chain are almost always the last series of bones to appear, and every bony element appears before skulls reach two-thirds of their expected adult size. Exceptions of note are related to the unique pharynx of rhinolophid/hipposiderid bats. The great expansion of their nasal passages has led to a local derangement of tissues, including the presence of a large fontanel between the nasal, maxillary and frontal bones that persists until well after birth. Such elemental translations and/or distortions to the midface are not uncommon in either developmental or evolutionary terms (Haines, 1940; Presley and Steel, 1976; Presley, 1981; Hanken, 1983, 1984; Pedersen, 1996, 2000). However, the unique coupling and suspension of the larynx and soft palate in rhinolophids manifests itself in a characteristic shift in the ossification sequence of the stylohyoid chain. These features do not appear to be shared with the phyllostomids, which exhibit a more generic ossification sequence (Pedersen, 1996).

14.4.4   Skull mechanics

Within each skull form, bat crania exhibit a stunning range of morphological diversity reflecting their diverse dietary specializations. The opposing jaws, teeth and muscles of mastication exist as a well-integrated functional unit, regardless of their relative position on the braincase (Starck, 1952; Czarnecki and Kallen, 1980; Freeman, 1984; Pedersen, 1993). Despite this generalization, does the dichotomy in skull form partition skull mechanics in any predictable manner?

During mastication, the rostrum and zygomatic arches brace the palate against the braincase posteriorly, which then transfers these forces to the occiput, cervical spine and thorax. The distribution of bone mass in the strongly ventro-flexed skulls of many rhinolophids/hipposiderids suggests a skull poorly designed to resist torsional/bending forces; the zygoma are relatively weak and the midface is attenuated dorsally – together leaving the pterygoid complex to be the primary reinforcement of the palate (Starck, 1952; Czarnecki and Kallen, 1980). As such, occlusal dynamics should help avoid structural failure in what can only be described as a flimsy rostrum, i.e., muscle, ligament and tendons must play a far more important role therein than do self-limiting features of the skeleton. In contrast, a generic nasal-emitting phyllostomid skull is more robust, exhibiting a solid midface and zygoma in all but the most extreme forms (e.g., glossophagines). Only the phyllostomines, wherein insects and vertebrates comprise the bulk of the diet, have retained tuberculosectorial teeth. It would be interesting to compare phyllostomine feeding mechanics with those of rhinolophids. Bat skulls that are strongly dorsi-flexed are typically constructed so that forces are passed directly from the rostrum to the braincase via the frontal bones (i.e., *Mormoops*) through a rather broad, robust interorbital midface.

One might predict that bats with extremely klinorhynchal (ventro-flexed) or aryrhynchal (dorsi-flexed) skulls cannot generate robust masticatory forces (Starck, 1952; Czarnecki and Kallen, 1980), rather, durophages must exhibit more moderate skull angulations within their oral- or nasal-emitting construct (*Cheiromeles* and *Vampyrum*, respectively; see Freeman, 1984). The inner dimensions of the nasal passages and the composition of the nasal septum are strangely absent from discussions concerning the evolution of echolocation or of skull mechanics in bats. Furthermore, given the spatial restrictions of the midface (Haines, 1940; Hanken, 1984; Hoyte, 1987), it seems unlikely that a nasal-emitting skull could exhibit both large olfactory fossae and resonating chambers within the interorbital midface. In the balance, phyllostomids may have retained olfaction at the cost of loudness of the call, while rhinolophids may have emphasized loudness of the call at the expense of olfaction and robust mastication (Pedersen, 1996, 2000).

14.4.5   Ultrasonic vocalization

The echolocation pulses emitted by the larynx in microchiropterans consist of a fundamental frequency (f_1) that may be accompanied by several overtones, or harmonics. Vocalizations are modified in the vocal tract by the differential filtering and amplification of various frequency combinations. Filtering (removal of harmonics) is affected by changes in the length and diameter of the vocal tract and discontinuities in the pharyngeal wall. Ultrasound produced by the larynx is subsequently emitted from either the mouth or the nostrils.

Regardless of the orifice, ultrasonic calls show striking variation in terms of emitted power and frequency structure (broad band, low-duty-cycle, frequency-modulated (FM) multiharmonic calls; narrow band, high-duty-cycle, "single" harmonic calls and everything in between). Depending on prey type, foraging strategy and habitat complexity, many species of bats will shift between broad- and narrow-band types of calls (e.g., *Eptesicus fuscus*; Surlykke and Moss, 2000), or modulate where they put energy into each type of call. Consequently, there is no fixed relationship/constraint between call structure and taxonomy, except perhaps in the rhinolophids and hipposiderids, where tuned nasal cavities impose significant restrictions on the emitted sound (Pedersen, 2000).

Call design has been mapped onto various molecular and morphological phylogenies (Eick *et al.*, 2005; Jones and Teeling, 2006; Jones and Holderied, 2007) and Jones and Teeling (2006) state explicitly that "Overall, our perspective on the evolution of echolocation is clouded by the diversity and plasticity of signals that we see in extant bats, suggesting that the animal's habitat is often more important in shaping its call design than is its evolutionary history." However, this particular statement neglects a body of work showing that the evolution of nasal-emitting heads has imposed significant mechanical restrictions on echolocation (Roberts, 1972, 1973; Hartley and Suthers, 1988; Suthers *et al.*, 1988; Pedersen, 1996). More specifically, the echolocation calls of oral-emitting bats vary from narrow-band calls without harmonics to broad-band calls with harmonics, and may combine the two. Some oral-emitting bats may employ narrow-band (CF)-type calls (emballonurids, molossids, mormoopids), but typically oral emitters produce multiharmonic, broad-band calls.

Conversely, nasal-emitting bats are faced with the problem of projecting sound through the restrictive nasal passages (Pedersen, 1993, 1995, 2000). Therefore, nycterids, megadermatids and most phyllostomid bats generate low-intensity multiharmonic calls of varying structure due primarily to the restrictions of the nasal cavity. The low-intensity calls of phyllostomid bats led Griffin (1958) to refer to them as "whispering bats," though recent work has shown these bats to be much louder than previously believed (Brinkløv *et al.*, 2009). Conversely, the nasal cavities of rhinolophid and hipposiderid bats are tuned to dramatically reduce the general impedance of the cavity (Roberts, 1972, 1973; Hartley and Suthers, 1988; Suthers *et al.*, 1988). These bats typically emit loud high-duty-cycle (CF) calls comparable to those of oral-emitting bats, but with one very important difference. The acoustic limitations imposed by the dimensions of the nasal passages in these rhinolophid and hipposiderid bats emphasize the second harmonic ($f2$) while reducing or removing the remaining overtones, including the fundamental ($f1$) (Roberts, 1972; Matsumura, 1979; Hartley and Suthers, 1988; Suthers et al., 1988).

In rhinolophids, neither the nasal passages nor the pinnae exhibit dimensions capable of sending or receiving the dominant spectral component of the adult call ($f2$) until well after birth (Matsumura, 1979; Konstantinov, 1989; Obrist et al., 1993; Pedersen, 1996). Infants must literally grow into their second harmonic – the use of $f2$ does not lie in some unique feature of the frequency itself, but rather that it is the only frequency permitted by the anatomical constraints imposed by the supra-glottal vocal tract in these bats (Roberts, 1973; Matsumura, 1979; Hartley and Suthers, 1987; Pedersen, 1996, 2000). Once established, the tuning of these systems is exquisite (Leonard et al., 2004; Zhuang and Müller, 2006, 2007).

Call structure does not correlate well with brain size, dentition or taxonomic diversity (Pedersen, 2000) and may be best seen as a behavioral response to clutter and selection of prey-type (Surlykke and Moss, 2000). So, what if any advantage comes from being a nasal-emitting bat? Could it be something so simple as how these animals use their mouths? After all, an echolocating predator would have difficulty flying and orienting simultaneously if it's mouth is full, or the food item requires a great deal of on-the-wing processing. Therein, the majority of carnivorous/durophagous microchiropterans are perch-hunting nasal-emitters (e.g., Nycteris, Chrotopterus etc.) that can echolocate with their mouths full; or are nasal-emitting nocturnal frugivores (e.g., Artibeus, Sturnira etc.) or are oral-emitting bats that must forage in open air well away from clutter whilst processing food items (e.g., Lasiurus, Molossus etc.).

14.4.6   Facial ornamentation: nose leaves

All nasal-emitting forms exhibit at least two taxonomically related features: reorganization of the skull about the nasal passages and flaps of skin projecting around the nostrils. Earlier in this chapter, we postulated that the carrying of food/prey items in the mouth by a primitive oral-emitting bat might suffice to favor the evolution of bats that emitted their calls from their nasal passages either primarily or secondarily.

The evolution of facial ornamentation may be a predictable response to nasal emission of the call, but this ornamentation is at least as varied as the differential packaging of the rostrum in the various nasal-emitting taxa.

Are mammalian nasal cartilages and facial musculature predisposed/ preadapted to generate a nose leaf (Göbbel, 2000, 2002)? This elaborate flap of skin is the final element of a bat's vocal tract and it reduces back pressure in the nasal cavity and may also help focus the call as it is emitted through the nostrils (Möhres, 1966a, 1966b; Simmons and Stein, 1980; Hartley and Suthers, 1987, 1988, 1990; Pye, 1988; Arita, 1990; Ghose, 2006; Zhuang and Müller, 2007). Such immediate physiological benefits would be sufficient to drive a standard issue mammalian nose into a rudimentary nose leaf or into something far more elaborate (Arita, 1990; Göbbel, 2000, 2002a, 2002b). Significantly, nose-leaf primordia appear before eyes and external ears are visible, indicating that such ornamentation is related to developmental timing and construction of the face, and not simply the product of ecological niche specialization (Göbbel, 2000, 2002a, 2002b; Yokoyama and Uchida, 2000; Chen et al., 2005; Cretekos et al., 2005, 2007; Giannini et al., 2006; Wyant and Adams, 2007; Nolte et al., 2008).

The homology of nose leaves is still debated (Yokoyama and Uchida, 2000; Springer et al., 2001a; Göbbel, 2002a, 2002b). What are we to make of nasal-emitting taxa that reduce the size and complexity of their nose leaves (brachyphyllines) or those oral-emitting taxa that exhibit incipient nose leaves (plecotines, Antrozous, Craseonycteris or Rhinopoma)? Certainly, the facial cleft of nycterids is unique, if not bizarre, leaving Pedersen (1995) to muse that the unique length/depth relationship in the facial cleft of nycterid bats might be a "resonating chamber outside the bony nasal cavity." Recently, our lab has worked out the anatomy of this cleft and its articulated muscular palps, leaving no question that these structures are: (1) homologous (though reduced in complexity) with the nose leaves of rhinolophids and (2) constructed so as to modify the nasally emitted call in these unique slit-faced bats.

It is naïve to think that nuclear or mitochondrial gene sequences will provide answers to these questions; however, there are several regulatory genes (Sumo, Irf, Bmp, Msx, Shox, Gabrb) associated with the integration and development of the frontonasal process/palate that might provide a better place to start looking at the evolution of these remarkable facial ornaments.

14.5   Beyond the genetic code: evolutionary developmental biology

The last decade has witnessed giant steps in the field of developmental genetics insofar as to exhaust the coding sequence as the impetus for morpho- logical evolution (Hanken and Thorogood, 1993; Carroll 2005; Chai and

Maxson, 2006; Radlanski and Renz, 2006; Young and Badyaev, 2007). Indeed, the key to understanding evolutionary change has now shifted focus to the study of regulatory gene function and the function of regulatory gene networks (Beddington and Robertson, 1989; Hanken and Thorogood, 1993; Keranen *et al.*, 1998, 1999; Acampora *et al.*, 1999; Merlo *et al.*, 2000; Trainor and Krumlauf, 2000; Alappat *et al.*, 2003; Trainor *et al.*, 2003; Blechschmidt, 2004; Meulemans and Bronner-Fraser, 2004; Carroll, 2005; Chai and Maxson, 2006; Evans and Noden, 2006; Radlanski and Renz, 2006; Hoekstra and Coyne, 2007; Young and Badyaev, 2007).

It is instructive to step back and remember that the vertebrate head is in itself a novelty and its genes, tissues and form are often cannibalized from old postcranial material (Gans and Northcutt, 1983). Events associated with the innovation of the vertebrate head include, but are not limited to, signaling cascades that effect pattern organogenesis, the establishment of tissue boundaries and tissue–tissue induction, and site-specific induction (Beddington and Robertson, 1989; Acampora *et al.*, 1999; Blechschmidt, 2004; Radlanski and Renz, 2006; Hoekstra and Coyne, 2007). Studies regarding cranial neural crest cell migration and *Hox* gene regulation have dramatically improved our understanding of morphogenetic plasticity and evolutionary novelty (Couly *et al.*, 1993; Keränen *et al.*, 1998, 1999; Acampora *et al.*, 1999; Merlo *et al.*, 2000; Trainor and Krumlauf, 2000; Brault *et al.*, 2001; Alapatt *et al.*, 2003; Trainor *et al.*, 2003; Meulemans and Bronner-Fraser, 2004; Wilson and Tucker, 2004; Gross and Hanken, 2005; Chai and Maxson, 2006; Evans and Noden, 2006). Additional studies of craniofacial development continue to evaluate this exceedingly complex network of mechanisms that contribute to morphological form (Hanken and Thorogood, 1993; Carroll, 2005; Creuzet *et al.*, 2005; Chai and Maxson, 2006; Radlanski and Renz, 2006; Young and Badyaev, 2007). Arguably, selection *in utero* (Katz *et al.*, 1981; Alberch, 1982; Katz, 1982; Müller, 1990) may well be more important than natural selection after parturition (Schmalhausen, 1949; Kuhn, 1987; Bonner, 1988; Maier, 1989).

Therefore, it seems apparent that the driving force behind morphological novelty will not be found in some mitochondrial gene sequence or in some nuclear gene with a limited/unknown connection with organogenesis; but rather from an understanding of developmental timing, regulatory genes and the ability of various anatomical structures to accommodate change (Lauder, 1982; Alberch, 1989; Klingenberg, 1998; Carroll, 2000; Chase *et al.*, 2002; Hilliard *et al.*, 2005; Peaston and Whitlaw, 2006; Radlanski and Renz, 2006; Hallgrimsson *et al.*, 2007, 2008; Salazar-Ciudad, 2007; Young and Badyaev, 2007).

14.6   Bat phylogeny: molecules, morphology and developmental mechanisms

Molecules have become the apparent gold standard of many recent reconstructions of bat phylogeny and several studies have sequenced a rather impressive collection of genes from a wide variety of locations within the genome (Hutcheon *et al.*, 1998; Springer *et al.*, 2001b; Teeling *et al.*, 2002; Eick *et al.*, 2005; Lim and Dunlop, 2008). This remains despite the fact that genes do not provide equally valid phylogenetic signals and the addition of more genes does not necessarily clarify the situation and may generate anomalous trees (Wiens and Hollingsworth, 2000; Wiens, 2004; Willa and Rubinoff, 2004; Degnan and Rosenberg, 2006; Rodriquez-Ezpeleta *et al.*, 2007; Belfiore *et al.*, 2008).

Regardless, the molecular lobby postulates that the evolution of echolocation must be far more complex than previously thought (Jones and Teeling, 2006). If they are correct, the evolution of echolocation is actually quite messy, with nasal emission having evolved independently four different times (Rhinolophidae + Hipposideridae; Megadermatidae; Nycteridae and Phyllostomidae; Figure 14.7).

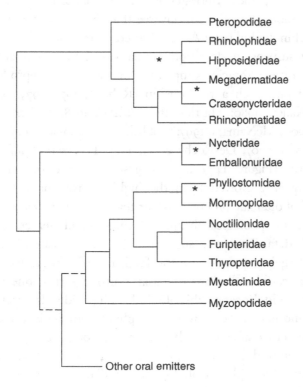

Figure 14.7 Molecular tree (after Teeling *et al.*, 2005). Asterisks indicate nose-leafed, nasal-emitting bats.

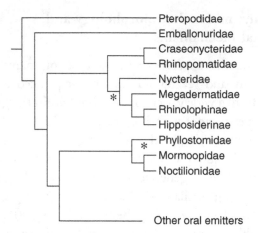

Figure 14.8 Morphology tree (after Simmons and Geisler, 1998). Asterisks indicate nose-leafed, nasal-emitting bats.

The molecular sequence data also suggest that the non-echolocating pteropodids and the highly sophisticated nasal-emitting rhinolophids are closely related (Teeling *et al.*, 2000, 2002, 2005; Springer *et al.*, 2001b; Eick *et al.*, 2005; Hutcheon and Kirsch, 2006). But is it conceivable that megachiropterans are a sister group related to rhinolophid bats? After all, there is little or nothing about their brains, skulls, jaw suspension, dentition, cranial vasculature, neuroacoustic systems, flight musculature, thoracic compliance or even their reproductive biology that would support such a relationship (Roberts, 1972, 1973; Kallen, 1977; Czarnecki and Kallen, 1980; Hartley and Suthers, 1988; Suthers *et al.*, 1988; Pedersen, 1996, 2000; McNamara, 1997; Göbbel, 2000; Leonard *et al.*, 2004; Zhuang and Müller, 2006, 2007; DesRoche *et al.*, 2007; Pedersen *et al.*, 2009). This molecular scenario (Figure 14.7) also suggests that the oral-emitting Rhinopomatidae subsequently evolved from the highly derived nasal-emitting rhinolophids (a reversal of dramatic proportions), or that *Craseonycteris* and *Megaderma* are sister groups (a morphological absurdity: Simmons and Geisler, 1998).

The morphological data, however, suggest a less complicated scenario (Figure 14.8) and are largely supported by the fossil record – there were two separate and rather successful radiations of nasal-emitting forms, one in the Old World during the mid Eocene (rhinolophids+hipposiderids, nycterids, megadermatids), and another in the New World (phyllostomids) in the early-mid Oligocene (Wetterer *et al.*, 2000; Baker *et al.*, 2003). In each case, nasal-emitting bats are derived from more primitive oral-emitting stock, without any reversals in form. Examples of phylogenetic radiation coincident with morphological innovation such as this are relatively common (Needham, 1933; Alberch *et al.*, 1979; Alberch and Alberch, 1981; Müller, 1990).

The inclusion of developmental mechanisms into evolutionary theory, during the last decade, represents an important shift from studies that focus on adaptation to those that include *emergence* (Müller and Newman, 2005; Salazar-Ciudad, 2007). Towards this end, numerous studies have tied regulatory genes to the development of the palate (*Msx, Bapx, Gsc, Emx, Sox, Hox, Prx* – Kaur *et al.*, 1992; ten Berge *et al.*, 1998; Bianchi *et al.*, 2000; Scapoli *et al.*, 2002; López *et al.*, 2008; Ji *et al.*, 2009), the ear (*Prx, Otx, Six, Eya* – ten Berge *et al.*, 1998; Morsli *et al.*, 1999; Schlosser, 2007), the cranial base (*Papps, Kena* – Hallgrímsson and Lieberman, 2008), the cervical vertebra (*Hox* – Kaur *et al.*, 1992; Galis *et al.*, 2006) and skeletal development in general (Young and Badyaev, 2007).

It is of great interest that several papers have recently pursued regulatory genes or karyological data that may provide more appropriate insight into the emergence of morphological innovation in bats themselves (Volleth *et al.*, 2002; Sears *et al.*, 2006; Cretekos *et al.*, 2007, 2008; Hockman *et al.*, 2008; Veselka *et al.*, 2010), even if we disagree with their definition of echolocation or their acceptance of the yinpterochiroptera (Li *et al.*, 2007, 2008).

14.7   Closing comments and future directions

As we wrote this chapter, many of our colleagues argued that because molecular sequence data is more readily collected than morphological data, the sheer volume of data (weight of the evidence) somehow equates to a more complete, possibly more correct, phylogenetic signal. However, this "more is better" approach is under debate (Wiens and Hollingsworth, 2000; Wiens, 2004; Willa and Rubinoff, 2004; Degnan and Rosenberg, 2006; Rodriquez-Ezpeleta *et al.*, 2007; Belfiore *et al.*, 2008; McDonough *et al.*, 2008). The often impenetrable language of molecular systematics can further obfuscate why various gene sequences were chosen, or how they relate to bat evolution. Not surprisingly, when morphology conflicts with molecules, most of our colleagues found it easier to point to inadequate taxon sampling (Heath *et al.*, 2008), or to even dismiss poorly known chiropteran taxa, rather than challenge the nature of the data. As Osborn pointed out as far back as 1932, "The geneticists are trying to make evolution fit the genes rather than to make the genes fit evolution."

Because many aspects of cranial form and function are well integrated and auto-correlated (Atchley and Hall, 1991; Lieberman *et al.*, 2004; Bulygina *et al.*, 2006; Bruner and Ripani, 2008), many phylogenetic studies have tried to argue that nasal emission in bats is a simple, easily coded character state during phylogenetic analyses. However, as we gain greater understanding of the epigenome and developmental mechanics (Peaston and Whitelaw, 2006), it is now clearly a dangerous gamble to divorce an innovative suite of character states from their developmental history, and to do so would reflect a gross ignorance of vertebrate embryology and what we are now learning about developmental genetics (Willa and Rubinoff, 2004; Müller and Newman, 2005; Davidson, 2006; Degnan and Rosenberg, 2006; Salazar-Ciudad, 2007). In this light, the evolution of nose leaves and the myriad of changes required during the reformulation of the chiropteran head (discussed earlier in this chapter) provides a magnificent opportunity to take our field beyond the shop-worn molecules-vs.-morphology debate.

In a draft proposal for this symposium, the convenors (Gunnell, Simmons and Eiting) stated that, "Molecular studies based on different genes and taxon samples often produce somewhat incongruent results, and morphological trees often conflict with molecular trees. However, these conflicts may be more superficial than previously thought, and new analyses of larger data sets are quickly converging on a new consensus of the pattern, if not the *process*, of bat evolution." ... and there's the rub.

It is our opinion that without the inclusion of the developmental process, phylogenetic inferences based in either adult morphology or molecular data (or a combination thereof) may be misleading or even specious. Are the genes that are currently being sequenced simply the low-hanging fruit that are at best casually correlated with evolutionary change, or should we shift our focus to developmental events and regulatory gene activity that directly relate to structural innovation and the emergence of novelty in a more causal manner?

So let us commit blasphemy and walk away from any pretense of *objectivity* and ignore the *populist* notion that more is better, but instead actively pursue causal mechanisms of novelty during phylogeny – the *cause célébre* in the burgeoning field of developmental genetics. We argue that our field must refocus our efforts on epigenetic mechanisms that act throughout ontogeny as a way to better understand the evolution of bats – because it would seem that we can no longer see the forest for all the gene trees.

14.8   REFERENCES

Acampora, D., Merlo, G. R., Paleari, L. *et al.* (1999). Craniofacial, vestibular and bone defects in mice lacking the Distal-less-related gene Dlx5. *Development*, **126**, 3795–3809.

Alappat, S., Zhang, Z. Y. and Chen, Y. P. (2003). Msx homeobox gene family and craniofacial development. *Cell Research*, **13**, 429–442.

Alberch, P. (1982). Developmental constraints in evolutionary processes. In *Evolution and Development: Dahlem Konferenzen, 1982*, ed. J. T. Bonner and I. Dawid. Berlin: Springer-Verlag, pp. 313–332.

Alberch, P. (1989). The logic of monsters: evidence for internal constraints in development and evolution. *Geobios (Lyon) mémoires spécial*, **12**, 21–57.

Alberch, P. and Alberch, J. (1981). Heterochronic mechanisms of morphological diversity and evolutionary change in the neotropical salamander, *Bolitoglossa occidentalis* (Amphibia: Plethodontidae). *Journal of Morphology*, **167**, 249–264.

Alberch, P., Gould, S. J., Oster, G. F. and Wake, D. B. (1979). Size and shape in ontogeny and phylogeny. *Paleobiology*, **5**, 296–317.

Arita, H. T. (1990). Noseleaf morphology and ecological correlates in phyllostomid bats. *Journal of Mammalogy*, **71**, 36–47.

Atchley, W. R. and Hall, B. K. (1991). A model for development and evolution of complex morphological structures. *Biological Review of the Cambridge Philosophical Society*, **66**, 101–115.

Baker, R. J., Hoofer, S. R., Porter, C. A. and Van Den Bussche, R. A. (2003). Diversification among New World Leaf-Nosed Bats: an evolutionary hypothesis and classification inferred from digenomic congruence of DNA sequence. *Occasional Papers, Museum of Texas Tech University*, **230**, 1–32.

Beddington, R. S. P. and Robertson, E. J. (1989). An assessment of the developmental potential of embryonic stem cells in the midgestation embyro. *Development*, **105**, 733–737.

Belfiore, N. M., Liu, L. and Moritz, C. (2008). Multilocus phylogenetics of a rapid radiation in the genus *Thomomys* (Rodentia: Geomyidae). *Systematic Zoology*, **57**, 294–310.

Bianchi, F., Calzolari, E., Ciulli, L. *et al.* (2000). Environment and genetics in the etiology of cleft lip and cleft palate with reference to the role of folic acid. *Epidemiology Preview*, **24**, 21–27.

Blechschmidt, E. (2004). *The Ontogenetic Basis of Human Anatomy: The Biodynamic Approach to Development from Conception to Adulthood*, ed. B. Freeman. Berkeley, CA: North Atlantic Books.

Bonner, J. (1988). *The Evolution of Complexity*. Princeton, NJ: Princeton University Press.

Brault, V., Moore, R., Kutsch, S. K. *et al.* (2001). Inactivation of the beta-catenin gene by Wnt1-Cre-mediated deletion results in dramatic brain malformation and failure of craniofacial development. *Development*, **128**, 1253–1264.

Brinkløv, S., Kalko, E. K. V. and Surlykke, A. (2009). Intense echolocation calls from two "whispering" bats, *Artibeus jamaicensis*. *Journal of Experimental Biology*, **212**, 11–20.

Bruner, E. and Ripani, M. (2008). A quantitative and descriptive approach to morphological variation of the endocranial base in modern humans. *American Journal of Physical Anthropology*, **137**, 30–40.

Buckland-Wright, J. C. (1978). Bone structure and the patterns of force transmission in the cat skull (*Felis catus*). *Journal of Morphology*, **155**, 35–62.

Bulygina, E., Mitteroecker, P. and Aiello, L. (2006). Ontogeny of facial dimorphism and patterns of individual development within one human population. *American Journal of Physical Anthropology*, **131**, 432–443.

Burdi, A. R. (1968). Morphogenesis of mandibular dental arch shape in human embryos. *Journal of Dental Research*, **47**, 50–58.

Byrd, K. (1988). Craniofacial sequelae of lesions to facial and trigeminal motor nuclei in growing rats. *American Journal of Physical Anthropology,* **76,** 87–103.

Carroll, S. B. (2000). Endless forms the evolution of gene regulation and morphological diversity. *Cell,* **101,** 577–580.

Carroll, S. B. (2005). Evolution at two levels: on genes and form. *PloS Biology,* **3,** e245, doi:10.1371/journal.pbio.0030245.

Chai, Y. and Maxson, R. E. (2006). Recent advances in craniofacial morphogenesis. *Developmental Dynamics,* **235,** 2353–2375.

Chase, K., Carrier, D. R., Adler, F. R. and Jarvik, T. (2002). Genetic basis for systems of skeletal quantitative traits: principal component analysis of the canid skeleton. *Proceedings of the National Academy of Sciences, USA,* **99,** 9930–9935.

Chen, C. H., Cretekos, C. J., Rasweiler, J. J. and Behringer, R. R. (2005). Hoxd13 expression in the developing limbs of the short-tailed fruit bat, *Carollia perspicillata. Evolution and Development,* **7,** 130–141.

Couly, G. F., Coltey, P. M. and Le Douarin, N. M. (1993). The triple origin of skull in higher vertebrates: a study in quail-chick chimeras. *Development,* **117,** 409–429.

Covey, D. S. G. and Greaves, W. S. (1994). Jaw dimensions and torsion resistance during canine biting in the Carnivora. *Canadian Journal of Zoology,* **72,** 1055–1060.

Cretekos, C. J., Weatherbee, S., Chen, C. *et al.* (2005). Embryonic staging system for the short-tailed fruit bat, *Carollia perspicillata,* a model organism for the mammalian order Chiroptera, based upon timed pregnancies in captive-bred animals. *Developmental Dynamics,* **233,** 721–738.

Cretekos, C. J., Deng, J.-M., Green, E. D. *et al.* (2007). Isolation, genomic structure and developmental expression of Fgf8 in the short-tailed fruit bat, *Carollia perspicillata. International Journal of Developmental Biology,* **51,** 333–338.

Cretekos, C. J., Wang, Y., Green, E., Martin, J., Rasweiler, J. and Behringer, R. (2008). Regulatory divergence modifies limb length between mammals. *Genes and Development,* **22,** 141–151.

Creuzet, S., Couly, G. and Le Douarin, N. M. (2005). Patterning the neural crest derivatives during development of the vertebrate head: insights from avian studies. *Journal of Anatomy,* **207,** 447–459.

Czarnecki, R. T. and Kallen, F. C. (1980). Craniofacial, occlusal, and masticatory anatomy in bats. *Anatomical Record,* **198,** 87–105.

Davidson, E. H. (2006). *The Regulatory Genome: Gene Regulatory Networks in Development and Evolution.* Burlingame: Academic Press.

de Beer, G. (1937). *Development of the vertebrate skull.* Oxford: Oxford University Press.

Dechmann, D. K. and Safi, K. (2009). Comparative studies of brain evolution: a critical insight from the Chiroptera. *Biological Review,* **84,** 161–172.

Degnan, J. H. and Rosenberg, N. A. (2006). Discordance of species trees with their most likely gene trees. *Public Library of Science, Genetics,* **2,** 762–768.

DesRoche, K., Fenton, M. B. and Lancaster, W. C. (2007). Echolocation and the thoracic skeletons of bats: a comparative morphological study. *Acta Chiropterologica,* **9,** 483–494.

Dumont, E. R. and Herrel, A. (2003). The effects of gape angle and bite point on bite force in bats. *Journal of Experimental Biology*, **206**, 2117–2123.

Eick, G., Jacobs, D. S. and Matthee, C. (2005). A nuclear DNA phylogenetic perspective on the evolution of echolocation and historical biogeography of extant bats (Chiroptera). *Molecular Biology and Evolution*, **22**, 1869–1886.

Eisenberg, J. F. and Wilson, D. E. (1978). Relative brain size and feeding strategies in the Chiroptera. *Evolution*, **32**, 740–751.

Evans, D. J. and Noden, D. M. (2006). Spatial relations between avian craniofacial neural crest and paraxial mesoderm cells. *Developmental Dynamics*, **235**, 1310–1325.

Fenton, M. B., Portfors, C. V., Rautenbach, I. L. and Waterman, J. M. (1998). Compromises: sound frequencies used in echolocation by aerial-feeding bats. *Canadian Journal of Zoology*, **76**, 1174–1182.

Freeman, P. W. (1984). Functional cranial analysis of large animalivorous bats (Microchiroptera). *Biological Journal of the Linnean Society*, **21**, 387–408.

Galis, F., Van Dooren, T. J. M., Feuth, J. D. *et al.* (2006). Extreme selection in humans against homeotic transformations of cervical vertebrae. *Evolution*, **60**, 2643–2654.

Gans, C. and Northcutt, R. G. (1983). Neural crest and the origin of vertebrates: a new head. *Science*, **220**, 268–274.

Gaunt, W. A. (1967). Observations upon the developing dentition of *Hipposideros caffer* (Microchiroptera). *Acta Anatomica*, **68**, 9–25.

Ghose, K. (2006). Sonar beam direction and flight control in an echolocating bat. Unpublished Ph.D. thesis, University of Maryland, College Park.

Giannini, N. P., Goswami, A. and Sanchez-Villagra, M. R. (2006). Development of integumentary structures in *Rousettus amplexicaudatus* (Mammalia: Chiroptera: Pteropodidae) during late-embryonic and fetal stages. *Journal of Mammalogy*, **87**, 993–1001.

Göbbel, L. (2000). The external nasal cartilages in Chiroptera: significance for intraordinal relationships. *Journal of Mammalian Evolution*, **7**, 167–201.

Göbbel, L. (2002a). Morphology of the external nose in *Hipposideros diadema* and *Lavia frons* with comments on its diversity and evolution among leaf-nosed Microchiroptera. *Cell Tissue Organ*, **170**, 39–60.

Göbbel, L. (2002b). Ontogenetic and phylogenetic transformations of the lacrimal-conducting apparatus among Microchiroptera. *Mammalian Biology*, **67**, 338–357.

Gould, S. J. and Vrba, E. S. (1982). Exaptation, a missing term in the science of form. *Paleobiology*, **8**, 4–15.

Griffin, D. R. (1958). *Listening in the Dark*. New Haven, CT: Yale University Press.

Griffioen, F. M. and Smit-Vis, J. H. (1985). The skull: mould or cast? *Acta Morphologica Neerlando-Scandinavica*, **23**, 325–335.

Gross, J. and Hanken, J. (2005). Cranial neural crest contributes to the bony skull vault in adult *Xenopus laevis*: insights from cell labeling studies. *Journal of Experimental Zoology*, **304** (B), 169–176.

Haines, W. (1940). The interorbital septum in mammals. *Journal of the Linnean Society, London*, **41**, 585–607.

Hallgrimsson, B. and Lieberman, D.E. (2008). Mouse models and the evolutionary developmental biology of the skull. *Journal of Integrative and Comparative Biology*, **48**, 373–384.

Hallgrimsson, B., Lieberman, D.E., Liu, W., Ford-Hutchinson, A.F. and Jirik, F.R. (2007). Epigenetic interactions and the structure of phenotypic variation in the cranium. *Evolution and Development*, **9**, 76–91.

Hanken, J. (1983). Miniaturization and its effects on cranial morphology in plethodontid salamanders, genus *Thorius* (Amphibia, Plethodontidae): II. The fate of the brain and sense organs and their role in skull morphogenesis and evolution. *Journal of Morphology*, **177**, 255–268.

Hanken, J. (1984). Miniaturization and its effects on cranial morphology in plethodontid salamanders, genus *Thorius* (Amphibia: Plethodontidae). I. Osteological variation. *Biological Journal of the Linnean Society*, **23**, 55–75.

Hanken, J. and Thorogood, P. (1993). Skull development during anuran metamorphosis: I. Early development of the first three bones to form – the exoccipital, the parasphenoid, and the frontoparietal. *Trends in Ecology and Evolution*, **8**, 9–15.

Hartley, D.I. and Suthers, R.A. (1987). The sound emission pattern and the acoustical role of the noseleaf in the echolocating bat, *Carollia perspicillata*. *Journal of the Acoustic Society of America*, **82**, 1892–1900.

Hartley, D.I. and Suthers, R.A. (1988). The acoustics of the vocal tract in the horseshoe bat, *Rhinolophus hildebrandti*. *Journal of the Acoustic Society of America*, **84**, 1201–1213.

Hartley, D.I. and Suthers, R.A. (1990). Sonar pulse radiation and filtering in the mustached bat, *Pteronotus parnellii*. *Journal of the Acoustic Society of America*, **87**, 2756–2772.

Heath, T.A., Hedtke, S.M. and Hillis, D.M. (2008). Taxon sampling and the accuracy of phylogenetic analyses. *Journal of Systematics and Evolution*, **46**, 239–257.

Herring, S. (1985). The ontogeny of mammalian mastication. *American Zoologist*, **25**, 339–349.

Herring, S. and Lakars, T.C. (1981). Craniofacial development in the absence of muscle contraction. *Journal of Craniofacial Genetics and Developmental Biology*, **1**, 341–357.

Hilliard, S.A., Yu, L., Gu, S., Zhang, Z. and Chen, Y.P. (2005). Regional regulation of palatal growth and patterning along the anterior-posterior axis in mice. *Journal of Anatomy*, **207**, 655–667.

Hockman, D., Cretekos, C.J., Mason, M.K., Behringer, R.R., Jacobs, D.S. and Illing, N. (2008). A second wave of Sonic hedgehog expression during the development of the bat limb. *Proceedings of the National Academy of Sciences, USA*, **105**, 16982–16987.

Hoekstra, H. and Coyne, J. (2007). The locus of evolution: evo devo and the genetics of adaptation. *Evolution*, **61**, 995–1016.

Hoyte, D.A.N. (1987). Muscles and cranial form. In *Mammalia Depicta: Morphogenesis of the Mammalian Skull*, ed. H. Kuhn and U. Zeller. Hamburg: Verlag Paul Parey, pp. 123–144.

Hutcheon, J.M. and Kirsch, J.A.W. (2006). A moveable face: deconstructing the Microchiroptera and a new classification of extant bats. *Acta Chiropterologica*, **8**, 1–10.

Hutcheon, J. M., Kirsch, J. A. W. and Pettigrew, J. D. (1998). Base-compositional biases and the bat problem. III. The question of microchiropteran monophyly. *Philosophical Transactions of the Royal Society of London Series B – Biological Sciences*, **353**, 607–617.

Hutcheon, J. M., Kirsch, J. A. W. and Garland Jr., T. (2002). A comparative analysis of brain size in relation to foraging ecology and phylogeny in the Chiroptera. *Brain, Behavior, and Evolution*, **60**, 165–180.

Ji, Q., Luo, Z.-X., Zhang, X., Yuan, C.-X. and Xu, L. (2009). Evolutionary development of the middle ear in Mesozoic therian mammals. *Science*, **326**, 278–281.

Jolicoeur, J., Pirlot, P. and Stephan, H. (1984). Brain structure and correlation patterns in Insectivora, Chiroptera, and Primates. *Systematic Zoology*, **33**, 14–33.

Jones, G. and Holderied, M. (2007). Bat echolocation calls: adaptation and convergent evolution. *Proceedings of the Royal Society of London Series B – Biological Sciences*, **274**, 905–912.

Jones, G. and Teeling, E. C. (2006). The evolution of echolocation in bats. *Trends in Ecology and Evolution*, **21**, 149–156.

Kallen, F. C. (1977). The cardiovascular systems of bats. In *Biology of Bats*, vol. 3, ed. W. Wimsatt. New York: Academic Press, pp. 289–483.

Kangas, A. T., Evans, A. R., Thesleff, I. and Jernvall, J. (2004). Nonindependence of mammalian dental characters. *Nature*, **432**, 211–214.

Katz, M. (1982). Ontogenetic mechanisms: the middle ground of evolution. In *Evolution and Development: Dahlem Konferenzen 1982*, ed. J. T. Bonner and I. Dawid. Berlin: Springer-Verlag, pp. 207–212.

Katz, M., Lasek, R. and Kaiser-Abramof, J. (1981). Ontophyletics of the nervous system: eyeless mutants illustrate how ontogenetic buffer mechanisms channel evolution. *Proceedings of the National Academy of Sciences, USA*, **7**, 397–401.

Kaur, S., Singh, G., Stock, J. L. *et al.* (1992). Dominant mutation of the murine Hox-2.2 gene results in developmental abnormalities. *Journal of Experimental Zoology*, **264**, 323–336.

Keränen, S. V. E., Åberg, T., Kettunen, P. and Thesleff, I. (1998). Association of developmental regulatory genes with the development of different molar tooth shapes in two species of rodents. *Development, Genes, and Evolution*, **208**, 477–486.

Keränen, S. V. E., Kettunen, P., Åberg, T. and Thesleff, I. (1999). Gene expression patterns associated with suppression of odontogenesis in mouse and vole diastema regions. *Development, Genes, and Evolution*, **209**, 495–506.

Klingenberg, C. P. (1998). Heterochrony and allometry: the analysis of evolutionary change in ontogeny. *Biological Reviews*, **73**, 79–123.

Konstantinov, A. I. (1989). The ontogeny of echolocation functions in horseshoe bats. In *Proceedings of the Fourth European Bat Research Symposium*. Prague: Charles University Press, pp. 271–280.

Kuhn, H. (1987). Introduction. In *Mammalia depicta: Morphogenesis of the Mammalian Skull*, ed. H. Kuhn and U. Zeller. Hamburg: Verlag Paul Parey, pp. 9–16.

Kylamarkula, S. (1988). Growth changes in the skull and upper cervical skeleton after partial detachment of neck muscles. An experimental study in the rat. *Journal of Anatomy*, **159**, 197–205.

Lakars, T. C. and Herring, S. (1980). Ontogeny and oral function in hamsters (*Mesocricetus auratus*). *Journal of Morphology*, **165**, 237–254.

Lauder, G. V. (1982). Historical biology and the problem of design. *Journal of Theoretical Biology*, **97**, 57–67.

Leonard, G., Flint, J. A. and Müller, R. (2004). On the directivity of the pinnae of the rufous horseshoe bat (*Rhinolophus rouxii*). *Proceedings of the Institute of Acoustics*, **26**, 29–37.

Lewin, R. (1986). Punctuated equilibrium is now old hat. *Science*, **231**, 672–673.

Li, G., Wang, J., Rossiter, S. J., Jones, G., Cotton, J. A. and Zhang, S. (2008). The hearing gene Prestin reunites echolocating bats. *Proceedings of the National Academy of Sciences, USA*, **105**, 13959–13964.

Li, G., Wang, J., Rossiter, S. J., Jones, G. and Zhang, S. (2007). Accelerated FoxP2 evolution in echolocating bats. *PloS ONE*, **2**, e900, doi: 10.1371/journal.pone.0000900.

Lieberman, D. E., Krovitz, G. E. and McBratney-Owen, B. (2004). Testing hypotheses about tinkering in the fossil record: the case of the human skull. *Journal of Experimental Zoology Part B Molecular and Developmental Evolution*, **302**, 284–301.

Lim, B. and Dunlop, J. (2008). Evolutionary patterns of morphology and behavior as inferred from a molecular phylogeny of New World emballonurid bats (Tribe Diclidurini). *Journal of Mammalian Evolution*, **15**, 79–121.

López, E. K. N., Stock, S. R., Taketo, M. M., Chenn, A. and Ravosa, M. J. (2008). A novel transgenic mouse model of fetal encephalization and craniofacial development. *Integrative and Comparative Biology*, **48**, 360–372.

Maier, W. (1989). Ala temporalis and alisphenoid in therian mammals. In *Trends in Vertebrate Morphology: Fortschritte der Zoologie, Band A*, vol. 35, ed. H. Splechtna and H. Hilgers. Stuttgart: Gustav Fischer Verlag, pp. 396–400.

Matsumura, S. (1979). Mother-infant communication in a Horseshoe bat (*Rhinolophus ferrumequinum nippon*): development of vocalization. *Journal of Mammalogy*, **60**, 76–84.

McDonough, M. M., Ammerman, L. K., Timm, R. M. *et al.* (2008). Speciation within bonneted bats (genus *Eumops*): the complexity of morphological, mitochondrial, and nuclear data sets in systematics. *Journal of Mammalogy*, **89**, 1306–1315.

McNamara, K. J. (1997). *Shapes of Time: The Evolution of Growth and Development*. Baltimore, MD: Johns Hopkins University Press.

Merlo, G., Zerega, B., Paleari, L., Trombino, S., Mantero, S. and Levi, G. (2000). Multiple functions of Dlx genes. *International Journal of Developmental Biology*, **44**, 619–626.

Meulemans, D. and Bronner-Fraser, M. (2004). Gene-regulatory interactions in neural crest evolution and development. *Developmental Cell*, **7**, 291–299.

Möhres, F. P. (1966a). Ultrasonic orientation in megadermatid bats. In *Animal Sonar Systems – Biology and Bionics, Proceedings NATO Advanced Study Institute*, ed. R. Busnel. Joury-en-Josas: Laboratoire de physiologie acoustique, Institut national de la recherche agronomique, pp. 115–128.

Möhres, F. P. (1966b). General characters of acoustic orientation sounds and performance of sonar in the order of Chiroptera. In *Animal Sonar Systems – Biology and Bionics:*

Proceedings NATO Advanced Study Institute, ed. R. Busnel. Joury-en-Josas: Laboratoire de physiologie acoustique, Institut national de la recherche agronomique, pp. 401–407.

Morsli, H., Tuorto, F., Choo, D., Postiglione, M. P., Simeone, A. and Wu, D. K. (1999). Otx1 and Otx2 activities are required for the normal development of the mouse inner ear. *Development*, **126**, 2335–2343.

Moss, M. (1958). Rotations of the cranial components in the growing rat and their experimental alteration. *Acta Anatomica*, **32**, 65–86.

Moss, M. (1976). Experimental alteration of basisynchondrosal cartilage growth in rat and mouse. In *Development of the Basicranium: Department of Health, Education, and Welfare Publication No. (NIH) 76–989*, ed. J. F. Bosma. Washington, DC: Government Printing Office, pp. 541–569.

Müller, G. B. (1990). Developmental mechanisms: a side-effect hypothesis. In *Evolutionary Innovations*, ed. M. H. Nitecki. Chicago, IL: University of Chicago Press, pp. 99–130.

Müller, G. B. and Newman, S. A. (2005). The innovation triad: an EvoDevo agenda. *Journal of Experimental Zoology B*, **304** (6), 487–503.

Needham, J. (1933). On the dissociability of the fundamental process in ontogenesis. *Biological Reviews*, **8**, 180–223.

Nolte, M., Hockman, D., Cretekos, C., Behringer, R. and Rasweiler, J. (2008). Embryonic staging system for the black mastiff bat, *Molossus rufus* (Molossidae), correlated with structure-function relationships in the adult. *Anatomical Record*, **292**, 155–168.

Obrist, M., Fenton, M. B., Eger, J. and Schlegel, P. (1993). What ears do for bats: a comparative study of pinna sound pressure transformation in Chiroptera. *Journal of Experimental Biology*, **180**, 119–152.

Osborn, H. F. (1932). The nine principals of evolution revealed by paleontology. *American Naturalist*, **66**, 52–60.

Ostyn, J. M., Maltha, J. C., Van't Hof, M. A. and Van Der Linden, F. P. G. M. (1995). The role of intercuspation in the regulation of transverse maxillary development in *Macaca fascicularis*. *Angle Orthodontist*, **65**, 215–222.

Peaston, A. E. and Whitelaw, E. (2006). Epigentics and phenotypic variation in mammals. *Mammalian Genome*, **17**, 365–374.

Pedersen, S. C. (1993). Cephalometric correlates of echolocation in the Chiroptera. *Journal of Morphology*, **218**, 85–98.

Pedersen, S. C. (1995). Cephalometric correlates of echolocation in the Chiroptera. II: Fetal development. *Journal of Morphology*, **225**, 107–123.

Pedersen, S. C. (1996). Skull growth and the presence of auxiliary fontanels in rhinolophoid bats (Microchiroptera). *Zoomorpholology*, **116**, 205–212.

Pedersen, S. C. (2000). Skull growth and the acoustical axis of the head in bats. In *Ontogeny, Functional Ecology, and Evolution of Bats*, ed. R. A. Adams and S. C. Pedersen. Cambridge: Cambridge University Press, pp. 174–213.

Pedersen, S., Riede, T., Nguyen, T. *et al.* (2009). Reconstruction of the rhinolophid vocal tract. *Bat Research News*, **50**, 131A.

Presley, R. (1981). Alisphenoid equivalents in placentals, marsupials, monotremes, and fossils. *Nature*, **294**, 668–670.

Presley, R. and Steel, F. (1976). On the homology of the alisphenoid. *Journal of Anatomy*, **121**, 441–459.

Price, T., Turelli, M. and Slatkin, M. (1993). Peak shifts produced by correlated response to selection. *Evolution*, **47**, 280–290.

Proffit, W. (1978). Equilibrium theory revisited: factors influencing position of the teeth. *Angle Orthodontics*, **48**, 175–186.

Pye, J. (1988). Noseleaves and bat pulses. In *Animal Sonar – Processes and Performance: Proceedings NATO Advanced Study Institute on Animal Sonar Systems*, ed. E. Nachtigall and P. W. B. Moore. New York: Plenum Press, pp. 791–796.

Radinsky, L. B. (1968). A new approach to mammalian cranial analysis, illustrated by examples of prosimian primates. *Journal of Morphology*, **124**, 167–180.

Radlanski, R. J. and Renz, H. (2006). Genes, forces, and forms: mechanical aspects of prenatal craniofacial development. *Developmental Dynamics*, **235**, 1219–1229.

Reep, R. L. and Bhatnagar, K. P. (2000). Brain ontogeny and ecomorphology in bats. In *Ontogeny, Functional Ecology, and Evolution of Bats*, ed. R. A. Adams and S. C. Pedersen. Cambridge: Cambridge University Press, pp. 93–136.

Roberts, L. (1972). Variable resonance in constant frequency bats. *Journal of Zoology (London)*, **166**, 337–348.

Roberts, L. (1973). Cavity resonances in the production of orientation cries. *Periodicum Biologorum*, **75**, 27–32.

Rodriquez-Ezpeleta, N., Brinkmann, H., Roure, B., Lartillot, N., Lang, B. F. and Philppe, H. (2007). Detecting and overcoming systematic errors in genome-scale phylogenies. *Systematic Biology*, **56**, 389–399.

Ross, C. F. and Henneberg, M. (1995). Basicranial flexion, relative brain size, and facial kyphosis in *Homo sapiens* and some fossil hominids. *American Journal of Physical Anthropology*, **98**, 575–593.

Ross, C. F. and Ravosa, M. J. (1993). Basicranial flexion, relative brain size, and facial kyphosis in nonhuman primates. *American Journal of Physical Anthropology*, **91**, 305–324.

Safi, K. and Dechmann, D. (2005). Adaptation of brain regions to habitat complexity: a comparative analysis in bats (Chiroptera). *Proceedings of the Royal Society of London, B*, **272**, 179–186.

Salazar-Ciudad, I. (2007). On the origins of morphological variation, canalization, robustness, and evolvability. *Integrative and Comparative Biology*, **47**, 390–400.

Scapoli, L., Martinelli, M., Pezzetti, F. *et al.* (2002). Linkage disequilibrium between GABRB3 gene and nonsyndromic familial cleft lip with or without cleft palate. *Human Genetics*, **110**, 15–20.

Schachner, O. (1989). Raising of the head and allometry: morphometric study on the embryonic development of the mouse-skeleton. In *Trends in Vertebrate Morphology: Fortschritte der Zoologie, Band A*, vol. 35, ed. H. Splechtna and H. Hilgers. Stuttgart: Gustav Fischer Verlag, pp. 291–298.

Schlosser, G. (2007). How old genes make a new head: redeployment of Six and Eya genes during the evolution of vertebrate cranial placodes. *Integrative and Comparative Biology*, **47**, 343–359.

Schmalhausen, I. (1949). *Factors of Evolution: The Theory of Stabilizing Selection.* Chicago, IL: University of Chicago Press.

Schneiderman, E. (1992). *Facial Growth in the Rhesus Monkey.* Princeton, NJ: Princeton University Press.

Schumacher, G. H. Von, Schoof, S., Fanghanel, J., Mildschlag, G. and Kannmann, F. (1986). Skull deformities following unilateral mandibular dysbalance. 1. General review of secondary changes. *Anatomische Anzeiger, Jena,* **161,** 105–111.

Schumacher, G. H. Von, Fanghanel, J., Koster, D. and Mierzwa, J. (1988). Craniofacial growth under the influence of blood supply; 11. Scoliosis of the skull. *Anatomische Anzeiger, Jena,* **165,** 303–309.

Sears, K. E., Behringer, R. R., Rasweiler, J. J. and Niswander, L. A. (2006). Development of bat flight: morphologic and molecular evolution of bat wing digits. *Proceedings of the National Academy of Sciences, USA,* **103,** 6581–6586.

Simmons, J. A. (1980). *Phylogenetic Adaptations and the Evolution of Echolocation in Bats. Proceedings of the Fifth International Bat Research Conference.* Lubbock, TX: Texas Tech. Press, pp. 267–278.

Simmons, J. A. and Stein, R. A. (1980). Acoustic imaging in bat sonar: echolocation signals and the evolution of echolocation. *Journal of Comparative Physiology A,* **135,** 61–84.

Simmons, N. B. and Geisler, J. H. (1998). Phylogenetic relationships of *Icaronycteris, Archaeonycteris, Hassianycteris,* and *Palaeochiropteryx* to extant bat lineages, with comments on the evolution of echolocation and foraging strategies in Microchiroptera. *Bulletin of the American Museum of Natural History,* **235,** 1–182.

Smit-Vis, J. H. and Griffioen, F. M. M. (1987). Growth control of neurocranial height of the rat skull. *Anatomische Anzeiger, Jena,* **163,** 401–406.

Solow, B. and Greve, E. (1979). Craniocervical angulation and nasal respiratory resistance. In *Naso-respiratory function and Craniofacial Growth: Craniofacial Growth Series, Monograph #9,* ed. J. A. McNamara. Ann Arbor, MI: University of Michigan Press, pp. 87–119.

Spatz, W. V. (1968). Die Bedeutung der Augen fur die sagittale Gestaltung des Schadels von *Tarsius* (Prosimiae, Tarsiiformes). *Folia Primatologica,* **99,** 22–40.

Sperry, T. P. (1972). Development of the specialized craniofacial complex in bats of the genera *Mormoops* and *Chilonycteris* (Family Mormoopidae). Unpublished Master' thesis, University of Illinois Medical Center, Chicago.

Springer, M. S., Teeling, E. C., Madsen, O., Stanhope, M. J. and Jong, W. W. (2001b). Integrated fossil and molecular data reconstruct bat echolocation. *Proceedings of the National Academy of Sciences, USA,* **98,** 6241–6246.

Springer, M. S., Teeling, E. C. and Stanhope, M. J. (2001a). External nasal cartilages in bats: evidence for microchiropteran monophyly? *Journal of Mammalian Evolution,* **8,** 231–236.

Springer, M. S., Meredith, R., Eizirik, E., Teeling, E. C. and Murphy, W. (2008). Morphology and placental mammal phylogeny. *Systematic Biology,* **57,** 499–503.

Spyropoulos, M. N. (1977). The morphogenetic relationship of the temporal muscle to the coronoid process in human embryos and fetuses. *American Journal of Anatomy,* **150,** 395–410.

Starck, D. (1952). Form und Formbildung der Schadelbasis bei Chiropteren. Ergänzung-sheft zum 99 Band; Ver. Anat. Ges., 50th Versammlung Marburg. *Anatomische Anzeiger, Jena*, **99**, 114–121.

Starck, D. (1954). Morphologische Untersuchungen am Kopf der Saugetiere, besonders der Priosimier, ein Beitrag zum Problem des Formwandels des Saugerschadels. *Zeitschrift für Wissenschaftliche Zoologie, Abteilung A*, **157**, 169–219.

Starck, D. (1989). Considerations on the nature of skeletal elements in the vertebrate skull, especially in mammals. In *Trends in Vertebrate Morphology: Fortschritte der Zoologie, band A*, vol. 35, ed. H. Splechtna and H. Hilgers. Stuttgart: Gustav Fischer Verlag, pp. 375–385.

Stephan, H., Nelson, J. and Frahm, H. (1981). Brain size in Chiroptera. *Zeitschrift für Zoologische Systematik und Evolutionsforschung*, **19**, 195–222.

Surlykke, A. and Moss, C. (2000). Echolocation behavior of big brown bats, *Eptesicus fuscus*, in the field and the laboratory. *Journal of the Acoustic Society of America*, **108**, 2419–2429.

Suthers, R. A., Hartley, D. I. and Wenstrup, J. (1988). The acoustic role of tracheal chambers and nasal cavities in the production of sonar pulses by the horseshoe bat, *Rhinolophus hildebrandti*. *Journal of Comparative Physiology A*, **162**, 799–813.

Teeling, E. C., Scally, D., Kao, D. *et al.* (2000). Molecular evidence regarding the origin of echolocation and flight in bats. *Nature*, **403**, 188–192.

Teeling, E. C., Madsen, O., Van Den Bussche, R. A. *et al.* (2002). Microbat paraphyly and the convergent evolution of a key innovation in Old World rhinolophoid microbats. *Proceedings of the National Academy of Sciences, USA*, **99**, 1431–1436.

Teeling, E. C., Springer, M., Madsen, O. *et al.* (2005). A molecular phylogeny for bats illuminates biogeography and the fossil record. *Science*, **307**, 580–584.

Tejada-Flores, A. E. and Shaw, C. A. (1984). Tooth replacement and skull growth in *Smilodon* from Rancho La Brea. *Journal of Vertebrate Paleontology*, **4**, 114–121.

ten Berge, D., Brouwer, A., Korving, J., Martin, J. F. and Meijlink, F. (1998). Prx1 and Prx2 in skeletogenesis: roles in the craniofacial region, inner ear and limbs. *Development*, **125**, 3831–3842.

Thilander, B. and Ingervall, B. (1973). The human spheno-occipital synchondrosis II. A histological and microradiographic study of its growth. *Acta Odontologica Scandinavica*, **31**, 323–336.

Trainor, P. A. and Krumlauf, R. (2000). Patterning the cranial neural crest: hindbrain segmentation and Hox gene plasticity. *Nature Reviews Neuroscience*, **1**, 116–124.

Trainor, P. A., Melton, K. R. and Manzanares, M. (2003). Origins and plasticity of neural crest cells and their roles in jaw and craniofacial evolution. *International Journal of Developmental Biology*, **47**, 541–553.

Veselka N., McErlain, D. D., Holdsworth, D. W. *et al.* (2010). A bony connection signals laryngeal echolocation in bats. *Nature*, **463**, 939–942.

Volleth, M., Heller, K. G., Pfeiffer, R. A. and Hameister, H. (2002). A comparative ZOO-FISH analysis in bats elucidates the phylogenetic relationships between Megachiroptera and five microchiropteran families. *Chromosome Research*, **10**, 477–497.

Washburn, S. L. (1947). The relation of the temporal muscle to the form of the skull. *Anatomical Record*, **99**, 239–248.

Waters, D. A. and Vollratch, C. (2003). Echolocation performance and call structure in the megachiropteran fruit-bat *Rousettus aegyptiacus*. *Acta Chiropterologica*, **5**, 209–219.

Wetterer, A. L., Rockman, M. V. and Simmons, N. B. (2000). Phylogeny of phyllostomid bats (Mammalia: Chiroptera): data from diverse morphological systems, sex chromosomes, and restriction sites. *Bulletin of the American Museum of Natural History*, **248**, 1–200.

Wiens, J. J. (2004). The role of morphological data in phylogeny reconstruction. *Systematic Biology*, **53**, 653–661.

Wiens, J. J. and Hollingsworth, B. D. (2000). War of the iguanas: conflicting molecular and morphological phylogenies and long-branch attraction in iguanid lizards. *Systematic Zoology*, **49**, 143–159.

Willa, K. W. and Rubinoff, D. (2004). Myth of the molecule: DNA barcodes for species cannot replace morphology for identification and classification. *Cladistics*, **20**, 47–55.

Wilson, J. and Tucker, A. S. (2004). Fgf and Bmp signals repress the expression of Bapx1 in the mandibular mesenchyme and control the position of the developing jaw joint. *Developmental Biology*, **266**, 138–150.

Wimberger, P. (1991). Plasticity of jaw and skull morphology in the neotropical cichlids *Geophagus brasiliensis*, and *G. steindachneri*. *Evolution*, **45**, 1545–1563.

Wyant, K. and Adams, R. A. (2007). Prenatal growth and development in the Angolan tree-tailed bat, *Mops condylurus* (Chiroptera: Molossidae). *Journal of Mammalogy*, **88**, 1248–1251.

Yokoyama, K. and Uchida, A. (2000). Evolutional implications of recapitulation concerning the round nose leaf seen at the middle prenatal period in the Japanese Lesser Horseshoe bat, *Rhinolophus cornutus cornutus*. *Annals of the Speleological Research Institute of Japan*, **19**, 19–31.

Young, R. and Badyaev, A. (2007). Evolution of ontogeny: linking epigenetic remodeling and genetic adaptation in skeletal structures. *Integrative and Comparative Biology*, **47**, 234–244.

Zhuang, Q. and Müller, R. (2006). Noseleaf furrows in a horseshoe bat act as resonance cavities shaping the biosonar beam. *Physical Review Letters*, **97**, 2187014.

Zhuang, Q. and Müller, R. (2007). Numerical study of the effect of the noseleaf on biosonar beam forming in a horseshoe bat. *Physical Review Letters*, **76**, 519020111.

15

Vertebral fusion in bats: phylogenetic patterns and functional relationships

DAWN J. LARKEY, SHANNON L. DATWYLER
AND WINSTON C. LANCASTER

15.1 Introduction

The general shape, function and development of vertebrae tend to be highly conserved among mammals (Vaughan, 1970; Simmons and Geisler, 1998; Hildebrand and Goslow, 2001; Buchholtz, 2007), where the vertebral column is divided into five distinct regions: cervical, thoracic, lumbar, sacral and caudal. Typical mammalian vertebrae consist of a centrum, a neural arch and two pairs of zygopophyses. On the centrum, a pair of dorsally directed pedicles fuse with the lamina to form the neural arch for protection of the spinal cord. An intervertebral disc separates each centrum; it facilitates multiaxial motion and acts as a cushion between adjacent centra (Hildebrand and Goslow, 2001). Although individual vertebrae separated by intervertebral discs typically remain distinctly separate bones throughout life, vertebral bodies may fuse into multibone units.

Characteristic fusions of vertebrae are well known in turtles and birds, but also occur to varying degrees in some mammals. Fusion of three or more vertebrae into a sacrum that articulates to the ilium is a primitive characteristic in mammals; its loss is considered a derived trait (e.g., Flower, 1885; Vaughan, 1970) and is usually seen only in obligate aquatic mammals. In contrast to the loss of sacral fusion, the cervical vertebrae of many cetaceans are craniocaudally compressed and often fuse into units of two to seven vertebrae, presumably to provide rigidity of the neck (Flower, 1885). In addition, some rodents, such as jerboas (*Dipus sagitta*), have fused cervical vertebrae. Jerboas use ricochetal locomotion (using only the hind feet for forward propulsion) and fused cervical vertebrae may provide increased surface area for muscle attachment and vertebral column strength to avoid whiplash injury (Hatt, 1932).

Although bats generally retain the independent vertebral elements typical of mammals, there are a number of exceptions. Some bats have vertebrae other than the sacrum that articulate so tightly that they form a rigid, apparently

Evolutionary History of Bats: Fossils, Molecules and Morphology, ed. G. F. Gunnell and N. B. Simmons. Published by Cambridge University Press. © Cambridge University Press 2012.

immobile unit. Still other bats have series of vertebrae fused into a synostosis (Miller, 1907; Vaughan, 1970). Dobson (1878) documented vertebral fusion in bats, but asserted that the fusion was due to age processes and provided no further data on its taxonomic occurrence. Walton and Walton (1970) identified three regions of vertebral fusion in bats: cervicothoracic, thoracolumbar and sacral. Vaughan (1970) stated that cervicothoracic and thoracolumbar fusion seen in some species of bats could reduce the required musculature that normally braces the vertebral column. Fenton and Crerar (1984) noted the occurrence of fusion between the last cervical and first thoracic vertebrae in some species of Rhinolophoidea and Molossidae, but did not comment on its functional significance. These studies have only provided cursory functional interpretations of vertebral fusion. They have focused only on the external description of vertebral fusion (e.g., Vaughan, 1970), while other studies only considered the presence or absence of vertebral fusion in phylogenetic analyses of the order Chiroptera (Simmons and Geisler, 1998; Simmons et al., 2008).

Simmons and Geisler (1998) concluded that the retention of independent vertebrae is the ancestral state for bats because all bats that occupy the basal branches of the phylogenetic tree lack vertebral fusion. However, some fossils of bats from the Middle Eocene can be ascribed to the families Hipposideridae and Rhinolophidae, partly due to the presence of vertebral fusion (Bogdanowicz and Owen, 1998).

Vertebral fusion in bats may represent a homoplastic adaptation, and may have arisen independently multiple times. Alternatively, it may have arisen only once, but occurs in multiple taxa through shared ancestry. In either case, fusion could offer a selective advantage, or could itself be a neutral accompaniment of another developmental process. Lancaster et al. (1995) proposed that vertebral fusion could contribute rigidity to the axial skeletons of bats, which could facilitate the production of energetically expensive biosonar calls. They suggested that fusion offered a passive constraint of the thoracic cage against expansion that could dissipate the force generated by contractions of abdominal muscles that power vocalization (termed paradoxical expansions; Estenne et al., 1990). Such a rigid axial skeleton could provide an energetic advantage in contrast to active constraint provided by axial musculature.

In this contribution, we catalog the systematic occurrence of vertebral fusion in bats and attempt to test the association of fusion with ecological or behavioral factors that may offer possible functional explanations. We test the hypothesis of Lancaster et al. (1995) that fusion is an adaptation to facilitate energetically demanding modes of biosonar call production. If this hypothesis is correct, we would predict that bats which use energetically costly modes of biosonar call production should have fusions in the thoracic vertebrae and ribs in order to constrain the thorax passively against paradoxical expansions of the thorax during call production. We would not expect to see these adaptations in bats that do not echolocate, bats that produce biosonar calls by the use of tongue clicks or bats that use biosonar calls of low intensity.

15.2 Methods

We obtained skeletal materials of bats from the American Museum of Natural History (AMNH), the California Academy of Sciences (CAS), the Natural History Museum of Los Angeles County (LACM) and the Museum of Vertebrate Zoology at the University of California, Berkeley (MVZ). We followed the taxonomy of Simmons (2005) with the exception of recognizing Miniopteridae as a distinct family following Miller-Butterworth *et al.* (2007). Data collection focused on sampling multiple genera, and not necessarily on sampling every available species due to variation in the number of species in each family. The families Craseonycteridae, Mystacinidae and Myzopodidae were analyzed based on vertebral fusion data reported by Simmons *et al.* (2008).

We selected 1–11 individuals of each species and examined their skeletons with a dissecting microscope, concentrating on specimens with the most intact vertebral columns. Adult status was established by complete epiphyseal fusion (Howell and Pylka, 1977) and/or by the presence of complete adult dentition. Vertebral fusion was determined by the absence of an intervertebral disc due to the ossified union of adjacent vertebral centra, fusion of zygopophyseal joints of neural arches or both.

The monophyly of Chiroptera is well supported, but recent studies have reorganized higher-level taxonomic relationships (Miller-Butterworth *et al.*, 2007; Simmons *et al.*, 2008). For the purposes of our phylogenetic analyses we followed these authors in recognizing Rhinolophoidea and Pteropodidae as sister groups which together form a larger clade (Yinpterochiroptera), with the remainder of extant families grouped as Yangochiroptera.

To analyze the occurrence of vertebral fusion at the family level, we used the morphological data set from Simmons *et al.* (2008). We analyzed the data using PAUP* (Phylogenetic Analysis Using Parsimony* and Other Methods; Swofford, 2003), except that we replaced the data on vertebral fusion from Simmons *et al.* (2008) with data from our study. We constrained trees with a molecular backbone scaffold from Simmons *et al.* (2008) and performed a heuristic search using tree bisection-reconnection (TBR) branch swapping and 100 addition sequence replicates. We obtained the resulting most parsimonious trees along with consistency and retention indices (CI and RI, respectively) as measurements of homoplasy (Simmons, 1993). Using the strict consensus of the most parsimonious trees, the number of independent origins of vertebral fusion were identified using character mapping as implemented in MacClade 4.06 (Maddison and Maddison, 2003).

To test for correlations between behavioral characteristics and vertebral fusion within a phylogenetic framework, we used the pairwise comparison technique outlined by Maddison (2000). We obtained the species-level data set from the supertree of bats (Jones et al., 2002) and constrained it using the molecular backbone scaffold tree derived from molecular studies (Simmons et al., 2008). Taxa for which we had no data for vertebral fusion were pruned from the constrained supertree.

Our analyses compared the distribution of two features of laryngeal echolocation and the method of feeding employed by bats to patterns of distribution of vertebral fusions within a phylogenetic framework. Intensity and duty-cycle of biosonar calls are usually consistent within families of Rhinolophoidea and Yangochiroptera. These parameters reflect the energy content of calls. A high-intensity call is defined as a signal greater than 110 decibels (dB) and a low-intensity signal is less than 60 dB when measured 10 cm from the source (Fenton, 1995). Intensity data were obtained from Fenton (1995), except Thyropteridae (Fenton et al., 1999a), Furipteridae (Fenton et al., 1999b) and some species of Vespertilionidae (Myotis septentrionalis, Faure et al., 1993; M. evotis, Faure and Barclay, 1994; M. grisescens, M. leibii, M. lucifugus and M. nigricans, Schnitzler and Kalko, 2001; Eptesicus furinalis, M. B. Fenton, personal communication, 2008), for a total of 157 species with call intensity data. Duty-cycle refers to the percentage of time of a respiratory cycle that a bat produces an echolocation signal. Most echolocating bats use low-duty-cycle signals, in which 20% or less of a respiratory cycle is used for call production. Hipposideridae, Rhinolophidae and one member of Mormoopidae, Pteronotus parnellii, use high-duty-cycle signals. These species produce signals that may consume 80% of a respiratory cycle (Fenton, 1995).

Bats use a range of methods to capture prey. Aerial hawking is performed in flight and relies on biosonar to detect flying insects. Speakman (2001) considered this to be the ancestral state for bats (but see Simmons and Geisler, 1998), but many other derived feeding methods are also used, including gleaning, perch hunting, trawling, quadrupedal terrestrial foraging and frugivory/nectarivory. For the purposes of analysis, we classified bats as either the ancestral condition (aerial hawkers) or any of the derived feeding methods. All comparisons were made with the pairwise comparison package for Mesquite using maximum pairs (Kalko and Schnitzler, 1998; Maddison, 2006; Maddison and Maddison, 2009).

15.3   Results

A total of 598 specimens of 178 species representing 16 families were of sufficient quality for evaluation of vertebral fusion (Table 15.1, Appendices 15.1

Table 15.1 The presence or absence of vertebral fusion loci per family. Vertebral fusion loci were defined as follows: FL – Fusion locus; o – No fusion at this locus; 1 – Fusion present at this locus; 1/o – At least one species with fusion at this locus; ? – Data were not available. Bold type indicates interspecific variation within families.

Family	Fusion Loci		
	FL1	FL2	FL3
Pteropodidae ($n = 75$)	o	o	o
Rhinolophidae ($n = 5$)	1	o	o
Hipposideridae ($n = 40$)	**1**	**1/o**	**1/o**
Aselliscus tricuspidatus ($n = 5$)	1	1	1
Hipposideros calcaratus ($n = 1$), *H. diadema* ($n = 6$) and *H. maggietaylorae* ($n = 5$)	1	o	o
Hipposideros caffer ($n = 3$), *H. cervinus* ($n = 5$), *H. cineraceus* ($n = 1$), *H. ruber* ($n = 5$) and *Rhinonicteris aurantia* ($n = 5$)	1	o	1
Megadermatidae ($n = 9$)	**1**	**o**	**1/o**
Cardioderma cor ($n = 2$) and *Lavia frons* ($n = 2$)	1	o	1
Macroderma gigas ($n = 5$)	1	o	o
Craseonycteridae*	1	?	1
Rhinopomatidae ($n = 5$)	o	o	o
Nycteridae ($n = 10$)	**1/o**	**o**	**o**
Nycteris aurita ($n = 2$) and *N. thebaica* ($n = 3$)	o	o	o
N. macrotis ($n = 5$)	1	o	o
Emballonuridae ($n = 35$)	o	o	o
Phyllostomidae ($n = 164$)	o	o	o
Mormoopidae ($n = 23$)	**1/o**	**o**	**1**
Mormoops blainvillei ($n = 1$) and *M. megalophylla* ($n = 3$)	1	o	1
Pteronotus davyi ($n = 4$), *P. macleayii* ($n = 2$), *P. parnellii* ($n = 5$), *P. personatus* ($n = 5$) and *P. quadridens* ($n = 3$)	o	o	1
Noctilionidae ($n = 10$)	o	o	o
Furipteridae ($n = 1$)	1	o	1
Thyropteridae ($n = 3$)	1	o	o
Mystacinidae*	o	?	o
Myzopodidae*	o	?	o
Miniopteridae ($n = 2$)	o	o	o
Vespertilionidae ($n = 139$)	o	o	o
Molossidae ($n = 74$)	1	o	o
Natalidae ($n = 5$)	1	o	1

* Data gathered using descriptions of vertebral fusion from Simmons *et al.* (2008).

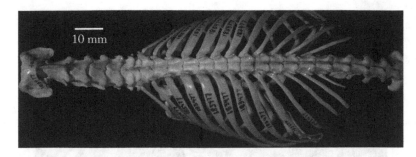

Figure 15.1 *Pteropus vampyrus* (MVZ 183477), dorsal view (anterior to the left). An example of a specimen with no vertebral fusion other than the sacrum.

and 15.2). Specimens with vertebrae were not available for Craseonycteridae, Mystacinidae and Myzopodidae. In addition to the sacrum, we found three distinct locations of vertebral fusion. Although areas of fusion were consistent, they showed both intraspecific and interspecific variability as to the vertebrae involved; therefore, we categorized them into three fusion loci. Fusion locus 1 (FL1) most often consisted of the last cervical vertebra (C7) united with the first thoracic vertebra (T1). At most, this fusion locus extended from C7 to the third thoracic (T3). Occasionally, only the first and second thoracic vertebrae are fused (T1–T2). Fusion locus 2 (FL2) involved only the mid-thoracic vertebrae and did not extend into the lumbar or cervical regions. Fusion locus 3 (FL3) nearly always involved lumbar vertebrae; it usually extended into the lower thoracic region, but never involved the last lumbar vertebra or the sacrum.

Many families showed no vertebral fusion apart from the sacrum: Pteropodidae (Figure 15.1), Emballonuridae, Noctilionidae, Phyllostomidae, Rhinopomatidae and Vespertilionidae. In three families, Rhinolophidae (Figure 15.2A, B), Molossidae and Thyropteridae, we found only FL1. Families Natalidae (Figure 15.3) and Furipteridae both exhibited FL1 and FL3. We examined only one species each of Natalidae and Thyropteridae; Furipteridae was represented by a single individual (Appendix 15.2).

Variation in patterns of fusion took two major forms: (1) individual variation within species and (2) variation within a family with respect to which fusion loci occurred. Within-species variation concerned the specific vertebrae involved in the fusion, or in two cases, the occurrence of entire loci in individuals that were not seen in conspecifics. We found individual variation in six of the nine families where fusion occurred. With respect to fusion locus 1 (FL1), of the three individuals of *Thyroptera tricolor* (Thyropteridae) that we examined, two had C7–T1 fused, whereas a third had C7–T2 fused. Although most species of Molossidae ($n = 74$) had only C7–T1 fused, we found individual variation in

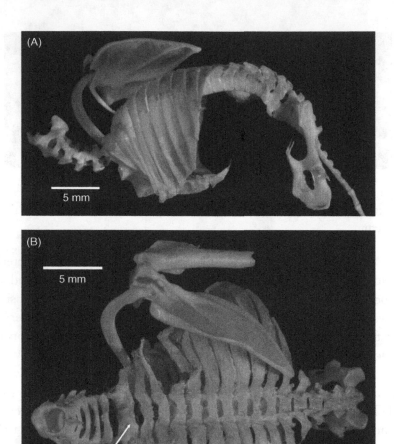

Figure 15.2 *Rhinolophus affinis* (AMNH 257199). (A) Lateral view (anterior to the left). Discrete, unfused lumbar vertebrae are visible. (B) Dorsal view (anterior to the left). An example of fusion locus 1 (FL1) between C7 and T1. Fusion includes the vertebral centra and neural arches as well as the adjacent laminae.

three species. We found one individual each of *Chaerephon jobensis* ($n = 5$) and *Nyctinomops laticaudatus* ($n = 5$) in which C7–T2 were fused. In *Nyctinomops femorosaccus* ($n = 5$), we found one individual with fusion from C7–T2, and two individuals with T1–T2 fused. The other individuals of these species followed the typical pattern of Molossidae. One individual of *Macroderma gigas* (Megadermatidae) added the T2 vertebra to the C7–T1 pair of FL1 found in four other

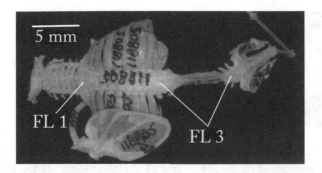

Figure 15.3 *Natalus stramineus* (MVZ 118805), dorsal view (anterior to the left). An example of fusion at FL1 involving the C7 and T1, and FL3 extending from T11 to L4. Individual vertebrae in this example of FL3 are nearly indistinguishable.

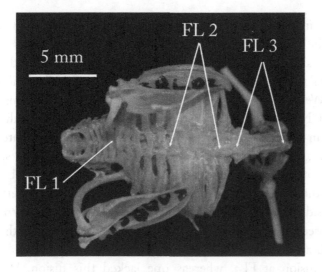

Figure 15.4 *Aselliscus tricuspidatus* (MVZ 138603), dorsal view (anterior to the left). FL 1, 2 and 3 are all present in this species. FL1 is not clearly visible here, but includes C7 through T3. Mid-thoracic fusion, FL2, spans from T6 to T9. FL3 includes only the L2 and L3. Both vertebral centra, neural arches and laminae are fused in all.

individuals. Otherwise, fusion at FL1 was remarkably consistent across a diverse range of taxa.

Individual variation was not limited to FL1. Fusion locus 2 appeared in only one individual of *Rhinonicteris aurantia* (Hipposideridae) of the five examined. This was one of the most unusual cases of individual variation in that it involved the presence of an entire fusion locus in one individual not seen in conspecifics (see also *Hipposideros caffer*, below). Variation at this locus in *Aselliscus tricuspidatus* (Figure 15.4) occurred only as a minor variation in the

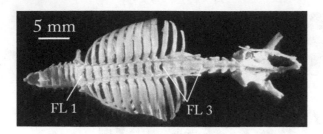

Figure 15.5 *Mormoops megalophylla* (MVZ 183506), dorsal view (anterior to the left). An example of fusion at FL1 and FL3. Fusion locus 3 (FL3) included the last two thoracic and the first lumbar vertebrae. Although there is complete fusion between the centra, discrete neural arches are still visible dorsally. FL1 did not occur in other species of Mormoopidae.

composition of FL2 in one of the five individuals examined; T5–T8 was fused in one individual, while the remaining four had T6–T9 fused.

Vertebral fusion at FL3 occurred in four families; we found individual variation in each of them, making this locus the most variable of the three (Appendix 15.2). We found variation in Hipposideridae, Megadermatidae, Mormoopidae and Natalidae (excepting species where only one individual was available, and in Furipteridae, where only one individual represented the entire family). Here, variation from the common pattern for various species consisted primarily of the inclusion of additional lower thoracic vertebrae on the cranial end of the fused unit, or inclusion of additional lumbar vertebrae caudally. In Hipposideridae, significant variation was seen in *Aselliscus tricuspidatus*, where the fused unit included as few as two or as many as five thoracic and lumbar vertebrae. Two of the three individuals of *Hipposideros caffer* we examined had fusion at FL3, whereas one lacked this fusion. We also found individual variation at FL3 in *Hipposideros ruber* and *Rhinonicteris aurantia*. Each of the five individuals of *Hipposideros ruber* examined had different vertebrae comprising the fused unit: T12–L3, L1–L4 and L1–L3. In *Rhinonicteris aurantia* two individuals had T12–L4 fused, and one individual had T12–L2 fused.

We found minor individual variation at FL3 in *Cardioderma cor* and *Lavia frons* (Megadermatidae). Of the two individuals of *Cardioderma cor* examined, one had T13–L3 fused, while the other had T12–L2 fused. In *Lavia frons*, one individual had T12–L2 fused, and the remaining individual had T11–L1 fused.

We found variation in FL3 in every species of Mormoopidae (Figure 15.5) that we examined (except *Mormoops blainvillei*, $n = 1$). *Pteronotus parnellii* showed the greatest degree of individual variation in FL3 of any species examined. The fused unit in this species ranged from only three vertebral bodies (T_{11}–L_1) to as many as seven (T_{10}–L_2). Part of this variation involved the relative numbers of thoracic and lumbar vertebrae, which complicated the variation at FL3 (Appendix 15.2). The total number of thoracic vertebrae in the *P. parnellii* examined ranged from 12 –14 and the lumbar count ranged from 4–5. Although the fusion between T_{10}–L_2 and T_{11}–L_1 appeared to include only one additional vertebra anteriorly, the difference between the total number of thoracic and lumbar vertebrae in each individual (12 thoracic and 5 lumbar compared to 14 thoracic and 5 lumbar vertebrae, respectively) was substantial.

Despite the variation among bats with FL3, this fusion locus consistently involved the lumbar series. We never found fusion between the penultimate and ultimate lumbar vertebrae or at the lumbosacral junction in any bat.

Intrafamilial variation, in which fusion was consistent within species, but varied among species within a family, was seen in three families. We found considerable variation among the species of Hipposideridae, Nycteridae and Mormoopidae that we examined. We found FL1, but not FL3, in all individuals of *Hipposideros calcaratus* ($n = 5$), *H. diadema* ($n = 6$) and *H. maggietaylorae* ($n = 5$). In contrast, both FL1 and FL3 were found in all *H. cervinus* ($n = 5$), *H. cineraceus* ($n = 1$) and *H. ruber* ($n = 5$). Within Nycteridae, we found no fusions in *Nycteris aurita* ($n = 2$) and *N. thebaica* ($n = 3$), but FL1 was consistently present in *N. macrotis* ($n = 5$). In the Mormoopidae, both FL1 and FL3 occurred in both species of *Mormoops* (Figure 15.5), whereas only FL3 was present in the five species of *Pteronotus* we examined (Appendix 15.2).

We identified three characters relating to vertebral fusion in the data matrix from Simmons *et al.* (2008) and excluded them from the analysis. Fusion loci 1, 2 and 3 were then added. Using the molecular backbone constraint from Simmons *et al.* (2008), the best composite tree including all three loci resulted in 741 steps. The consistency and retention indices (CI and RI) were 0.398 and 0.616, respectively, values that vary only slightly from those obtained using only Simmons *et al.* (2008) data (748 steps, CI = 0.400 and RI = 0.616). Separate phylogenetic trees were constructed for each fusion locus (Figure 15.6A–C). Optimization of character states on the tree for FL1 indicates that cervico-thoracic vertebral fusion has evolved independently between four and eight times within Chiroptera (Figure 15.6A). Fusion locus 2 has only arisen once (Figure 15.6B). Thoracolumbar fusion (FL3) has evolved independently on at least three occasions or possibly as many as five (Figure 15.6C).

Pairwise comparisons using the backbone scaffold of the molecular tree resulted in no statistically significant relationships between vertebral fusion, including individual fusion loci, when compared with intensity, duty-cycle and feeding methods. We found no relationship between the occurrence of vertebral fusion and the functional or behavioral characters (Table 15.2).

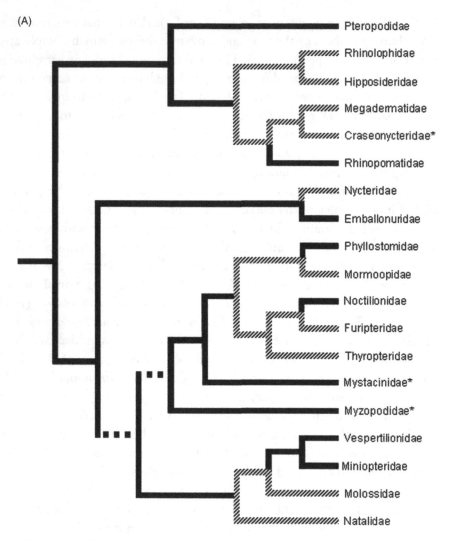

(A)

Figure 15.6 Phylogenetic occurrence of vertebral fusion in the extant families of order Chiroptera. Occurrence of vertebral fusion among families were constrained by using a backbone scaffolding tree *sensu* Simmons *et al.* (2008). Hatched lines indicate the presence of vertebral fusion in at least one species of that family. Asterisks indicate where data were gathered from Simmons *et al.* (2008). (A) Fusion locus 1. Cervicothoracic fusion evolved independently on at least three occasions and possibly as many as eight. (B) Fusion locus 2. Gray lines indicate data for this fusion locus was not available. This fusion locus was isolated to the mid-thoracic vertebrae and was restricted to a single species. (C) Fusion locus 3. Thoracolumbar fusion arose independently on at least three separate occasions and possibly as many as five. Fusion at this locus showed the greatest interspecific variability.

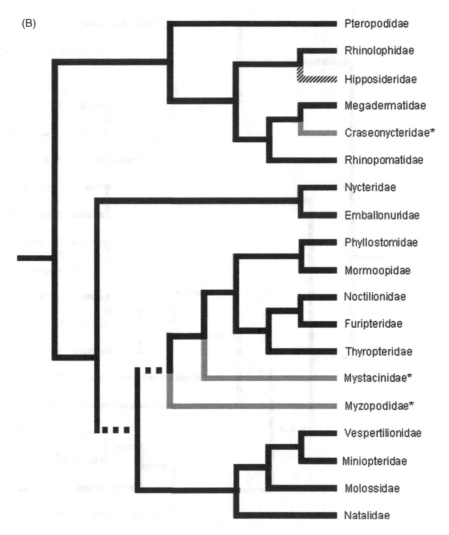

Figure 15.6 (*cont.*)

15.4 Discussion

Our examination of vertebral columns from most families of bats presents clear evidence that fusion of vertebrae in addition to the sacrum has arisen independently on at least four or as many as eight occasions in Chiroptera. Based on character mapping and correlation analyses, we reject the hypotheses of Lancaster *et al.* (1995); our tests show no association of vertebral fusion patterns with biosonar behavior that is independent of phylogeny.

Vertebral fusion in bats is a recurrent phenomenon that follows patterns among and within families. However, clear patterns also underlie the variation

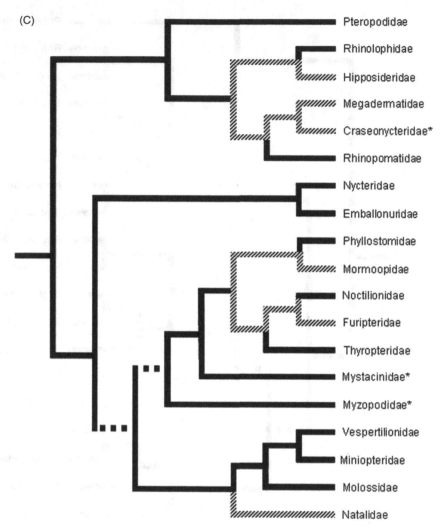

(C)

Pteropodidae
Rhinolophidae
Hipposideridae
Megadermatidae
Craseonycteridae*
Rhinopomatidae
Nycteridae
Emballonuridae
Phyllostomidae
Mormoopidae
Noctilionidae
Furipteridae
Thyropteridae
Mystacinidae*
Myzopodidae*
Vespertilionidae
Miniopteridae
Molossidae
Natalidae

Figure 15.6 (*cont.*)

seen within species. To organize these patterns, we developed the concept of
fusion loci, focused on identifying consistent regions of fusion as opposed to
categorization of fusions by the exact vertebrae involved. Previous studies have
described vertebral fusions and categorized them in different ways. Miller
(1907) recognized vertebral fusion in the cervicothoracic region and noted its
occurrence in several families (Nycteridae, Megadermatidae, Rhinolophidae,
Hipposideridae, Thyropteridae, Natalidae and Molossidae). He also noted
lumbar fusion in families where we documented it as well. Vaughan (1970)
similarly described the cervicothoracic and thoracolumbar fusion in several
families. Walton and Walton (1970) stated that vertebral fusion can occur in

Table 15.2 Results of the pairwise comparisons. Presence or absence of vertebral fusion and each fusion locus compared in succession to biosonar call intensity (high, low or non-echolocating), duty-cycle (high, low or non-echolocating) and feeding method (aerial hawking or other methods) using Mesquite (Maddison, 2006; Maddison and Maddison, 2009).

Fusion	Comparison	p range		best tail p
Vertebral Fusion	Intensity	0.25	1.0	0.5
	Duty-cycle	0.5	1.0	0.5
	Feeding Method	0.5	1.0	0.75
Fusion Locus 1	Intensity	0.5	0.75	0.75
	Duty-cycle	0.5	1.0	0.75
	Feeding Method	0.5	1.0	0.75
Fusion Locus 2	Intensity	1.0	1.0	1.0
	Duty-cycle	1.0	1.0	1.0
	Feeding Method	1.0	1.0	1.0
Fusion Locus 3	Intensity	0.125	0.5	0.25
	Duty-cycle	0.5	1.0	1.0
	Feeding Method	0.5	1.0	0.5

three places, including the sacral region (the latter is ubiquitous in mammals; Flower, 1885; and was therefore disregarded here). Neither Miller (1907) nor Walton and Walton (1970) mentioned fusion of the mid-thoracic vertebrae consistently found in *Aselliscus tricuspidatus* (Hipposideridae).

Our observations differ from previous studies in several details. Thyropteridae, Molossidae, Mormoopidae and Nycteridae differed from previous studies (Peffley, 1995; Simmons *et al.*, 2008). With regard to cervicothoracic fusion, these authors considered C7–T1 fusion as distinctly different from T1–T2 fusion. Our data do not support this conclusion. Fusion locus 1 varied from C7–T1, C7–T2, C7–T3 and T1–T2, and varied among individuals within species (Appendix 15.2). Simmons *et al.* (2008) classified Thyropteridae as having C7–T1 and T1–T2 fused; however, the three individuals of *Thyroptera tricolor* that we examined showed intraspecific variation, two individuals exhibited C7–T2 fused and one with C7–T1 fused. Previous studies (Miller, 1907; Vaughan, 1970; Simmons *et al.*, 2008) described all species of Molossidae as having C7–T1 fused, but we found several exceptions. Contrary to our data, Miller (1907) characterized all species of the family Nycteridae as having C7–T1 fused. *Aselliscus tricuspidatus* ($n = 5$) was the only species to consistently have FL1, FL2 and FL3 in the observed individuals (Figure 15.5), and to our knowledge the thoracic fusion in this species (what we termed FL2) has not been previously described.

Although the consistency of the patterns in vertebral fusion was more striking than the variability, we nonetheless found significant patterns in the nature of intraspecific variations. Variation at FL1 was minor and consisted simply of inclusion of vertebrae further caudal from the C7–T1 fusion or occasionally exclusion of C7 from the zone of fusion. We saw more variation at FL3, and it often accompanied variation in the total numbers of thoracic and lumbar vertebrae. This finding is supported by Lancaster *et al.* (1995), who mentioned both the presence of fusion and variation in the number of lumbar vertebrae involved in *Pteronotus parnellii*.

Our data offer no explanation as to the functional significance of vertebral fusion in bats. Lancaster *et al.* (1995) proposed vertebral fusion as an adaptation for high-intensity, high-duty-cycle echolocation, as fusions are seen in all species that use this style of biosonar vocalization. However, the results of the pairwise comparisons indicate that the phylogenetic relationships of the species that use these behaviors (with the exception of *Pteronotus parnellii*) are a more plausible explanation for the association between the morphology and behavior. Among bats that use high-intensity, high-duty-cycle echolocation, only hipposiderids have fusion at both FL1 and FL3. Rhinolophids, which use a similar form of biosonar, have no fusion at FL3 and *P. parnellii* lacks fusion at FL1. Finally, several groups with significant fusion at FL1, FL3 or both sites use low-intensity, low-duty-cycle echolocation. With regard to species that use the more commonly seen high-intensity, low-duty-cycle echolocation, at least one major family shows no extra-sacral fusion (Vespertilionidae), whereas a closely related family (Molossidae) uses similar echolocation, but has consistent fusion at FL1. We found no association between vertebral fusion and these aspects of biosonar behavior, and therefore reject all predictions based on the hypothetical association between the two.

Miller (1907) included cervicothoracic fusion in his description of morphological characteristics of the sternum and shoulder girdle. Although no direct function was ascribed to the vertebral fusion that frequently accompanied fusion of the first or second ribs to the sternum, Miller (1907) suggested that it could contribute to the stability of the pectoral girdle. However, he ended his discussion stating, "The mechanical need for this remarkable strengthening in bats of this size is by no means apparent" (Miller, 1907, p. 20).

Miller (1907) also noted that vertebral fusion occurred in the closely related families Nycteridae, Megadermatidae, Rhinolophidae and Hipposideridae using the accepted phylogenetic relationships of that time. Our data and those of Simmons *et al.* (2008) concur, although Nycteridae is now considered to be the sister group of Emballonuridae and is group with Yangochiroptera rather than Rhinolophoidea (Figure 15.6). Craseonycteridae is now included in

Yinpterochiroptera; Thyropteridae and Furipteridae form another clade with vertebral fusion. Although Molossidae, Natalidae and Mormoopidae exhibit vertebral fusion, their sister groups (Vespertilionidae plus Miniopteridae and Phyllostomidae, respectively) do not. Because the results of the pairwise comparisons indicate no relationship between the occurrence of vertebral fusion, and functional and behavior characters that we examined, we are left to conclude that the distribution of fusion may be explained in part by inheritance from common ancestors, or some other untested factor.

The finding that vertebral fusion has arisen independently on at least four and possibly eight separate occasions in bats suggests an association with the lifestyles of bats that we have not considered. Other instances of consistent non-pathological fusion of vertebrae in mammals (other than the sacrum) are plausibly correlated with unusual functional adaptations (e.g., stabilization of the neck against water pressure in cetaceans, Buchholtz and Schur, 2004, and the ricochetal locomotion of jerboas, Hatt, 1932). We have no data to relate vertebral fusion to flight. Although it could conceivably relate to specializations that we did not test, occurrence of fusion in groups such as rhinolophoids, molossids, natalids, thyropterids and furipterids indicate that it is compatible both with highly maneuverable flight in closed spaces as well as fast, straight flight in open spaces (Norberg and Rayner, 1987); likewise, species without fusion use every variety of styles of flight. Neither can fusion be reasonably associated with any of the other unusual characteristics that are common to all bats. To our knowledge, no other group of mammals exhibits the tendency to develop fused vertebral segments in multiple regions, yet it has arisen and stabilized independently in this one monophyletic order on multiple occasions. We must, therefore, consider the possible genetic origins of fusion and speculate on why it recurs in this group.

Hox genes orchestrate the pattern of much of the vertebrate body plan, and specifically the axial skeleton. Development of vertebrate embryos proceeds in general from anterior to posterior (cranial to caudal), and during development groups of specific *Hox* genes are expressed sequentially with discrete expression boundaries along the anteroposterior axis. *Hox* genes belong to a class of transcription factors that contain a homeodomain that binds to DNA and is involved in transcription regulation. The temporal and spatial expression (e.g., anterior to posterior boundary limits) of these genes control many downstream events, such as cell patterning and differentiation (Wellik, 2007).

Specifically, Wellik (2007) suggested that patterning of cervical vertebrae 6 and 7 is influenced not only by the anterior expression limit of *Hox6*, but also by other *Hox* genes posterior to it. Therefore, variations in the function of the *Hox6* group and others could explain fusion at FL1 that does not extend farther anteriorly. As this was the most consistent of the three loci, specifically in the case of the Molossidae, Rhinolophidae and Hipposideridae, we suggest that this character transformation became stabilized early in the histories of these groups. The inclusion of ribs and the manubrium in FL1 in the latter two families (Miller, 1907; Vaughan, 1970) is of interest as the ventral portions of ribs and the sternum form not from somitic mesoderm, but from lateral plate mesoderm. The picture is further complicated by the finding that *Hox* genes pattern the sternum in a non-colinear manner that is independent from the somite-derived structures of the vertebral column and may require the effects of two or more *Hox* groups (McIntyre *et al.*, 2007). The rarity of FL2 may be explained by the observation that mutations of *Hox* genes and loss of paralogous *Hox* genes have not been associated with changes in the mid-thoracic vertebrae, likely due to the redundancy or plasticity of their functional properties (Wellik, 2007).

Fusion locus 3, the area of greatest individual and intrafamilial variability, is also the vertebral region associated with the greatest effects of *Hox* mutation described in the literature. The caudal-most thoracic and the lumbar vertebrae, and the sacrum are considered to be patterned by *Hox10* and *Hox11*. Without the function of *Hox11*, sacral vertebrae will fail to develop and will assume the form of lumbar vertebrae (anterior homeotic transformation; Wellik and Capecchi, 2003; Wellik, 2007). The opposite effect, posterior homeotic transformation, could explain fusion at FL3. Gérard *et al.* (1997) experimentally produced such an effect through an interspecies exchange of an enhancer of the *Hoxd11* gene. In comparison to controls, their experimental animals activated *Hoxd11* transcription prematurely, resulting in a cranial transposition of the sacrum. The fused lumbar vertebrae seen in Natalidae (Figure 15.3) and Furipteridae bear a remarkable resemblance to the sacrum, and could be conceivably explained by posterior homeotic transformation of sacral vertebral fusion.

The extraordinary adaptations of bats have recently captured the attention of developmental geneticists, as evidenced by numerous recent studies on the genetic control of limb development (e.g., Chen *et al.*, 2005; Hockman *et al.*, 2008; Ray and Capecchi, 2008). Some of these studies examined *Hox* gene activity, along with that of *Sonic hedgehog*, *Gremlin*, *Fgf8*, *Bmp2* and others. Although *Hox* gene expression along the anteroposterior axis exerts a considerable effect on patterning of the axial skeleton, bilateral and dorsoventral patterning as seen in limb development may also play a role in patterning of the axial skeleton, and the actions of these downstream genes cannot be excluded.

Regardless of the developmental explanations for the origins of vertebral fusion, we found no functional or behavioral characters that related to fusion independent of phylogeny. Our findings do not explain the mode of stabilizing selection that maintained these mutations for 50 million years in some lineages. The similarity of structure among vertebrae attributable to serial homology belies the functional differences between regions of the vertebral column. Differences in the shapes and orientations of zygapophyseal joints as well as the structure of muscles and ligaments govern the range of motion that occurs between successive vertebrae and within regions. Functions of cervical, thoracic and lumbar vertebrae are markedly different as unfused segments. We cannot, therefore, assume that vertebral fusions in different areas share any functional similarity simply because they appear to be similar due to the anatomical coincidences of serial homology. The discrete nature of the three fusion loci is further borne out by their distinct phylogenetic histories, in which FL2 is monoplyletic, and FL1 and FL3 co-occur in some instances, but separately in others (Figure 15.6).

Further studies of this phenomenon should seek to separate the functional significance of fusions in different regions. Functional significance could also be clarified by relating variations in vertebral fusion and vertebral counts to populations and consider genetic relationships (e.g., Mormoopidae). Examination of embryonic development of species from families with the greatest variation could shed light on the genetic control of fusion (e.g., Nycteridae). Finally, an examination of the biomechanical consequences of fusion could elucidate the function of this adaptation in groups where it is found.

15.5   Summary

Fusion between the vertebrae of bats occurred at three distinct places in the vertebral column other than the sacrum, and we refer to these as fusion loci. Fusion locus 1 (FL1) occurred between the caudal-most cervical vertebra and the adjacent one to three thoracic vertebrae. Fusion locus 2 (FL2) included only middle thoracic vertebrae. Fusion locus 3 (FL3) included both thoracic and lumbar vertebrae, but never extended to, or included the sacrum. Fusion locus 2 showed the least variability and was found consistently in only one species. Fusion locus 1 showed variability mostly in the number of thoracic vertebrae involved. Fusion locus 3 showed the greatest variability, including variations in the total number of thoracic and lumbar vertebrae. Our phylogenetic analysis and pairwise comparisons indicate no relationship between patterns of vertebral fusion and biosonar call intensity, duty-cycle or method of feeding. Our data suggest that fusion arose independently from four to eight times. Vertebral development is under the control of *Hox* genes which are expressed as sequential groups with discrete expression boundaries along the anteroposterior axis

(Wellik, 2007). The normal function of *Hox* genes posteriorly are responsible for the fusion of the sacrum (Wellik and Capecchi, 2003). Disruptions of the borders of expression of these genes could transform the normal sacral fusion anteriorly into the lumbar region. Future studies should examine the variation and biomechanical function of vertebral fusion.

15.6 Acknowledgments

We greatly appreciate the access to the museum collections used in this study and thank Chris Conroy (MVZ), Jim Dines (LACM), Maureen Flannery (CAS) and Eileen Westwig (AMNH) for their support and cooperation. We thank the editors, Gregg Gunnell and Nancy Simmons, for the opportunity to contribute to this volume. We thank Brock Fenton, Andrea Wetterer, Nancy Simmons and Gregg Gunnell for their critical reviews of the manuscript and gratefully acknowledge the contributions of Thomas Peavy and Thomas Landerholm to our understanding of the genetic control of development. Thanks to Ted Cranford for his help with the photographic illustrations.

Appendix 15.1

Species without vertebral fusion arranged by family. Abbreviations: AMNH – American Museum of Natural History, CAS – California Academy of Sciences, LACM – Natural History Museum of Los Angeles County, MVZ – Museum of Vertebrate Zoology.

Species	Catalog number
Pteropodidae	
Cynopterus brachyotis	MVZ 183470
Dobsonia minor	MVZ 138497, 138498, 140206, 141097, 141102
Dobsonia moluccensis	MVZ 138495, LACM 67792, 74681, 74682
Eidolon helvum	AMNH 86245, MVZ 196219
Epomophorus wahlbergi	AMNH 187274, 187275, 187276, 187277
Epomops buettikoferi	AMNH 207007, 207008
Macroglossus minimus	MVZ 138558, 140232, 140236, 140238, 140239
Nyctimene aello	MVZ 138524, 140319, 141171
Nyctimene albiventer	LACM 65795, MVZ 140322, 140324, 140326, 140327, 140329
Paranyctimene raptor	MVZ 140343, 140362, 140363, 140365, 141129, 141137
Pteropus alecto	LACM 74685

Species	Catalog number
Pteropus conspicillatus	MVZ 140203, 141010, 141016, 141017, 141029
Pteropus giganteus	LACM 90801, 90802, 90803
Pteropus mariannus	MVZ 200079, 200144, 200185, 200193, 200205
Pteropus molossinus	MVZ 182511, 182515, 182526, 182527, 182528
Pteropus neohibernicus	MVZ 138492, 138493
Pteropus vampyrus	MVZ 116835, 183475, 183477
Rousettus aegyptiacus	MVZ 196220, 196221, 196222
Rousettus amplexicaudatus	MVZ 138502, 138504, 140186, 140198, 140199
Syconycteris australis	MVZ 140279, 140280, 140281, 140705, 140707
Emballonuridae	
Balantiopteryx plicata	MVZ 109959, 109960, 109965, 109967, 109968
Coleura afra	LACM 55542
Cormura brevirostris	AMNH 266002, 267071, 267389, 267391, 267823
Emballonura raffrayana	LACM 65826, 65828
Peropteryx kappleri	AMNH 265992, 267393, 267834
Peropteryx leucoptera	AMNH 266010
Rhynchonycteris naso	AMNH 209198, 209199, 209208, 209209, 210454
Saccolaimus flaviventris	AMNH 197201, 197202
Saccopteryx bilineata	AMNH 210459, 210464, 210465, MVZ 98187, 98188
Taphozous georgianus	AMNH 197175, 197182, 197560, 197760
Taphozous melanopogon	MVZ 183484
Taphozous theobaldi	MVZ 179919
Rhinopomatidae	
Rhinopoma microphyllum	AMNH 244385
Rhinopoma muscatellum	AMNH 244329, 244390, 244393, 244394
Nycteridae	
Nycteris aurita	AMNH 187293, 187294
Nycteris thebaica	AMNH 169164, 187321, LACM 14485
Phyllostomidae	
Anoura geoffroyi	LACM 55376, 53604, 55370, MVZ 148037, 183520
Ardops nichollsi	MVZ 166214, 166216, 166219, 172394, 172395
Artibeus anderseni	MVZ 165995, 165996

Species	Catalog number
Artibeus aztecus	LACM 53558, 53595
Artibeus fimbriatus	MVZ 144442, 144444
Artibeus fraterculus	MVZ 168895, 168896, 168903, 168912, 168913, 168914
Artibeus jamaicensis	MVZ 144445, 154906, 154907, 154908, 154909, 157727, 166011, 166115, 166116, 166181, 166224
Artibeus lituratus	MVZ 157724, 166553, 166554, 166556, 183527
Artibeus obscurus	MVZ 154894, 154895, 157728, 157730
Artibeus phaeotis	LACM 15484, 19369, 19371
Artibeus toltecus	LACM 55647, 55648
Brachyphylla cavernarum	MVZ 166201, 173792
Carollia brevicauda	LACM 53572, 53602, MVZ 154828, 157683, 157685, 160048, 160049, 160067
Carollia castanea	MVZ 154827, 157680, 157682, 165968, 165969
Carollia perspicillata	MVZ 114402, 118781, 160046, 160066, 183530
Centurio senex	CAS 15485
Chiroderma trinitatum	AMNH 264075, 264076, 264077, 268531
Choeronycteris mexicana	MVZ 100080, 119067, 186389, 181845, 181846
Chrotopterus auritus	AMNH 209353, 256824, 261373, 267444, 267852
Desmodus rotundus	LACM 53577, MVZ 113503, 113504, 183539
Diphylla ecaudata	MVZ 98248
Erophylla bombifrons	MVZ 166160, 166171
Glossophaga leachii	MVZ 183541
Glossophaga soricina	MVZ 148036, 154802, 160044, 160045
Leptonycteris curasoae	CAS 22534
Leptonycteris nivalis	CAS 10112, 10113, 12030, 12031, 12796, MVZ 80326
Lonchophylla thomasi	AMNH 266103, 266109, 267940, 267943
Lonchorhina aurita	AMNH 230122
Lophostoma silvicolum	AMNH 267107, 267108, 267923
Macrotus californicus	MVZ 44194, 106340, 106343, 106345, 106348
Mesophylla macconnelli	AMNH 209579, 267281
Micronycteris megalotis	MVZ 118759, LACM 55322
Mimon crenulatum	MVZ 160043
Monophyllus plethodon	MVZ 166188, 166189, 166193, 172397, 172398

Species	Catalog number
Monophyllus redmani	MVZ 166165, 166166
Phylloderma stenops	AMNH 205371
Phyllostomus hastatus	MVZ 114230, 114386, 135840, 166647, 183555
Platyrrhinus infuscus	MVZ 116632, 154886, 154887, 154888, 154889
Pygoderma bilabiatum	AMNH 248340, 261758, 261760, 261762
Rhinophylla pumilio	MVZ 165973
Sphaeronycteris toxophyllum	AMNH 262637, 209740
Sturnira lilium	MVZ 144406, 144407, 160076, 160079, 173796
Sturnira ludovici	CAS 14977
Trachops cirrhosus	AMNH 209348, 209351, 210579, 210680, 261378, MVZ 98198
Uroderma bilobatum	MVZ 116758, 118791, 118792, 118794, 166572
Vampyrum spectrum	AMNH 261379, MVZ 154782
Noctilionidae	
Noctilio albiventris	AMNH 209310, MVZ 116739, 144328, 118749, 118750
Noctilio leporinus	AMNH 265974, LACM 12956, MVZ 166218, 144345, 183516
Miniopteridae	
Miniopterus minor	CAS 26843
Miniopterus tristis	MVZ 138608
Vespertilionidae	
Antrozous pallidus	MVZ 21122, 104233, 110718, 114349, 183561
Corynorhinus townsendii	MVZ 19215, 112913, 146647, 182470, 182473
Eptesicus furinalis	MVZ 144879, 144880
Eptesicus fuscus	MVZ 136800, 148681, 182363
Euderma maculatum	LACM 13856, MVZ 102744, MVZ 139209
Falsistrellus mackenziei	AMNH 197218, 197220, 197221
Glischropus tylopus	AMNH 235586
Histiotus macrotus	MVZ 150965, 152156
Histiotus montanus	MVZ 150930
Idionycteris phyllotis	LACM 67102, MVZ 167524
Lasionycteris noctivagans	MVZ 181859, 181860, 181862, 182376, 182484
Lasiurus borealis	MVZ 138995, 182393, 182497, 182500, 183575
Lasiurus cinereus	MVZ 112947, 163913, 182406, 182413, 182423
Lasiurus ega	LACM 93997, 93998, 94005, MVZ 144988, 144989
Lasiurus intermedius	LACM 56062
Lasiurus xanthinus[†]	MVZ 181891, 181928, 181935, 181939, 182508

Species	Catalog number
Myotis adversus	LACM 66902
Myotis albescens	MVZ 144572, 144573
Myotis austroriparius	MVZ 77506, 77507, 77508, 77509
Myotis californicus	MVZ 146523, 146525, 182438, 182439, 182440
Myotis ciliolabrum	MVZ 31555, 78377
Myotis dominicensis	MVZ 173791, 173793
Myotis evotis	MVZ 182441, 182443, 182442, 206880, 206886
Myotis grisescens	MVZ 63099, 104707
Myotis leibii	LACM 91139, 91140, 91142, 91143, 91144
Myotis lucifugus	MVZ 109457, 109458, 182444, 182447, 182448
Myotis nigricans	MVZ 144673
Myotis septentrionalis	MVZ 109452, 109453
Myotis sodalis	MVZ 136773, 136775, 136779, 136782, 136783
Myotis thysanodes	MVZ 146555, 146556, 146558, 146559, 206881
Myotis velifer	MVZ 106171, 186397
Myotis vivesi	LACM 28273, 28274, 28275, 70219, 72640, 119076, MVZ 119077
Myotis volans	MVZ 78357, 78361, 78368, 78369, 206882
Myotis yumanensis	MVZ 148658, 148660, 148661, 148663, 148665
Nyctophilus timoriensis	MVZ 138609
Pipistrellus hesperus	MVZ 198302, 198303, 198304, 198314, 198319
Pipistrellus papuanus	MVZ 140411, 140413, 140414, 140415, 140416
Pipistrellus stenopterus	MVZ 116841, 116842
Pipistrellus subflavus	MVZ 104451, 104454, 104483, 136799
Vespadelus caurinus	AMNH 197699, 197710

[†] *Lasiurus xanthinus* was combined with *L. ega* in the pairwise comparisons; these were previously considered the same species (Simmons, 2005).

Appendix 15.2

Composition and count (in parentheses) of fused vertebrae at each fusion locus of species examined in this study.

Abbreviations: AMNH – American Museum of Natural History, CAS – California Academy of Sciences, LACM – Natural History Museum of Los Angeles County, MVZ – Museum of Vertebrate Zoology, N – number of individuals, FL – fusion locus, C – cervical, T – thoracic, L – lumbar. Note that family Nycteridae appears both in Appendix 15.1 and 15.2.

Species	Catalog no.	N	FL1	FL2	FL3
Rhinolophidae					
Rhinolophus affinis	AMNH 257199, 256200	2	C7–T1 (2)		
Rhinolophus darlingi	AMNH 257157	1	C7–T1 (2)		
Rhinolophus pusillus	AMNH 257198	1	C7–T1 (2)		
Rhinolophus stheno	AMNH 235557	1	C7–T1 (2)		
Hipposideridae					
Aselliscus tricuspidatus	MVZ 140407	1	T1–T2 (2)	T5–T8 (4)	T12–L3 (4)
	MVZ 138603	1	C7–T3 (4)	T6–T9 (4)	L2–L3 (2)
	MVZ 138604	1	C7–T3 (4)	T6–T9 (4)	L1–L3 (3)
	MVZ 138606	1	C7–T3 (4)	T6–T9 (4)	L1–L5 (5)
	MVZ 138607	1	C7–T3 (4)	T6–T9 (4)	T12–L4 (5)
Hipposideros caffer	LACM 56251	1	C7–T2 (3)		
	AMNH 257174, 216220	2	C7–T2 (3)		L1–L3 (3)
Hipposideros calcaratus	MVZ 140376	1	C7–T1 (2)		
	MVZ 140380, 140381, 140382, 140383	4	C7–T2 (3)		
Hipposideros cervinus	MVZ 138594, 138595, 138596, 140400, 140403	5	C7–T2 (3)		L1–L3 (3)
Hipposideros cineraceus	AMNH 235578	1	C7–T2 (3)		L1–L3 (3)
Hipposideros diadema	MVZ 138571, 138572, 138573, 140391, 140392, 183558	6	C7–T2 (3)		
Hipposideros maggietaylorae	MVZ 138577, 138578, 140385, 140386, 140388	5	C7–T2 (3)		
Hipposideros ruber	CAS 26848	1	C7–T2 (3)		T12–L3 (4)
	CAS 26851	1	C7–T2 (3)		L1–L4 (4)
	CAS 26850, 26852, 26853	3	C7–T2 (3)		L1–L3 (3)
Rhinonicteris aurantia	LACM 74725	1	C7–T2 (3)	T6–T7 (2)	T12–L4 (5)
	AMNH 197213	1	C7–T2 (3)		T12–L2 (3)
	LACM 74726, 74727, 85755	3	C7–T2 (3)		T12–L4 (5)

Species	Catalog no.	N	FL1	FL2	FL3
Megadermatidae					
Cardioderma cor	AMNH 187327	1	C7–T1 (2)		T13–L3 (4)
	AMNH 187329	1	C7–T1 (2)		T12–L2 (4)
Lavia frons	AMNH 187343	1	C7–T1 (2)		T12–L2 (4)
	AMNH 187340	1	C7–T1 (2)		T11–L1 (4)
Macroderma	LACM 74699	1	C7–T2 (3)		
gigas	LACM 74697, 74698, 74700, 74701	4	C7–T1 (2)		
Nycteridae					
Nycteris macrotis	AMNH 187304, 187305, 187312, 187317, 187319	5	C7–T1 (2)		
Mormoopidae					
Mormoops blainvillei	MVZ 166152	1	C7–T1 (2)		T11–L1 (3)
Mormoops megalophylla	AMNH 190138	1	C7–T1 (2)		T11–L1 (4)
	MVZ 183506, 183507	2	C7–T1 (2)		T11–L1 (3)
Pteronotus davyi	MVZ 159447	1			T11–L1 (3)
	AMNH 203565	1			T12–L1 (3)
	AMNH 189596	1			T12–L1 (3)
	AMNH 203566	1			T12–L2 (3)
Pteronotus macleayii	AMNH 176188	1			T11–L2 (5)
	AMNH 60917	1			T11–L3 (6)
Pteronotus parnellii	AMNH 267285	1			T10–L2 (7)*
	MVZ 183508, AMNH 267288	2			T11–L1 (4)
	MVZ 160042	1			T11–L1 (3)
	CAS 12021	1			T11–T14 (4)*
Pteronotus personatus	LACM 55106	1			T10–L2 (5)
	LACM 55108, 55316	2			T10–L2 (6)
	LACM 55107, 55109	2			T11–L2 (5)
Pteronotus quadridens	MVZ 166139	1			T10–L2 (5)
	MVZ 166140	1			T11–L3 (5)
	MVZ 166141	1			T10–L2 (6)
Furipteridae					
Furipterus horrens	AMNH 265975	1	C7–T1 (2)		T8–L5 (10)

Species	Catalog no.	N	FL1	FL2	FL3
Thyropteridae					
Thyroptera	AMNH 267216	1	C7–T2 (3)		
tricolor	AMNH 266361, 267217	2	C7–T1 (2)		
Molossidae					
Chaerephon ansorgei	AMNH 257447, 257448, 257449, 257450	4	C7–T1 (2)		
Chaerephon jobensis	AMNH 197156	1	C7–T2 (3)		
	AMNH 197159, 197160, 197164, 197166	4	C7–T1 (2)		
Cheiromeles parvidens	AMNH 241942	1	C7–T1 (2)		
Cheiromeles torquatus	AMNH 247583	1	C7–T1 (2)		
Cynomops paranus	AMNH 267535	1	C7–T1 (2)		
Eumops auripendulus	AMNH 248210, 248222, 248223, 248225, 248226	5	C7–T1 (2)		
Eumops bonariensis	AMNH 235959, 235960, 235961	3	C7–T1 (2)		
Eumops glaucinus	AMNH 179948	1	C7–T1 (2)		
Eumops perotis	MVZ 109569, 109572, 109573, 182348, 182476	5	C7–T1 (2)		
Eumops trumbulli[†]	AMNH 209901, 209902	2	C7–T1 (2)		
Eumops underwoodi	MVZ 82154, 82155, 82156	3	C7–T1 (2)		
Molossops temminckii	AMNH 234460, 248362	2	C7–T1 (2)		
Molossus aztecus	AMNH 190179	1	C7–T1 (2)		
Molossus barnesi[†]	AMNH 269105	1	C7–T1 (2)		
Molossus coibensis[†]	AMNH 217447	1	C7–T1 (2)		
Molossus molossus	AMNH 211365, 267244, 267246,	5	C7–T1 (2)		

Species	Catalog no.	N	FL1	FL2	FL3
	267250, MVZ 183494				
Molossus rufus	AMNH 203946, 209904, 263285, MVZ 98903, 183489	5	C7–T1 (2)		
Molossus sinaloae	AMNH 269110, 269112	2	C7–T1 (2)		
Mops mops	AMNH 235590	1	C7–T1 (2)		
Nyctinomops femorosaccus	MVZ 181963, 181979	2	C7–T1 (2)		
	MVZ 181972, 181974	2	T1–T2 (2)		
	MVZ 181975	1	C7–T2 (3)		
Nyctinomops laticaudatus	AMNH 209878	1	C7–T2 (3)		
	AMNH 209881, 209882, 209884, 209888	4	C7–T1 (2)		
Nyctinomops macrotis	MVZ 97153, 97154, 97155, 97157, 97158	5	C7–T1 (2)		
Promops centralis	AMNH 269114, MVZ 145138, 145139	3	C7–T1 (2)		
Tadarida australis	MVZ 129297, 129298	2	C7–T1 (2)		
Tadarida brasiliensis	MVZ 146579, 146583, 146584, 146585, 146591	5	C7–T1 (2)		
Natalidae					
Natalus stramineus	MVZ 110225	1	C7–T1 (2)		T10–L3 (6)
	MVZ 110227	1	C7–T1 (2)		T10–L4 (6)
	MVZ 110223	1	C7–T1 (2)		T10–L5 (8)
	MVZ 118805	1	C7–T1 (2)		T11–L4 (6)
	MVZ 110222	1	C7–T2 (3)		T11–L5 (7)

[†] *Eumops trumbulli* was combined with *E. perotis* and *Molossus barnesi* and *M. coibensis* was combined with *M. molossus* in the pairwise comparisons; these were previously considered the same species (Simmons, 2005).

[*] Individuals with 14 thoracic vertebrae.

15.7   REFERENCES

Bogdanowicz, W. and Owen, R. D. (1998). In the minotaur's labyrinth: phylogeny of the bat family Hipposideridae. In *Bat Biology and Conservation*, ed. T. H. Kunz and P. A. Racey. Washington, DC: Smithsonian Institution Press, pp. 27–42.

Buchholtz, E. A. and Schur, S. A. (2004). Vertebral osteology in Delphinidae (Cetacea). *Zoological Journal of the Linnean Society*, **140**, 383–401.

Buchholtz, E. A. (2007). Modular evolution of the Cetacean vertebral column. *Evolution and Development*, **9**, 278–289.

Chen, C.-H., Cretekos, C. J., Rasweiler IV, J. J. and Behringer, R. R. (2005). *Hoxd13* expression in the developing limbs of the short-tailed fruit bat, *Carollia perspicillata*. *Evolution and Development*, **7**, 130–141.

Dobson, G. E. (1878). *Catalogue of the Chiroptera in the Collection of the British Museum edn.* London: British Museum of Natural History.

Estenne, M., Zocchi, L., Ward, M. and Macklem, P. T. (1990). Chest wall motion and expiratory muscle use during phonation in normal humans. *Journal of Applied Physiology*, **68**, 2075–2082.

Faure, P. A. and Barclay, R. M. R. (1994). Substrate-gleaning versus aerial-hawking: plasticity in the foraging and echolocation behaviour of the long-eared bat, *Myotis evotis*. *Journal of Comparative Physiology A*, **174**, 651–660.

Faure, P. A., Fullard, J. H. and Dawson, J. W. (1993). The gleaning attacks of the northern long-eared bat, *Myotis septentrionalis*, are relatively inaudible to moths. *Journal of Experimental Biology*, **178**, 173–189.

Fenton, M. B. (1995). Natural history and biosonar signals. In *Hearing by Bats*, ed. A. N. Popper and R. R. Fay. New York: Springer-Verlag, pp. 37–86.

Fenton, M. B. and Crerar, L. M. (1984). Cervical vertebrae in relation to roosting posture in bats. *Journal of Mammalogy*, **65**, 395–403.

Fenton, M. B., Rydell, J., Vonhof, M. J., Eklöf, J. and Lancaster, W. C. (1999a). Constant-frequency and frequency-modulated components in the echolocation calls of three species of small bats (Emballonuridae, Thyropteridae, and Vespertilionidae). *Canadian Journal of Zoology*, **77**, 1891–1900.

Fenton, M. B., Whitaker, J. O., Jr., Vonhof, M. J. *et al.* (1999b). The diet of bats from Southeastern Brazil: the relation to echolocation and foraging behaviour. *Revista Brasileira de Zoologica*, **16**, 1081–1085.

Flower, W. H. F. (1885). *An Introduction to the Osteology of the Mammalia*, 2nd edn. London: Macmillan and Co.

Gérard, M., Zákány, J. and Duboule, D. (1997). Interspecies exchange of a *Hoxd* enhancer *in vivo* induces premature transcription and anterior shift of the sacrum. *Developmental Biology*, **190**, 32–40.

Hatt, R. T. (1932). The vertebral columns of ricochetal rodents. *Bulletin of the American Museum of Natural History*, **68**, 599–738.

Hildebrand, M. and Goslow, G. E., Jr. (2001). *Analysis of Vertebrate Structure*, 5th edn. New York: John Wiley & Sons, Inc.

Hockman, D., Cretekos, C. J., Mason, M. K. *et al.* (2008). A second wave of *Sonic hedgehog* expression during the development of the bat limb. *Proceedings of the National Academy of Sciences, USA*, **105**, 16982–16987.

Howell, D. J. and Pylka, J. (1977). Why bats hang upside down: a biochemical hypothesis. *Journal of Theoretical Biology*, **69**, 625–631.

Jones, K. E., Purvis, A., MacLarnon, A., Binida-Emonds, O. R. P. and Simmons, N. B. (2002). A phylogenetic supertree of the bats (Mammalia: Chiroptera). *Biological Reviews*, **77**, 223–259.

Kalko, E. K. V. and Schnitzler, H.-U. (1998). How echolocating bats acquire food. In *Bat Biology and Conservation*, ed. T. H. Kunz and P. A. Racey. Washington, DC: Smithsonian Institution Press, pp. 197–204.

Lancaster, W. C., Henson, O. W., Jr. and Keating, A. W. (1995). Respiratory muscle activity in relation to vocalization in flying bats. *Journal of Experimental Biology*, **198**, 175–191.

Maddison, W. P. (2000). Testing character correlation using pairwise comparison on a phylogeny. *Journal of Theoretical Biology*, **202**, 195–204.

Maddison, W. P. (2006). Pairwise comparisons package for Mesquite, version 1.1. http://mesquiteproject.org.

Maddison, W. P. and Maddison, D. R. (2003). MacClade, version 4.04.

Maddison, W. P. and Maddison, D. R. (2009). Mesquite: a modular system for evolutionary analysis, version 2.6. http://mesquiteproject.org.

McIntyre, D. C., Rakshit, S., Yallowitz, A. R. *et al.* (2007). Hox patterning of the vertebrate rib cage. *Development*, **134**, 2981–2989.

Miller, G. S., Jr. (1907). The families and genera of bats. *Bulletin of the United States National Museum*, **57**, 1–282.

Miller-Butterworth, C. M., Murphy, W. J., O'Brien, S. J. *et al.* (2007). A family matter: conclusive resolution of the taxonomic position of the long-fingered bats, *Miniopterus*. *Molecular Biology and Evolution*, **24**, 1553–1561.

Norberg, U. M. and Rayner, J. M. V. (1987). Ecological morphology and flight in bats (Mammalia; Chiroptera): wing adaptations, flight performance, foraging strategy and echolocation. *Philosophical Transactions of the Royal Society of London B*, **316**, 335–427.

Peffley, A. L. (1995). Vertebral fusion in bats: phylogenetic interpretation. *Bat Research News*, **36**, 99–100.

Ray, R. and Capecchi, M. (2008). An examination of the Chiropteran *HoxD* locus from an evolutionary perspective. *Evolution and Development*, **10**, 657–670.

Schnitzler, H.-U. and Kalko, E. K. V. (2001). Echolocation by insect-eating bats. *BioScience*, **51**, 557–569.

Simmons, N. B. (1993). The importance of methods: Archontan phylogeny and cladistic analysis of morphological data. In *Primates and Their Relatives in Phylogenetic Perspective*, ed. R. D. E. MacPhee. New York: Plenum Press, pp. 1–61.

Simmons, N. B. (2005). Order Chiroptera. In *Mammal Species of the World: A Taxonomic and Geographic Reference*, ed. D. E. Wilson and D. M. Reeder, vol. 1. Baltimore, MD: Johns Hopkins University Press, pp. 312–529.

Simmons, N. B. and Geisler, J. H. (1998). Phylogenetic relationships of *Icaronycteris, Archaeonycteris, Hassianycteris*, and *Palaeochiopteryx* to extant bat lineages, with comments on the evolution of echolocation and foraging strategies in Microchiroptera. *Bulletin of the American Museum of Natural History*, **235**, 1–182.

Simmons, N. B., Seymour, K. L., Habersetzer, J. and Gunnell, G. F. (2008). Primitive early Eocene bat from Wyoming and the evolution of flight and echolocation. *Nature*, **451**, 818–821.

Speakman, J. R. (2001). The evolution of flight and echolocation in bats: another leap in the dark. *Mammal Review*, **31**, 111–130.

Swofford, D. L. (2003). *PAUP* (Phylogenetic Analysis Using Parsimony *and Other Methods), Version 4*.

Vaughan, T. A. (1970). The skeletal system. In *Biology of Bats*, ed. W. A. Wimsatt. vol. 1, New York: Academic Press, pp. 98–139.

Walton, D. W. and Walton, G. M. (1970). Post-cranial osteology of bats. In *About Bats: A Chiropteran Biology Symposium*, ed. B. H. Slaughter and D. W. Walton. Dallas, TX: Southern Methodist University Press, pp. 93–126.

Wellik, D. M. (2007). *Hox* patterning of the vertebrate axial skeleton. *Developmental Dynamics*, **236**, 2454–2463.

Wellik, D. M. and Capecchi, M. R. (2003). *Hox10* and *Hox11* genes are required to globally pattern the mammalian skeleton. *Science*, **301**, 363–367.

16

Early evolution of body size in bats

NORBERTO P. GIANNINI, GREGG F. GUNNELL,
JÖRG HABERSETZER AND NANCY B. SIMMONS

16.1 Introduction

Size is the single most important factor affecting physiology, locomotion, ecology and behavior of mammals (MacNab, 2007 and citations therein). Understanding evolution of size is important in all organisms, but especially so in cases like bats which exhibit many energetically expensive behaviors (e.g., powered flight, echolocation, long-distance migration), as well as characteristics that represent extreme energy-saving mechanisms (e.g., torpor and hibernation). Most bat species are small: from data in Smith *et al.* (2004), the central tendency in size in extant bats, as estimated by the median value, is around 14 g (Figure 16.1). However, size in bats as a group spans three orders of magnitude, ranging from 2–3 g (e.g., *Craseonycteris*, *Thyroptera*, *Furipterus*, some vespertilionids; Smith *et al.*, 2004) to a few species exceeding 1 kg (e.g., *Acerodon jubatus*, *Pteropus vampyrus*; Kunz and Pierson, 1994). This variation in size scales a number of fundamental traits in bats, including physiological features (e.g., basal metabolic rate; McNab and Bonaccorso, 2001; MacNab, 2003, Speakman and Thomas, 2003); aerodynamic performance (Norberg, 1986, 1990; Rayner, 1986; Watts *et al.*, 2001); dimensions of limb bones and their biomechanical properties (Swartz, 1997, 1998; Swartz and Middleton, 2008); behaviors (e.g., extreme dietary habits like carnivory; Norberg and Fenton, 1988); echolocation call parameters (Jones, 1999); and most life-history traits (e.g., litter mass; Hayssen and Kunz, 1996). These traits likely have an important phylogenetic component of variation, as has been shown, for instance, for the relationship of basal metabolic rate to body mass (Cruz-Neto *et al.*, 2001; cf. MacNab, 2007). Besides the many dependent variables responding to body mass in various ways, size is a fundamental trait that should be understood by itself as an evolving character in bat lineages.

Variation in size is unevenly distributed across Chiroptera, and the three major clades of bats (Megachiroptera, Yinochiroptera and Yangochiroptera, with their alternative definitions) differ in important ways with respect to size.

Evolutionary History of Bats: Fossils, Molecules and Morphology, ed. G. F. Gunnell and
N. B. Simmons. Published by Cambridge University Press. © Cambridge University Press 2012.

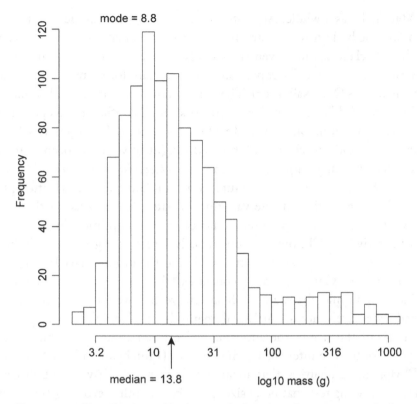

Figure 16.1 Distribution of body mass in bats based on 1071 records for 905 extant species from Smith *et al.* (2004). Mode and median values are indicated.

Just considering ranges of body sizes, bats in the Pteropodidae vary in size from an average of *c.* 14 g in *Balionycteris maculata* (Smith *et al.*, 2004) to the largest bats, with individuals of *Pteropus vampyrus* and *Acerodon jubatus* reaching 1.5 kg (Kunz and Pierson, 1994). Species in Yinochiroptera (*sensu* Eick *et al.*, 2005 and Teeling *et al.*, 2005) vary from an average of *c.* 2 g in *Craseonycteris thonglongyai* to an average of *c.* 126 g in the megadermatid *Macroderma gigas* (Smith *et al.*, 2004). Yangochiroptera, comprising all other extant families (Eick *et al.*, 2005; Teeling *et al.*, 2005; Hermsen and Hendricks, 2008), range from an average of *c.* 3 g in thyropterids, furipterids and some vespertilionids, to an average of *c.* 169 g (and maximum weight of 250 g) in the phyllostomid *Vampyrum spectrum*, the largest microbat (Smith *et al.*, 2004). However, size ranges are poor indicators of the importance of size variation across major bat groups because ranges overlap extensively, thereby masking significant within-clade variation that should be studied with reference to the phylogenetic history of the group.

Despite the ranges in size seen in extant families, it seems clear that ancestral bats were relatively small mammals, especially when considered in the context of Mammalia as a whole. Accurate estimations of mass are needed for under-standing the biology of ancestral bats and the role of size in the early evolution of the bat clade because even small mass differences in this segment of size variation (<100 g) are disproportionately important for homeotherms, particu-larly mammals (MacNab, 2007). The oldest known bats are Early Eocene in age, and by the Middle Eocene there were bats on all continents except Antarctica (Gunnell and Simmons, 2005; Tejedor *et al.*, 2005; Simmons *et al.*, 2008). Habersetzer and Storch (1987) estimated weight of various Eocene bat species and noted that they varied from 7–10 g in *Palaeochiropteryx tupaiodon* to up to 65 g in *Hassianycteris magna*. Hutcheon and Garland (2004) attempted to quantify the evolution of size variation in bats with a focus on the question of whether megabats (= Pteropodidae), the group including the largest representatives of Chiroptera, were actually larger as a group than other bats (collectively, the microbats) when considering the evolutionary history of bats. Hutcheon and Garland (2004) concluded that megabats indeed tended to be larger than microbats, although specific results were rather ambiguous and varied extensively across analytical methods applied. They also gave point estimates of the size of the ancestral bat between 20 and 23 g, but with wide (9–51 g) confidence intervals for these values (Hutcheon and Garland, 2004).

Phylogenetic analyses that include fossils, particularly recent discoveries, compellingly suggest that body size, just like any other evolving trait, can best be understood in the light of paleontological evidence and the early history of the group. More than 30 fossil taxa of undisputed Paleogene bats have been described (Gunnell and Simmons, 2005; Eiting and Gunnell, 2009), including spectacularly preserved specimens from the Early Eocene of North America and Middle Eocene of Europe. These now include a primitive bat, *Onychonycteris finneyi* from the Green River Formation (Early Eocene of Wyoming; 52.5 mya), which was recovered as sister to all other known bats in recent phylogenetic analyses (Simmons *et al.*, 2008). Other well-preserved taxa include *Icaronycteris index* (also from the Green River Formation) and several genera from Messel, Germany, each of which appears to represent a different family of bats (Simmons and Geisler, 1998).

In this chapter we first present new estimates of body mass of key fossil taxa based on allometric relationships of osteological dimensions in extant bats. Next, we integrate those estimates with bat phylogenies that explicitly included the fossil taxa. Our aim is to explore patterns of size evolution beginning with the early history of bats to the onset of extant familial diversity. We also compare the results with those made by mapping body mass onto molecular phylogenetic trees that include only extant bats to single out the contribution of fossils to the understanding of early size evolution in the group. In addition, this approach allows us to explore the evolution of bat body size under the possible scaling constraints of laryngeal echolocation (Barclay and Brigham, 1991; Jones, 1999). Several lines of evidence suggest that fundamental aspects of echolocation (particularly peak frequency, pulse duration and pulse repetition rate) may relate to body mass in different ways, acting as effective limiting factors on interspecific mass variation among bats (Barclay and Brigham, 1991; Jones, 1999). We investigate implications of these constraints in the context of recent findings suggesting that laryngeal echolocation was absent from primitive bats (Simmons *et al.*, 2008, 2010; cf. Veselka *et al.*, 2010).

16.2   Materials and methods

16.2.1   Estimation of size in Early Eocene fossils

In our analyses, we included several well-preserved Early Eocene bats, from which we obtained linear measurements of long-bone diameters, specifically the humerus, radius, femur and tibia. We used least mid-shaft diameter of these bones as an estimator of body mass, as determined for extant mammals, including bats (see Calder, 1996; Swartz and Middleton, 2008). Least diameter of long bones is expected to scale body size accurately, likely responding to one of three major scaling principles: geometric similarity (scaling exponent $b_1 = 0.33$), elastic similarity ($b_1 = 0.375$) or static-stress similarity ($b_1 = 0.40$; McMahon, 1975; Niklas, 1994; Calder, 1996).

We obtained least mid-shaft diameter of long bones from a sample of 105 extant species of bats representing all 18 extant bat families and used body mass data from Smith *et al.* (2004) for the corresponding species (their data were reported as single values per species representing a simple mean between male mean and female mean values). These data were used as our primary source for species size, with minor corrections and some additions from other sources. We incorporated new data as ranges per species whenever possible. We calculated ranges trimming the *m* largest and *m* smallest data (with *m* varying from one to three, depending on species) in order to avoid biases due to inclusion of subadults and pregnant females, on the low and high ends of the reported range, respectively. If age and reproductive condition were not available, we took the arithmetic average and used a single value per species.

Least squares regression (LS) was used to estimate the fit of least mid-shaft diameter of humerus, radius, femur and tibia (all in mm) to body mass variation (in kg) in extant species, using averages per species as sampling units with all data values transformed to logarithms (base 10). We fitted LS regressions for each bone, but for femur and tibia we analyzed data with and without *Desmodus rotundus*. This species is known to possess more robust hindlimb bones than is typical for bats of its size due to extensive quadrupedal capabilities (Altenbach, 1979; Riskin *et al.*, 2005; Schutt and Simmons, 2005), which result in femora and tibiae with bone diameters similar to non-volant mammals of comparable size (Swartz *et al.*, 2005). As a result, we fitted a total of six models, including humerus, radius, femur and tibia with *Desmodus*, plus femur and tibia without *Desmodus*. Of those models, we selected the one with the highest fit as estimated by the coefficient of determination, r^2, to reverse-estimate body size in fossils from the mid-shaft diameter of the corresponding long bone. We calculated body mass M from least mid-shaft diameter data D taken from fossil specimens using the equation:

$$\log_{10} M = \left(\log_{10} D - \log_{10} b_0\right)/b_1 \tag{16.1}$$

where b_0 is the y-intercept and b_1 is the slope parameter of the corresponding equation from extant bats.

We incur some potential error with this procedure. First, we used unmatched data from diverse specimens to compose a mean value for each variable (mean least diameter of humerus and mean body mass) per species. Inevitably, some variation is masked by using averages instead of using matched measurements from each individual. However, paired data of this nature are rarely available from museum specimens. In our sample of 1160 museum specimens, only eight had both skeleton elements to measure bone diameters *and* size (body mass) taken at the field capture site for the same individual. Second, by using ordinary LS, we assume that relevant variation occurred only in least mid-shaft diameter data and ignore measurement error and other sources of variation in mass data. The LS slope parameter b_1 can be directly converted to an orthogonal parameter such as the reduced major axis (RMA) slope by dividing b_1 by the correlation coefficient r (i.e., the square root of the coefficient of determination reported here). RMA allows variation and measurement error in both y and x variables (Niklas, 1994). However, the relationship involved is such that bone diameters effectively depend functionally on mass at both intra- and interspecific levels (Calder, 1996). A dependence model like LS, as opposed to an interdependence model such as RMA, preserves the functional relationship of mass to other variables (y_i dependent on x = mass) and we judged this method more appropriate for the reverse-estimation of size from bone diameter variables. Third, ordinary LS ignores phylogenetic dependence among data points when species are taken as sampling units, thus possibly incurring an inflated type I error rate and erroneous parameter estimation. Therefore we also compared our estimates with those from models given by Swartz and Middleton (2008) using an independent data set with parameters calculated under the Ornstein–Uhlenbeck model of character evolution fitted on a composite phylogenetic tree of extant Chiroptera (Swartz and Middleton, 2008).

All our estimates for fossil taxa depend on the availability of well-preserved specimens. This was a significant limiting factor in spite of the abundance of specimens for certain taxa because most of the specimens were crushed during the preservation process and actual measurements of sections of bones may be inaccurate. We measured specimens from the following institutions: American Museum of Natural History, Fossil Mammal collection (AMNH FM); Royal Ontario Museum (ROM); Forschungsinstitut Senckenberg, Frankfurt, Germany (SMF-ME); Tanzanian National Museum (TNM MP); University of Michigan Museum of Paleontology (UM); and Yale Peabody Museum (YPM). The taxa and specimens examined include *Onychonycteris finneyi* (ROM 55351A (holotype); original specimen of paratype (private collection, casts = AMNH FM 142467, ROM 55055, UM 12405), *Icaronycteris index* (AMNH FM 125000; YPM 18150), *Archaeonycteris trigonodon* (SMF-ME 663, 1504), *Palaeochiropteryx tupaiodon* (AMNH FM 142649, 142650, SMF-ME 492, 1038, 1469, 1540a/b, 1919a/b), *Hassianycteris messelensis* (AMNH 141310) and *Tanzanycteris mannardi* (TNM MP-207). Images of selected specimens are shown in Figures 16.2–16.4.

16.2.2   Optimization of size

Our approach to investigating the early evolution of size in bats consisted of parsimoniously optimizing body mass on several different bat phylogenies that differed in taxonomic sampling and the hierarchical level of focus (see below). Body mass is an evolving character best expressed as a continuous, metric variable, and as such it can be optimized as an ordered character on a phylogenetic tree. Goloboff *et al.* (2006) described the optimization of continuous characters as a natural extension of Farris's (1970) optimization of additive (= ordered) characters. In their implementation (Goloboff *et al.*, 2008), the optimization proceeds as typical down- and up-pass routines for additive characters. Terminal values can be either single-value estimates of the continuous character (e.g., body mass), or ranges (e.g., minimum–maximum adult body mass), or a mix in which some terminals are represented by single values while others are represented by ranges. Ambiguous changes are represented as an interval of minimum to maximum possible values for the node. Steps (here, one step = 1 g) are counted every time a net, unambiguous change is estimated for a given node, or an interval must be formed or changed, either by contraction or expansion of the range of assigned values. The result of this optimization is a set of ancestral nodal

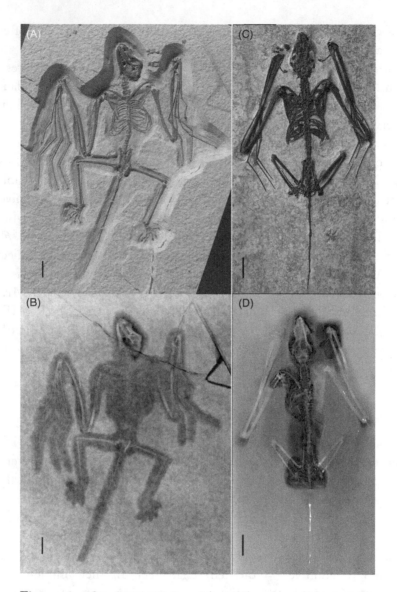

Figure 16.2 *Onychonycteris finneyi* (uncataloged specimen in private collection), ventral view of specimen (A) and X-ray of slab (B), and *Icaronycteris index* (YPM-PU 18150, holotype), dorsal view of specimen (C) and corresponding X-ray of slab (D), both from Green River Formation, Wyoming. Scale = 1 cm.

assignments mapped onto the chosen tree that minimize the total sum of steps in the continuous variable for that tree.

We were interested in the location, magnitude and sign of evolutionary changes in body mass tracked locally on each of the trees analyzed. Location

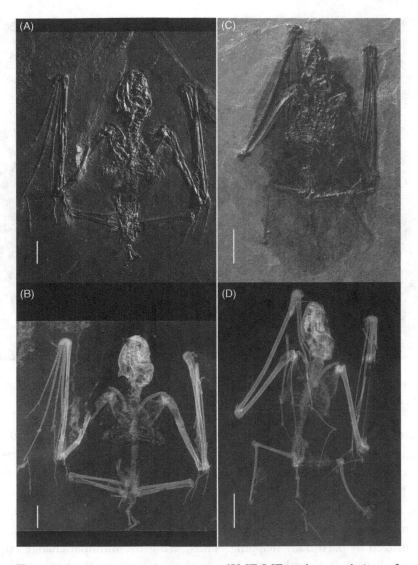

Figure 16.3 *Archaeonycteris trigonodon* (SMF-ME 963), ventral view of specimen (A) and corresponding X-ray of slab (B), and *Palaeochiropteryx tupaiodon* (SMF-ME 1790), dorsal view of specimen (C) and corresponding X-ray of slab (D), from Messel, Germany. Scale = 1 cm.

of a change was recorded simply as the node on the tree where a net change in size was reconstructed. The magnitude was the absolute value of change along a branch, i.e., the difference between values reconstructed for two consecutive nodes. If a branch with a change involved an ambiguous node, the local value taken was the minimum possible. For instance, if a node was assigned 20 g, and its immediate ancestral node was assigned the range 15–18 g, the net change

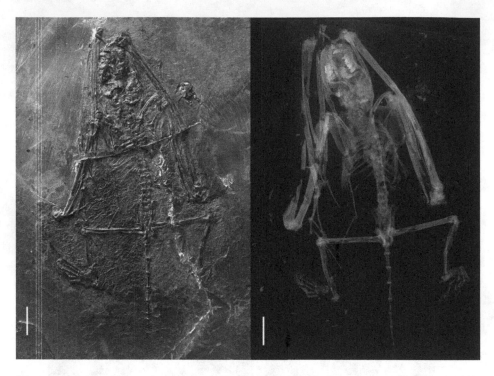

Figure 16.4 *Hassianycteris messelensis* (SMF-ME 1469b), dorsal view of specimen (A) and corresponding X-ray of slab (B), from Messel, Germany. Scale = 1 cm.

assumed was 20 g − 18 g = 2 g (i.e., the minimum difference). This is equivalent to choosing one particular reconstruction of ancestral values, in this case the one in which the ancestral value was larger (18 g). Selecting the reconstruction in which the ancestral value was smaller (15 g) implies a change of 5 g (20−15). The additional three steps required by this choice are subtracted from neighboring nodes in order to balance the fixed total sum of steps which is globally minimized by parsimony mapping. Thus, using the minimal change required by each branch (rather than the larger value or a mean value) represented the most conservative approach. Finally, the sign of the change (increases [+] or decreases [−]), which is generally irrelevant in optimization because absolute values of changes are summed in the calculation of total steps, was key in interpreting the evolutionary trend in a continuous character on the rooted tree. We recorded the proportion of increases vs. decreases in each tree analyzed, the dominant values along the backbone of each tree, and the percentage of mass change, if any, involved in selected nodes. For our analyses, we chose nodes in which important changes in diet or habit were unambiguously reconstructed. These categories included dominant dietary items: animalivory (from insectivory to carnivory, with piscivory as a special case of the latter; see

Giannini and Kalko, 2005), sanguivory, frugivory and nectarivory, with the latter two categories treated jointly as phytophagy. We recorded the size difference, expressed as percentage with respect to the ancestral value, estimated by optimization between a node at which a change of interest (e.g., in diet) occurred and the preceding node in the tree (e.g., phytophagy derived from insectivory). All analyses were done using the program TNT (Goloboff *et al.*, 2008).

16.2.3 Bat phylogenies

We explored evolution of body size in bats as described above using a series of phylogenetic trees that differed in both breadth and depth of taxonomic coverage, thus allowing us to explore questions at a variety of taxonomic levels. First, we optimized body size on three different phylogenies that included Early Eocene fossil taxa. The first tree (Simmons *et al.*, 2008, Suppl. fig. 5) was obtained from an unconstrained parsimony analysis of morphological data. The second tree (Simmons *et al.*, 2008, Suppl. fig. 6) was based on a constrained parsimony analysis of morphological data, using a molecular backbone tree from Teeling *et al.* (2005). A third tree (Hermsen and Hendricks, 2008, fig. 1) was obtained from an unconstrained parsimony analysis of combined morphological data from Gunnell and Simmons (2005) and aligned nucleotide sequence data from Teeling *et al.* (2005). In these phylogenies, the terminals were defined at the family level. As such, these phylogenies are inadequate for analysis of body size based on species values. We followed Hermsen and Hendricks (2008) in inserting the set of terminals from the molecular phylogeny of Teeling *et al.* (2005) as representatives of specific representatives of each bat family. Then we optimized size of terminals on the trees from Simmons *et al.* (2008) and Hermsen and Hendricks (2008). Finally, we optimized size as a continuous character on molecular phylogenies that were aimed at resolving interfamilial relationships among extant bats. The purpose of this exercise was to compare the effect of fossils (or the lack thereof) on our interpretation of early evolution of bat body size. The molecular phylogenies chosen for analysis were those of Teeling *et al.* (2005) and Eick *et al.* (2005), each of which included a wide representation of major bat groups with each group represented by a few exemplar members. Teeling *et al.* (2005) included representatives of all bat families but Miniopteridae, with 30 bat terminals of which 5 were chimeric at the level of genus and composed of DNA data from 2–3 distinct species. The Teeling *et al.* (2005) phylogenetic hypothesis was based on 13.7 kbp from 17 nuclear coding genes. In turn, Eick *et al.* (2005) generated

Table 16.1 Linear size estimators based on a sample of extant bats, and size estimations of Eocene fossil taxa (see text).

	Linear size estimator (least mid-shaft diameter of bone)					
	Humerus	Radius	Femur	Femur*	Tibia	Tibia*
Parameters						
y-intercept	−0.273	−0.242	−0.455	−0.455	−0.536	−0.536
slope	0.363	0.318	0.355	0.353	0.342	0.340
r^2	0.991	0.986	0.981	0.987	0.972	0.977
Estimated range of weight in grams (n)						
Archaeonycteris trigonodon (2)	29–31	15–17	34–42	35–43	34–45	34–46
Hassianycteris messelensis (1)	90	59	72	74	82	84
Onychonycteris finneyi (2)	39–41	25–32	44–49	45–50	63–76	65–78
Icaronycteris index (2)	24–27	7–8	34–39	34–40	24–38	24–39
Palaeochiropteryx tupaiodon (5)	9–16	6–12	6–11	7–11	8–11	8–11
Tanzanycteris mannardi (1)	14	?	?	?	?	?

*Excluding *Desmodus*.

4 kbp from 4 nuclear introns from 54 bat terminals, none of which were chimeric, representing all bat families but Craseonycteridae.

16.3 Results

16.3.1 Size estimates for Early Eocene bats

In all six models, least mid-shaft diameter of the long bones fitted body mass in extant bats remarkably well (adjusted $r^2 > 0.97$; Table 16.1). Note that with r^2 values this high LS and RMA slope estimates b are nearly identical because $b_{RMA} = b_{LS}/r$ and $r \sim 1$ (e.g., for humerus, $b_{LS} = 0.363$ and $b_{RMA} = 0.365$). In spite of the high fit, size estimates in fossils varied across models (Table 16.1). However, least mid-shaft diameter of humerus was the absolute best estimator with r^2 as high as 0.991 (Table 16.1); therefore, we used this model to reverse-estimate size in Eocene fossils. As expected, *Hassianycteris messelensis* was the largest Eocene bat in our taxonomic sample with an estimated mass of ~90 g (with a minimum size of 59 g when estimated using radius diameter). Based on humerus diameter, the archaic North American bats *Onychonycteris finneyi* and *Icaronycteris index* weighed an estimated 38.7–40.9 g and 24.0–27.3 g, respectively.

16.3.2 Evolution of size from Eocene fossils to the ancestors of extant families

The patterns of size evolution were very similar in the morphology tree, the constrained tree and the combined tree, in spite of key topological differences. Figures 16.5 and 16.6 show the optimization of size in the later two trees.

In the constrained morphology tree (Figure 16.5), the basal node of Chiroptera was assigned states 39–41 g (the estimated size of *Onychonycteris finneyi*). This range of mass represented a reduction of nearly 120 g with respect to the previous node in the laurasiatherian clade of the tree. Bats further decreased in size along the backbone of their subtree in a pattern we identify as phyletic nanism, *sensu* Gould and MacFadden (2004; see Discussion). The next node up the chiropteran subtree exhibited a reduction of 9 g, with no net change in subsequent backbone nodes until the common ancestor of *Palaeochiropteryx* and extant bats. This node was assigned 14–16 g, i.e., a net decrease of 14 g, and a cumulative decrease up to this node of at least 141 g from the hypothetical chiropteran ancestor. Two contrasting changes were assigned to Eocene bats, a small decrease of 3 g in the branch leading to *Icaronycteris index*, and a net increase of 60 g in *Hassianycteris messelensis*. This represents a 250% increase in body mass with respect to the reconstructed ancestor. The ancestral megachiropteran node exhibited an increase of 13 g (or an 81% increase) in this topology, and several large net increases were reconstructed inside the pteropodid family subtree, which is consistent with phyletic giantism, *sensu* Gould and MacFadden (2004; see Discussion). No net changes were reconstructed along the backbone microbat branches, with all important net increases or decreases reconstructed within currently recognized families. In this tree, the width of the interval reconstructed at the backbone also changed, first increasing at the lower end (from 14–16 g to 11–16 g), then shrinking at the upper end (from 11–16 g to 11–12 g; represented by horizontal arrows in Figure 16.6).

In general, increases were of higher magnitude than decreases, which may reflect the general phenomenon of variance positively scaling with size.

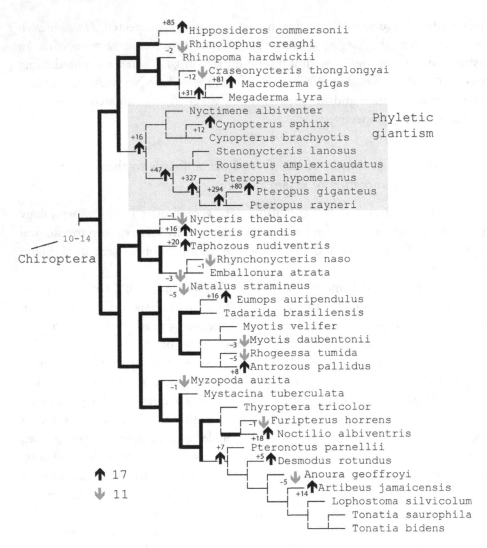

Figure 16.5 Optimization of body mass on the phylogenetic tree of Teeling *et al.* (2005), based on molecular data. Signed changes (increments and decrements) are indicated by arrows and values at or near corresponding internal nodes. These values are minimal changes in body mass required by the tree (see text). Branches in bold show the backbone of the tree in which no changes are reconstructed. The gray box indicates pteropodids.

The largest increase in any bat branch was reconstructed at the *Pteropus* node (+327 g, or a 417% increase with respect to the reconstructed ancestor), whereas the largest decrease (−118 g) was located at the ancestral bat node. Among microbats, the largest increases were reconstructed on the branches leading to the hipposiderid *Hipposideros commersoni* (+83 g) and the megadermatid

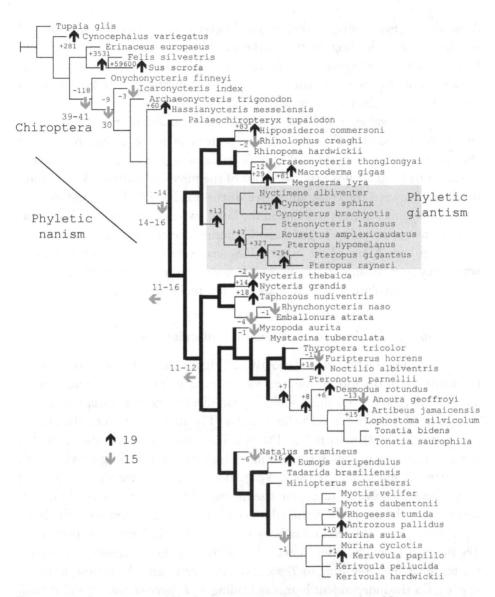

Figure 16.6 Optimization of body mass on the phylogenetic tree of Simmons *et al.* (2008), based on morphological data constrained by a molecular scaffolding. Signed changes (increments and decrements) are indicated by arrows and values at or near corresponding internal nodes. These values are minimal changes in body mass required by the tree (see text). Branches in bold show the backbone of the tree in which no changes are reconstructed. The gray box indicates pteropodids. Horizontal arrows at backbone nodes.

Macroderma gigas (+81 g), whereas the largest decreases were reconstructed on the branches leading to the phyllostomid *Anoura geoffroyi* (−13 g) and *Craseonycteris thonglongyai*, the smallest bat and one of the smallest mammals (−12 g). This represents a sixfold decrease in body mass, a finding which is similar to our results using a molecular tree (see below).

As expected, values and details were not identical among trees with fossils. In the combined tree (Figure 16.6) the pattern of phyletic nanism in stem chiropteran nodes converge to the single value of 14 g at Microchiroptera and remains unaltered all along the backbone of the microbat subtree. A significant difference between the constrained tree discussed above and both the morphology and combined trees was that in the latter two trees the ancestral megabat was not reconstructed as larger than the immediate common ancestor shared with other bats. Many of the other values were very similar among trees, and when they were dissimilar, they never affected our interpretation of the general pattern of the evolution of body mass in Chiroptera (but see Discussion).

16.3.3   Optimization of size on molecular trees

On the Teeling *et al.* (2005) tree (Figure 16.7), the body mass value reconstructed at the root of the tree was 10.1–14.2 g. There was no net change along the backbone (branches connecting families) of the tree, except for a +7 g change in Mormoopidae + Phyllostomidae. The first change (i.e., the change closest to the root) occurred in Pteropodidae (+15 g) which also exhibited phyletic giantism in this tree. In Yinochiroptera *sensu* Teeling *et al.* (2005), important changes included a *c.* 85 g increase in the terminal branch of *Hipposideros commersoni*, successive size increases in carnivorous megadermatids and a −12 g change in the terminal branch leading to *Craseonycteris*, the latter representing a sixfold decrease in body mass. Patterns in Yangochiroptera *sensu* Teeling *et al.* (2005) were less remarkable, with the largest increase at *c.* 20 g (in the terminal branch leading to *Taphozous nudiventris*) and the largest decreases at *c.* 5 g (in the independent branches leading to *Rhogeessa tumida* and *Anoura geoffroyi*; Figure 16.7).

On the Eick *et al.* (2005) tree, the body mass value reconstructed at the root was 12–14 g (not shown). The only backbone branch exhibiting change (albeit a small change of −2 g) was the one joining *Myzopoda* and vespertilionoids (Natalidae, Miniopteridae, Molossidae and Vespertilionidae). Again, the first large change occurred in Pteropodidae with a net increase of 29 g. No other major changes appeared in Pteropodidae because large species (e.g., of the genus *Pteropus*) were not included in the analysis of Eick *et al.* (2005). In Yinochiroptera, only three changes were notable, increases in size along the

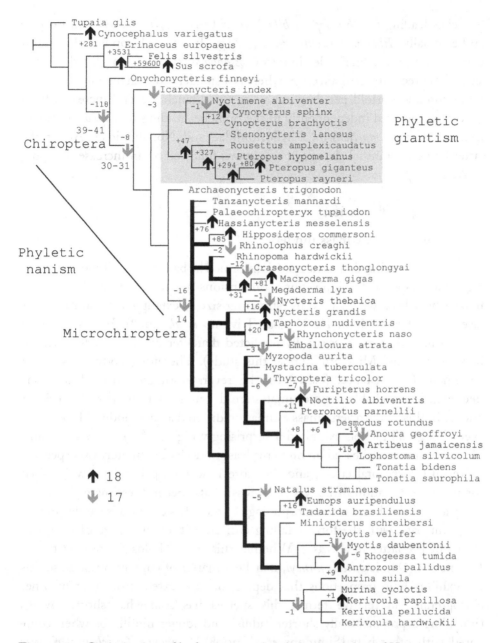

Figure 16.7 Optimization of body mass on the phylogenetic tree of Hermsen and Hendrick (2008), based on a combined analysis of molecular and morphological data. Signed changes (increments and decrements) are indicated by arrows and values at or near corresponding internal nodes. These values are minimal changes in body mass required by the tree (see text). Branches in bold show the backbone of the tree in which no changes are reconstructed. The gray box indicates pteropodids.

branches leading to *Rhinolophus hildebrandti* (+13 g), *Megaderma lyra* (+18 g) and especially *Hipposideros commersoni* (+87 g). In Yangochiroptera, the largest decrease was in the branch leading to *Glossophaga soricina* (−19 g), halving the size of the reconstructed ancestor, whereas a smaller −9 g change in *Furipterus* represented a fourfold net decrease along this branch. Several increases from 13 to 18 g appeared independently in the branches leading to *Taphozous mauritianus*, *Otomops martienseni*, Phyllostomidae and Noctilionidae. Along the latter branch, an initial change of +14 g was followed by an increase of +30 g in *Noctilio leporinus*.

16.4   Discussion

16.4.1   Body mass models

Interspecific allometric models informed by basic mechanical properties – e.g., the response of limb bone dimensions to forces exerted by body mass – have been widely used in estimating size of extant and fossil mammals (see a recent example in De Esteban-Trivigno *et al.*, 2008). Bats are not an exception: body mass successfully predicted diameter of limb bones in extant bats (Swartz and Middleton, 2008; this study). Therefore, body mass of bat specimens (for instance, fossils) can be reverse-estimated from limb bone diameter. How reliable these estimates are depends on a number of considerations. In this study, reverse mass estimation differed across models (Table 16.1), even though all models exhibited a surprisingly high fit ($0.972 < r^2 < 0.991$). However, one model based on humerus least mid-shaft diameter was especially powerful ($< 1\%$ total error), and therefore it seemed particularly well suited for reverse estimation of body size in fossil bats. Second, estimates in fossils likely are reliable because the morphology of most Eocene taxa is similar to that of extant bats, with the former falling entirely within the range of measurements of the latter in all cases. Whether this model holds for one particular Eocene bat, *Onychonycteris finneyi*, may be a matter of contention because this bat exhibits limb proportions that depart to some extent from that of other chiropterans; in relative terms, this species has somewhat shorter wings (including a proportionally shorter radius) and longer hindlimbs when compared with other bats (Simmons *et al.*, 2008). However, *Onychonycteris* was clearly capable of powered flight (Simmons *et al.*, 2008), and therefore its skeleton was likely subjected to aerodynamic forces comparable to those experienced by more typically proportioned bats. Moreover, our mass estimates are consistent with previously reported morphometric data (Simmons *et al.*, 2008), which indicate that this bat was clearly larger than the other Eocene taxa included in our study, except *Hassianycteris*. We conclude that humerus least mid-shaft diameter is an extremely accurate proxy of size in extant bats and a reliable predictor of body mass in fossils.

16.4.2   Patterns of early size evolution in bats

Mapping body mass onto two different molecular trees (from Teeling *et al.*, 2005 and Eick *et al.*, 2005) resulted in a congruent estimate of mass at the root of the chiropteran tree, between 10 and 14 g. Interestingly, this range includes the median mass of extant bats – 13.8 g (Figure 16.1), as calculated from data in Smith *et al.* (2004) for all bats reported. This interval was conserved with very little variation along the backbone nodes of the molecular trees, from the root to the ancestor of each bat family. Departures from stasis were nested within familial clades and could be associated with dietary specialization (see below).

This remarkable pattern of size stability in the backbone of the crown clade was also reconstructed in the three phylogenies that included Eocene fossils. However, mapping body mass onto any of these trees revealed a trend of phyletic nanism, i.e., a pattern of pronounced decreasing size along successive branches (Gould and MacFadden, 2004), at the base of the chiropteran tree. A cumulative decrease of >280% was reconstructed until stabilizing around 14 g at the crown clade *Palaeochiropteryx* + microbats (using the morphology tree), or *Palaeochiropteryx* + *Hassianycteris* + microbats (i.e., Microchiropter-aformes, using the combined tree), or *Palaeochiropteryx* + all extant bats (using the constrained tree topology). This trend is key to understanding evolution of size in bats and is missed altogether in phylogenies not explicitly including Eocene fossils. These trees also indicated that the central tendency of bat size was established during the Early Middle Eocene (around 49 mya, the age of Messel bats) and maintained along all bat lineages. This timing differs by several million years from estimates based on molecular dating. In the molecular phylogenies, the median value of *c.* 14 g was reconstructed at the basal split of major chiropteran groups, which has been dated to 65 Ma (close to the KT-boundary; Teeling *et al.*, 2005) or 62 Ma (Early Paleocene; Eick *et al.*, 2005). However, fossil bats older than 49 Ma (*Icaronycteris* and *Onycho-nycteris*) were considerably (170–292%) larger.

Outgroup taxa differed across source trees, so the magnitude of change from the various outgroups to the chiropteran root was tree dependent. However, in all cases the outgroup taxa were much larger (≫100 g) than the reconstructed chiropteran roots (*c.* 14 g in molecular trees, *c.* 40 g in trees with fossils). This means that essentially the same interpretation holds for all trees used: early bats were remarkably smaller than any of the available outgroups. However, although it seems clear that bats belong in Laurasiatheria (e.g., Murphy *et al.*, 2001), there is uncertainty regarding the sister taxon of bats. Among the laurasiatherian candidates are small insectivores in the Erinaceomorpha (e.g., Prasad *et al.*, 2008); if such mammals are finally recovered as sister to bats, then all the evolution toward the acquisition of powered flight occurred within the 100 g boundary.

Our study reveals a slight trend in the directionality of change with the frequency of size increases between nodes being greater than that of decreases. However, we found no evidence of a steady increase in size through the history of the bat lineage, as might be predicted by Cope's Rule. Such a trend, which in recent terminology corresponds with "phyletic giantism" *sensu* Gould and MacFadden (2004), is certainly present in some restricted parts of the bat tree, most conspicuously in the megabat clade. For instance, in our topologies *Pteropus giganteus* achieves an average mass of *c.* 800 g (from Smith *et al.*, 2004) accumulating several net increases along successive branches from the chiropteran backbone, which represents a total increase of 5700% (Figures 16.5–16.7). However, stasis and reductions in size (nanism) are even more common. Size reductions were most frequently manifested as apomorphic changes, i.e., decreases in size reconstructed along a single branch, as opposed to size reduction along several successive branches (as in stem Chiroptera). One such case of apomophic nanism is represented by the >600% decrease in size in Craseonycteridae. Conversely, however, cases of apomorphic giantism were also found, even among Eocene bats (e.g., *Hassianycteris*). Several cases of sister lineages showing opposite size trends were found; these include the pairs Craseonycteridae vs. Megadermatidae, *Nycteris thebaica* vs. *Nycteris grandis*, *Taphozous* vs. other emballonurids, Furipteridae vs. Noctilionidae, *Anoura* vs. *Artibeus* and *Rhogeessa* vs. *Antrozous*. It is premature to draw conclusions from these data because these patterns are heavily dependent on a small taxonomic sample, but it is certainly intriguing and these patterns clearly call for a deeper exploration with a more comprehensive taxonomic sampling.

16.4.3   Functional constraints on evolution of body size in bats

Flight is relatively easier, i.e., less expensive energetically, for small-sized organisms (Pennycuick, 1986). Small size reduces induced and inertial power (Norberg, 1994). Small size also enhances maneuverability because smaller organisms require a small radius of turn (Norberg, 1994). Power margin, or the ratio of the power available from muscles to the power required to fly horizontally at minimum power speed, varies inversely with body mass and the power-to-mass relationship is best below 100 g (Pennycuick, 1986). Thus, the trend of phyletic nanism we reconstruct at the base of the chiropteran tree likely reflects an improvement in power relationships (decrease in total power, increase in power margin) and maneuverability that funneled early bats into a new aerial foraging niche with the advent of sophisticated laryngeal echolocation.

Onychonycteris apparently lacked the ability to detect and track airborne prey using echolocation (Simmons *et al.*, 2008, 2010; *cf.* Veselka *et al.*, 2010). Laryngeal echolocation likely appeared at some node soon after the split of *Onychonycteris* and other bats, after flight evolved. Analyses of wing morphologies and stomach contents indicate that aerial foraging was established early in the chiropteran lineage, at least by the Early Middle Eocene (Simmons and Geisler, 1998). Evolution of laryngeal echolocation would have brought a series of new constraints to bear on the evolution of bat body size, and probably represented a strong stabilizing force that kept the majority of aerial foraging, echolocating bats within a narrow range of body size (Barclay and Brigham, 1991; Jones, 1999).

Echolocation is a short-range sensory device, so echolocating bats are apparently limited by several factors. These include the trade-off between flight speed and detection range, the double constraints of scaling of echolocation call parameters (pulse duration and peak frequency) for detection of small airborne prey (Jones, 1999), and wingbeat frequency (Norberg, 1998) and its direct influence in pulse repetition rate (coordinated with wingbeat frequency; Speakman and Racey, 1991). Flight speed and pulse duration scale positively with body size, whereas frequency parameters, as expected, scale inversely with body size (Calder, 1996; Jones, 1999). This set of conditions apparently compromises the ability of large bats to detect small airborne targets, such as flying insects, using echolocation (see details in Jones, 1999). Nocturnal small insects are extremely common prey in most habitats, but catching them on the wing using echolocation requires high-frequency calls with returning echoes that should be strong enough to carry reliable information on location and movement of prey. Aerial insectivores likely need to be small in order to: (1) fly slowly, with sufficient maneuverability within detection range; (2) generate high-frequency calls capable of producing detectable echoes returning from a small target; and (3) generate enough information about a small moving target via a high call repetition rate (see Jones, 1999). Therefore, departures from stasis in size evolution would be predicted to involve dietary and behavioral specializations that would free bats from the scaling constraints of the demanding combination of flight and echolocation. Deviations from this physically constrained system are possible in several ways.

One potential way to overcome the constraints of flight and echolocation is shifting from preying on small flying insects to preying on larger organisms, e.g., very large insects or small vertebrates. In our trees, the largest size increase within microbat clades corresponded with the branch leading to *Hipposideros commersoni*. This is a 95 g bat specialized in hunting large (up to 15 g) dung beetles (Vaughan, 1977). Dung beetles probably are large enough to be detected and traced using low-frequency, low-repetition-rate calls. Another way is to avoid aerial foraging altogether and use echolocation for orientation only, as gleaners do. Large size is more frequent among gleaners as compared to aerial hawkers, and large size, in combination with low wing loading and aspect ratio, and hence high carrying capacity, represents morphofunctional signatures of carnivorous bats (Norberg and Fenton, 1988). Giannini and Kalko (2005) have shown that carnivory is only an extreme in a continuous gradient of animal-ivory. In line with this, it is interesting to note that most of the increases (Figure 16.2) were directly associated with either animalivores that are partial-to-obligate gleaners or perch-hunters (e.g., *Hipposideros*, *Macroderma*, *Megaderma*, *Nycteris*, *Antrozous*), with animalivores exhibiting special habits (e.g., *Noctilio*), or with large frugivores.

Pteropodids deserve special consideration. These bats are obviously not constrained by laryngeal echolocation (the only echolocators in this clade are species of *Rousettus* that use tongue clicks; Roberts, 1975), and this phytophagous family spans the largest size range (*c.* 14–1500 g) among natural bat groups. This can be interpreted in two different ways, depending on the phylogenetic hypothesis used. If pteropodids diverged before laryngeal echolocation was established in the bat lineage, then these bats never experienced the penalties imposed on large-bodied aerial echolocators by the physical limitations of the echolocation system; therefore, they realized nearly the whole range of size that is allowed for bats from an exclusively aerodynamic perspective (not exceeding 2 kg; U. M. Norberg, personal communication). Alternatively, if the pteropodid lineage evolved from a clade nested within echolocating bats, as strongly suggested by current molecular phylogenies (Eick *et al.*, 2005; Teeling *et al.*, 2005), then echolocation must have been lost in this lineage. The loss of echolocation initially might have represented a significant adaptive disadvantage, but in giving up echolocation these bats might have simultaneously been freed from the constraints of echolocation on the evolution of size. Therefore pteropodids may have experienced an ecological release, rapidly expanding their size range and diversifying as phytophagous volant mammals in the Old World tropics. They could have taken advantage of reducing cost of transport and so improving on commuting time and resource exploitation range. For instance, one of the largest bats, *Pteropus vampyrus*, can travel the astonishing maximum foraging distance of 87 km and has between-roost commuting distance of up to 364 km, with a maximum home range as large as 128 000 km² (see details in Epstein *et al.*, 2009).

16.5 Conclusions

Size is a fundamental factor influencing most aspects of bat biology and likely evolved in response to complex interactions between developmental control and environment. Reliable estimations of mass in fossils are available from interspecific models fitted in extant bats based on the response of limb bone dimensions to varying body size, particularly least mid-shaft diameter of the humerus. Size trends were robust to differences in phylogeny reconstruction and revealed rich patterns of size variation among bats, which contradict Cope's Rule. Early in their history bats experienced phyletic nanism probably associated with flight power demands and the need for increased maneuverability to effectively exploit the rich night insect food resource. Our reconstructions indicate that the median value of mass in extant bats was achieved by the Early Middle Eocene and was maintained through their phylogenetic history, probably due to physical constraints on size imposed by the evolution of echolocation. This supports Jones's (1999) hypothesis that requirements of echolocation for aerial hawking tend to drive reductions in body size. The subsequent prediction, that lineages that do not use that foraging strategy should be the ones showing increases in body size, is also supported given that bat lineages that seem to have escaped from these constraints exhibit divergent dietary and behavioral specializations. These include giantism associated with gleaning and perching behavior in extreme insectivory and carnivory, and with frugivory. Megabats either evolved before those constraints on size were enforced by echolocation, or their size evolution reflects an ecological release from the loss of echolocation.

16.6 Acknowledgments

We thank Dr. Philip Myers and Mr. Steven Hinshaw (University of Michigan, Museum of Zoology), Dr. Judith Eger (Royal Ontario Museum), Dr. John Flynn and Ms. Ruth O'Leary (AMNH, Division of Vertebrate Paleontology) and Ms. Eileen Westwig (AMNH, Department of Mammalogy) for providing access to extant and fossil bat specimens and body mass data. NPG thanks K. M. Helgen who graciously provided additional measurements. We thank one anonymous reviewer and Daniel Riskin, for suggestions that led to important improvements of the manuscript. In addition, Daniel

Riskin encouraged us to explore further our results in connection with echolocation constraints. This research was supported in part by CONICET, Argentina, American Museum of Natural History and National Science Foundation AToL Grant 0629959.

16.7   REFERENCES

Altenbach, J. C. (1979). Locomotor morphology of the Vampire Bat, *Desmodus rotundus*. *American Society of Mammalogists, Special Publication*, **6**, 1–137.

Barclay, R. M. R. and Brigham, R. M. (1991). Prey detection, dietary niche breadth and body size in bats: why are aerial insectivorous bats so small? *American Naturalist*, **137**, 693–703.

Calder, W. A. (1996). *Size, Function, and Life History*. New York: Dover Publications.

Cruz-Neto, A. P. and Jones, K. E. (2007). Exploring the evolution of the basal metabolic rate in bats. In *Functional and Evolutionary Ecology of Bats*, ed. A. Zubaid, G. F. McCracken and T. H. Kunz. Oxford: Oxford University Press, pp. 56–89.

Cruz-Neto, A. P., Garland, T. and Abe, A. S. (2001). Diet, phylogeny and basal metabolic rate in phyllostomid bats. *Zoology*, **104**, 49–58.

De Esteban Trivigno, S., Mendoza, M. and De Renzi, M. (2008). Body Mass Estimation in Xenarthra: a predictive equation suitable for all quadrupedal terrestrial placentals? *Journal of Morphology*, **269**, 1276–1293.

Eick, G. N., Jacobs, D. S. and Matthee, C. A. (2005). A nuclear DNA phylogenetic perspective on the evolution of echolocation and historical biogeography of extant bats. *Molecular Biology and Evolution*, **22**, 1869–1886.

Eiting, T. P. and Gunnell, G. F. (2009). Global completeness of the bat fossil record. *Journal of Mammalian Evolution*, **16**, 151–173.

Epstein, J. H., Olival, K. J., Pulliam, J. R. C. *et al.* (2009). *Pteropus vampyrus*, a hunted migratory species with a multinational home-range and a need for regional management. *Journal of Applied Ecology*, doi: 10.1111/j.1365-2664.2009.01699.x.

Farris, J. S. (1970). Methods for computing Wagner trees. *Systematic Zoology*, **19**, 83–92.

Giannini, N. P. and Kalko, E. K. V. (2005). The guild structure of animalivorous leaf-nosed bats of Barro Colorado Island, Panama, revisited. *Acta Chiropterologica*, **7**, 131–146.

Goloboff, P. A., Mattoni, C. I. and Quinteros, A. S. (2006). Continuous characters analyzed as such. *Cladistics*, **22**, 589–601.

Goloboff, P. A., Farris, J. S. and Nixon, K. (2008). TNT, a free program for phylogenetic analysis. *Cladistics*, **24**, 774–786.

Gould, G. C. and MacFadden, B. J. (2004). Gigantism, dwarfism, and Cope's Rule: "Nothing in evolution makes sense without a phylogeny". In *Tributes to Malcolm C. McKenna: His Students, His Legacy*, ed. G. C. Gould and S. K. Bell. *Bulletin of the American Museum of Natural History*, **285**, 219–237.

Gunnell, G. F. and Simmons, N. B. (2005). Fossil evidence and the origin of bats. *Journal of Mammalian Evolution*, **12**, 209–246.

Habersetzer, J. and Storch, G. (1987). Klassifikation und funktionelle Flügelmorphologie paläogener Fledermäuse (Mammalia, Chiroptera). *Courier Forschungsinstitut Senckenberg*, **91**, 11–150.

Hayssen, V. and Kunz, T. H. (1996). Allometry of litter mass in bats: comparisons with maternal size, wing morphology, and phylogeny. *Journal of Mammalogy*, **77**, 476–490.

Hermsen, E. J. and Hendricks, J. R. (2008). W(h)ither fossils? Studying morphological character evolution in the age of molecular sequences. *Annals of the Missouri Botanical Garden*, **95**, 72–100.

Hutcheon, J. M. and Garland, Jr., T. (2004). Are megabats big? *Journal of Mammalian Evolution*, **11**, 257–277.

Jones, G. (1999). Scaling of echolocation call parameters in bats. *Journal of Experimental Biology*, **202**, 3359–3367.

Kunz, T. H. and Pierson, E. D. (1994). Bats of the world: an introduction. In *Walker's Bats of the World*, ed. R. M. Nowak. Baltimore, MD: Johns Hopkins University Press, pp. 1–46.

MacNab, B. K. (2003). Standard energetics of phyllostomid bats: the inadequacies of phylogenetic-contrast analyses. *Comparative Biochemistry and Physiology*, **135A**, 357–368.

MacNab, B. K. (2007). The evolution of energetics in birds and mammals. In *The Quintessential Naturalist: Honoring the Life and Legacy of Oliver P. Pearson*, ed. D. A. Kelt, E. P. Lessa, J. Salazar-Bravo and J. L. Patton. *University of California Publications, Zoology*, **134**, 67–110.

MacNab, B. K. and Bonaccorso, F. J. (2001). The metabolism of New Guinean pteropodid bats. *Journal of Comparative Physiology B*, **171**, 201–214.

McMahon, T. A. (1975). Allometry and biomechanics: limb bones in adult ungulates. *American Naturalist*, **109**, 547–563.

Murphy, W. J., Eizirik, E., O'Brien, S. J. *et al.* (2001). Resolution of the early placental mammal radiation using Bayesian phylogenetics. *Science*, **294**, 2348–2351.

Niklas, K. J. (1994). *Plant Allometry. The Scaling of Form and Process*. Chicago, IL: University of Chicago Press.

Norberg, U. M. (1986). On the evolution of flight and wing forms in bats. In *Bat Flight/Fledermausflug. BIONA Report 5*, ed. W. Nachtigall. Stuttgart: Gustav Fischer, pp. 13–26.

Norberg, U. M. (1990). *Vertebrate Flight: Mechanics, Physiology, Morphology, Ecology and Evolution*. Berlin: Springer-Verlag.

Norberg, U. M. (1994). Wing design, flight performance, and habitat use in bats. In *Ecological Morphology: Integrative Organismal Biology*, ed. P. C. Wainright and M. Reill. Chicago, IL: University of Chicago Press, pp. 205–239.

Norberg, U. M. (1998). Morphological adaptations for flight in bats. In *Bat Biology and Conservation*, ed. T. H. Kunz and P. A. Racey. Washington, DC: Smithsonian Institution Press, pp. 93–108.

Norberg, U. M. and Fenton, M. B. (1988). Carnivorous bats? *Biological Journal of the Linnean Society*, **33**, 383–394.

Prasad, A. B., Allard, M. W., NISC Comparative Sequencing Program and Green, E. D. (2008). Confirming the phylogeny of mammals by use of large comparative sequence datasets. *Molecular Biology and Evolution*, **25**, 1795–1808.

Pennycuick, C. J. (1986). Mechanical constraints on the evolution of flight. In *The Origin of Birds and the Evolution of Flight*, ed. K. Padian. *Memoirs of the California Academy of Sciences*, **8**, 83–98.

Rayner, J. (1986). Vertebrate flapping mechanics and aerodynamics, and the evolution of flight in bats. In *Bat Flight/Fledermausflug. BIONA Report 5*, ed. W. Nachtigall. Stuttgart: Gustav Fischer, pp. 27–74.

Riskin, D. K., Bertram, J. E. A. and Hermanson, J. W. (2005). Testing the hindlimb-strength hypothesis: non-aerial locomotion by Chiroptera is not constrained by the dimensions of the femur or tibia. *Journal of Experimental Biology*, **208**, 1309–1319.

Roberts, L. H. (1975). Confirmation of the echolocation pulse production mechanism of *Rousettus*. *Journal of Mammalogy*, **56**, 218–220.

Simmons, N. B. and Geisler, J. H. (1998). Phylogenetic relationships of *Icaronycteris*, *Archaeonycteris*, *Hassianycteris*, and *Palaeochiropteryx* to extant bat lineages, with comments on the evolution of echolocation and foraging strategies in Microchiroptera. *Bulletin of the American Museum of Natural History*, **235**, 1–182.

Simmons, N. B., Seymour, K. L., Habersetzer, J. and Gunnell, G. F. (2008). Primitive early Eocene bat from Wyoming and the evolution of flight and echolocation. *Nature*, **451**, 818–821.

Simmons, N. B., Seymour, K. L., Habersetzer, J. and Gunnell, G. F. (2010). Inferring echolocation in ancient bats. *Nature*, **466**, E8–E10.

Schutt, W. A. and Simmons N. B. (2005). Quadrupedal bats: form, function, and evolution. In *Functional and Evolutionary Ecology of Bats*, ed. A. Zubaid, G. F. McCracken and T. H. Kunz. Oxford: Oxford University Press, pp. 145–159.

Smith, F. A., Lyons, S. K., Morgan, E. K. *et al.* (2004). Body mas of Late Quaternary mammals. Ecological Archives E084–094. *Ecology*, **84**, 3403.

Speakman, J. R. and Thomas, D. W. (2003). Physiological ecology and energetics of bats. In *Bat Ecology*, ed. T. H. Kunz and M. B. Fenton. Chicago, IL: University of Chicago Press, pp. 430–490.

Speakman, J. R. and Racey, P. A. (1991). No cost of echolocation for bats in flight. *Nature*, **350**, 421–423.

Swartz, S. M. (1997). Allometric patterning in the limb skeleton of bats: implications for the mechanics and energetics of powered flight. *Journal of Morphology*, **234**, 277–294.

Swartz, S. M. (1998). Skin and bones: the mechanical properties of bat wing tissues. In *Bats: Phylogeny, Morphology, Echolocation, and Conservation Biology*, ed. T. H. Kunz and P. A. Racey. Washington, DC: Smithsonian Institution Press, pp. 109–126.

Swartz, S. M. and Middleton, K. M. (2008). Biomechanics of the bat limb skeleton: scaling, material properties and mechanics. *Cell Tissues Organs*, **187**, 59–84.

Swartz, S. M., Bishop, K. and Aguirre, M.-F. I. (2005). Dynamic complexity of wing form in bats: implications for flight performance. In *Functional and Evolutionary*

Ecology of Bats, ed. A. Zubaid, G. F. McCracken and T. H. Kunz. Oxford: Oxford University Press, pp. 110–130.

Teeling, E. C., Springer, M. S., Madsen, O. *et al.* (2005). A molecular phylogeny for bats illuminates biogeography and the fossil record. *Science*, **307**, 580–584.

Tejedor, M. F., Czaplewski, N. J., Goin, F. J. and Aragón, E. (2005). The oldest record of South American bats. *Journal of Vertebrate Paleontology*, **25**, 990–993.

Vaughan, T. A. (1977). Foraging behavior of the giant leaf-nosed bat (*Hipposideros commersoni*). *East African Wildlife Journal*, **15**, 237–249.

Veselka, N., McErlain, D. D., Holdsworth, D. W. *et al.* (2010). A bony connection signals laryngeal echolocation in bats. *Nature*, **463**, 939–942.

Watts, P., Mitchell, E. J. and Swartz, S. M. (2001). A computational model for estimating the mechanics of horizontal flapping flight in bats: model description and validation. *Journal of Experimental Biology*, **204**, 2873–2898.

Index